FLORA ZAMBESIACA

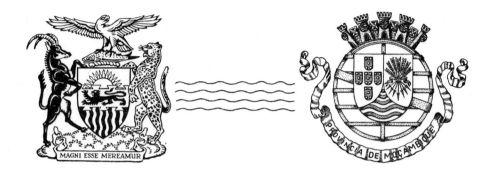

Flora terrarum Zambesii aquis conjunctarum

To the memory of
our Pioneer Botanical Collectors

PADRE JOÃO DE LOUREIRO (1710-1791)

THOMAS BAINES (1822-1875)

SIR JOHN KIRK (1832-1922)

CLEOME DIANDRA Burch.

Adapted by Miss L. M. Ripley from a water-colour sketch by Mrs H. E. Lugard

FLORA ZAMBESIACA

MOZAMBIQUE
FEDERATION OF RHODESIA AND NYASALAND
BECHUANALAND PROTECTORATE

VOLUME ONE: PART ONE

Edited by

A. W. EXELL and H. WILD

on behalf of the Editorial Board:

J. P. M. BRENAN
Royal Botanic Gardens, Kew

A. W. EXELL
British Museum (Natural History)

F. A. MENDONÇA
Junta de Investigações do Ultramar, Lisbon

H. WILD
Federal Department of Agriculture, Salisbury

Published on behalf of the Governments of
Portugal
The Federation of Rhodesia and Nyasaland
and the United Kingdom by the
Crown Agents for Oversea Governments and Administrations,
4, Millbank, London, S.W.1
January 28, 1960

*Printed at the University Press Glasgow
by Robert MacLehose & Company Limited*

PREFACE

WHEN THE *Flora of Tropical Africa* began publication in 1868 as part of a series of Floras initiated by Sir William Hooker and based mainly on the collections from British overseas territories then rapidly accumulating at Kew, the area drained by the River Zambezi was little known botanically although collections of plants had been made by a few pioneers like Loureiro, Chapman, Baines, Kirk and Peters. Indeed at that time many other parts of tropical Africa were no better known, and the need for an amplified and modernized treatment of the flora, especially that part of it covered by the earlier volumes of the *Flora of Tropical Africa*, has long been obvious. To meet this requirement it has been decided, for practical reasons, to produce regional Floras rather than to attempt an up-to-date version of the *Flora of Tropical Africa*. Publication of some of these has already begun, and with this Preface we introduce another.

Although Portuguese settlements on the coast of Mozambique date from the early 16th Century, European occupation of what is now the Rhodesias did not begin until 1890, so that it is understandable that the Zambezi area as a whole has only recently become sufficiently investigated botanically to make the production of a regional Flora practicable.

Southern tropical Africa—the southern half of the Sudano-Zambezian phytogeographical region—ought perhaps to be treated as a floristic unit but, before the present Flora was envisaged, a regional work of a floristic nature on the Portuguese territory of Angola (*Conspectus Florae Angolensis*), initiated by the late Prof. Carrisso, had already been started in 1937. This continued a long tradition of Anglo-Portuguese collaboration and facilitated negotiations for a joint work on the flora of the Federation of Rhodesia and Nyasaland and its eastern neighbour Mozambique. It was felt that it would be wrong to neglect such a considerable area as the Bechuanaland Protectorate, and the decision has been made to include this also, as well as, for convenience, the Caprivi Strip of South West Africa which connects Bechuanaland with Northern Rhodesia. A title is needed and, to avoid a long and unwieldy one, *Flora Zambesiaca* has been chosen, appropriate for an international project and for territories linked by one of the greatest of African rivers.

Flora Zambesiaca has meant collaboration between the Governments of the Federation of Rhodesia and Nyasaland, Portugal and the United Kingdom, who are sharing the cost between them ; as well as the felicitous co-operation of a number of botanical institutions : Department of Botany, British Museum (Natural History) ; Royal Botanic Gardens, Kew ; Centro de Botânica, Junta de Investigações do Ultramar, Lisbon ; Government Herbarium, Salisbury, Federation of Rhodesia and Nyasaland.

It is natural that, owing to the size of their collections, the possession of much type-material and the immense library facilities available, a great part of the work involved in the production of *Flora Zambesiaca* will be done at the British Museum and Kew, and it is a great pleasure to us to put our resources at the disposal of this project and to welcome to our respective institutions members of the *Flora Zambesiaca* staff and those botanists from the Federation and from Portugal engaged in the work.

Much of the demand and support for the project has come from the Agricultural and Forestry Departments of the rapidly developing territories of the area covered, and it is our hope that *Flora Zambesiaca* will satisfy their needs and at the same time stimulate local interest in the botany of the region by providing the inhabitants as a whole with the means of recognizing the plants which grow around them.

J. E. DANDY

Department of Botany, British Museum
(Natural History)

G. TAYLOR

Royal Botanic Gardens, Kew

1 January 1960

PREFÁCIO

QUANDO EM 1868 começou a publicação da *Flora of Tropical Africa*, fazendo parte da série de Floras iniciada por Sir William Hooker, fundamentadas nas colecções que então se acumulavam ràpidamente em Kew, provenientes mormente de territórios britânicos ultramarinos, a área da bacia do Zambeze era botânicamente pouco conhecida, apesar de terem sido feitas colecções de plantas por um pequeno número de pioneiros : Loureiro, Chapman, Baines, Kirk, Peters, etc. É verdade que naquele tempo muitas outras partes da África tropical não eram melhor conhecidas e desde há muito se fazia sentir a necessidade de estudo mais amplo e modernizado da flora, especialmente a parte abrangida pelos volumes mais antigos da *Flora of Tropical Africa*. Para satisfazer este objectivo decidiu-se, por razões de ordem prática, que se fizessem Floras regionais, de preferência a tentar uma nova versão da *Flora of Tropical Africa*. Já se começou a publicação de algumas daquelas e com este Prefácio apresentamos outra.

Se bem que o estabelecimento dos Portugueses na costa de Moçambique date do começo do século dezasseis, a ocupação europeia do que é agora território rodesiano não começou senão em 1890. É, assim, compreensível que a área do Zambeze em toda a sua extensão só nos tempos recentes tenha sido suficientemente estudada do ponto de vista botânico, de modo a ser praticável a publicação de uma Flora regional.

A África austro-tropical—a metade meridional da região fitogeográfica sudano-zambeziana—devia, talvez, ser tratada como uma unidade florística, mas antes da presente Flora ter sido encarada, já tinha começado em 1937, por iniciativa do falecido Prof. Carrisso, uma obra regional de natureza florística concernente o território português de Angola—*Conspectus Florae Angolensis*. Esta obra continuava a velha tradição de colaboração anglo-portuguesa e facilitou as negociações para um trabalho conjunto sobre a flora da Federação das Rodésias e Niassalândia e o seu vizinho oriental, Moçambique. Pensou-se que seria erro deixar de lado uma tão consideravel área como a do Protectorado de Bechuanalândia, e tomou-se a decisão de incluir esta também, assim como, por conveniência, o Caprivi Strip do Sudoeste africano que liga a Bechuanalândia com a Rodésia do Norte. Precisava-se de um título e, para evitar um muito longo e deselegante, foi adoptado *Flora Zambesiaca*, muito apropriado para um projecto internacional e territórios ligados por um dos rios mais imponentes da África.

A *Flora Zambesiaca* envolve a colaboração dos governos da Federação das Rodésias e Niassalândia, de Portugal, e do Reino Unido que entre si custeiam os encargos da obra, e a feliz cooperação das instituições botânicas : Department of Botany, British Museum (Natural History) ; Royal Botanic Gardens, Kew ; Centro de Botânica da Junta de Investigações do Ultramar, Lisboa ; Government Herbarium, Salisbury, Federation of Rhodesia and Nyasaland.

É natural que, devido à riqueza das suas colecções, à posse de tipos e às imensas facilidades bibliográficas de que dispõem, uma grande parte do trabalho relativo à elaboração da *Flora Zambesiaca* seja feito no British Museum e em Kew, e para nós é um grande prazer pôr os nossos recursos à disposição deste projecto e de receber nas nossas instituições o pessoal da *Flora Zambesiaca* e os botânicos de Portugal e da Federação que trabalham na obra.

Os incitamentos e o apoio ao projecto vieram em grande parte dos Serviços de Agricultura e Florestas dos territórios da nossa área em rápido desenvolvimento e o que nós esperamos é que a *Flora Zambesiaca* corresponda às suas necessidades e ao mesmo tempo estimule o interesse local pela botânica da região, dando a todos os habitantes os meios de reconhecerem as plantas que crescem à sua volta.

J. E. DANDY

Department of Botany, British Museum
(Natural History)

G. TAYLOR

Royal Botanic Gardens, Kew

1 de Janeiro de 1960

CONTENTS

LIST OF NEW NAMES PUBLISHED IN THIS WORK

LIST OF FAMILIES INCLUDED IN VOL. I, PART 1

GYMNOSPERMAE

1. Cycadaceae
2. Podocarpaceae
3. Cupressaceae

ANGIOSPERMAE

4. Ranunculaceae
5. Dilleniaceae
6. Annonaceae
7. Menispermaceae
8. Berberidaceae
9. Cabombaceae
10. Nymphaeaceae
11. Papaveraceae
12. Fumariaceae
13. Cruciferae
14. Capparidaceae
15. Resedaceae
16. Violaceae
17. Bixaceae
18. Flacourtiaceae (incl. Samydaceae)
19. Pittosporaceae
20. Polygalaceae

INTRODUCTION

THE *Flora Zambesiaca* area is situated in south-central Africa between latitudes 8° 10′ and 26° 50′ S. and longitudes 20° and 40° 50′ E. It is roughly 2,740,000 square kilometres (1,069,000 square miles) in area and, apart from the Lourenço Marques District and part of the Sul do Save Province in Mozambique, and the southern part of Bechuanaland Protectorate, lies within the tropics. In addition to the territories of the Federation, Mozambique and Bechuanaland Protectorate, the Caprivi Strip is included by arrangement with Dr. R. A. Dyer, Chief of the Division of Botany, Pretoria, since it projects as a narrow wedge into our area. Apart from areas draining into the Sabi, Limpopo, Okovango, Congo, Lake Tanganyika and Rovuma systems the Flora falls within the drainage basin of the Zambezi River, hence the title of the Flora. Altitudes vary from sea-level to almost 3,000 m. on Mt. Mlanje. Annual rainfall varies between about 25 cm. in the drier parts of Bechuanaland and S. Rhodesia and more than 250 cm. on Mt. Mlanje and the Chimanimani Mts. The vegetation varies in accordance with the rainfall between a very dry type of bush or scrub in south-western Bechuanaland and evergreen forest on the slopes of our higher mountains with ericoid scrub and grassland at the highest altitudes.

Phytogeographically the region is quite complex with a number of widely divergent influences but perhaps the most striking feature of the vegetation is the very large area covered by species of *Brachystegia* which dominate the woodlands of our plateau country between 1,000 and 2,000 m. as well as the watersheds and certain coastal areas of Mozambique. Extending widely in the north into Tanganyika, the Belgian Congo, Angola and beyond they stop abruptly in the south just north of the Limpopo R., no species of *Brachystegia* ever having been recorded south of our area. Second in importance are the large areas covered by the tree *Colophospermum mopane*, often interchanging with areas dominated by *Acacia*, which stretch from near the coast of Angola through all the low-lying areas of our region below about 1,000 m. to the coastal areas of Mozambique. Along our southern boundary a very close affinity naturally exists between our flora and that of the Union of S. Africa and particularly with the floras of Natal and the Transvaal, but of special interest is the affinity between the mountain flora of the border country between S. Rhodesia and Mozambique and that of the Nyasaland mountains and the flora of the Cape. This Cape element in our flora is associated with an Afro-montane element along the mountains of East Africa and extending northwards to Ethiopia. Both elements as they relate to our flora have been exhaustively analysed by Weimarck in his *Groups, Centres and Intervals within the Cape Flora* (Lund Univ. Arssk., N.F.Avd. 2, **37** (1951)). In addition, there is a strong representation of SW. African species in Bechuanaland Protectorate which spreads further east at times into the relatively dry Kalahari sand regions of N. Rhodesia (Barotseland) and S. Rhodesia. It is here also that affinities with the Angolan flora are most marked. The SW. African element is predominantly xerophytic in type but the Angolan element is often relatively mesophytic. The occurrence of evergreen forest is usually associated with the high rainfall of our mountain areas. Such forests are usually small and of a relic nature but they are sometimes rich in species. One of the few well-developed evergreen forests in S. Rhodesia is at Chirinda but they are more frequent on the mountains of Nyasaland and Mozambique. There is also a tendency towards evergreen forest development in the Western and Northern

Provinces of N. Rhodesia because of a relatively high annual rainfall of about 150 cm. These latter forests are naturally very similar to those of the nearby Belgian Congo and Angola and indeed all our forest patches have a high proportion of species with W. African affinities, and genera like *Lovoa*, *Entandrophragma*, *Homalium*, *Strombosia*, *Cola*, etc. are well represented. There is also a small but interesting Madagascan element in our flora. Such genera as *Neopalissya*, *Faurea*, *Bivinia*, *Coleotrype*, *Pseudocalyx* and *Dianella* offer examples of this type of distribution and are usually found on the mountains of S. Rhodesia, Nyasaland and Mozambique. Finally the coastal area of Mozambique shows the affinities one would expect with the coastal floras of Tanganyika and Natal and, in addition, a certain lowland element of Madagascan affinity (e.g. *Brexia madagascariensis*).

The sequence of families that is being followed is that of Bentham and Hooker except that the position of a few families has been changed in the light of more modern research into their relationships with other families. The key to the families has been prepared for us by Mr. J. E. Dandy and we are most grateful to him for his assistance in this regard.

New taxa described in the course of preparing this volume were published in a series entitled *New and little known Species from the Flora Zambesiaca Area* in the *Boletim da Sociedade Broteriana*. This arrangement was made possible through the great kindness of Professor A. Fernandes of the University of Coimbra, editor of this journal. Future new taxa will also be published in the *Anais da Junta de Investigações do Ultramar*.

The taxonomic treatment of the families themselves is as uniform as possible without interfering too much with the liberty of the individual authors and allowing for the fact that recent research has concentrated on the elucidation of some families more than others.

The full synonymy of the individual species has not been given but only that which concerns our area or which is necessary to explain the nomenclature and the concept of the species which has been adopted. The same criteria have been adopted in the choice of bibliographical references. Reference to the type specimen or specimens (syntypes) of a species or of its synonyms is made in detail if they come from our area but, where the types have been collected outside our area, only the country of origin is given, as in such cases the decision as to what constitutes the type may present difficulties best investigated by the authorities on those regions. In spite of this latter proviso every effort is being made to trace and see all the types of the species included in the Flora. The descriptions of the species are not meant to be complete but are mainly designed to help distinguish any one species from the related ones occurring in our area. The terms employed have been chosen with a view to their being internationally understood. This is desirable in any case but in particular will facilitate the use of the Flora by the inhabitants of Mozambique.

The specimens chosen for citation are arranged topographically by territories. Each territory is divided into divisions according to the map which can be found in the pocket of the back cover and in general one specimen is cited for each division in which the species occurs. The size of the divisions is necessarily rather large in order to save space but in choosing their boundaries some attempt has been made to divide the territories into divisions which are fairly uniform ecologically. At the same time, to make the rapid location of a specimen easier, existing provincial or magisterial district boundaries have been made use of where possible. These divisions are therefore of rather hybrid origin and, especially as provincial boundaries have a tendency to change from time to time, the map in the end cover must be consulted if specimens are to be accurately localized. A list of these divisions together with the territories to which they relate is given on page 21.

Following on the abbreviation for the division, the specimen is more exactly localized and then follows an indication of the condition of the material (fl. = flowering ; fr. = fruiting ; st. = sterile) on the date collected. Next follows the collector's name with the number of the specimen if known and the standard abbreviations of the herbaria in which specimens have been seen. The general distribution of the species in the rest of the world is given followed by a note on its ecology in our area. This last item depends largely on the information given by collectors which is often very brief or may be lacking altogether. Every effort has been made, however, to make this information reliable and the terminology has been standardized as far as possible by adopting with a few modifications the scheme devised by Dr. P. J. Greenway and used in the *Flora of Tropical East Africa*. We are most grateful for Dr. Greenway's permission to do this. Further explanatory botanical or nomenclatural notes may follow here, if necessary. Although it is not practicable to figure every species, most of the genera represented are illustrated in the black and white plates.

In conclusion the editing of this volume has been the responsibility of those members of the editorial board most concerned with the writing of the particular families included herein. It is likely that later volumes will be edited by different members of this board as is convenient and taxonomically desirable. Meanwhile the present editors wish to acknowledge with deep gratitude all the assistance they have received in the production of this volume. Firstly, we must thank all those administrators and officials whose understanding and encouragement have led to the provision of the funds necessary to carry out the work and in particular to the President of the Executive Board of the Junta de Investigações do Ultramar, Professor Carrington da Costa, Comandante G. Teixeira, Governor-General of Mozambique, and Mr. J. M. Caldicott, until recently Minister of Agriculture in the Federation of Rhodesia and Nyasaland, who gave us their whole-hearted support from the very instigation of the project in 1955. On the organizational and administrative side, Dr. G. R. Bates, Chief Botanist and Plant Pathologist in the Federal Department of Agriculture, Salisbury, has borne much of the burden through the years when we were trying to get the project started and has been of constant help to us ever since.

Secondly, we are much indebted to the Directors of those two great institutions the British Museum (Natural History) and the Royal Botanic Gardens, Kew, for acceding to the request made by our governments to assist us in the preparation of this Flora.

To Mr. A. W. Exell, co-editor of this first volume, we owe our greatest thanks for all his help throughout the whole history of the undertaking. Although Mr. Exell has played such a notable part in the production of this first volume, he has left it to us, as representing the territories for whom the work is being done, to write this introduction.

Mr. J. E. Dandy of the British Museum and Mr. E. Milne-Redhead of Kew deserve our very warm thanks for their invaluable assistance in editorial matters. We have also had, of course, the constant support and assistance of our own *Flora Zambesiaca* staff. To them and to our colleagues at the British Museum and Kew, who have given us generous help on many occasions we are much indebted. We are most grateful to the Directors of the institutions which have lent us material and particularly for the loan of type specimens. Large amounts of material have also been lent to Kew and the British Museum by those herbaria officially linked with the *Flora Zambesiaca* project but in addition we must specially thank Dr. R. A. Dyer and his staff of the National Herbarium, Pretoria, for the loan of their very large collections from our area and for frequent collaboration in the elucidation of problems common to the flora of S. Africa and the *Flora Zambesiaca* and Professor A. Fernandes, Director of the Instituto Botânico " Dr. Julio Henriques ", Coimbra, for his much valued advice and co-operation in the loan of material from the Coimbra Herbarium. Professor Quintanilha, Director of the Centro de Investigação Científica do Algodão, Lourenço Marques, and his staff and Eng.º Martins da Silva, Chief of Serviços Técnicos de Agricultura, have also lent us large amounts of material from the herbaria of their institutions and are collaborating actively in the collection of further specimens.

We have also been very fortunate in that a number of foreign specialists have contributed to this volume those families on which they are authorities. As these families are among those offering particular difficulties we are especially grateful to the authors, namely, Dr. G. Cufodontis, Professor P. Duvigneaud, and Dr. G. Troupin.

The great majority of the illustrations in this volume are by Miss L. M. Ripley and Miss G. W. Dalby. We congratulate them on the accuracy and attractiveness of their work.

The cost of the coloured frontispiece has been provided by the National Publications Trust of Salisbury, S. Rhodesia. We thank them for this handsome addition to our Flora.

The map showing the geographical divisions used in citing specimens for the Flora was prepared for us through the kind offices of the Director of Overseas Surveys to whom we are most grateful.

Finally, the work of writing a Flora would be impossible without the help of numerous collectors in the field. We thank them for all their efforts in the past and hope for many interesting and well-collected specimens from them in the future.

                              F. A. MENDONÇA and H. WILD

INTRODUÇÃO

A ÁREA DA *Flora Zambesiaca* situa-se entre 8° 10′ e 26° 50′ de latitude Sul e e 20° e 40° 50′ de longitude Este. Tem aproximadamente 2,740,000 quilómetros quadrados de superfície e, a não ser a parte meridional do Protectorado da Bechuanalândia e em Moçambique os distritos de Lourenço Marques e Gaza, fica compreendida na zona tropical. Em adição aos territórios de Moçambique, Federação e Protectorado, incluímos o do Caprivi Strip, mediante entendimento com o Dr. R. A. Dyer, Chief of the Division of Botany, Pretória, visto que aquele penetra como estreita cunha na nossa área. Com excepção de certas áreas periféricas que drenam para os sistemas do Save e Limpopo, Ocovango, Congo, Lago Tanganica e Rovuma, a Flora implanta-se na bacia do Zambeze—e daí o título de *Flora Zambesiaca*. As altitudes sobem do nivel do mar até quase 3,000 m. no Mt. Mlanje. A precipitação anual de chuva varia entre cerca de 25 cm. nos lugares mais áridos da Bechuanalândia e Rodésia do Sul e mais de 250 cm. em Mt. Mlanje e Chimanimani. A vegetação varia, conforme a pluviosidade, desde o tipo muito seco arbustivo ou sufruticoso no sudoeste da Bechuanalândia, até à floresta sempervirente das encostas das nossas mais altas montanhas com sufrutices ericóides e relvados nas altitudes mais elevadas.

Do ponto de vista fitogeográfico a região é assaz complexa por virtude de múltiplas influências amplamente divergentes, mas a vegetação que mais vivamente impressiona é, porventura, a vastíssima área coberta por *Brachystegia* cujas espécies dominam nos bosques do planalto, entre 1,000 e 2,000 m., vertentes e certas áreas litorais de Moçambique, estendendo-se largamente para o norte no Tanganica, Congo Belga e Angola, enquanto que do lado sul termina abruptamente quase a tocar no Limpopo. Nenhuma espécie de *Brachystegia* foi jamais registada ao sul da nossa área. Segunda em importância é a grande área de *Colophospermum mopane*, de ordinário entremeada de áreas em que as espécies de *Acacia* dominam, a qual se estende encostada à precedente, desde o sublitoral do sul de Angola, através das áreas da nossa região subjacentes à curva dos 1,000 m., até o sublitoral de Moçambique. Em toda a extensão da fronteira meridional da mesma área existe naturalmente muito estreita afinidade da nossa flora com a da União da Africa do Sul, particularmente com as floras do Natal e do Transval. É, porém, de especial interesse a afinidade das floras das serras limítrofes de Moçambique e Rodésia do Sul, da Niassalândia e norte de Moçambique, com a flora do Cabo. O elemento capense da nossa flora está associado com o afromontano das montanhas da África Oriental que se estende para o norte até a Abissínia. Ambos têm sido exaustivamente estudados nas suas relações com a nossa flora, nomeadamente por Weimarck, em *Groups, Centres and Intervals within the Cape Flora* (Lund Univ. Arssk., N.F.Avd. 2, **37** (1951)). Por outro lado, há uma forte representação de espécies sudoesteafricanas na flora do Protectorado da Bechuanalândia, as quais às vezes se expandem mais para leste em regiões relativamente secas de areias do Calaari da Rodésia do Norte (Barotseland) e Rodésia do Sul. É tambem aqui que as afinidades com a flora de Angola são mais acentuadas. O elemento sudoesteafricano é predominantemente do tipo xerofítico, enquanto que o angolano é muitas vezes relativamente mesofítico. A presença da floresta sempreverde está ordinariamente relacionada com a elevada precipitação de chuvas nas áreas montanhosas. Estas florestas são, em regra, de natureza residual ou relíquias de pequena extensão, mas às vezes ricas de espécies. São em pequeno número as florestas sempreverdes bem desenvolvidas, como a de Chirinda, na Rodésia do Sul, e poucas mais nas montanhas de Moçambique e Niassalândia. É, porém, notória a tendência para o desenvolvimento da floresta sempreverde nas províncias do norte e oeste da Rodésia do Norte, por virtude da elevada precipitação de chuvas, de cerca de 150 cm. Estas últimas florestas são naturalmente muito semelhantes às dos territórios vizinhos do Congo Belga e de Angola, e não há dúvida que todas as nossas manchas de floresta têm elevada proporção de espécies de afinidade ocidental-africana, e géneros tais como *Lovoa*, *Entandrophragma*, *Homalium*, *Strombosia*, *Cola* etc. são bem representados. Há também que considerar o elemento malgaxe, pouco representado mas interessante na nossa flora. Géneros tais como *Neopalissya*, *Faurea*, *Bivinia*, *Coleotrype*, *Pseudocalyx* e *Dianella* exemplificam este tipo de distribuição e encontram-se, usualmente, nas montanhas de Moçambique, Niassalândia e Rodésia do Sul. Finalmente, a flora da área litoral de Moçambique mostra naturalmente estreitas afinidades com as floras costeiras da Tanganica e do Natal e, adicionalmente, certo elemento litoral de afinidade madagascariana (e.g. *Brexia madagascariensis*).

A sequência das famílias é a do sistema de Bentham e Hooker, salvo quando a posição dé algumas foi alterada à luz de modernas investigações conducentes ao conhecimento de relações mais íntimas com outras famílias. A clave das famílias foi elaborada por Mr. J. E. Dandy, a quem estamos muito reconhecidos pela valiosa ajuda nesta matéria. Os taxa novos descritos no decurso da preparação deste volume foram publicados em uma série intitulada *New and little known Species from the Flora Zambesiaca Area*, no *Boletim da Sociedade Broteriana*. Este arranjo foi possível por amável oferecimento do Prof. Abílio Fernandes, da Universidade de Coimbra, editor do dito jornal. De futuro, taxa novos serão publicados também nas *Memorias da Junta de Investigações do Ultramar*.

O tratamento das famílias é o mais uniforme possível, sem demasiada interferência com a liberdade individual dos autores e tendo em atenção o facto de que estudos recentes têm incidido na elucidação de algumas famílias mais que em outras. Não é dada a sinonímia completa das espécies, mas apenas a que diz respeito à nossa área, ou a que é necessária para elucidar a nomenclatura da espécie adoptada. O mesmo critério foi seguido no que respeita à selecção de referências bibliográficas. A citação do tipo ou dos sintipos de uma espécie ou dos seus sinónimos é feita em pormenor se são provenientes da nossa área, mas, se provém de outros áreas, apenas é citado o país de origem, para não termos de tocar em problemas que podem ter melhor solução tratados pelas autoridades das respectivas regiões. Não obstante estas limitações, todos os esforços foram feitos para ver os tipos das espécies incluídas na Flora. As descrições não são pormenorizadas, mas delineadas principalmente para facilitar a distinção de cada espécie de outras da mesma área entre si correlacionadas.

A terminologia usada foi escolhida com vista a ser internacionalmente compreensível. Isto é sempre desejável, mas especialmente neste caso, para facilitar o uso da Flora às populações de Moçambique.

Os espécimes escolhidos para citação são ordenados topogràficamente sob os nomes dos territórios (Moçambique, Rodésia, etc.) e estes repartidos em divisões especificadas adiante e marcadas no mapa que se encontra na bolsa interna da capa detrás da primeira parte do volume. Em regra citamos apenas um espécime para cada divisão, onde a espécie foi encontrada. As áreas das divisões são necessàriamente bastante grandes, de modo a permitir poupar espaço, mas na escolha dos seus limites tentamos fazer a partilha dos territórios em divisões ecològicamente tão uniformes quanto possível. Do mesmo passo, para facilitar a localização rápida dos espécimes, usamos, sempre que possível, os limites tradicionais de províncias ou dos distritos administrativos. Estas divisões são, por consequência, de origem assaz híbrida, e, visto que os limites dos distritos tendem a mudar de vez em quando, é preciso consultar o mapa para localizar correctamente a proveniência dos espécimes. Damos na página 21 a lista destas divisões, sob o nome dos territórios a que dizem respeito.

Depois da abreviatura topográfica, segue-se a localidade mais exacta, e logo a indicação do estado do espécime (fl. = em flor ; fr. = em fruto ; st. = esteril), a data da colheita, o nome do colector, o número do espécime (quando o tem), e as abreviaturas internacionais dos herbários onde os espécimes foram vistos. A distribuição da espécie no mundo é seguida de uma nota acerca da sua ecologia na nossa área. Esta última depende principalmente das informações dadas pelos colectores as quais muitas vezes são breves, deficientes, ou faltam. Fizemos todos os esforços para verificar esta informação, e a terminologia foi uniformizada tanto quanto possível pela adopção, com pequenas modificações, do sistema do Dr. P. J. Greenway, usado na *Flora of East Tropical Africa*. Estamos muito gratos ao Dr. Greenway por nos ter permitido isto. Quando necessário são adicionadas notas explicativas, botânicas ou nomenclaturais. Se bem que seja impossível ilustrar todas as espécies, quase todos os géneros são iconografados em estampas a preto e branco.

Em conclusão, a edição do presente volume é da responsabilidade dos membros da Comissão Editorial que mais a fundo trabalharam na elaboração das famílias aqui tratadas. Como é desejàvel, e conveniente do ponto de vista taxonómico, os volumes ulteriores serão muito provàvelmente editados por outros membros da mesma Comissão Editorial. Entretanto, os actuais editores desejam agradecer calorosamente toda a ajuda recebida na preparação deste volume. Em primeiro lugar, agradecemos às entidades oficiais a alta compreensão e encorajamento que se traduz no provimento dos fundos necessários à realização da obra, particularmente ao Presidente da Comissão Executiva da Junta de Investigacões do Ultramar, Prof. J. Carrington da Costa ; Governador-geral de Moçambique, Commandante Gabriel Teixeira ; e Mr. J. M. Caldicott, até recentemente Minister of Agriculture in the Federation of Rhodesia and Nyasaland, que nos deram pleno apoio desde o início da execução do projecto, em 1955. No âmbito da organização administrativa, o Dr. G. R. Bates, Chief Botanist and Plant Pathologist of the Federal Dept. of Agriculture, Salisbury, apoiou o projecto em multiplas diligências durante anos, enquanto tentávamos pô-lo em obra, e desde então tem sido constante o seu auxílio.

Por outro lado estamos muito obrigados aos Directores das duas grandes instituições, British Museum of Natural History e Royal Botanic Gardens, Kew, por terem acedido ao pedido dos nossos Governos de nos darem ajuda na preparação desta Flora.

Mr. A. W. Exell, co-editor deste primeiro volume, é credor do nosso mais vivo reconhecimento pelo auxílio constante desde o início do empreendimento. Se bem que Mr. Exell tenha desempenhado papel tão notável na preparação deste volume, deixou-nos a nós, como representantes dos territórios a que a Flora diz respeito, o encargo de elaborar a presente introdução.

Mr. J. E. Dandy, do British Museum, e Mr. E. Milne-Redhead, de Kew, são credores dos nossos mais calorosos agradecimentos pela sua inestimável ajuda em questões editoriais. Tivemos tambem, naturalmente, constante apoio e auxílio do pessoal próprio da *Flora Zambesiaca*. A estes e aos nossos colegas do British Museum e Kew que em muitas ocasiões nos deram o seu generoso auxílio, estamos muito obrigados. Estamos muito gratos aos Directores das instituições que nos emprestaram materiais, particularmente pelo empréstimo de tipos. Grande massa de materiais foram tambem recibidos por empréstimo, em Kew e no British Museum, dos herbários oficialmente interessados no projecto da *Flora Zambesíaca*. Especialmente, cumpre-nos agradecer ao Dr. R. A. Dyer e seu pessoal, do National Herbarium, Pretória, o empréstimo das suas vastas colecções da nossa área e a frequente colaboração no estudo de problemas comuns á flora sul-africana e *Flora Zambesiaca* e ao Professor A. Fernandes, Director do Instituto Botânico " Dr. Julio Henriques ", Coimbra, pela sua muito apreciada opinião e pela co-operação no empréstimo de material do Herbário que dirige. Ao Prof. A. Quintanilha, Director do Centro de Investigação Científica do Algodão, e Eng.º Martins da Silva, Chefe da Repartição Técnica de Agricultura, Lourenço Marques, agradecemos o empréstimo dos materiais dos herbários das suas instituições e a colaboração activa do seu pessoal na colheita de novos materiais de herbário.

Tivemos também a boa sorte de certo número de especialistas estrangeiros terem contribuído para este volume com o estudo de famílias de que são autoridades. Como estas famílias são as que em regra oferecem especiais dificuldades, estamos muito gratos a estes autores : Dr. G. Cufodontis, Prof. P. Duvigneaud, e Dr. G. Troupin.

A grande maioria dos desenhos deste volume foram feitos por Miss L. M. Ripley e Miss G. W. Dalby. Felicitamo-las pela perfeição e elegância daqueles. O frontispício colorido foi custeado pelo National Publications Trust of Salisbury, Rodésia do Sul. Agradecemos-lhes esta bela adição à nossa Flora.

O mapa das divisões geográficas usadas para citação de espécimes da Flora foi para nós preparado por amável deferência do Director of Overseas Surveys, a quem estamos muito obrigados.

Finalmente, elaborar a Flora teria sido impossível sem o auxílio do trabalho de campo de numerosos colectores. A todos agradecemos os seus esforços do passado, e de futuro esperamos deles novos espécimes, interessantes e bem preparados.

F. A. MENDONÇA e H. WILD

ABBREVIATIONS

The following abbreviations are used for works frequently cited:

C.F.A. : Conspectus Florae Angolensis.
F.C. : Flora Capensis.
F.C.B. : Flore du Congo Belge et du Ruanda-Urundi.
F.P.F.T. : A Manual of the Flowering Plants and Ferns of the Transvaal and Swaziland.
F.T.A. : Flora of Tropical Africa.
F.T.E.A. : Flora of Tropical East Africa.
F.W.T.A. : Flora of West Tropical Africa.
N.C.L. : Check-lists of the Forest Trees and Shrubs of the British Empire No. 2, Nyasaland Prot. (1936).
T.T.C.L. : Check-lists of the Forest Trees and Shrubs of the British Empire No. 5, Tanganyika Territory, Part II (1949).

Citation of specimens

Collectors' names have been abbreviated in two instances :
E.M. & W. = Exell, Mendonça and Wild.
F.N. & W. = Fries, Norlindh and Weimarck.

The following sequence of geographical divisions has been used : N. W. C. E. S. signifying north, west, central, east and south respectively unless otherwise indicated below :

Caprivi Strip.
Bechuanaland Protectorate : N ; SW ; SE.
Northern Rhodesia : B (Barotseland) ; N ; W ; C ; E ; S.
Southern Rhodesia : N ; W ; C ; E ; S.
Nyasaland : N ; C ; S.
Mozambique : N (Niassa) ; Z (Zambezia) ; T (Tete) ; MS (Manica e Sofala) ; SS (Sul do Save) ; LM (Lourenço Marques).

Citation of Herbaria

Herbaria are cited according to the internationally accepted *Index Herbariorum* ed. 3 (1956) but for those to whom the international index is not available a list is given below of the abbreviations used in *Flora Zambesiaca* Vol. 1, part 1.

B : Berlin-Dahlem, Botanisches Museum.
BM : London, British Museum (Natural History).
BOL : Capetown, Bolus Herbarium.
BOLO : Bologna, Istituto ed Orto Botanico.
BR : Brussels, Jardin Botanique de l'État.
BRLU : Brussels, Laboratoire de Botanique Systématique, Université Libre de Bruxelles.
CN : Caen, Institut Botanique.
COI : Coimbra, Instituto Botânico " Dr. Julio Henriques ".
E : Edinburgh, Royal Botanic Garden.
EA : Nairobi, East African Herbarium.

FHO :	Oxford, Forest Herbarium.
FI :	Florence, Herbarium Universitatis Florentinae.
G :	Geneva, Conservatoire et Jardin Botaniques.
GB :	Göteborg, Botaniska Trädgård.
GRA :	Grahamstown, Herbarium of the Albany Museum.
IMI :	Kew, Commonwealth Mycological Institute.
J :	Johannesburg, Moss Herbarium.
K :	Kew, Royal Botanic Gardens.
L :	Leiden, Rijksherbarium.
LD :	Lund, Botanical Museum and Herbarium.
LISC :	Lisbon, Centro de Botânica da Junta de Investigações do Ultramar.
LISJC :	Lisbon, Jardim e Museu Agrícola do Ultramar.
LISU :	Lisbon, Instituto Botânico da Universidade de Lisboa (formerly " Escola Polytéchnica ").
LM :	Lourenço Marques, Repartição Técnica de Agricultura de Moçambique.
LMJ :	Lourenço Marques, Junta de Algodão.
M :	Munich, Botanische Staatssammlung.
MO :	St. Louis, Missouri Botanical Garden.
NDO :	Ndola, Forestry Department Herbarium (now at Kitwe).
NH :	Durban, Natal Herbarium.
NY :	New York, Botanical Garden.
P :	Paris, Muséum National d'Histoire Naturelle.
PO :	Oporto, Instituto de Botânica " Dr. Gonçalo Sampaio ".
PRE :	Pretoria, National Herbarium.
S :	Stockholm, Naturhistoriska Riksmuseum.
SAM :	Capetown, South African Museum Herbarium.
SRGH :	Salisbury, Government Herbarium.
UPS :	Uppsala, Institute of Systematic Botany.
US :	Washington, U.S. National Museum.
W :	Vienna, Naturhistorisches Museum.
WAG :	Wageningen, Laboratory for Plant Taxonomy.
WU :	Vienna, Botanisches Institut der Universität.
Z :	Zürich, Botanischer Garten und Museum.
ZOM :	Zomba, Department of Agriculture.

HISTORY OF BOTANICAL COLLECTING IN THE FLORA ZAMBESIACA AREA

By A. W. Exell

Information is gradually being collected about the botanical exploration of our area and it is hoped some day to publish a more complete account than it is possible to give here. An attempt has been made, however, to deal fairly fully with the earlier collectors and the itinerary of Kirk (by Miss G. Hayes) has been appended since the information is otherwise not easily obtainable. As well as for the Kirk itinerary I am indebted to Miss Hayes for much of the biographical details about the collectors, dates, etc. In spite of the conveniences of a straight-forward alphabetical list, I have decided on the more interesting method of trying to give a picture of the various phases to which it seems possible to allocate the collectors. There is naturally some overlap both geographical and chrono-logical. In such cases the collectors have been listed under the headings to which they seem most characteristically to belong. Although our area was, in general, one of the last parts of the world to be botanically explored, nevertheless the list of collectors is a long one and it will be realized that many modern collectors (especially since 1930) have had to be omitted in order to include the historically interesting pioneers.

1. *18th-Century Collectors*

The *Flora Zambesiaca* area was botanically unknown until towards the end of the 18th Century. Although the Mozambique coast was partially explored by Vasco da Gama early in 1498 and Portuguese settlements made at Sofala, Quelimane and Mozambique early in the 16th Century, there is no specimen in the Sloane Herbarium and the flora was unknown to Linnaeus and his con-temporaries.

João de Loureiro (1710–1791) a Portuguese priest and missionary of the Society of Jesus made small collections on the east coast of Africa in 1781–82 in the course of voyages to the East. Some of these specimens are unlocalized : 9 of them are known to have been collected in Mozambique. Loureiro is botanically famous for his *Flora Cochinchinensis* (1790) and in that work specimens from our area are cited for the first time in botanical literature. Such specimens as survive are in BM, LISU and P (see list of herbarium abbreviations, p. 21). An account of his plants has been published by E. D. Merrill (in Trans. Amer. Phil. Soc., New Ser. **24**, 2 : 1 (1935)).

Two other collections were made before 1800 by *da Silva* and *Surcouf.*

Manuel Galvão da Silva, Portuguese, head of a Portuguese scientific mission sent to India and Mozambique in 1783. He collected in Mozambique from 1784–88 principally in the provinces of Manica e Sofala and Tete, the collections amounting in all to 256 specimens. These were sent to Lisbon and in 1808 removed to Paris, at the order of St.-Hilaire, during the French occupation. This was a tragedy, for a collection of that size made at that period would have been interesting. Some at least of the specimens are still in the Paris Herbarium but no systematic search has been made for them. A complication is that they were mixed or confused with collections from Brazil. The type of *Rhexia princeps* Bonpl. (now *Dissotis princeps* (Bonpl.) Triana) is probably one of da Silva's specimens (see A. & R. Fernandes, Contrib. Conhec. Melast. Moçamb. :

42 (1955)). Owing to this accident of war and cupidity, Portugal lost the credit for the first official scientific expedition to our area.

Robert Surcouf (1773–1827) French, born at St. Malo, was a naval officer who collected in Mauritius and Madagascar and is said to have visited the coast of Mozambique in 1795. Specimens (K, P, PRE).

2. Early 19th-Century Collectors in Mozambique

After the three 18th-Century collectors there was a great gap. During this period the eastern part of tropical Africa was largely in the grip of Arab slave-traders. It was a period, however, when the Horticultural Society of London (now the Royal Horticultural Society) was sending collectors to many parts of the world and *John Forbes* (1798–1823) was sent in 1822 on H.M.S. *Leven*, which had been commissioned to make a survey of the east coast of Africa. Forbes died at Sena; specimens at BM and K, many of them cited in the *Flora of Tropical Africa*. He was the first British collector in our area.

Towards the end of the first half of last century came the first two " foreign " collectors in the sense of their not belonging to the countries sponsoring *Flora Zambesiaca*: Cavaliere *Carlo Antonio Fornasini* (1805–68), Italian, and *Dr. Wilhelm Carl Hartwig Peters* (1815–83), German, whose collections were approximately contemporaneous.

Fornasini collected in and around Inhambane from 1839 onwards and sent his plants back to *Giuseppe Bertoloni* of Bologna who described and figured many of them in a series of dissertations entitled *Illustrazione di Piante Mozambicesi* (1850–55) first published in Mem. Accad. Sci. Inst. Bologn. The plates, mostly drawn and lithographed by C. Bettini, were probably the first illustrations made from specimens from our area. His specimens are in BOLO and FI.

Peters made his celebrated voyage to Mozambique in 1842–48 and was collecting in that province from 1843–47. The results were published in *Reise nach Mossambique*, Bot. (1861–64). This substantial and valuable collection was preserved in the Berlin Herbarium and largely destroyed during the last war. It is fortunate that in most cases modern collecting from the type-localities makes identification of the species based on the destroyed types reasonably certain.

3. Chapman and Baines

Up to now the story has been entirely of Mozambique. To whom should be awarded the honour of collecting the first specimen from the inland (British) territories of our area it is difficult to say. Chapman was certainly in Bechuanaland and Southern Rhodesia before 1849, somewhat earlier than Livingstone, but I have not been able to find out which of them first collected plants. It is certain, however, that the *Chapman and Baines* collection was far more important. The various expeditions made separately and jointly by these two cannot be given in detail: some of them were in no way connected with Botany.

James Chapman (1831–72), described as a Commission Agent, was an explorer and adventurer.

Thomas Baines (1822–75), born at King's Lynn, was an artist of repute and an explorer by nature. It was the joint expedition of Chapman and Baines to Lake Ngami in Bechuanaland (1861–63) which provided most of the specimens from our area cited in the earlier volumes of the *Flora of Tropical Africa*. Baines had been appointed as artist to Livingstone's Zambezi Expedition (1858) on the recommendation of the Royal Geographical Society but left it after two years owing to a disagreement with Livingstone. Later he was in Southern Rhodesia during 1869–71. (See J. P. R. Wallis, *Thomas Baines of King's Lynn*, 1941, which gives a list of plants.) Specimens in K.

4. *Early English and Scottish Missionaries*

David Livingstone (1813–73) was the moving spirit. His life and journeys are too well known to need notice here. He collected but few specimens. These are at Kew. His companions on various journeys, to whom the main botanical collections were due, were : *Baines* (see above), *Kirk*, *Meller*, *Stewart* and *Waller*.

Sir John Kirk (1832–1922), born at Barrie, near Arbroath, Scotland, surgeon, accompanied Livingstone on the Zambezi Expedition (1858–63). His botanical collections (at K) are by far the most important of those made on the Livingstone expeditions. Kirk's specimens were mostly dated and as it is often important to localize them accurately a summarized version of his itinerary is appended from which most dated specimens can be localized fairly accurately (see p. 35).

Dr. Charles James Meller (1836?–69) was with Livingstone (1860–63) and collected in Mozambique and Nyasaland, specim. (K).

Rev. Horace Waller (1833–1896), born in London, joined the Universities Mission, coll. (1861–62) in Mozambique and Nyasaland, specim. (K).

Rev. Dr. James Stewart (1831–1905), born in Edinburgh, of the Livingstone Mission, coll. (1862–63) in Mozambique (Manica e Sofala) and Nyasaland, specim. (BM, K).

5. *Mozambique Collections 1850–99*

Mozambique continued almost to the end of the 19th Century to be visited far more frequently and to be collected more intensively than the rest of our area. In addition to the missionaries already mentioned, belonging to the Livingstone orbit, the following is a summary of the principal collectors of the second half of the 19th Century :

Dr. Manuel Rodrigues Pereira de Carvalho (1848–1909), Portuguese, coll. " Mozambique " (1875–86) specim. (COI) ; *Joaquim John Monteiro*, Portuguese, and *Rose Monteiro* (1833–78), British, coll. Lourenço Marques (1876), specim. (K) ; *E. Durand*, French, coll. (1881) Quelimane, Mopeia, Sena, Chemba, Tete, Machinga, Gorongosa and Manica, 250 specim. (P) ; *Henry O'Neill*, British, Consul in Mozambique from 1879, coll. Mozembe (?) and Inhambane, specim. (K) ; *Harry Bolus* (1834–1911), British, born at Nottingham, coll. Lourenço Marques (1886), specim. (B, BM, K, PRE, Z) ; *Rev. Henri Alexandre Junod* (1863?–1934), Swiss, born Chezard, coll. Delagoa Bay (1889–96), specim. (G, Z), wrote *Le Climat de la Baie de Delagoa* (in Bull. Soc. Sci. Nat. Neuch. 25 : 76 (1897)), and *Zur Kenntnis Pflanzenw. Delagoa-Bay* with Schinz ; *Rev. Ladislau Menyhart* (1849–97), Hungarian, born at Szarvas, coll. (1890–) Boroma (Tete), specim. (Z) ; *Joseph Thomas Last* (1847–1933), British, born at Tuddenham, Suffolk, coll. " Zambezia " (1886), specim. (K) ; *Dr. Frederic Wilms*, German, pharmacist living in the Transvaal, coll. Lourenço Marques (c. 1886), specim. (BM, K, PRE, Z) ; *William Lawrence Scott*, British, coll. (1887–88) Nyasaland and Mozambique, specim. (K) ; *Francisco Joaquim Dias Quintas*, Portuguese, coll. Lourenço Marques (1893), specim. (COI) ; *Prelado de Moçambique D. Antonio Barroso* (1854–1918), Portuguese, born at Remelhe, coll. Mozambique (1894), specim. (COI) ; *Dr. Antonio José Rodrigues Braga* (1859–1913), Portuguese, coll. " Mozambique " (1894–95), specim. (COI) ; *Otto Kuntze* (1843–1907), German, coll. (1894) Lourenço Marques, Beira, Mozambique, specim. (K, NY) ; *Friedrich Richard Rudolf*

Schlechter (1872–1925), German, a professional collector, coll. Lourenço Marques (1897), Beira (1898), Inhambane (1898), specim. (BM, COI, K, PRE, Z) ; *Reginald Charles Fulke Maugham*, British, coll. Quelimane (1898), Beira (1902), Lourenço Marques (1908–11), specim. (K) ; *Hon. Evelyn Cecil* (*Lord Rockley*) (1865–1941) British, and his wife *Alicia Margaret, Lady Rockley* (née *Amhurst*) (1865–1941), coll. Manica e Sofala (1899) and S. Rhodesia (1899–1900) specim. (K); *Morais Sarmento*, Portuguese, coll. Mozambique (1899), specim. (COI) ; *Dr. Stefan Paulay* (1839–1913), Austrian, ship's doctor, coll. Niassa (" cap Cubeiro " which should probably be " Cabaceira grande ") date uncertain, specim. (W) ; *Johann Maria Hildebrandt* (1847–81), German, coll. (date uncertain) Beira and Vila Machado, specim. (BM, K).

Specimens collected by *Franz Ludwig Stuhlmann* (1863–) have been cited as coming from Mozambique but there is evidence that he did not collect these himself ; specim. (B).

6. *Journeys of Major Serpa Pinto and Capello and Ivens*

These two Portuguese journeys across the continent from west to east are of much geographical interest but only a few specimens were collected from our area.

Major Alexandre Alberto da Rocha de Serpa Pinto (1846–1900), Portuguese, crossed Africa in 1877 passing through Angola, N. Rhodesia and Mozambique. Most of his specimens which I have seen have been from Rio Ninda in the Moxico Distr. of Angola although not far from the Rhodesian frontier and from an almost unexplored region but a few are from Samuco (or Simuco), lat. 14° S., on the coast of Mozambique, where he collected with *Augusto Cardoso*. Specimens (LISU).

The journey of the two Portuguese, *Guilherme Augusto de Brito Capello* (1839–1926) and *Roberto Ivens* was in 1884–85 but in this instance the only specimens I have seen have been from Angola. Very briefly, they entered N. Rhodesia between Sept. 1–8, 1884 and proceeded in a NE. direction reaching as far north as Lat. 10° 23′ S., Long. c. 27° E. by Nov. 23–29, 1884, well into territory now the Belgian Congo. They then turned SE. back through Rhodesian territory, entering Mozambique at Zumbo on 4th May, 1885. Specimens (LISU, a few in BM).

7. *British Central Africa Collections* (*Nyasaland and N. Rhodesia*)

British occupation of the territory in our area now forming part of the British Commonwealth took place in two distinct phases and from two different directions. The one, which provides a distinct period of collecting, originated with the foundation of the Universities Mission followed by the Scottish Mission and resulted in the formation of the African Lakes Corporation in 1878 and the proclamation in 1889 of a protectorate over the Shire Highlands in Nyasaland and in 1891 of the creation of British Central Africa, which included parts of what are now N. Rhodesia and Tanganyika Territory, administered from Blantyre and Zomba. Two years later the area was reduced to the present Nyasaland and the title changed to Nyasaland Protectorate in 1907. The lines of penetration were up the Shire valley, across to Fort Manning and Fort Jameson via Lilongwe, up the western side of Lake Nyasa to Livingstonia, Fort Hill and northwards to Abercorn. Botanical collecting roughly followed the same routes. The principal collectors were : *Archdeacon William Percival Johnson* (1854–1928) British, coll. Nyasaland (1876–1919), specim. (K) ; *Alexander Whyte* (1834–1908), British, born at Fettercairn, Kincardineshire, assistant to Sir Harry Johnston, coll. (1890–97) especially Mt. Mlanje where he was one of the first to make botanical collections, specim. (K); *John Buchanan* (1855–96), British, gardener at Blantyre, coll. Nyasaland (1891–96) especially in the Shire Valley, specim. (BM, K); *George Francis Scott Elliot* (1862–1934), British, born at Calcutta, travelled and collected widely, coll. Nyasaland (1893–94) especially Shire Highlands, a few specim. from Mozambique and in N. Rhodesia (1893–94) usually labelled " Ruwenzori Expedition " and localized " Stevenson Road ", specim. (BM, K); *Kenneth J. Cameron*, British, coll. Nyasaland (1896–99, 1905) especially Namasi, specim. (B, K); *Alexander Carson* (1850–96), British, engineer, born at Stirling, coll. N. Rhodesia, Fwambo (1892), specim. (B, K); *W. H. Nutt* coll. in N. Rhodesia with *Carson*, specim. (K); *G. Adamson*, coll. Nyasaland (1906–12), specim. (BM, E, P), also with *Buchanan*, specim. (K).

8. *The SW. route to the Bechuanaland Protectorate and Matabeleland*

The other penetration leading to eventual British political occupation, naturally accompanied by botanical collecting, was the route followed by Livingstone and now approximately followed by the main line of the Rhodesia Railways to the Congo. Expeditions and collectors of various nationalities followed more or less in the same track.

Frank Oates (1840–75), British, coll. (1873–75) in Bechuanaland Prot. and S. Rhodesia north to Victoria Falls, specim. (K); *Dr. Emil Holub* (1847–1902), Austrian, coll. N. and S. Rhodesia (1875), Bechuanaland Prot. (1876), specim. (K, W); *Major Edward James Lugard* (1865–1944) and *Charlotte Eleanor* (" *Nell* ") *Lugard* (–1939) his wife, British, coll. Bechuanaland, Kwebe Hills (1898), our frontispiece adapted from a coloured sketch signed " Nell Lugard " made on this expedition, specim. (K); *Dr. Hans Schinz* (1858–1941), Swiss, coll. (1885–87) in the Kalahari, north as far as Lake Ngami, specim. (Z); *Dr. Siegfried Passarge* (1867–), German, geologist, coll. (1896–98) Bechuanaland Prot., 178 specim. (B).

9. *Period of the Rhodesian Pioneers (1890–1930)*

The Pioneer Column to Mashonaland sponsored by Cecil Rhodes reached Salisbury on the 12th September, 1890 and the modern history of Southern Rhodesia began. The country was administered by the British South Africa Company (the " Chartered Company ") whose territory extended into what later became known as N. Rhodesia when S. Rhodesia was granted self-government in 1923. During this period there was a gradual change in the nature of the collecting. As well as expeditions and collectors remaining only during comparatively brief periods of residence, we have local collectors living permanently in the country and often devoting much of their lives to a study of its flora. Our knowledge of the Rhodesian flora thus began to catch up with that of Mozambique and Nyasaland. This may also be called the " British Museum period ". It will be noted that up to the end of the 19th Century nearly all the collections from British territory went to Kew and most of the resulting publications were also naturally from there. During the earlier decades of the 20th Century important collections from *Eyles, Rand, Swynnerton* and others were received by the Botany Department of the British Museum and three publications (see " Selected Bibliography " p. 38) by L. S. Gibbs (1906), C. F. M. Swynnerton (1911–12) and F. Eyles (1916) were mainly based on identifications supplied by the staff of the department.

The principal collectors of this period were : *Dr. Richard Frank Rand* (1856–

1937), born at Plaistow, Essex, medical officer to the Chartered Company's Police, went to Mashonaland with the Pioneer Column of 1890, pioneer of botanical collecting in Mashonaland, specim. (BM), published *Wayfaring Notes* (in Journ. of Bot. 1898–1926), the last of which was appropriately linked with the first incursion into Rhodesian botany of one of the present editors of *Flora Zambesiaca*; *Henry George Flanagan* (1861–1919), born at Komgha, Cape Prov., S. Africa, coll. S. Rhodesia (1892–98), specim. (K, PRE); *Frederick Eyles* (–1938), coll. (1900–37) in many parts of S. Rhodesia, about 8000 specim. (BM, IMI, K, MO, PRE, SAM, SRGH); *Claude Frederick Hugh Monro* (1863–1918), British, coll. S. Rhodesia, Fort Victoria and district (1900–16), specim. (BM), wrote *Some Indigenous Trees of S. Rhodesia* (Proc. Rhod. Sci. Ass. **8** (1908)); *Archdeacon Frederick Arundel Rogers* (1876–1944), Chaplain to the South African Railways, made extensive collections (1904–25) in N. & S. Rhodesia and Mozambique, specim. (BM, K, US); *Lilian Suzette Gibbs* (1870–1925), British, coll. (1905), Matopos and Victoria Falls, specim. (BM); *Joseph Burtt Davy* (1870–1940), British, coll. Victoria Falls (1905); *Heinrich Gustav Adolf Engler* (1844–1930), German, coll. S. Rhodesia (1905), specim. (B); *Hermann Wilhelm Rudolf Marloth* (1855–1931), coll. S. Rhodesia (1905), specim. (K, PRE); *Alfred Barton Rendle* (1865–1938), British, Keeper of Botany, British Museum, made small collections in S. Rhodesia (1929), specim. (BM), was probably instrumental in directing interest of British Museum towards the Rhodesian flora; *C. E. F. Allen* of the Rhodesian Forestry Dept., coll. (1904–06) Barotseland and S. Rhodesia, specim. (K, SRGH) and Mozambique (1912) specim. (K); *E. C. Chubb*, British, Curator of Bulawayo Museum, coll. Bulawayo (1908) specim. (BM); *Mrs. W. S. Craster*, British, coll. Salisbury Distr., (1910–20) specim. (K); *Miss F. Edith Cheesman*, British, coll. S. Rhodesia (1930) specim. (BM, K); *T. Kassner*, German, coll. N. Rhodesia (1908), specim. (BM); *Mrs. Arthur Shinn* (–1938), British, coll. Mlanje Distr. of Nyasaland (1913), specim. (BM); *Miss M. A. Pocock* went across Barotseland and Angola on foot with porters (1924), specim. (PRE); *A. J. Teague*, mining engineer, coll. S. Rhodesia (E), Odzani R. (1914), specim. (BM, BOL, K); *Mrs. M. A. Macaulay* (née *Gairdner*), coll. N. Rhodesia, Mumbwa, specim. (K); *D. C. M. Jelf*, coll. N. Rhodesia, Fort Rosebery (1924) specim. (K); *Dr. John Hutchinson* (1884–), British, born at Wark, Northumberland, works at Kew Herbarium, coll. S. & N. Rhodesia (1928–30) with *General J. C. Smuts* (see p. 32) and *Jan Bevington Gillett* (1911–), specim. (BM, K); *Charles Francis Massey Swynnerton* (1877–1938) British, born in India, coll. S. Rhodesia (E) and Mozambique (Manica e Sofala), specim. (BM, K), farmer and naturalist in S. Rhodesia, later Director of Tsetse Research in Tanganyika where he also collected, died with B. D. Burtt in an air crash; *O. B. Miller*, British, coll. Bechuanaland Prot. and S. Rhodesia, specim. (BM, FHO, K, SRGH); *van Son*, S. African, coll. Bechuanaland Prot. (1930), specim. (BM, K, PRE, SRGH)).

10. *Early 20th-Century Collectors in Mozambique (1900–30)*

In Mozambique this period was somewhat unproductive of important collections. Probably many other small ones could be added to the following list: *Guillaume Vasse*, French, coll. (1904–06), Beira, Buzi, Macequece, R. Pungue, Gorongosa, Chamba, Mopeia, Chinde, specim. (P); *W. Tiesler*, German?, coll. Chifumbazi (1905–07), 106 specim. (B); *J. Stocks*, British, of Kew Gardens, coll. (1906–07), Ibo, Mocimboa da Praia, Lago Nangadi, R. Rovuma, Mêdo, specim. (K); *William Howard Johnson* (1875–) British, coll. Mozambique

(1906–10), specim. (K); *M. T. Dawe* (1880–), British, coll. Mozambique (1911–12), specim. (BR, K); *Dr. Américo Pires de Lima* (1886–), Portuguese, coll. Niassa (1916–17), specim. (PO); *Thomas Honey* (1872–1937), British, born in Northumberland, Director of Gardens etc. in Lourenço Marques, coll. (1919–26) Beira and Lourenço Marques, specim. (K); *Mrs. J. M. Borle* (1880–), Swiss, at American Mission at Chicuque, coll. Mozambique (1920), specim. (BR, COI, NY, PRE, S, UPS, WU) also in N. & S. Rhodesia (1921), specim. (BR, K, NY, PRE, S, SRGH, UPS, WU); *Charles Edward Moss* (1870–1930), British, Prof. of Botany at Johannesburg, coll. Lourenço Marques, specim. (PRE); *Thomas Robertson Sim* (1858–1938) specim. (PRE).

11. *Two Swedish Expeditions*

Swedish interest in the botany of the area of our Flora has been considerable and the two great Swedish expeditions merit special mention both for the importance and excellence of the collections made and the publications resulting from them.

Robert Elias Fries (1876–), Swedish, went on the Swedish Rhodesia-Congo Expedition of 1911–12 led by Eric Graf von Rosen, coll. N. Rhodesia, specim. (B, BR, G, K, UPS) results published in Wiss. Ergebn. Schwed. Rhod.-Kongo-Exped. 1911–12, Bot. Untersuch. (1914–16); *Thore Christian Elias Fries* (1886–1931), Swedish, Prof. of Botany at Lund, went to S. Rhodesia (1930–31) with *Tycho Norlindh* (1906–) and *Henning Weimarck*, specim. (BM, GB, K, LD, S, SRGH, UPS). In *Flora Zambesiaca* this collection is cited *F.N. & W.* up to the unfortunate death of Thore Fries. The later numbers are cited normally as *Norlindh & Weimarck*. The results of the expedition are being published in Bot. Notis. Lund (1932–).

12. *Modern Collections from the Federation (1930–)*

Collecting has speeded up enormously in recent years and since the foundation of the Government Herbarium at Salisbury and especially since it has been under the able direction of my friend and co-editor, Dr. Hiram Wild, the collection has grown to some 100,000 specimens and collectors have become so numerous that only a selection can be cited. Meanwhile in N. Rhodesia and Nyasaland much useful work has been done by officers of the forestry and agricultural services usually in liaison with Kew and the Imperial Forestry Institute, Oxford. In the following much abbreviated list I have concentrated on the " outside " collectors somewhat to the detriment of the " local " collectors, about whom the Government Herbarium, Salisbury, will usually have sufficient records, and references to nationality have been omitted : *F. B. Armitage* (1921–) of the Forestry Commission, S. Rhodesia, coll. (1954–), specim. (FHO, K, SRGH); *Mrs. Florence Mary Benson* (1909–), born Kimberley, S. Africa, coll. (1943–) Nyasaland and N. Rhodesia, specim. (BM, K, PRE); *C. K. Brain*, coll. S. Rhodesia (1930–42), 4000 specim. (K, SRGH); *John Patrick Micklethwait Brenan* (1917–), born at Chislehurst, works at Kew, coll. N. Rhodesia (1947) with Greenway (*Brenan & Greenway* specimens form part of Brenan's enumeration, *Greenway & Brenan* specimens form part of Greenway's enumeration) also coll. with *Keay*, specim. (FHO, K). (See also Vernay Lang Exped. p. 32); *Arthur Allman Bullock* (1906–), works at Kew, coll. N. Rhodesia (1949–51), specim. (K, SRGH); *Bernard D. Burtt* (1902–1938), coll. N. Rhodesia, Abercorn (1936), specim. (BM, K) and largely in Tanganyika where he was unfortunately killed with Swynnerton in an air disaster ; *James D.* and *Elizabeth Chapman*, coll.

Nyasaland, specim. (BM, FHO, K); *Norman Centlivres Chase* (1888–), born at Seven Oakes, Uitenhage, now resident in S. Rhodesia, coll. (1945–) S. Rhodesia (E) and Mozambique (MS), one of the finest recent collections, over 6000 specimens (BM, K, COI, LISC, MO, PRE, SRGH) first set to SRGH, private set later presented to BM; *R. M. Davies* (1899–) Chief of Native Agricultural Dept., S. Rhodesia, coll. (1953–), specim. (K, SRGH); *Mrs. G. Dehn*, born in Germany, coll. S. Rhodesia (1941–52) 1000 specim. (M, SRGH), described by Suessenguth and Merxmüller (in Trans. Rhod. Sci. Ass. **43** (1951))); *Dennis Basil Fanshawe* (1915–), forestry officer in N. Rhodesia, coll. N. Rhodesia, specim. (K; NDO); *Miss A. H. Gamwell*, coll. N. Rhodesia, Abercorn, (1931), specim. (BM); *Hamish Boyd Gilliland*, coll. S. Rhodesia (1931–37), 1500 specim. (BM, J, K, LD); *B. Goldsmith* (1924–), of Forestry Commission, S. Rhodesia, coll. Shangani (1954–), specim. (FHO, K, SRGH); *Percy James Greenway* (1897–), born at Germiston, near Johannesburg, coll. N. & S. Rhodesia and Nyasaland (1938) with *Trapnell*, N. Rhodesia (1947) and with *Brenan* (see p. 29) and N. & S. Rhodesia and Nyasaland (1954), specim. (EA, K, PRE), elaborated the classification (ined.) of vegetational units which is being used, with a few modifications, in *Flora Zambesiaca*; *J. C. F. Hopkins* (1898–), formerly Chief Botanist and Plant Pathologist, Salisbury, concerned with preliminary stages in implementing the Flora Zambesiaca project, coll. S. Rhodesia (1938–), 1500 specim. (SRGH); *George Jackson* (1927–), born at Sheffield, Government Ecologist, Zomba, coll. Nyasaland, specim. (BM, K, SRGH), accompanied and greatly assisted " Iter Zambesiacum 1955 " in Nyasaland; *Ronald William John Keay* (1920–) born at Richmond, Surrey, coll. (1947) N. (with *Brenan*) and S. Rhodesia, specim. (FHO, SRGH); *J. D. Martin* of Forestry Dept., N. Rhodesia, coll. mainly Victoria Falls and Zambezi Valley, specim. (FHO, K, NDO); *G. M. McGregor* (1904–) of Forestry Commission, S. Rhodesia, coll. (1938–), specim. (FHO); *Edgar Wolston Bertram Handsley Milne-Redhead* (1906–) born near Frome, Somerset, works at Kew, coll. N. Rhodesia (1930–38) especially Mwinilunga adding many new records to the flora, specim. (BM, BR, K, PRE) coll. also Angola, Tanganyika etc.; *R. C. Munch* of Rusape, coll. S. Rhodesia and Mozambique, specim. (SRGH); *Mrs. Edna J. Nash*, coll. N. Rhodesia (1951–), specim. (BM); *A. A. Pardy* (1902–) formerly of Forestry Commission, S. Rhodesia, coll. (1932–), specim. (FHO, K. SRGH); *Darrel Charles Herbert Plowes*, born at Estcourt, Natal, coll. S. Rhodesia, especially Matopos and Nyamandhlovu, 1600 specim. (K, PRE, SRGH); *I. B. Pole Evans*, coll. N. & S. Rhodesia (with *J. Erens*) and Mozambique, specim. (PRE, SRGH); *James McFarlane Rattray*, Pasture Research Dept., S. Rhodesia, coll. S. Rhodesia, mainly *Gramineae* and an authority on this group, specim. (SRGH) (see also *S. M. Stent*); *Mrs. H. M. Richards* (1885–) coll., N. Rhodesia, Abercorn etc. (1952–), magnificent collections with many new records, specim. (B, BM, BR, EA, K, LISC, LMJ, NDO, P, PRE, S, SRGH); *E. A. Robinson*, teacher at Mapanza Mission, coll. N. Rhodesia, specim. (K, SRGH); *Sydney M. Stent*, coll. S. Rhodesia (1933–), specim. (SRGH), wrote (with *J. M. Rattray*) the first checklist of the grasses of S. Rhodesia; *Kathleen E. Sturgeon* (*Mrs. Bennett*), born at Wankie, formerly assistant in Salisbury Herbarium, coll. S. Rhodesia, specim. (SRGH) wrote *A Revised List of the Grasses of S. Rhodesia* published in 11 parts in Rhod. Agric. Journ. (1953–56); *P. Topham* coll. Nyasaland, specim. (FHO, K), wrote " Some forest types in Nyasaland " in the Nyasaland Check-list (1936); *Colin Graham Trapnell*, coll. N. Rhodesia, specim. (K, SRGH), made ecological survey of N. Rhodesia (1937–40); *Oliver West* (1910–) Chief Pasture

Research Officer, S. Rhodesia, coll. Matabeleland, Zambezi Valley and Caprivi Strip, specim. (SRGH) ; *J. A. Whellan*, Chief Entomologist, Salisbury, coll. S. Rhodesia (1947–), specim. (SRGH) ; *P. O. Wiehe*, coll. Nyasaland, 600 specim. (K, SRGH) ; *Hiram Wild* (1917–), born at Sheffield, appointed Government Botanist, Salisbury, 1945, coll. widely and thoroughly in S. Rhodesia (1945–) etc. *(*see " Iter Zambesiacum, 1955 " p. 32), specim. (BM, K, PRE, SRGH) ; *Mrs. Jessie Williamson*, coll. Nyasaland (1938–), specim. (BM) ; *R. G. N. Young*, coll. S. Rhodesia (1928), specim. (SRGH), N. Rhodesia (1932) Ndola etc., en route for Angola, specim. (BM).

13. *African Collectors in the Federation*

The principal African collectors who have contributed to *Flora Zambesiaca* are : *Elias A. Banda* (1936–), Assistant to Government Ecologist, Zomba, coll. Nyasaland, specim. (BM, K) ; *Juliassi*, Native Assistant, Dept. of Pasture Research, coll. S. Rhodesia (1939) with *Newton*, specim. (SRGH) ; *D. Kafuli*, coll. N. Rhodesia (1955), specim. (K) ; *J. Kantikana* coll. Nyasaland, Lilongwe (1952), specim. (BM) ; *Newton*, Native Assistant, Dept. of Pasture Research, coll. S. Rhodesia (1939) with *Juliassi*, specim. (SRGH) ; *Wilfred Siame*, Native Assistant, International Red Locust Control Service, coll. N. Rhodesia, Abercorn (1952), specim. (BM, K).

14. *Missão Botânica de Moçambique*

This Mission was created in 1942 for phytogeographical investigation of Mozambique under *Francisco d'Assensão Mendonça* (1889–), Portuguese, born at Faro, Algarve, Director of the Centro de Botânica of the Junta de Investigações do Ultramar, Lisbon. (See also " Iter Zambesiacum, 1955 " p. 32). Very large collections were made throughout Mozambique, specim. (BM, K, LISC, SRGH). Publications include *Itinerário Fitogeográfico da Campanha de 1942 da Missão Botânica de Moçambique* by F. A. Mendonça (in An. Junt. Miss. Geogr. e Invest. Colon. **3**, 1 (1948)), and various taxonomic works by Mendonça and others in Contrib. Conhec. Fl. Moçamb. (1950–). In addition to local Mozambique botanists mentioned later, Mendonça's principal assistants were : *A. L. Cavaco* (1916–), Portuguese, coll. Mozambique (1947–48), specim. (LISJC, LISU) ; *J. G. Garcia* (1904–), Portuguese, born at Guarda, coll. Mozambique (1948), specim. (BM, COI, LISC) and *António Rocha da Torre* (1904–), Portuguese, born at Viana do Castelo, coll. Mozambique (1934–48), specim. (BM, COI, K, LISC, SRGH), coll. also in Angola.

15. *Modern Collectors in Mozambique (1930–)*

Apart from the " Missão Botânica " mentioned above the following are among the principal modern collectors in Mozambique ; *E. C. Andrada*, Portuguese, coll. Niassa (1948), specim. (BM, COI, LISC, LM) ; *Luis Augusto Grandvaux Barbosa*, Portuguese, specim. (COI, K, LISC, LM, MO, SRGH), wrote *Esboço da Vegetação da Zambézia*, 1952 (publ. orig. in Documentario Moçambique, No. 69 (1952)) with coloured vegetation map, coll. also with *M. F. Carvalho* ; *Mrs. H. G. Faulkner*, British, coll. Mozambique (1944–49) specim. (COI, K, PRE, SRGH) also coll. Angola ; *António de Figueiredo Gomes e Sousa*, Portuguese, coll. Mozambique (1930–50), specim. (B, BM, COI, K, LISJC) ; *A. J. W. Hornby* (1893–), coll. Mozambique, specim. (K, LISC, LM, PRE) ; *H. E.* (1890–) *and Mrs. R. M. Hornby*, (1893–), coll. S. Rhodesia, specim. (SRGH) and Mozambique, specim. (K, MO, PRE, SRGH) ; *Francisco Leal de Lemos* (1915–), born at Coimbra, Portugal, formerly collector for the Botanical Institute at Coimbra, is making excellent collections in Mozambique, specim. (BM, COI, LM) ; *Georges Le Testu* (1877–), French, born at Caen, coll. Mozambique, specim. (BM, CN, P), coll. also French W. Africa ; *José Gomes Pedro* (1915–), Portuguese, coll. Mozambique, specim. (COI, LMJ) also with *José Pedrógão* ; *José Alves Pereira* (1912–), born at Aguedas, Portugal, coll. Zambezia, (1951–), specim. (COI) ; *J. Simão*, Portuguese, coll. Manica e Sofala and Tete, (1947), specim. (LISC, LM, PRE, SRGH) ; *General Jan C. Smuts* (1870–1950) ; S. African, coll. Mozambique, Nyasaland and also in N. and S. Rhodesia (with *Hutchinson*), specim. (K, PRE) ; *Père Charles Tisserant* (1886–), French, coll. Mozambique, specim. (BM, CN, COI, P), coll. also French W. Africa ; *Lieut.-Colonel Jack Vincent*, coll. Mozambique (1931–32), specim. (BM).

16. *Caprivi Strip Collections*

Our knowledge of the Caprivi Strip is due mainly to South African collections kindly lent to us by Pretoria. *Seiner*, coll. in Bechuanaland Prot. and the Caprivi Strip (1905–06), specim. (Z) ; *L. E. Codd*, South African, of Pretoria Herbarium, coll. recently in N. Rhodesia and Caprivi Strip, specim. (PRE) ; *Curson*, coll. Bechuanaland Prot. (1930) and Caprivi Strip (1945), specim. (PRE).

17. *Vernay Nyasaland Expedition (1946)*

This important expedition deserves special mention both for the admirable collection made and for the account published by J. P. M. Brenan (see p. 29) and collaborators (in Mem. N.Y.Bot. Gard. **8**, 3 (1953) et seq.) with an interesting introduction by the collector, *Leonard John Brass* (1900–) Australian, 2000 specim. (BM, K, MO, NY, SRGH).

18. *Imperial Forestry Institute Expedition to N. Rhodesia (1951–53)*

This expedition by *Frank White* and *A. Angus* to N. Rhodesia and Nyasaland (1951–53) under the auspices of the Imperial Forestry Institute, Oxford, in connexion with the *N. Rhodesia Forest Flora*, in course of publication, resulted in a large and excellent collection mainly of woody plants, specim. (BM, FHO, K). The expedition traversed almost the whole of N. Rhodesia and the collection is still being worked out.

19. *Iter Zambesiacum 1955*

This survey of the *Flora Zambesiaca* area for the purpose of agreeing upon the terms to be used for the vegetational units etc. was made by *Arthur Wallis Exell* (1901–), British, born at Handsworth, Staffordshire, *Francisco d'Ascenção Mendonça* (see p. 31) and *Hiram Wild* (see p. 31). The expedition made a rapid but extensive journey through S. Rhodesia, Transvaal, Mozambique (LM and SS), Nyasaland (C and S), N. Rhodesia and Bechuanaland Prot. (N). The specimens (BM, LISC, SRGH) are cited *E.M. & W.* in the *Flora Zambesiaca*.

20. *Cambridge University Mt. Mlanje Expedition (1956)*

This expedition made good collections on Mt. Mlanje. The botanists were : *Edward I. Newman* (1935–) British and *T. C. Whitmore* (1935–) British, specim. (BR, BM, NY, SRGH, WAG).

21. *Iter Zambesiacum 1958–59*

At the time of going to press *Norman Keith Bonner Robson* (1928–), born at Aberdeen, member of the *Flora Zambesiaca* staff, has recently returned from collecting in N. Rhodesia and Nyasaland.

22. *Notes for future collecting*

It may be useful to conclude this summary of botanical collecting in our area by giving some indication of the regions already reasonably well known and those in which further collecting is highly desirable.

Bechuanaland Prot. Various collections have been made around Lake Ngami and at different points along the railway. Otherwise our knowledge of this territory is botanically still very incomplete.

N. Rhodesia. Collections are still needed from almost everywhere. The least-known regions are : Barotseland (especially west of the R. Zambezi, almost unexplored) and the Eastern Province. Very few collections have been made in the Fort Jameson district.

Victoria Falls. Easily accessible localities near the falls have been thoroughly collected. The gorges below the falls are very difficult of access and are almost uncollected.

S. Rhodesia. Fairly well known near the main centres, especially the environs of Salisbury, Bulawayo (incl. Matopos) and Umtali. The northern district is still insufficiently known, especially the valley of the Zambezi. In the south there is scope for more collecting in the valley of the Limpopo and the escarpment to the north of it. The eastern border mountains are becoming fairly well known but their flora is certainly not yet exhausted.

Nyasaland. Nearly all collections still produce new records. Neither the Vipya nor the Nyika Plateau is yet sufficiently known. Further collections would be welcome from Dedza Mt. and even Zomba Mt. The extreme north and the lower Shire Valley in the extreme south are also insufficiently known.

Mozambique. Further collections are needed from the R. Rovuma in the extreme north, from many parts of the Niassa Province, from considerable areas in Sul do Save and from many of the less accessible coastal regions. The environs of Lourenço Marques have been well collected and also Inhaca Island. Collections along the railway route between Beira and Umtali have also been numerous. The flora of Mozambique is a rich one and there are certainly many discoveries still to be made.

Fernandes, A. & R., 23
Flanagan, H. G., 28
Forbes, J., 24
Fornasini, C. A., 24
Fries, R. E., 29
Fries, T. C. E., 29

Gairdner, Miss M. A., 28
Galvão da Silva, M., 23
Gamwell, Miss A. H., 30
Garcia, J. G., 31
Gibbs, Miss L. S., 27
Gillett, J. B., 28
Gilliland, H. B., 30
Goldsmith, B., 30
Gomes e Sousa, A. de F., 31
Greenway, P. J., 30

Hildebrandt, J. M., 26
Holub, E., 27
Honey, T., 29
Hopkins, J. C. F., 30
Hornby, A. J. W., 31
Hornby, H. E., 31
Hornby, Mrs. R. M., 31
Hutchinson, J., 28

Ivens, R., 26

Jackson, G., 30
Jelf, D. C. M., 28
Johnson, W. H., 28
Johnson, W. P., 26
Johnston, H., 27
Juliassi, 31
Junod, H. A., 25

Kafuli, D., 31
Kantikana, J., 31
Kassner, T., 28
Keay, R. W. J., 30
Kirk, J., 25
Kuntze, O., 25

Last, J. T., 25
Lemos, F. L. de, 31
Le Testu, G., 32
Livingstone, D., 25
Loureiro, J. de, 23

Lugard, Mrs. C. E., 27
Lugard, E. J., 27

Macaulay, Mrs. M. A., 28
Marloth, H. W. R., 28
Martin, J. D., 30
Maugham, R. C. F., 26
McGregor, G. M., 30
Meller, C. J., 25
Mendonça, F. A., 31
Menyhart, L., 35
Miller, O. B., 28
Milne-Redhead, E. W. B. H., 30
Monro, C. F. H., 28
Monteiro, J. J., 25
Monteiro, Mrs. R., 25
Moss, C. E., 29
Munch, R. C., 30

Nash, Mrs. E. J., 30
Newman, E. I., 32
Newton, 31
Norlindh, T., 29
Nutt, W. H., 27

Oates, F., 27
O'Neill, H., 25

Pardy, A. A., 30
Passarge, S., 27
Paulay, S., 26
Pedro, J. G., 32
Pedrógão, J., 32
Pereira, J. A., 32
Peters, W. C. H., 24
Pires de Lima, A., 29
Plowes, D. C. H., 30
Pocock, Miss M. A., 28

Quintas, F. J. D., 25

Rand, R. F., 27
Rattray, J. M., 30
Rendle, A. B., 28
Rhodes, C., 27
Richards, Mrs. H. M., 30
Robinson, E. A., 30
Robson, N. K. B., 32

Rockley, Lady A. M., 26
Rockley, Lord, 26
Rogers, F. A., 28

Sarmento, M., 26
Schinz, H., 27
Schlechter, F. R. R., 26
Scott, W. L., 25
Seiner, F., 32
Serpa Pinto, A. A. da R. de, 26
Shinn, Mrs. A., 28
Siame, W., 31
Sim, T. R., 29
Simão, J., 32
Smuts, J. C., 32
Stent, Miss S. M., 30
Stewart, J., 25
Stocks, J., 28
Stuhlmann, F. L., 26
Sturgeon, Miss K. E., 30
Surcouf, R., 24
Swynnerton, C. F. M., 28

Teague, A. J., 28
Tiesler, W., 28
Tisserant, C., 32
Topham, P., 30
Torre, A. R. da, 31
Trapnell, C. G., 30

van Son, 28
Vasse, G., 28
Vincent, J., 32

Waller, H., 25
Wallis, J. P. R., 25
Weimarck, H., 29
West, O., 30
Whellan, J. A., 31
White, F., 32
Whitmore, T. C., 32
Whyte, A., 26
Wiehe, P. O., 31
Wild, H., 31
Williamson, Mrs. J., 31
Wilms, F., 25

Young, R. G. N., 31

ITINERARY OF DR. JOHN KIRK

By Gwendolen A. Hayes

(Extracted from " Kirk on the Zambesi ", by R. Coupland (1928))

For abbreviations of the geographical divisions see p. 21.

1858

March 3rd :	Left London.
March 25th–30th :	Freetown.
April 21st :	Capetown.
May 14th :	Mouth of W. Luabo (said by Coupland to be " R. Luawe ", probably the Rio Mungari of modern Portuguese maps). M(MS)
May 15th–18th :	Charting estuary (Kirk and Stead). M(MS)
June 3rd :	Kongone. M(MS)
?	On island in mouth of R. Zambezi (Expedition Island). M(MS)
June 26th :	Expedition Island (first botanical specimens sent to Sir William Hooker).
July 9th :	Expedition Island.
July 19th :	Expedition Island (over a month spent on island).
August 2nd :	Up R. Zambezi.
August 4th :	Shupanga (Chupanga). M(MS)
August 5th :	Down river to Expedition Island.
August 9th :	Up R. Zambezi.
August 23rd (approx) :	2 miles SE. of Shupanga—Lake Bove. M(MS)
August 28th :	Shupanga.
September 2nd :	Shupanga.
?	Left Shupanga for mouth of R. Zambezi.
September 11th :	Shupanga.
September 14th :	Expedition Island (charting the estuary).
October 7th :	Up R. Zambezi.
October 10th :	Shupanga.
October 17th :	Sena. M(MS)
October 22nd :	Left Sena to go up R. Zambezi.
November 1st :	Lupata Gorge. M(MS) and M(T)
November 3rd :	?
November 4th :	Tete. M(T)
November 8th :	Up towards Kebrabasa (Quebrabasa) Gorge. M(T)
November 10th :	Kebrabasa Gorge.
November 13th :	Tete.
December 1st–3rd :	R. Zambezi near Kebrabasa. M(T)
December 6th :	Returned to Tete.
December 20th :	Left Tete.
December 25th :	Sena. M(MS)
December 29th :	Confluence of R. Zambezi and R. Shire. M(MS)

1859

January 3rd :	Up R. Shire. M(T) and M(Z)
January 4th :	Marsh (Elephant Marsh). M(T) and M(Z)

January 8th :	Chibisa. N(S)
January 9th :	Murchison Cataract. N(S)
January 10th :	Down R. Shire.
January (middle) :	Shupanga. M(MS)
January 31st :	Tete. M(T)
March 19th :	Up R. Shire.
March 29th :	Chibisa.
April 3rd :	Chibisa.
April 16th :	Near Lake Shirwa (Chilwa). N(S)
June 23rd :	Tete. M(T)
July 29th :	Expedition Island. M(MS)
August 28th :	Overland from Chibisa. N(S) and M(T)
September 14th :	Lake Pamalombe. N(S)
September 17th :	Lake Nyasa. N(S)
September 18th :	Returned towards Chibisa.
October 8th :	Chibisa. N(S)
October 25th :	Tete. M(T)
November 22nd :	Kongone Mouth to Tete.

1860

March 17th :	Kongone Mouth. M(MS)
March 25th :	Below Shupanga.

G. Rae here left the Expedition to return to England with four cases of Kirk's plants consigned to Sir William Hooker. These never arrived and in 1883 Sir Joseph Hooker received a note from Portsmouth Dockyard saying that four cases addressed to Sir William had been deposited there in 1870 and asking him to remove them! (Coupland, op. cit. : 169).

March 30th :	Shupanga. M(MS)
April 20th :	Tete. M(T)
May 15th :	Left Tete.
?	Kebrabasa Hills. M(T)
?	Chicova Plain. M(T)
June 26th :	Zumbo—confluence of R. Zambezi and R. Loangwa (Luangwa). M(T)
?	Mburuma Pass. NR(C) and SR(N)
July 9th :	Confluence of R. Zambezi and R. Kafue. NR(C), NR(S) and SR(N)
?	Kariba Gorge. NR(S) and SR(N)
?	R. Zangwe. NR(S)
?	Ascended Batoka Plateau. NR(S)
August 4th :	Sighted Victoria Falls. NR(S)
August 7th :	Left Victoria Falls and Garden Island. NR(S) and SR(W)
August 18th :	Shesheke (Sesheke). NR(B)
September 17th :	Left Shesheke.
November 11th :	Kebrabasa Rapids. M(T)
November 23rd :	Tete. M(T)
December 3rd :	Left Tete.
December 6th–21st :	R. Zambezi.
December 23rd :	Camped by R. Zambezi.

1861

January 1st :	Sena. M(MS)
?–March 17th :	Returned down R. Rovuma (no dates given for outward journey). M(N)
March 23rd :	Mouth of R. Rovuma. M(N)
April 1st–22nd :	Sailing—R. Rovuma—Johanna (Comoro Is.) thence—
May 1st :	Mouth of R. Zambezi.
July 8th :	Chibisa. N(S)
July 13th :	Near Chibisa. N(S)
July 29th :	Magomera. N(S)
August 1st :	Chibisa.
August 6th :	Up R. Shire. N(S)
August 26th :	R. Shire, passed Mt. Zomba. N(S)
August 31st :	Lake Pamalombe. N(S)
September 2nd, 3rd :	Lake Nyasa.
November 8th :	R. Shire. N(S)
November 14th :	Left Chibisa.
December 7th :	Mouth of R. Ruo. N(S)

1862

January 11th :	Entered R. Zambezi.
January 23rd :	Luabo Mouth. M(Z)
January 30th :	Mouth of R. Zambezi.
February 17th :	Left Shupanga. M(MS)
March 4th :	Chibisa. N(S)
March 5th :	Left Chibisa.
March 12th :	R. Ruo. N(S)
April 11th :	Up R. Zambezi.
April 14th :	Shupanga. M(MS)
April 27th :	Shupanga.
May 18th :	Tete. M(T)
?	Shupanga. M(MS)
June 3rd–23rd :	Shupanga.
August ?	On way to Johanna.
September 9th :	Mouth of R. Rovuma. M(N)
September 13th–26th :	Up R. Rovuma.
September 27th–9th :	Down R. Rovuma to estuary. M(N)
October 20th :	Johanna.

1863

January 1st :	Shupanga. M(MS)
January 11th :	Down R. Zambezi.
January 15th :	Mt. Morumbala. M(Z)
April 21st :	Chibisa. N(S)
May 19th :	Kirk left expedition.

SELECTED BIBLIOGRAPHY OF GENERAL WORKS DEALING MAINLY WITH THE BOTANY OF THE FLORA ZAMBESIACA REGION AND ADJACENT TERRITORIES

Taxonomic works of restricted scope will be found cited in their appropriate places in the general text.

Atlas de Portugal Ultramarino (1948) with a phytogeographical map of Mozambique by F. A. Mendonça.

E. G. Baker, A. W. Exell & S. Moore, *Notes on Dr. R. F. Rand's Rhodesian Plants.* (Journ. of Bot. **64** (1926)).

L. A. G. Barbosa, *Esboço da Zambezia.* Lourenço Marques (1952) (reprinted from Documentario Moçambique No. 69, 1952) with coloured vegetation map.

G. Bertoloni, *Illustrazione di Piante Mozambicesi.* Diss. 1–3 (Mem. Accad. Sci. Bol. **2–5** (1850–55)).

L. J. Brass, *Vegetation of Nyasaland. Report on the Vernay Nyasaland Expedition of 1946.* (Mem. N.Y.Bot. Gard. **8,** 3 (1953)).

C. E. B. Bremekamp & A. A. Obermeyer, *Scientific Results of the Vernay-Lang Kalahari Expedition, 1930: Sertum Kalahariense, a List of the Plants Collected.* (In Ann. Transv. Mus. **16** (1935)).

J. P. M. Brenan, *Check-lists of the Forest Trees and Shrubs of the British Empire No. 5, Tanganyika Territory,* Part II (1949).

J. P. M. Brenan, *Plants Collected by the Vernay Nyasaland Expedition of 1946.* (Mem. N.Y.Bot. Gard. **8,** 3 (1953) et seq.).

I. H. Burkill, *List of the Known Plants occurring in British Central Africa, and the British Territory north of the Zambezi.* (Published as Appendix II to H. H. Johnston, *British Central Africa* (1897)). The *Serpa Pinto* specimens listed were all from Angola.

J. Burtt Davy, *A Manual of the Flowering Plants and Ferns of the Transvaal with Swaziland, South Africa* (1926–32).

J. Burtt Davy & A. C. Hoyle, *Check-lists of the Forest Trees and Shrubs of the British Empire No. 2, Nyasaland Protectorate* (1936).

P. Duvigneaud, *La Flore et la Végétation du Congo Méridional.* (Lejeunia **16**: 95 (1953)); *La Végétation du Katanga et de ses Sols Metallifères.* (Bull. Soc. Roy. Bot. Belg. **90**: 127 (1958)).

H. G. A. Engler, *Die Pflanzenwelt Ost-Afrikas,* C (1895).

H. G. A. Engler, *Beiträge zur Kenntniss der Pflanzenformationen von Transvaal und Rhodesia.* (Sitz. K. Preuss. Akad. Wiss. Berl., 1906).

I. B. Pole Evans, *A Reconnaissance Trip through the Eastern Portion of the Bechuanaland Protectorate April 1931 and an Expedition to Ngamiland, 1937.* (Bot. Surv. S. Afr. Mem. 21 (1948)); *Roadside Observations on the Vegetation of East and Central Africa.* (Bot. Surv. S. Afr. Mem. 22 (1948)).

A. W. Exell & F. A. Mendonça, *Conspectus Florae Angolensis* (1937–).

A. W. Exell, N. K. B. Robson, H. Wild (and others), *New and Little Known Species from the Flora Zambesiaca Area.* (Bol. Soc. Brot., Sér. 2, **30** et seq. (1956–)).

F. Eyles, *A Record of Plants Collected in S. Rhodesia.* (Trans. Roy. Soc. S. Afr. **5**: 273 (1916)).

Conde de Ficalho, *Plantas Uteis da Africa Portugueza* (1884).

Flora Capensis (see Harvey).

Flora of Tropical Africa (see Oliver).

Flora of Tropical East Africa (see Turrill).

Flora of West Tropical Africa (see Hutchinson).

Flore du Congo belge et du Ruanda-Urundi (1948–) published by the Institut National pour l'Étude Agronomique du Congo Belge.

R. E. Fries, *Wissenschaftliche Ergebnisse der Schwedischen Rhodesia-Kongo-Expedition (1911–1912)* I. *Botanische Untersuchungen* (1914–16).

L. S. Gibbs, *A Contribution to the Botany of S. Rhodesia.* (Journ. Linn. Soc., Bot. **37** : 425 (1906)).

H. B. Gilliland, *A Proposed Delimitation of Botanical Counties for S. Rhodesia.* (Journ. S. Afr. Bot. **4** : 65 (1938)) ; *The Vegetation of Rhodesian Manicaland.* (tom. cit. : 73 (1938)).

A. F. Gomes e Sousa, *Subsidios para o estudo da flora do Niassa Português.* (Bol. Soc. Est. Col. Moçamb. No. 26 : 71 (1935) with a list of plants ; *Plantas Menyharthianas* (with H. Schinz and others, op. cit. No. 29 (1936) et. seq.) ; *Exploradores e Naturalistas da Flora de Moçambique,* Documentario Moçambique Nos. 18, 19, 20, 23, 27, 30, (1939–42) ; *Dendrology of Mozambique I (1948).*

W. H. Harvey & O. W. Sonder, *Flora Capensis* (1860–1933).

J. S. Henkel, *Types of Vegetation in Southern Rhodesia.* (Proc. & Trans. Rhod. Sci. Ass. **30** : 1 (1931)) with coloured maps.

J. C. Hopkins, A. L. Bacon & L. M. Gyde, *Look Around You Series, No. 1, Common Veld Flowers,* Salisbury (1940).

J. Hutchinson, *A Botanist in Southern Africa* (1946).

J. Hutchinson & J. M. Dalziel, *Flora of West Tropical Africa* (1927–36), ed. 2 (revised by R. W. J. Keay, 1954–).

G. Jackson, *Preliminary Ecological Survey of Nyasaland.* (Proc. 2nd. Inter-Afr. Soils Conf., Leopoldville (1954)).

H. A. Junod, *Le Climat de la Baie de Delagoa.* (Bull. Soc. Sci. Nat. Neuch. **25** : 76 (1897)) with an account of the vegetation.

J. de Loureiro, *Flora Cochinchinensis* (1790) with descriptions of the first plants collected in Mozambique.

W. Macnae & M. Kalk, *A Natural History of Inhaca Island, Moçambique* (1958) with botanical chapter and list of plants by A. O. D. Mogg.

R. A. S. Martineau, *Rhodesian Wild Flowers* (dated 1953 but published 1954) with descriptions and coloured plates.

F. A. Mendonça, *Itinerário Fitogeográfico da Campanha de 1942 da Missão Botânica de Moçambique.* (Anais Junta Inv. Ultr. **3**, 1 : 7 (1948)) ; *The Vegetation of Mozambique.* (Lejeunia, **16** : 127 (1953)).

F. A. Mendonça & A. R. Torre, *Contribuições para o Conhecimento da Flora de Moçambique-I.* (Estud., Ens. e Docum. Junta Inv. Ultr. (1950)).

F. A. Mendonça (and others), *Contribuições para o Conhecimento da Flora de Moçambique-II.* (op. cit. (1954)).

O. B. Miller, *Check-lists of the Forest Trees and Shrubs of the British Empire No. 6, Bechuanaland Protectorate,* (1948) ; *The Woody Plants of the Bechuanaland Protectorate.* (Journ. S. Afr. Bot. **18** : 1 (1952), **19** : 177 (1953)).

C. F. H. Monro, *Some Indigenous Trees of Southern Rhodesia.* (Proc. & Trans. Rhod. Sci. Ass. **8** (1908)).

E. I. Newman & T. C. Whitmore, *Cambridge Mlanje (Nyasaland) Expedition,
      1956* (1957) with a good map of Mt. Mlanje and some notes on the
      vegetation.
T. Norlindh & H. Weimarck, *Beiträge zur Kenntnis der Flora von Süd-Rhodesia.*
      (Bot. Notis. **1932** : 1 (1932) et seq.).
D. Oliver (and others) *Flora of Tropical Africa*, (1868–).
K. C. Palgrave, *Trees of Central Africa* (1956).
A. A. Pardy, *Notes on Indigenous Trees & Shrubs of S. Rhodesia.* (Rhod. Agric.
      Journ. (1951–56)).
S. Passarge, *Die Kalahari* (1904).
J. G. Pedro, *Contribuições para o Inventário Florístico de Moçambique.* (Bol. Soc.
      Estud. Moçamb. No. 87 (1954) et seq.).
J. G. Pedro & L. A. G. Barbosa, *Esboço do Reconhecimento Ecológico-Agrícola de
      Moçambique, II (Vegetação)* (1955).
W. C. H. Peters, *Reise nach Mossambique* (1861–64).
E. P. Phillips, *The Genera of South African Flowering Plants* (1926), ed. 2 (1951).
J. M. Rattray, *The Grasses and Grass Associations of Southern Rhodesia.* (Rhod.
      Agric. Journ. **54** : 197 (1957)).
J. M. Rattray & H. Wild, *Report on the Vegetation of the Alluvial Basin of the Sabi
      Valley and Adjacent Areas.* (Rhod. Agric. Journ. **52** : 484 (1955)) with
      map.
W. Robyns, *Flore des Spermatophytes du Parc National Albert* (1948–1955).
H. L. Shantz & C. F. Marbut, *The Vegetation and Soils of Africa* (1923) with
      map.
F. Seiner, *Ergebnisse einer Bereisung des Gebiets zwischen Okawango und Sambesi
      (Caprivi-Zipfel) in den Jahren 1905 und 1906.* (In Mitt. Deutsch. Schutz.
      22 (1909)).
T. R. Sim, *Forest Flora & Forest Resources of Portuguese East Africa* (1909).
E. C. Steedman, *Some Trees, Shrubs & Lianes of Southern Rhodesia* (1933)
      (expanded from a series of papers in Proc. & Trans. Rhod. Sci. Ass. (1925–)).
K. Suessenguth & H. Merxmüller, *A Contribution to the Flora of the Marandellas
      District, Southern Rhodesia.* (Proc. & Trans. Rhod. Sci. Ass. **43** : 175
      (1951)).
C. F. M. Swynnerton (and others), *A Contribution to our Knowledge of the Flora
      of Gazaland.* (Journ. Linn. Soc., Bot. **40** : 1 (1911)).
F. Thonner, *Die Blütenpflanzen Afrikas* (1908), English ed. *The Flowering
      Plants of Africa* (1915).
C. G. Trapnell, *Vegetation-Soil Map of Northern Rhodesia* (1950).
C. G. Trapnell, *Soils, Vegetation and Agriculture of North-Eastern Rhodesia*
      (1953).
C. G. Trapnell & J. N. Clothier, *Soils, Vegetation and Agricultural Systems of
      North Western Rhodesia* (1937).
W. B. Turrill & E. Milne-Redhead, *Flora of Tropical East Africa* (1952–).
Vegetation Map of Africa (1959) compiled by members of the Association pour
      l'Étude Taxonomique de la Flore d'Afrique Tropicale.
O. Warburg, *Kunene-Sambesi-Expedition, H. Baum* (1903).
H. Wild, *Some Records of Phytogeographic Interest from Southern Rhodesia* (Proc.
      & Trans. Rhod. Sci. Ass. **43** (1951)) ; *Southern Rhodesian Aquatic Plants*
      (Rhod. Agric. Journ. **49** : 111 (1952)) ; *The Vegetation of Southern Rhodesian
      Termitaria* (tom. cit. : 280 (1952)) ; *A Southern Rhodesian Botanical
      Dictionary of Native and English Plant Names* (1952) ; *A Guide to the Flora*

of the Victoria Falls (1953) ; *Observations on the Vegetation of the Sabi-Lundi Junction Area* (Rhod. Agric. Journ. **52** : 533 (1955)) ; *Common Rhodesian Weeds* (1955) ; *The Principal Phytogeographic Elements of the Southern Rhodesian Flora* (Proc. & Trans. Rhod. Sci. Ass. **44** : 53 (1956)).

J. Williamson, *Useful Plants of Nyasaland* (1955).

F. White, *Trees, Shrubs and Man in Northern Rhodesia* (Oxf. Univ. For. Soc. Journ., Ser. 4, **5** : 38 (1957)).

F. White & A. Angus, *Forest Flora of Northern Rhodesia* (in press).

PROVISIONAL GLOSSARY OF BOTANICAL TERMS

(Reprinted from the *Flora of West Tropical Africa* with additions, emendations and Portuguese equivalents.)

abaxial (*abaxial*), applied to the side or face away from the axis.

abortion (*aborto*), suppression of parts usually present.

abruptly pinnate (*paripinado*), a pinnate leaf without an odd terminal leaflet, the same as paripinnate (see fig. 6A).

acaulescent (*acaule*), becoming stemless or seemingly so.

accrescent (*acrescente*), increasing in size, e.g. the calyx of some plants in fruit.

accumbent (*acumbente*), lying against, as the cotyledons (in some *Cruciferae*) with their edges against the radicle.

achene (*aquénio*), a small dry fruit, not splitting when ripe, and containing a single seed, as in *Clematis*.

acicular (*acicular*), very narrow, stiff, and pointed.

actinomorphic (*actinomórfico*) (regular), applied to flowers which may be bisected in more than one vertical plane.

aculeate (*aculeado*), armed with prickles as distinct from spines ; *aculeolate* (*aculeolado*) with small prickles.

acuminate (*acuminado*) (see fig. 4A).

acute (*agudo*) (see fig. 4B).

adaxial (*adaxial*), applied to the side or face next the axis.

adnate (*adnado*), united with another organ or series, e.g. the ovary and calyx-tube.

adventitious buds (*gomos adventícios*), those produced elsewhere than in the axils of the leaves or the extremity of the branch.

aestivation (*estivação*), the manner in which the sepals and petals are arranged in bud (see fig. 1).

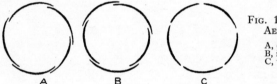

Fig. 1.—Types of Aestivation

A, contorted.
B, imbricate.
C, valvate.

A   B   C

alternate (*alterno*), applied to leaves, etc., inserted at different levels along the branch (as distinct from *opposite*).

amentiferous (*amentífero*), bearing catkins.

amplexicaul (*amplexicaule*), stem-clasping, as when the base of the leaf is dilated and embraces the stem.

anastomosis (*anastomose*), union of one vein with another, the connections forming a network.

anatropous (*anatrópico*), the ovule reversed, with micropyle close to the side of the hilum, and the chalaza at the opposite end.

androecium (*androceu*), the stamens and accessories.

androgynophore (*androginóforo*), a stalk supporting both androecium and gynoecium.

androphore (*andróforo*), a stalk supporting the androecium.

anemophilous (*anemófilo*), wind-pollinated.

anisophyllous (*anisofilo*), with the two leaves of a pair of different size or shape.

annular (*anular*), used of any organs arranged in a circle.

anterior (*anterior*), in position most remote from the axis.

anther (*antera*), the part of the stamen which contains the pollen, usually divided into two pouches or thecae.

anthesis (*ântese*), the time when the flower is expanded.

apetalous (*apétalo*), without petals.

apiculate (*apiculado*), ending abruptly in a short point (see fig. 4G).

apocarpous (*apocárpico*), with carpels free from one another (see fig. 7A).

appressed (*adpresso*), lying close and flat along the surface.

arcuate (*arqueado*), curved like a bow.

areolate (*areolado*), divided into distinct spaces.

aril (*arilo*), an appendage covering or partly enclosing the seed and arising from the funicle (or stalk) of the seed.

aristate (*aristado*), with a long, bristle-like point.

articulated (*articulado*), jointed or separating at a certain point, and leaving a clean scar.

asperous (*áspero*), rough, harsh to the touch.

attenuate (*atenuado*), tapering gradually.

auriculate (*auriculado*), having an ear-like lobe or appendage at the base of a leaf or other organ (see fig. 3E).

awn (*arista*), a fine bristle usually terminating an organ (usually applied in the case of the flowers of grasses).

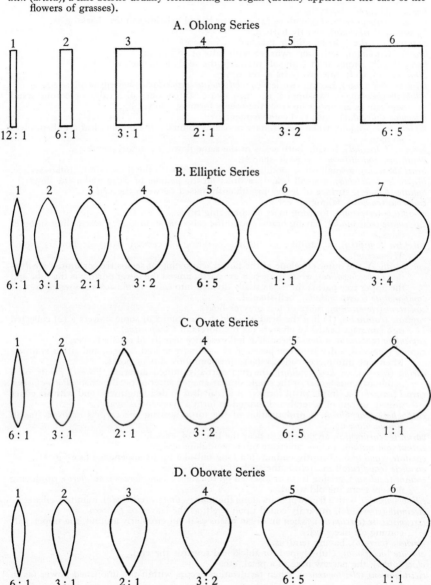

A. Oblong Series

1	2	3	4	5	6
12 : 1	6 : 1	3 : 1	2 : 1	3 : 2	6 : 5

B. Elliptic Series

1	2	3	4	5	6	7
6 : 1	3 : 1	2 : 1	3 : 2	6 : 5	1 : 1	3 : 4

C. Ovate Series

1	2	3	4	5	6
6 : 1	3 : 1	2 : 1	3 : 2	6 : 5	1 : 1

D. Obovate Series

1	2	3	4	5	6
6 : 1	3 : 1	2 : 1	3 : 2	6 : 5	1 : 1

FIG. 2.—SHAPES OF LEAVES

A. Oblong Series. 1, linear; 2, cultrate or lorate; 3, narrowly oblong; 4, oblong; 5, broadly oblong; 6, very broadly oblong.

B. Elliptic Series. 1, very narrowly elliptic; 2, narrowly elliptic; 3, elliptic; 4, broadly elliptic; 5, rotund; 6, orbicular; 7, oblate.

C. Ovate Series. 1, narrowly lanceolate; 2, lanceolate; 3, narrowly ovate; 4, ovate; 5, broadly ovate; 6, very broadly ovate.

D. Obovate Series. 1, narrowly oblanceolate; 2, oblanceolate; 3, narrowly obovate; 4, obovate; 5, broadly obovate; 6, very broadly obovate.

With acknowledgement to Mr. W. T. Stearn and Dr. W. B. Turrill

axil (axila), the angle between the leaf and the branch; *axillary (axilar)*, arising from the axil.

axile (axial), used of the attachment of ovules to the axis or to the inner angle of the cells of a syncarpous ovary (fig. 7D).

axis (eixo), of inflorescence, that part of the stem or branch on which the individual flowers are borne.

axis (eixo), of syncarpous ovary, the central column.

baccate (bacaceo), berry-like.

barbed (retrorso-celheado), with rigid points or lateral bristles pointing backwards.

barbellate (barbelado), shortly barbed.

basifixed (basifixo), of an anther attached by its base.

bearded (barbado), with long awns, or with tufts of stiff hairs.

berry (baga), a juicy fruit with soft pericarp, the seeds immersed in pulp.

bifid (bifido), cleft into two parts at the tip.

bifoliate (bifoliado), having two leaves ; *bifoliolate (bifoliolado)*, having two leaflets.

bilabiate (bilabiado), two-lipped, as when two or three lobes of a calyx or corolla stand separate as an upper lip from the others forming a lower lip.

bilocular (bilocular), with two compartments.

bipinnate (bipenado), when the primary divisions (pinnae) of a pinnate leaf are themselves pinnate.

bisexual (bisexual), having both sexes in the same flower or inflorescence.

botuliform (botuliforme), sausage-shaped.

bract (bráctea), a small often modified leaf associated with the flower or the inflorescence.

bracteole (bractéola), a small bract ; a bract on the pedicel or close under the flower.

bullate (bulado), surface of leaves prominently raised between the veins.

caducous (caduco), falling off early.

caespitose (cespitoso), forming mats or spreading tufts.

calyculate (caliculado), having bracts round the calyx, or an involucre resembling an outer calyx.

calyptra (caliptra), a cap-like or lid-like covering of certain fruits or flowers (some *Myrtaceae*).

calyx (cálice), the outer envelope of the flower, consisting of sepals free or united ; *calyx-tube (tubo do cálice)*, when the sepals are partly united the lower portion is the tube and the upper free part is the *limb* usually divided into *calyx-teeth, -lobes* or *-segments*.

campanulate (campanulado), bell-shaped.

canescent (canescente), more or less grey or hoary.

capitate (capitado), (1) like the head of a pin as the stigma of some flowers ; (2) collected into compact head-like clusters (as the florets of *Compositae*).

capitulum (capítulo), a dense head-like inflorescence usually of sessile flowers.

capsule (cápsula), a dry fruit composed of two or more united carpels, and either splitting when ripe into pieces called valves, or opening by slits or pores.

carina (carina), or *keel (quilha)*, the two partially united anterior (lowest) petals of a papilionaceous flower or the single similar-shaped anterior petal in many *Polygalaceae*.

carpel (carpelo), a simple pistil formed of a fruit-leaf folded lengthwise and with its edges united, or one of several such united to form a syncarpous ovary.

carpophore (carpóforo), a prolongation of the torus bearing the carpels or ovary (as in *Ranunculaceae*).

caruncle (carúncula), an outgrowth near the hilum of a seed.

catkin (amentilho), a close bracteate often pendulous spike.

caudate (caudado), abruptly ending in a long tail-like tip or appendage (see fig. 4I).

caudicle (caudículo), in orchids the stalk of a pollen-mass.

cauline (caulinar), arising from or inserted on the stem ; *cauliflorous (caulifloro)*, producing flowers from the old wood.

ciliate (ciliado), with a fringe of hairs along the edge ; *ciliolate (ciliolado)*, minutely ciliate.

circinate (circinado), inwardly coiled upon itself, as the leaves of *Drosera*.

circumscissile (circunciso), when an organ opens as if cut circularly around, the upper part coming off like a lid.

cirrhose (cirroso), having tendrils.

clavate (aclavado), club-shaped, or thickened towards the end.

claw (unha), the narrow base of a petal, etc.

cleistogamous (cleistogámico), when fertilization occurs within the unopened flower (as in *Viola*).

coccus (coca), a separate part of a lobed fruit (*Euphorbiaceae*).

column (coluna), (1) in orchids the adnate stamen and style forming a solid central body ; (2) the tube of connate filaments (as in *Malvaceae*, etc.).

coma (coma), a tuft of hairs at the end of some seeds.

commissure (comissura), the faces of cohering carpels (*Umbelliferae*).

compound (composto), the opposite of simple ; composed of several similar parts, as a leaf of several leaflets ; *compound fruit (fruto composto)*, when the fruits of separate flowers become united into a mass, as in *Morinda (Rubiaceae)* ; *compound umbel (umbela composta)*, where each ray again bears an umbel.

compressed (*comprimido*), flattened lengthwise from side to side (*laterally*), or from front to back (*dorsally*).

conduplicate (*conduplicado*), folded together lengthwise.

confluent (*confluente*), of parts of organs which are blended without fusion.

connate (*conato*), when parts of the same series are united so closely that they cannot be separated without tearing.

connective (*conectivo*), the part of an anther which connects its two thecae.

connivent (*conivente*), when parts converge so as to be nearer together above than below.

contorted (*contorcido*), of sepals or petals in the bud when each overlaps an adjoining one on one side, and is overlapped by the other adjoining one on the other side (see fig. 1A).

convolute (*convoludo*), rolled.

cordate (*cordiforme*), applied to the base of a leaf when it is more or less deeply notched (see fig. 3B).

coriaceous (*coriáceo*), firm, tough, of the consistence of leather.

corm (*cormo*), a tuberous bulb-like root stock.

corolla (*corola*), the inner envelope of the flower, consisting of petals free or united.

corona (*coroa*), a circle of appendages between the corolla and stamens, often united in a ring or crown (as in *Passifloraceae* and *Asclepiadaceae*).

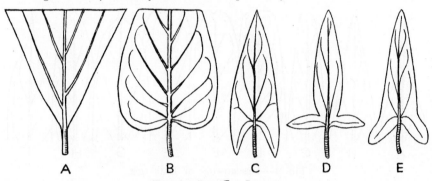

FIG. 3.—BASES OF LEAVES

A, cuneate. B, cordate. C, sagittate. D, hastate. E, auriculate.

corticate (*encorticado*), covered with bark or a bark-like substance.

corymb (*corimbo*), a more or less flat-topped inflorescence in which the branches or pedicels start from different points but all reach to about the same level.

cotyledon (*cotilédone*), seed-leaf.

crenate (*crenado*), the margin notched with regular blunt or rounded teeth ; *crenulate* (*crenulado*), crenate with very small teeth (see fig. 5C).

crispate (*encrespado*), curled.

crustaceous (*crustáceo*), of brittle texture.

cucullate (*cuculado*), hooded or hood-shaped.

culm (*colmo*), the stem of sedges and grasses.

cultrate (*cultriforme*), the shape of a knife-blade.

cuneate (*acunheado*), of the base of a leaf when tapering gradually, i.e. wedge-shaped (fig. 3A).

cuspidate (*cuspidado*), abruptly tipped with a sharp rigid point (see fig. 4I).

cymbiform (*cimbiforme*), boat-shaped.

cyme (*cimeira*), an inflorescence in which the central flower opens first (centrifugal), and the first branches at least are usually forked or opposite.

cystolith (*cistolito*), mineral concretion.

deciduous (*decíduo*), falling off at end of season of growth ; not evergreen.

declinate (*declinado*), bent or curved downwards or forwards.

decumbent (*decumbente*), prostrate-ascending.

decurrent (*decorrente*), as when the edges of the leaf are continued down the stem or petiole as raised lines or narrow wings.

decussate (*decussado*), in pairs alternately at right angles.

deflexed (*deflexo*), bent abruptly downwards.

dehiscent (*deiscente*), opening spontaneously when ripe, as capsules and anthers.

deltoid (*deltóide*), shaped like an equal-sided triangle.

dentate (*dentado*), the margin prominently toothed, the teeth directed outwards (see fig. 5E).

denticulate (*denticulado*), finely toothed.

depressed (*deprimido*), more or less flattened from above downwards or at least at the top.

diadelphous (*diadelfo*), in two bundles.

dichotomous (*dicótomo*), forking regularly into two.

dicotyledon (*dicotilédone*), a plant having two cotyledons.

didymous (*dídimo*), of anthers distinctly two-lobed, but with almost no connective.

didynamous (*didinâmico*), in two pairs of unequal length.

diffuse (*difuso*), loosely and openly branching or spreading.

digitate (*digitado*), a compound leaf whose leaflets diverge from the same point (the apex of the petiole), like the fingers of a hand (*Bombax*) (see fig. 6C).

dimorphic (*dimórfico*), of two forms.

dioecious (*dióico*), with unisexual flowers, the male flowers on one individual and the female flowers on another.

discoid (*discóide*), (1) like a disk or plate ; (2) applied to heads of *Compositae* without ray-flowers.

disk (*disco*), (1) an enlargement of the receptacle within the calyx or within the corolla or stamens, usually in the form of a ring, cup, or cushion, often lobed or even cut up into so-called *glands* ; (2) the central florets as compared with the ray-florets in a head of the *Compositae*.

dissepiment (*dissepimento*), a partition in an ovary or fruit.

distal (*distal*), farther from the place of attachment, the converse of *proximal*.

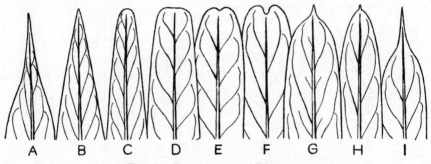

FIG. 4.—DESCRIPTION OF LEAF-TIPS

A, acuminate. B, acute. C, obtuse. D, truncate. E, retuse. F. emarginate. G, apiculate. H, mucronate. I, caudate or cuspidate.

distichous (*dístico*), regularly arranged one above another in two opposite rows, one on each side of the stem or rhachis

divaricate (*divaricado*), extremely divergent, or spreading in different directions.

dorsal (*dorsal*), of a carpel or leaf, the back, or the face turned away from the axis ; *dorsal suture* (see *suture*).

dorsifixed (*dorsifixo*), of an anther attached by its back to the filament.

drupe (*drupa*), a fleshy fruit such as plum, cherry, etc. (*Flacourtia, Ximenia*), with hard endocarp enclosing the seed.

ebracteate (*ebracteado*), without bracts.

echinate (*equinado*), covered with prickles like a hedgehog.

efoliate (*efoliado*), without leaves.

ellipsoid (*elipsóide*), an elliptic solid ; *elliptic* (*elíptico*) (see fig. 2B).

emarginate (*emarginado*), notched at the extremity (see fig. 4F).

embryo (*embrião*), the rudimentary plant still enclosed in the seed, consisting of the radicle from which the root develops, the cotyledons (one, two, rarely more), which become the earliest leaves, and the *plumule*, the bud from which the stem and further leaves develop.

endemic (*endémico*), confined to a region or country and not native anywhere else.

endocarp (*endocarpo*), the innermost layer of the pericarp.

endosperm (*endosperma*), the nutritive material (mealy, oily, fleshy or horny) stored within the seed, and often surrounding the embryo (formerly called *albumen*).

entire (*inteiro*), with an even margin without teeth, lobes, etc. (see fig. 5A).

epicalyx (*epicálice*), an involucre of bracts below the flower resembling an extra calyx as in some *Malvaceae*.

epigynous (*epiginico*), applied to the flower when the sepals, petals and stamens are apparently above the ovary, the latter being enclosed in an adnate receptacle or calyx-tube (see fig. 8C).

epipetalous (*epipétalo*), on the petals.

epiphyte (*epífita*), a plant which grows on another plant but without deriving nourishment from it, i.e. not parasitic, as some ferns and orchids growing on trees.

exocarp (*exocarpo*), the outer layer of the pericarp (of fruit).

exserted (*exserto*), projecting beyond, as the stamens from the tube of the corolla.

exstipulate (*sem estípulas*), without stipules.

extra-axillary (*extra-axilar*), beyond or outside the axil ; *extra-floral* (*extra-floral*), away from the flower.

extrorse (*extrorso*), of an anther which opens outwardly towards the circumference of the flower.

facultative (*facultativo*), occasional or incidental, as opposed to essential or necessary.

falcate (*falciforme*), curved like a scythe.

farinose (*farináceo*), covered with a meal-like powder.

fascicle (*fascículo*), a cluster of flowers, leaves, etc., arising from about the same point.

faveolate (*faveolado*), honeycombed.

ferruginous (*ferrugíneo*), rust-coloured.

filament (*filete*), the stalk of a stamen supporting the anther.

filiform (*filiforme*), slender, thread-like.

fimbriate (*fimbriado*), with the margin bordered by long slender processes.

flabellate (*flabelado*), fan-like.

flexuose or *flexuous* (*flexuoso*), zig-zag, or bent alternately in opposite directions.

floccose (*flocoso*), covered with woolly hairs.

foliaceous (*foliáceo*), leaf-like.

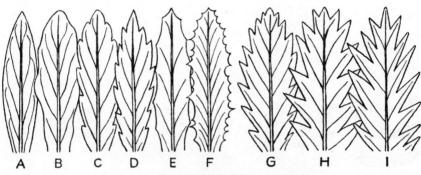

FIG. 5.—DESCRIPTION OF LEAF-MARGINS

A, entire. B, undulate. C, crenate. D, serrate. E, dentate. F, doubly dentate. G, pinnatifid. H, pinnatilobed. I, pinnatipartite.

follicle (*folículo*), a fruit formed from a single carpel opening usually only along the inner (i.e. ventral) suture to which the seeds are attached.

foveolate (*foveolado*), marked with small pitting.

free (*livre*), neither adhering nor united ; *free-basal placenta* (*placenta basal livre*), one in which the ovules are attached to a central column arising from the base of the ovary-cavity and not reaching the top, e.g. *Primulaceae* (see fig. 7F) ; *free-central placenta* (*placenta central livre*), the same, but reaching to the top of the cavity, e.g. most *Caryophyllaceae* (see fig. 7E).

frutescent (*frutescente*) or *fruticose* (*fruticoso*), having the characters of a shrub.

fugacious (*fugaz*), falling off early.

fulvous (*fulvo*), tawny.

funicle (*funículo*), the stalk which attaches the ovule to the placenta.

fuscous (*fosco*), dusky, brown rather than grey.

fusiform (*fusiforme*), spindle-shaped, thick but tapering towards each end.

gamopetalous (*gamopétalo*), when the petals are united either entirely or at the base into a tube, cup, or ring.

gamophyllous (*gamofilo*), when the perianth members are united.

gamosepalous (*gamosépalo*), when the sepals are united.

geniculate (*geniculado*), bent like a knee.

gibbous (*giboso*), slightly pouched.

glabrous (*glabro*), devoid of hairs ; *glabrescent* (*glabrescente*), becoming glabrous or nearly so.

gland (*glândula*), (1) a secreting structure on the surface or embedded in the substance of a leaf, flower, etc., or raised on a small stalk (*glandular hairs* or *stipitate glands*) ; (2) a warty protuberance or fleshy excrescence (often on petiole, inflorescence, or within the flower).

glaucous (*glauco*), pale bluish-green, or with a pale bloom.

glochidiate (*gloquidiado*), with barbed bristles or hooked hairs.

glomerate (*glomerado*), compactly clustered.

glomerule (*glomérulo*), a small compact cluster.

glumes (*glumas*), bracts, usually chaffy, in the spikelets of Grasses and Sedges.

gynoecium (gineceu) (pistil), the female part of the flower, consisting when complete of ovaries, styles, and stigmas.

gynobasic (ginobásico), when the style arises apparently from the base of the ovary (as in *Boraginaceae* and *Labiatae*).

gynophore (ginóforo), a stalk supporting the gynoecium and formed by an elongation of the receptacle.

hastate (hastado), of the base of a leaf when it has two more or less triangular lobes diverging laterally (see fig. 3D).

heterocarpous (heterocárpico), carpels of more than one kind.

heterogamous (heterogámico), when a flower-head, as in some *Compositae*, has two kinds of florets, those of the ray being neuter or unisexual, and those of the disk bisexual.

hilum (hilo), the scar left on the seed where it was attached to the funicle or placenta.

hirsute (hirsuto), with rather coarse, stiff hairs.

hispid (híspido), bristly-pubescent.

homogamous (homogámico), when a flower-head has all the flowers of the same kind.

hyaline (hialino), almost transparent.

hypanthium (hipanto), part of the flower between the base of the calyx-lobes and the insertion of the ovary, often called calyx-tube.

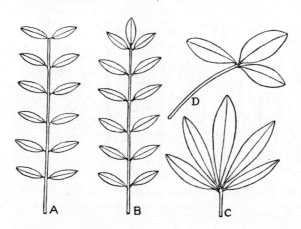

Fig. 6.—Compound Leaves

A, paripinnate.
B, imparipinnate.
C, digitate.
D, trifoliolate.

hypogynous (hipogínico), applies to the flower when the petals and stamens are inserted on the receptacle below the ovary and free from it, the ovary being thus *superior* (see fig. 8A).

imbricate (imbricado), (1) overlapping like tiles ; (2) in a flower-bud when one sepal or petal is wholly internal and one wholly external and the others overlapping at the edge only (see fig. 1B).

imparipinnate (imparipinado), pinnate with an odd terminal leaflet (see fig. 6B).

incised (inciso), cut rather deeply.

included (incluso), not projecting, the opposite of exserted.

incumbent (incumbente), lying on, as the cotyledons (in some *Cruciferae*) with the radicle against their lateral surface.

indefinite (indefinido), numerous (of stamens, etc.).

indehiscent (indeiscente), not opening when ripe.

indumentum (indumento), any covering to a surface, such as hairs, wool, scales, etc.

induplicate (induplicado), the margins (of petals or sepals) folded inwards but not overlapping.

indusium (indúsio), a cup covering the stigma, as in *Goodeniaceae*.

inferior (ínfero), of a calyx which is below the ovary, the latter being then *superior* ; of an ovary which appears to be below the calyx, the latter being adherent to the ovary.

inflorescence (inflorescência), the arrangement of the flowers ; *infructescence (infrutescencia)*, the inflorescence in the fruiting stage.

inserted (inserto), attached, included.

insertion (inserção), place of attachment.

internode (entrenó), the portion of stem between two nodes.

interpetiolar (interpeciolar), of stipules placed between the petioles of opposite leaves (often connate).

intrapetiolar (intrapeciolar), between the petiole and the stem.

introrse (introrso), of an anther when it opens towards the centre of the flower.

involucre (invólucro), a number of bracts surrounding the base of a flower-head or of an umbel ; *involucel*, the involucre of a secondary umbel.

involute (*involuto*), when the edges of the leaves are rolled inwards.

irregular flowers (*flores irregulares*), asymmetric flowers which cannot be divided into two equal halves in any vertical plane (see also *zygomorphic*).

jugate (*jugado*), coupled or yoked together ; applied especially to the leaflets of a pinnate leaf.

keel, see *carina*.

keeled (*aquilhado*), ridged along the middle of a flat or convex surface.

labellum (*labelo*), the lowest petal of an orchid, usually enlarged and different in form from the two lateral ones.

laciniate (*laciniado*), cut into slender lobes.

lamina (*limbo*), the limb, blade or expanded part of a leaf, sepal, or petal.

lanate (*lanoso*), woolly.

lanceolate (*lanceolado*) (see fig. 2, C2).

leaflet (*folíolo*), the ultimate member of a compound leaf.

legume (*vagem*), the fruit-pod of the family *Leguminosae*, consisting of a single carpel, usually opening along both sutures into two halves.

lemma (*lema*), the flowering glume of *Gramineae*.

lenticels (*lentículas*), corky spots on the bark.

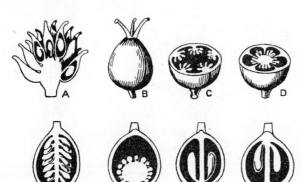

Fig. 7

A, apocarpous pistil.
B, syncarpous pistil.
C, parietal placentation.
D, axile placentation.
E, free-central placentation.
F, free-basal placentation.
G, basal erect ovules.
H, apical pendulous ovules.

lenticular (*lenticular*), shaped like a doubly convex lens.

lepidote (*lepídoto*), clothed with scales.

liana or *liane* (*liana*), a woody climber with rope-like stems.

ligulate (*ligulado*), strap-shaped, used of the ray-florets in many *Compositae*.

ligule (*lígula*), (1) a thin membranous appendage at the top of the leaf-sheath as in grasses ; (2) the limb of ray-florets in *Compositae*.

limb (*limbo*), the upper usually expanded part of the calyx, or corolla.

linear (*linear*), long and narrow with parallel edges.

lip (*lábio*), (1) one of the two divisions of a gamosepalous calyx or a gamopetalous corolla when it is cleft into an upper (posterior) and a lower (anterior) portion (see under *bilabiate*) ; (2) of an orchid (see *labellum*).

lobulate (*lobulado*), having small lobes.

locellate (*locelado*), divided into small secondary compartments.

locular (*locular*), having chambers, thus : *unilocular* (*unilocular*), one-chambered ; *bilocular* (*bilocular*), two-chambered.

loculicidal (*loculicida*), opening into the loculi, when a ripe capsule splits along the back (i.e. along the midrib or dorsal suture and not at the line of junction) of the carpels, e.g. in most *Liliaceae*.

loculus (*lóculo*), the chamber or chambers of an ovary or fruit containing the ovules or seeds

lodicule (*lodícula*), a small scale outside the stamens in flowers of *Gramineae*.

lorate, or *loriform* (*loriforme*), strap-shaped.

lyrate (*lirado*), of a leaf which is pinnately lobed or cut into small segments below, but with a much larger terminal lobe.

membranous, or *membranaceous* (*membranáceo*), of thin translucent texture.

mericarp (*mericarpo*), one of the separate halves or parts of a fruit (*Umbelliferae*, etc.).

mesocarp (*mesocarpo*), when the pericarp consists of three different layers, the middle one is the mesocarp ; it is often fleshy or succulent.

micropyle (*micrópilo*), a minute opening in the ovule through which the pollen-tube enters.

midrib (*nervura media*), the principal usually central nerve of a leaf or leaf-like part.

monadelphous (*monadelfo*), in one bundle (stamens in *Malvaceae*).

moniliform (*moniliforme*), like a string of beads.

monochlamydeous (*monoclamídeo*), of a flower having a perianth of one whorl.
monocotyledon (*monocotilédone*), a plant having a single cotyledon, or seed-leaf.
monoecious (*monóico*), when the male and female flowers are separate, but borne on the
    same individual plant.
mucronate (*mucronado*), ending abruptly in a short stiff point which is a continuation of
    the midrib (see fig. 4H).
muricate (*muricado*), rough, with short hard tubercles or pointed protuberances.
muticous (*mútico*), blunt, without a point.
nerves (*nervuras*), the principal or more conspicuous ribs of a leaf which start from the
    midrib and diverge or branch throughout the blade ; the smaller branches are veins ;
    nervose, with prominent nerves.
net-veined (*reticulado*), when the smaller veins are connected like the meshes of a net
    (=*reticulate*).
node (*nó*), the place on the stem at which a leaf or leaves and accompanying organs arise.
nut (*noz*), properly a one-seeded indehiscent fruit, with a hard, dry pericarp (the shell).
nutlet (*núcula*), a little nut.
obcordate (*obcordiforme*), more or less heart-shaped, but with the narrow end below and the
    broad end deeply notched.

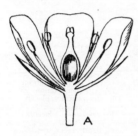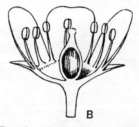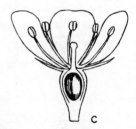

FIG. 8.—TYPES OF FLOWERS

A, hypogynous, i.e. ovary superior ; sepals, petals and stamens *below* the ovary. B, perigynous, i.e. ovary
    superior ; sepals, petals and stamens *around* the ovary. C, epigynous, i.e. ovary inferior ; sepals, petals
    and stamens *above* the ovary.

oblanceolate (*oblanceolado*), inverse of lanceolate (see fig. 2D).
oblate (*oblato*), transversely broadly elliptic (see fig. 2B).
oblique (*oblíquo*), of a leaf when the two sides of the blade are unequal at the base.
oblong (*oblongo*) (see fig. 2A).
obovate (*obovado*), ovate with the broadest part above.
obovoid (*obovóide*), solid shape of obovate outline.
obtuse (*obtuso*), blunt or rounded at the end (see fig. 4C).
ocreate (*ocreado*), with an *ocrea* (*ócrea*), i.e. a tubular stipule sheathing the stem (e.g. in
    Polygonum, see fig. 51).
opposite (*oposto*), (1) of leaves or branches when two are borne at the same node on oppo-
    site sides of the stem ; (2) of other organs, e.g. stamens, when opposite or placed in
    front of the petals instead of alternating with them.
orbicular (*orbicular*), flat with a more or less circular outline.
oval (*largamente elíptico*), broadly elliptic.
ovary (*ovário*), that part of the pistil (the usually enlarged base) which contains the ovules
    and eventually becomes the fruit.
ovate (*ovado*), egg-shaped, of a flat surface which is scarcely twice as long as broad, with
    the broader end below the middle (see fig. 2C).
ovoid (*ovóide*), solid shape of ovate outline.
ovule (*óvulo*), the immature seed in the ovary before fertilization.
pale, or *palea* (*pálea*), (1) one of the chaffy scales or thin colourless bracts amongst the
    flowers on the receptacle (*Compositae*) ; (2) of grasses, the inner of two bracts enclosing
    the flower.
palmate (*palmado*), shaped like the palm and fingers of a hand.
palmatifid (*palmatifido*), with the margin palmately cleft to less than halfway to the base.
palmatilobed (*palmatilobado*), palmately divided to about half-way to the midrib.
palmatipartite (*palmatipartido*), palmately divided almost to the midrib.
palmatisect (*palmatissecto*), palmately divided down to the midrib.
pandurate (*panduriforme*), fiddle-shaped.
panicle (*panícula*), an inflorescence in which the axis is divided into branches bearing
    several flowers.
papilionaceous (*papilionácea*), applied to flowers of the pea-flower shape.
papillose (*papiloso*), covered with minute, nipple-like protuberances.

pappus (*papilho*), applied to the ring of hairs or scales round the top of the fruit (as in *Compositae*) and representing the calyx-limb.

parietal placentation (*placentação parietal*), when the ovules are attached to the inner surface of the wall of the ovary (see fig. 7C).

paripinnate (*paripinado*) (see *abruptly pinnate* and fig. 6A).

partite (*partido*), cleft nearly but not quite to the base.

patent (*patente*), spreading.

pectinate (*pectinado*), like a comb.

pedate (*apedado*), resembling *palmate*, but the side-lobes further divided.

pedicel (*pedicelo*), the stalk of each individual flower of an inflorescence.

peduncle (*pedúnculo*), the general name for a flower-stalk bearing either a solitary flower or a cluster, or the common stalk (rhachis, or axis) of several pedicellate or sessile flowers.

pellucid (*perlúcido*), translucent.

peltate (*peltado*), of a leaf of which the stalk is attached to its under-surface instead of to its edge.

pendulous ovule (*óvulo pêndulo*), an ovule hanging down (see fig. 7H).

penicillate (*penicilado*), with a tuft of hairs at the end.

perfoliate (*perfoliado*), when the stem apparently passes through the blade of a leaf.

perianth (*perianto*), the floral envelope, consisting of calyx or corolla or both.

pericarp (*pericarpo*), the wall of the ripened ovary ; its layers may be fused into one, or more or less divisible into epicarp, mesocarp and endocarp.

perigynous (*periginico*), applied to the flower when the sepals, petals and stamens are inserted on an open receptacle surrounding the ovary but not adnate to it (see fig. 8B).

perulate (*perulado*), of buds covered with scales.

petiole (*peciolo*), the stalk of a leaf ; *petiolule* (*peciólulo*), the stalk of a leaflet.

phyllode (*filódio*), a flattened petiole or leaf-rhachis with the form and functions of a leaf.

pilose (*piloso*), hairy with rather long simple hairs.

pinna (*pina*), a primary division of a pinnate leaf ; *pinnate* (*pinado*), of a compound leaf with the leaflets arranged along each side of a common rhachis.

pinnatifid (*pinatífido*), with the margin pinnately cleft (see fig. 5G).

pinnatilobed (*pinatilobado*), pinnately divided to about half-way to the midrib (see fig. 5H).

pinnatipartite (*pinatipartido*), pinnately divided almost to the midrib (see fig. 5I).

pinnatisect (*pinatisecto*), pinnately divided down to the midrib.

pinnule (*pinula*), the secondary or tertiary division of a leaf which is twice or thrice pinnate.

pistil (*pistilo*), see *gynoecium*.

placenta (*placenta*), the part of the ovary, sometimes but not always thickened or raised, to which the ovules are attached.

plicate (*plicado*), folded.

plumose (*plumoso*), feathered, as bristles which have fine hairs on each side, e.g. the pappus of some *Compositae*.

pod (*vagem*), see *legume*.

pollen-mass (*massa polinica*), pollen-grains cohering into a single body (=*pollinium*).

polygamous (*poligámico*), when a species has unisexual and bisexual flowers on the same or on different individuals.

polypetalous (*polipétalo*), of flowers with petals free from each other.

polystichous (*polístico*), when leaves are borne in many series.

posterior (*posterior*), in position nearest to the axis.

precocious (*precoce*), appearing or developing early, often used of flowers which appear before the leaves.

prickle (*acúleo*), a sharp outgrowth from the bark or surface layer.

proliferous (*prolifero*), bearing adventitious buds on the leaves or in the flowers, capable of rooting and forming separate plants.

proterandrous (*proterandro*), when the anthers ripen before the stigmas.

proterogynous (*proteroginico*), when the stigmas are receptive before the anthers open.

proximal (*proximal*), nearer to the place of attachment, the converse of *distal* (*distal*).

pruinose (*pruinoso*), covered with a frost-like powder or bloom.

puberulous (*pubérulo*), slightly pubescent.

pubescent (*pubescente*), covered with short soft hairs.

pulverulent (*pulverulento*), powdery.

pulvinus (*pulvino*), a swollen base or apex of the petiole or petiolule.

punctate (*pontuado*), marked with dots or translucent glands.

punctiform (*punctiforme*), reduced to a mere point.

pungent (*pungente*), ending in a sharp rigid point.

pustulate (*pustuloso*), having slight elevations like pimples or blisters.

pyrene (*pireno*), a nutlet, or kernel, the stone of a drupe or similar fruit.

pyriform (*piriforme*), pear-shaped.

raceme (*racimo ou cacho*), an inflorescence in which the flowers are borne on pedicels along an unbranched axis or peduncle, the terminal flowers being the youngest and last to open (centripetal development).

radiate (*radiado*), of flower-heads of *Compositae* with ray-florets.

radical (*radical*), of leaves which arise so close to the base of the stem as to appear to proceed from the root.

radicle (*radículo*) (see *embryo*).

raphides (*ráfides*), mineral substances in the form of needle-shaped crystals, within the cells of plants.

ray (*raio*), (1) the florets of the margin of a flower of the *Compositae* when different from those of the centre, or disk ; (2) one of the radiating branches of an umbel.

receptacle (*receptáculo*) or *torus* (*toro*), the extremity of the axis on which the parts of the flower, sepals, petals, stamens and pistil are inserted.

regular (*regular*), (see *actinomorphic*).

reniform (*reniforme*), kidney-shaped.

repand (*repando*), when the margin is uneven or wavy, with shallow undulations ; not so deep as *sinuate*.

replum (*replo*), a frame-like placenta or false septum, as in *Cruciferae*.

reticulate (*reticulado*), net-veined, when the smallest veins of a leaf are connected together like the meshes of a net.

retrorse (*retrorso*), bent abruptly backwards.

retuse (*retuso*), notched (see fig. 4E).

revolute (*revoluto*), of leaves having the margins rolled backwards towards the midrib.

rhachis (*ráquis*), (1) the principal axis of an inflorescence above the peduncle ; (2) the axis on which the leaflets of a compound leaf are inserted ; *rhachilla* (*raquila*), the rhachis of the spikelet in grasses and sedges.

rhizome (*rizoma*), a root-stock or root-like stem prostrate on or under the ground, sending rootlets downwards and branches, leaves or flowering shoots upwards ; always distinguished from a true root by the presence of buds, leaves or scales.

rostellum (*rostelo*), in orchids a projection of the upper edge of the stigma in front of the anthers.

rostrate (*rostrado*), beaked.

rosulate (*rosetado*), when the leaves are in a circle or rosette.

rotate (*rodado*), wheel-shaped, of a corolla with a very short tube and spreading limb.

rotund (*arredondado*), with a shape between orbicular and broadly elliptic.

rufous (*rufo*), rusty or brownish-red.

rugose (*rugoso*), wrinkled.

ruminate (*ruminado*), of endosperm marked by transverse lines or divisions due to infolding of the inner layer of the seed-coat into the paler endosperm causing a marbled or mottled appearance (e.g. *Annonaceae*, *Myristicaceae*).

runcinate (*roncinado*), pinnatifid with the lobes pointing towards the base.

saccate (*saquiforme*), pouched.

sagittate (*sagitado*), of the base of a leaf with two acute straight lobes directed downwards (fig. 3C).

samara (*sâmara*), an indehiscent one-seeded fruit or mericarp provided with a wing (i.e. *Securidaca, Malpighiaceae*).

sarmentose (*sarmentoso*), with long whip-like branches.

scabrid or *scabrous* (*escábrido*), rough to the touch, usually from the presence of very short harsh hairs.

scales (*escamas*), (1) reduced leaves, usually sessile and scarious and seldom green ; (2) a kind of indumentum in the form of small, flat disks attached by the centre.

scape (*escapo*), a naked flower-stalk arising from the ground.

scarious (*escarioso*), thin and dry, not green.

scorpioid (*escorpióide*), when the main axis of an inflorescence is coiled in bud, the flowers being usually two-ranked, i.e. single flowers alternately right and left.

secund (*secundo*), when parts or organs (branches, leaves or flowers) are all directed to one side.

segment (*segmento*), a part or division of an organ.

septate (*septado*), divided by one or more partitions.

septicidal (*septicida*), when a ripe capsule splits along the lines of junction of the carpels (i.e. the lines of the placentas or dissepiments), each carpel then itself usually splitting down its ventral suture.

sericeous (*seríceo*), silky, with closely appressed soft, straight hairs.

serrate (*serrado*), toothed like a saw, with regular pointed teeth (see fig. 5D).

serrulate (*serrulado*), serrate with minute teeth.

sessile (*séssil*), without a stalk.

seta (*seda*), a bristle, or stiff hair.

setose (*setoso*), beset with bristles.

setulose (*setuloso*), a diminutive of setose.

sheath (*bainha*), the lower portion of the leaf clasping the stem.

silique (*silíqua*), a dry fruit divided into two compartments by a thin partition, opening by two valves which fall away from a frame on which the seeds are borne (e.g. *Cruciferae*); *silicule* (*silícula*), a short silique, not much longer than broad.

simple (simples), the opposite of compound ; *simple leaf*, of one blade, not divided into leaflets.

sinuate (sinuado), when the margin is uneven, with rather deep undulations.

sinus (sino), (1) recess between the teeth or lobes on a margin ; (2) angle formed by the basal lobes of a leaf or other organ.

spadix (espádice), a flower-spike with a fleshy or thickened axis (e.g. *Araceae* and some palms).

spathe (espata), a large bract enclosing a spadix, or one or more bracts enclosing a flower or inflorescence.

spathulate (espatulado), spoon-shaped, broadly rounded above and long and narrow below.

spike (espiga), an inflorescence with the flowers sessile along a simple undivided axis or rhachis ; *spicate*, arranged in a spike.

spikelet (espigueta), in grasses and sedges, a small spike composed of one or more flowers enclosed by glumes.

spine (espinho), a sharp-pointed hardened structure modified from another organ (leaf, branch, stipule etc.) or from part of an organ.

spinose (espinhoso), spiny or having spines.

spur (esporão), a slender usually hollow extension of some part of the flower.

staminode (estaminódio), an abortive or vestigial stamen without a perfect anther.

standard (estandarte), the large posterior petal (outside in the bud) of a papilionaceous corolla.

stellate hairs (pêlos estrelados), hairs with several arms radiating horizontally.

stigma (estigma), the point or surface of the pistil which receives the pollen, either sessile (when there is no style) or on the top or surface of the style or its branches.

stipe (estipe), the stalk supporting a carpel or gynoecium.

stipels (estipúlulas), two small secondary stipules at the base of a leaflet.

stipitate (estipitado), supported on a special stalk, or stipe, as an ovary (e.g. *Capparidaceae*).

stipules (estípulas), leaf-like or scale-like appendages of a leaf, usually at the base of the petiole.

stolon (estolho), a runner which roots.

stoma, pl. *stomata (estoma*, pl. *-s)*, pores in the epidermis.

stramineous (estramíneo), straw-like or straw-coloured.

striate (estriado), marked with parallel longitudinal lines, grooves or ridges.

strigose (estrigoso), with short stiff hairs lying close along the surface.

strobilus (estróbilo), a spike covered with imbricate bracts or sporophylls.

strophiole (estrofíolo), an appendage to the hilum of some seeds.

style (estilete), narrow upper part of an ovary supporting the stigma.

sub (sub), a prefix implying almost, e.g. *subacute*, almost acute, *subentire*, having a very slightly uneven margin.

subulate (subulado), awl-shaped.

suffrutescent (sufrutescente), like a suffrutex.

suffrutex (sufrútice), a shrublet, usually producing leafy and flowering shoots each year from a woody underground stock.

suffruticose (sufruticoso), shrubby.

sulcate (sulcado), grooved.

superior (súpero), of an ovary when the sepals, petals and stamens are inserted below it (hypogynous) ; also when the receptacle bearing the calyx, corolla and stamens is prolonged so as to be separate from the ovary, but forms a cup surrounding it (perigynous).

supra-axillary (supra-axilar), growing above an axil.

suture (sutura), the line of junction, or seam of union, commonly used of the line of opening of a carpel ; *dorsal* (outer or anterior) *suture (sutura dorsal)*, of a carpel represents the midrib of the carpellary leaf ; *ventral* (inner) *suture (sutura ventral)*, represents the united margins on which the ovules and placentas are borne.

synandrium (sinândrio), an androecium of a male flower with united anthers.

syncarpous (sincárpico), composed of two or more united carpels (see fig. 7B).

tepal (tépala), a division of an undifferentiated perianth.

terete (cilíndrico), cylindrical, circular in transverse section.

ternate (ternado), arranged in a whorl or cluster of three.

terrestrial (terrestre), on or in the ground.

testa (tegumento), the outer coat of a seed.

tetradynamous (tetradinâmico), of six stamens when four are longer and two shorter, as in Cruciferae.

thecae (tecas), anther-lobes.

thyrse (tirso), a panicle with the secondary and ultimate axes cymose.

thyrsoid (tirsoide), like a thyrse.

tomentellous (tomentelo), shortly tomentose.

tomentose (tomentoso), densely covered with short soft felted hairs.

torulose (toruloso), cylindrical with contractions or swellings at intervals.

torus (toro) (see *receptacle*).

trifoliate (*trifoliado*), three-leaved ;   *trifoliolate* (*trifoliolado*), having three leaflets (see fig. 6D).

trigonous (*trigonal*), obtusely three-angled.

trimerous (*trímero*), in threes, e.g. of a flower with three sepals, three petals, etc.

triquetrous (*tríquetro*), with three sharp angles.

tristichous (*trístico*), arranged one above another in three vertical rows.

truncate (*truncado*), cut off more or less squarely at the end (see fig. 4D).

tuber (*tubérculo*), (1) a thickened branch of an underground stem, which produces buds ; (2) a swollen root or branch of a root acting as a reserve store of nourishment.

tuberculate (*tuberculado*), covered with wart-like protuberances or knobs.

tumid (*túmido*), inflated, swollen.

tunicated bulb (*bolbo entunicado*), a bulb covered with complete enveloping coats, as an onion.

turbinate (*turbinado*) or *turbiniform* (*turbiniforme*), shaped like a top.

turgid (*túrgido*), swollen.

umbel (*umbela*), an inflorescence in which the divergent pedicels or rays spring from the same point ; *compound umbel* (*umbela composta*), where each ray itself bears an umbel, each of the latter being called a *partial umbel* (*umbela parcial*), *simple umbel*, where each ray terminates in a flower.

umbonate (*umbonado*), bearing a small boss (umbo) or elevation in the centre.

uncinate (*uncinado*), hooked.

undulate (*ondulado*), wavy on the margin (see fig. 5B).

unguiculate (*unguiculado*), contracted at the base into a claw.

unisexual (*unisexual*), having functional organs appertaining to only one sex.

urceolate (*urceolado*), urn-shaped, with a short swollen tube contracted near the top and then slightly expanded in a narrow rim.

utricle (*utrículo*), applied (1) to the traps of *Utricularia* ; (2) to the fruits of *Carex*, etc.

valvate (*valvado*), when the edges of petals or sepals meet without overlapping (see fig. 1C).

valve (*valva*), one of the parts produced by the splitting of a capsule when ripe.

valvule (*válvula*) (see *palea*).

velutinous (*velutino*), velvety.

venation (*nervação*), the arrangement of the veins of a leaf or other organ.

venose (*venoso*), having veins.

ventral (*ventral*), the inner face, or the surface towards the axis ; *ventral suture* (see *suture*).

ventricose (*ventricoso*), swollen or bulging on one side.

vernation (*vernação* ; *prefoliação*), folding of leaves in bud.

verrucose (*verugoso*), warty.

versatile (*versátil*), of an anther attached by its back to the very tip of the filament so as to swing loosely.

verticil (*verticilo*), a whorl or arrangement of similar parts in a circle at the same level ; *verticillate*, of leaves in a whorl or several arising at the same node arranged regularly around the stem.

vexillum (*vexilo*), the uppermost (or posterior) petal of a papilionaceous flower (see *standard*).

villous (*viloso*), beset with long weak hairs.

virgate (*virgado*), long, slender and straight as applied to stems.

viscid (*víscido*), sticky.

viscous (*viscoso*), glutinous, or very sticky.

vittae (*canais secretores oleíferos*), aromatic oil tubes in the fruit of some *Umbelliferae*.

viviparous (*vivíparo*), when the seeds germinate on the parent plant.

whorl (*verticilo*) (see *verticil*).

wing (*asa*), (1) any flat membranous expansion ; (2) one of the two lateral petals of a papilionaceous flower ; (3) one of the petaloid sepals of the flower of *Polygalaceae*.

zygomorphic (*zigomórfico*), applied to flowers which are bilaterally symmetric and may be bisected in only one vertical plane.

KEY TO FAMILIES AND HIGHER GROUPS

by J. E. Dandy

This key is designed to assist users of *Flora Zambesiaca* in preliminary identification down to family level. It is basically similar to the artificial keys to families published by F. W. Andrews, *The Flowering Plants of the Anglo-Egyptian Sudan*, vol. 1 (1950), which were adapted to the flora of the Sudan from the keys to the families of the world constructed by J. Hutchinson, *The Families of Flowering Plants*, vols. 1 and 2 (1926, 1934). The families included in it are those known to be represented in the *Flora Zambesiaca* area by indigenous or naturalized species. So far as can be foreseen, this key will cover all the phanerogamous plants to be dealt with in *Flora Zambesiaca* ; but it is obvious that some modifications may become necessary as the result of the discovery of additional plants within the area, or owing to changes in the delimitation and nomenclature of families (allowance for some of these is made in the index which follows the key). It is unlikely, however, that such modifications will impair the general usefulness of the key.

Ovules naked, not enclosed in ovaries formed of infolded or united carpels ; flowers (strobili) unisexual, without a perianth - - - GYMNOSPERMAE :
Palm-like plants with pinnate leaves forming a crown at the top of the stem
CYCADALES (see below)
Trees with simple leaves arranged along the branches - CONIFERAE (see below)
Ovules enclosed in ovaries formed of infolded or united carpels ; flowers bisexual or unisexual, with or without a perianth - - - - ANGIOSPERMAE :
Embryo with 2 (rarely more) cotyledons ; vascular bundles usually arranged in a cylinder in the stem ; leaves usually reticulate-veined ; flowers usually 5- or 4-merous
DICOTYLEDONES (see below)
Embryo with 1 cotyledon ; vascular bundles scattered in the stem ; leaves usually parallel-veined ; flowers usually 3-merous - - MONOCOTYLEDONES (see p. 72)

CYCADALES

Only family - - - - - - - - - - - **Cycadaceae**

CONIFERAE

Leaves well developed, strap-shaped, mostly alternate ; ovules solitary on the carpels ; seeds 1–2, large and drupe-like, borne on a receptacle (often swollen and fleshy) formed from the base of the strobilus - - - - - - - **Podocarpaceae**
Leaves small, scale-like, decussate (or subulate and ternate or alternate on young growths) ; ovules solitary or 2 or more on the carpels ; seeds enclosed in an indehiscent berry-like fruit or in a woody dehiscent cone - - - - - - **Cupressaceae**

DICOTYLEDONES

Gynoecium composed of 2 or more free carpels with separate styles and stigmas (sometimes, in *Nymphaeaceae*, the carpels immersed in the tissue of the expanded torus) :
  Petals present, free from each other - - - - - *Group 1* (see p. 56)
  Petals present, ± united - - - - - - - *Group 2* (see p. 56)
  Petals absent - - - - - - - - - *Group 3* (see p. 56)
Gynoecium composed of 1 carpel or of 2 or more united carpels with free or united styles, or if carpels free below then the styles or stigmas united :
  Ovules 2 or more in the gynoecium, attached to the wall of the ovary (placentation parietal) :
    Ovary superior :
      Petals present, free from each other - - - - *Group 4* (see p. 57)
      Petals present, ± united - - - - - - *Group 5* (see p. 58)
      Petals absent - - - - - - - - *Group 6* (see p. 58)
    Ovary ± inferior :
      Petals present, free from each other - - - - *Group 7* (see p. 59)
      Petals present, ± united - - - - - - *Group 8* (see p. 59)
      Petals absent - - - - - - - - *Group 9* (see p. 59)

Ovules 1 or more in the gynoecium, if more than 1 then attached to the central axis or
  the base or apex of the ovary :
  Ovary superior :
    Petals present, free from each other   -   -   -   -   *Group 10* (see p. 59)
    Petals present, ± united   -   -   -   -   -   -   *Group 11* (see p. 63)
    Petals absent   -   -   -   -   -   -   -   *Group 12* (see p. 66)
  Ovary ± inferior :
    Petals present, free from each other   -   -   -   -   *Group 13* (see p. 69)
    Petals present, ± united   -   -   -   -   -   -   *Group 14* (see p. 71)
    Petals absent   -   -   -   -   -   -   -   *Group 15* (see p. 72)

Group 1

Leaves opposite, without stipules ; carpels as many as the petals ; herbs, often succulent
      Crassulaceae
Leaves alternate, sometimes all radical :
  Aquatic herbs with long-petiolate floating leaves, the lamina ± peltate ; flowers
    solitary, long-pedunculate, often large and conspicuous :
    Carpels immersed in the tissue of the expanded torus ; petals numerous ; leaf-
      lamina cordate as well as peltate   -   -   -   -   - **Nymphaeaceae**
    Carpels not immersed in the tissue of the torus ; petals 3 ; leaf-lamina peltate, not
      cordate   -   -   -   -   -   -   -   - **Cabombaceae**
  Plants terrestrial ; leaf-lamina not peltate :
    Stamens perigynous, inserted at the mouth of the calyx-tube ; leaves pinnate or
      digitate, with stipules ; trees or shrubs or herbs, sometimes scrambling with
      prickles   -   -   -   -   -   -   -   -   - **Rosaceae**
    Stamens hypogynous, not inserted on the calyx :
      Sepals 6–24 in 2 or more series, free or the inner ones united ; shrublets or woody
        climbers with small dioecious flowers   -   -   -   - **Menispermaceae**
      Sepals 2–5 in 1 series, free or united :
        Herbs ; leaves crenate or lobed or much divided ; sepals free, herbaceous or
          petaloid, the median (posterior) one sometimes spurred   **Ranunculaceae**
        Trees or shrubs, sometimes climbing :
          Leaves pinnate or 3-foliolate ; stamens twice as many as the petals ; carpels 5
            Connaraceae
          Leaves simple :
            Sepals 2–3, valvate, free or united ; anthers often longer than the filaments,
              commonly with a broad prolongation of the connective above the thecae ;
              flowers solitary or fasciculate or in few-flowered cymes   **Annonaceae**
            Sepals 5, imbricate, free ; anthers much shorter than the filaments, with ±
              convergent thecae and no prolongation of the connective ; flowers in
              terminal cymes or panicles   -   -   -   -   - **Dilleniaceae**

Group 2

Leaves opposite ; petals united into an elongated tube ; succulent herbs or shrublets
      Crassulaceae
Leaves alternate ; petals shortly united at or above the base ; trees or shrubs, sometimes
  climbing :
  Leaves 3-foliolate ; sepals 5, free or almost so ; petals 5 -   -   -   **Connaraceae**
  Leaves simple ; sepals 2–3, free or united ; petals 6 or 4   -   -   **Annonaceae**

Group 3

Sepals united below ; stamens perigynous, inserted at the mouth of the calyx-tube ;
  leaves pinnate or digitate or palmately lobed, with stipules ; trees or shrubs or herbs,
  sometimes scrambling with prickles   -   -   -   -   -   - **Rosaceae**
Sepals free ; stamens hypogynous :
  Leaves compound, sometimes decompound ; sepals petaloid ; herbs or soft-wooded
    climbing or trailing plants   -   -   -   -   -   - **Ranunculaceae**
  Leaves simple :
    Flowers dioecious ; woody climbers with alternate leaves :
      Sepals 6–18 in 2 or more series ; stamens 3–6, free or united into a synandrium ;
        flowers in cymes or racemes ; leaves entire or palmately lobed
          Menispermaceae
      Sepals 5 in 1 series ; stamens 10–15, free ; flowers in elongated racemes ; leaves
        entire   -   -   -   -   -   -   -   - **Phytolaccaceae**
    Flowers bisexual ; herbs with opposite or subopposite leaves :
      Leaves entire ; calyx herbaceous ; flowers small, in cymes or panicles
          Aizoaceae
      Leaves toothed or lobed ; calyx petaloid ; flowers large and conspicuous, solitary
        or few together -   -   -   -   -   -   -   - **Ranunculaceae**

Group 4

Stamens 6, tetradynamous (inner 4 long, outer 2 short) ; sepals 4 ; petals 4 or fewer ; leaves without stipules ; herbs - - - - - - - **Cruciferae**
Stamens not tetradynamous :
  Gynoecium composed of 1 carpel, thus with only 1 placenta in the ovary :
    Flowers zygomorphic ; petals 5 or fewer, sometimes 1 ; leaves simple or compound ; trees or shrubs or herbs, sometimes climbing - - - - **Leguminosae**
    Flowers actinomorphic :
      Leaves 2-pinnate, with stipules ; trees or shrubs or herbs, sometimes climbing ; flowers in spikes or heads - - - - - - - **Leguminosae**
      Leaves simple or 1-foliolate, without stipules ; trees or shrubs, sometimes climbing :
        Sepals 5 ; petals 5 ; flowers in few-flowered racemes - - **Connaraceae**
        Sepals 3 ; petals 6 ; flowers solitary, axillary - - - **Annonaceae**
  Gynoecium composed of 2 or more united carpels, thus with 2 or more placentas :
    Leaves opposite or verticillate, simple, not all radical ; herbs :
      Stamens more than twice as many as the petals, often ± grouped into 3 or more bundles ; plants terrestrial ; leaves opposite, dotted with pellucid glands **Guttiferae**
      Stamens as many as the petals, free ; plants aquatic ; leaves verticillate, bladdery **Droseraceae**

    Leaves alternate or all radical :
      Flowers unisexual :
        Stamens as many as the petals ; herbaceous or woody climbers with tendrils ; leaves sometimes digitate or palmately lobed - **Passifloraceae**
        Stamens twice as many as the petals or more numerous ; trees or shrubs without tendrils :
          Leaves palmately 3–7-lobed ; petals free in the female flowers, united in the male flowers - - - - - - - - **Caricaceae**
          Leaves not palmately 3–7-lobed ; petals free in all the flowers **Flacourtiaceae**

      Flowers bisexual or sometimes polygamous :
        Ovary borne on a distinct gynophore :
          Flowers with a fimbriate or hairy corona outside the stamens ; styles 3–6 ; leaves simple ; trees or shrubs, sometimes climbing - **Passifloraceae**
          Flowers without a corona ; style 1 or stigma sessile ; leaves simple or digitate ; trees or shrubs or herbs, sometimes climbing - - **Capparidaceae**
        Ovary sessile or subsessile :
          Flowers ± zygomorphic :
            Leaves 2–3-pinnate ; trees ; flowers in axillary panicles ; stamens 5, alternating with 5 staminodes - - - - - - **Moringaceae**
            Leaves simple or digitate ; herbs or sometimes shrubs or small trees :
              Connective of the anthers prolonged above the thecae into an appendage ; flowers mostly solitary ; fruit a 3-valved capsule ; leaves simple **Violaceae**
              Connective of the anthers not prolonged above the thecae ; flowers in terminal racemes or spikes :
                Fruit a capsule dehiscing throughout its length by 2 valves ; leaves simple or digitate - - - - - - **Capparidaceae**
                Fruit a capsule gaping at the apex ; leaves simple, sometimes pinnately lobed or divided - - - - - **Resedaceae**
        Flowers actinomorphic :
          Stamens more numerous than the petals or as many as and opposite to them:
            Anthers opening by short apical pore-like slits ; sepals 5, caducous, leaving large persistent glands at the base of the flower ; trees or shrubs **Bixaceae**

            Anthers opening by longitudinal slits :
               Flowers with a fimbriate corona outside the stamens ; climbing shrublets with axillary tendrils - - - - **Passifloraceae**
              Flowers without a corona ; plants without tendrils :
                Prickly herbs ; leaves pinnately lobed ; sepals 2–3, caducous ; fruit a capsule dehiscing by 4–6 short valves at the top **Papaveraceae**

              Trees or shrubs ; leaves not pinnately lobed :
                Fruit indehiscent, surrounded by the 5 persistent accrescent wing-like sepals ; leaves entire - - **Dipterocarpaceae**
                Fruit capsular or baccate, not surrounded by persistent wing-like sepals ; leaves entire or toothed - - **Flacourtiaceae**

Stamens as many as and alternating with the petals :
　Styles 2–5, free or shortly united at the base :
　　Leaves bearing numerous sticky stipitate glands ; insectivorous herbs
　　　　　　　　　　　　　　　　　　　　　　　　　Droseraceae
　　Leaves without sticky stipitate glands ; plants not insectivorous :
　　　Flowers with a conspicuous fimbriate or hairy corona outside the
　　　　stamens ; herbs or shrubs, often climbing with tendrils
　　　　　　　　　　　　　　　　　　　　　　　Passifloraceae
　　　Flowers without a corona, or corona very small and inconspicuous ;
　　　　herbs without tendrils　-　　-　　-　　-　　-　　**Turneraceae**
　Style 1 :
　　Stamens accompanied by 2 series of staminodes, the outer filiform,
　　　the inner petaloid ; herbs ; leaves with pectinate stipules
　　　　　　　　　　　　　　　　　　　　　　　　　　Ochnaceae
　　Stamens not accompanied by staminodes ; trees or shrubs :
　　　Connective of the anthers prolonged above the thecae into an append-
　　　　age ; leaves with stipules　-　　-　　-　　-　　-　**Violaceae**
　　　Connective of the anthers not prolonged above the thecae ; leaves
　　　　without stipules　-　　-　　-　　-　　-　　-　**Pittosporaceae**

Group 5

Gynoecium composed of 1 carpel, thus with only 1 placenta in the ovary ; leaves simple or
　compound, with stipules or stipular spines ; trees or shrubs or herbs, sometimes
　climbing　-　-　-　-　-　-　-　-　-　-　-　**Leguminosae**
Gynoecium composed of 2 or more united carpels, thus with 2 or more placentas ; leaves
　without stipules :
　Flowers zygomorphic ; stamens fewer than the corolla-lobes, 4 or 2, inserted on the
　　corolla-tube :
　　Leaves pinnate ; trees ; stamens 4 ; fruit large, subcylindric, indehiscent
　　　　　　　　　　　　　　　　　　　　　　　Bignoniaceae
　　Leaves simple, sometimes reduced to scales ; herbs :
　　　Plants parasitic on roots ; leaves reduced to scales ; stamens 4　**Orobanchaceae**
　　　Plants not parasitic ; leaves (or solitary leaf) well developed ; stamens 2
　　　　　　　　　　　　　　　　　　　　　　　　Gesneriaceae
　Flowers actinomorphic ; stamens as many as the corolla-lobes or more numerous :
　　Flowers dioecious, the petals united in the male flowers but free in the female flowers
　　　leaves palmately 3–7-lobed ; trees　-　　-　　-　　-　　-　**Caricaceae**
　　Flowers bisexual ; leaves not palmately lobed :
　　　Stamens more numerous than the corolla-lobes, free from the corolla ; corolla-
　　　　lobes twice as many as the sepals ; trees or shrubs　-　　-　**Annonaceae**
　　　Stamens as many as the corolla-lobes, inserted on the corolla-tube ; corolla-lobes
　　　　as many as the sepals :
　　　　Leaves alternate, the lamina cordate-orbicular ; corolla-lobes induplicate-
　　　　　valvate ; aquatic herbs　-　　-　　-　　-　　-　　**Menyanthaceae**
　　　　Leaves opposite ; corolla-lobes contorted ; plants terrestrial :
　　　　　Shrubs or woody climbers ; fruit a large berry or prickly capsule
　　　　　　　　　　　　　　　　　　　　　　　　Apocynaceae
　　　　　Herbs ; fruit a septicidal capsule, not prickly　-　　-　　-　**Gentianaceae**

Group 6

Submerged aquatic herbs, fern-like in habit owing to the pinnately much-divided leaves ;
　flowers dioecious, in elongated pedunculate spikes ; calyx absent　**Hydrostachyaceae**
Plants terrestrial, not fern-like in habit :
　Leaves pinnate or 2-foliolate ; trees　-　　-　　-　　-　　-　**Leguminosae**
　Leaves simple or digitately 3-foliolate, sometimes reduced to scales :
　　Flowers dioecious ; leaves simple ; trees or shrubs :
　　　Calyx absent ; flowers in catkins ; fruit a 2-valved capsule ; seeds with a basal tuft
　　　　of long fine hairs　-　　-　　-　　-　　-　　-　　-　**Salicaceae**
　　　Calyx present ; flowers solitary or in fascicles or short racemes ; fruit a berry ;
　　　　seeds glabrous or covered with a woolly indumentum　-　　-　**Flacourtiaceae**
　　Flowers bisexual :
　　　Stamens as many as and opposite to the sepals ; flowers with a fimbriate corona
　　　　outside the stamens ; herbs or shrubs, often climbing with tendrils
　　　　　　　　　　　　　　　　　　　　　　　Passifloraceae
　　　Stamens more numerous than the sepals, or as many as and alternating with them ;
　　　　trees or shrubs, without tendrils :
　　　　Ovary borne on a distinct gynophore　-　　-　　-　　-　　-　**Capparidaceae**
　　　　Ovary sessile or subsessile　-　　-　　-　　-　　-　　-　**Flacourtiaceae**

Group 7

Flowers unisexual ; stamens 2–5, the anthers often curved or flexuous or folded ; herbs or shrubs, usually trailing or climbing with tendrils ; leaves often palmately lobed or deeply divided - - - - - - - - - **Cucurbitaceae**
Flowers bisexual ; stamens with small straight anthers ; plants without tendrils :
　Leaves absent or reduced to scales ; succulent shrubs, sometimes epiphytic **Cactaceae**
　Leaves present, well developed :
　　Trees or shrubs ; leaves alternate ; stamens arranged singly or in bundles opposite to the petals ; petals 5–9, persistent - - - - - **Flacourtiaceae**
　　Succulent herbs ; leaves opposite ; stamens numerous, not arranged in bundles ; petals numerous - - - - - - - - - **Aizoaceae**

Group 8

Flowers unisexual ; leaves alternate, often palmately lobed or deeply divided ; stamens 2–5, sometimes united, the anthers often curved or flexuous or folded ; trailing or climbing herbs or shrubs with tendrils - - - - - - **Cucurbitaceae**
Flowers bisexual ; leaves opposite or verticillate, entire, with interpetiolar stipules ; stamens with straight anthers ; trees or shrubs, sometimes climbing but without tendrils **Rubiaceae**

Group 9

Herbs or shrubs, often climbing, not parasitic ; leaves well developed ; flowers bisexual ; calyx with an elongated tube inflated below, the mouth 3-lobed or ± produced unilaterally into a single lobe - - - - - - **Aristolochiaceae**
Plants parasitic on branches of trees ; leaves reduced to scales ; flowers unisexual, solitary or sometimes paired, sessile on the bark of the host, small, surrounded by the scale-leaves and with a calyx of 4 or more free sepals - - - **Cytinaceae**

Group 10

Ovary 1-locular, sometimes septate towards the base :
　Sepals 1–2, free, sometimes caducous :
　　Flowers dioecious ; sepals 1–2 ; petals 1–4 ; stamens united into a synandrium ; leaf-lamina peltate or subpeltate ; woody climbers - **Menispermaceae**
　　Flowers bisexual ; sepals 2 ; petals 4–5 ; stamens free or with united filaments ; leaf-lamina not peltate ; herbs :
　　　Flowers zygomorphic ; ovary with 1 ovule ; fruit a nut ; leaves much divided **Fumariaceae**
　　　Flowers actinomorphic ; ovary with numerous ovules ; fruit a capsule ; leaves entire, fleshy - - - - - - - - **Portulacaceae**
　Sepals (or calyx-lobes) 3 or more :
　.Leaves opposite or verticillate, not all radical :
　　Leaves with stipules ; herbs or shrublets :
　　　Ovary with 2 ovules ; stamens 1–2 ; fruit indehiscent, surrounded by fleshy bracts - - - - - - - - - **Illecebraceae**
　　　Ovary with 3 or more ovules ; stamens up to twice as many as the petals ; fruit a 3–5-valved capsule - - - - - **Caryophyllaceae**
　　Leaves without stipules :
　　　Petals and stamens perigynous, inserted on the calyx-tube ; style 1 :
　　　　Ovary with 1 apical pendulous ovule ; stamens as many as the petals ; heath-like shrubs - - - - - - - - **Thymelaeaceae**
　　　　Ovary with numerous ovules on a free-basal or free-central placenta ; stamens twice as many as the petals or fewer ; herbs or shrubs or trees **Lythraceae**
　　　Petals and stamens hypogynous or only slightly perigynous, not inserted on the calyx :
　　　　Herbs ; ovary with numerous ovules on a free-basal or free-central placenta ; fruit a capsule ; styles 2–5, free - - - **Caryophyllaceae**
　　　　Trees or shrubs ; ovary with 1–2 basal ovules ; fruit drupaceous ; stamens as many as and alternating with the petals :
　　　　　Ovules 2 ; style 1 ; filaments united at the base ; flowers bisexual **Salvadoraceae**
　　　　　Ovule 1 ; styles 3, united at the base ; filaments free ; flowers polygamous or dioecious - - - - - - - **Anacardiaceae**
　Leaves alternate, sometimes all radical :
　　Leaves compound with 3 or more leaflets ; trees or shrubs, sometimes climbing :
　　　Leaves 2-pinnate ; flowers bisexual ; fruit winged, indehiscent **Leguminosae**
　　　Leaves 1-pinnate or 3-foliolate ; flowers polygamous or dioecious ; fruit drupaceous - - - - - - - - **Anacardiaceae**

Leaves simple or 1-foliolate :
 Ovary with 1 ovule :
  Flowers zygomorphic, the inner 2 sepals larger than the others, the lowest (median) petal forming a carina, the upper 2 petals vestigial or absent ; fruit long-winged, indehiscent ; trees or shrubs, sometimes climbing **Polygalaceae**
  Flowers actinomorphic ; fruit unwinged but sometimes surrounded by persistent wing-like sepals :
   Leaves with stipules ; stamens as many as the petals ; herbs **Illecebraceae**
   Leaves without stipules :
    Sepals united into an elongated tube ; stamens twice as many as the petals, inserted on the calyx-tube ; shrubs, sometimes heath-like **Thymelaeaceae**
    Sepals free or almost so :
     Stamens united into a synandrium ; leaf-lamina peltate or subpeltate ; woody climbers with dioecious flowers   -   **Menispermaceae**
     Stamens free or almost so ; leaf-lamina not peltate :
      Flowers polygamous or dioecious ; stamens 1–10 ; trees or shrubs **Anacardiaceae**
      Flowers bisexual ; stamens as many as and opposite to the petals ; woody climbers   -   -   -   -   -   **Opiliaceae**
 Ovary with 2 or more ovules :
  Anthers opening by 2 upcurving valves ; shrubs with 3-partite spines at the nodes ; fruit a berry   -   -   -   -   -   -   **Berberidaceae**
  Anthers opening by longitudinal slits :
   Leaves with stipules :
    Stamens as many as and opposite to the petals, united at the base ; petals and stamens hypogynous ; herbs or shrublets with stellate hairs **Sterculiaceae**
    Stamens more numerous than the petals, free ; petals and stamens perigynous, inserted at the mouth of the calyx-tube ; trees or shrubs :
     Ovary lateral, inserted on one side of the calyx-tube ; style arising from near the base of the ovary   -   -   -   **Chrysobalanaceae**
     Ovary central, not inserted on the calyx-tube ; style terminal **Rosaceae**
   Leaves without stipules ; trees or shrubs :
    Petals imbricate ; stamens as many as and opposite to the petals ; ovules on a free-basal placenta   -   -   -   -   -   **Myrsinaceae**
    Petals valvate :
     Ovules 2, pendulous from the apex of the ovary ; stamens as many as and alternating with the petals   -   -   -   -   **Icacinaceae**
     Ovules 2–5, pendulous from the apex of a central placenta ; stamens as many as and opposite to the petals, or up to twice as many, or reduced to 3 and accompanied by staminodes   -   - **Olacaceae**
Ovary 2- or more-locular :
 Stamens as many as and opposite to the petals :
  Filaments ± united into a tube or cup, sometimes alternating with staminodes ; calyx-lobes valvate ; herbs or shrubs, often with stellate hairs **Sterculiaceae**
  Filaments free, not alternating with staminodes :
   Ovary with numerous ovules in each loculus ; fruit a loculicidal capsule ; flowers in terminal panicles ; trees with gland-dotted leaves   -   **Heteropyxidaceae**
   Ovary with 1–2 ovules in each loculus ; fruit drupaceous or baccate :
    Inflorescences leaf-opposed ; herbaceous or woody plants, often climbing with tendrils ; leaves simple (often lobed or divided) or digitate ; ovules 2 in each loculus ; fruit a berry   -   -   -   -   -   -   -   **Vitaceae**
    Inflorescences axillary ; trees or shrubs, often spiny, without tendrils ; leaves simple, undivided ; ovules 1 in each loculus ; fruit drupaceous **Rhamnaceae**
 Stamens as many as and alternating with the petals, or more numerous or fewer :
  Leaves compound, with 2 or more leaflets :
   Inflorescences bearing tendrils ; climbing shrubs or herbs ; leaves pinnate or 2-ternate   -   -   -   -   -   -   -   -   **Sapindaceae**
   Inflorescences without tendrils :
    Herbs ; stamens twice as many as the petals :
     Leaves opposite (often unequal in a pair), pinnate, with stipules ; style 1 ; fruit spiny   -   -   -   -   -   -   -   **Zygophyllaceae**
     Leaves alternate (sometimes all radical), pinnate or 3-foliolate, without stipules ; styles 5 ; fruit not spiny   -   -   -   -   **Oxalidaceae**
    Trees or shrubs, sometimes climbing :
     Anthers 1-thecous ; leaves digitate, 3–9-foliolate ; stamens 10 or more, the filaments united below   -   -   -   -   -   -   **Bombacaceae**

Anthers 2-thecous ; leaves pinnate or 2-pinnate or 2–3-foliolate :
  Leaves opposite or subopposite, paripinnate ; flowers polygamous or
    dioecious ; stamens as many as the petals ; fruit a compressed capsule
    with winged seeds -   -   -   -   -   -   -   **Meliaceae**
  Leaves alternate :
    Leaves with lateral or intrapetiolar stipules, imparipinnate ; flowers ±
      zygomorphic, in racemes ; stamens 4–5 -   -   - **Melianthaceae**
    Leaves without stipules :
      Filaments united into a tube ; leaves pinnate or 2-pinnate
                                                          Meliaceae
      Filaments free or only shortly united at the base :
        Leaflets dotted with pellucid glands ; leaves pinnate or 3-foliolate
                                                          Rutaceae
        Leaflets without pellucid glands :
          Ovary with 2 or more ovules in each loculus :
            Ovules numerous in each loculus ; styles 5 ; fruit baccate ;
              leaves pinnate -   -   -   -   - **Averrhoaceae**
            Ovules 2 in each loculus ; style 1 ; fruit drupaceous, the
              exocarp sometimes dehiscent ; leaves pinnate or 3-foliolate ;
              plants secreting resin -   -   -   - **Burseraceae**
          Ovary with 1 ovule in each loculus :
            Styles 3–5, free and separated at the base ; flowers polygamous
              or dioecious -   -   -   -   -   - **Anacardiaceae**
            Style or styles central or terminal, not separated at the base :
              Ovules erect or ascending ; leaves paripinnate or 2-pinnate
                or 2–3-foliolate -   -   -   -   - **Sapindaceae**
              Ovules pendulous :
                Leaves 2-foliolate ; branches armed with straight spines
                                                          Balanitaceae
                Leaves imparipinnate or 3-foliolate ; branches unarmed
                  or armed with short hooked prickles
                                                          Simaroubaceae
Leaves simple or 1-foliolate, sometimes deeply divided :
  Perianth strongly zygomorphic ; sepals 3 or sometimes 5, the median one spurred ;
    petals 3 ; stamens 5, the anthers united round the ovary ; fruit an explosively
    dehiscent capsule ; herbs -   -   -   -   -   - **Balsaminaceae**
  Perianth actinomorphic or only slightly zygomorphic :
    Leaves opposite or verticillate, not all radical :
      Stamens more than twice as many as the petals :
        Leaves with stipules :
          Trees ; stipules interpetiolar ; petals fringed at the apex ; stamens
            15–45, the filaments not united into bundles - **Rhizophoraceae**
          Herbs ; stipules not interpetiolar ; petals not fringed ; stamens 15, the
            filaments united below into 5 bundles -   -   - **Geraniaceae**
        Leaves without stipules :
          Sepals united below ; stamens and petals perigynous, inserted at the
            mouth of the calyx-tube ; filaments free ; style 1, elongated ; leaves
            without glands ; trees -   -   -   -   - **Sonneratiaceae**
          Sepals free ; stamens and petals hypogynous ; filaments irregularly
            arranged or ± united into bundles ; styles 1–5 or absent ; leaves
            usually dotted or streaked with pellucid or opaque glands ; trees
            or shrubs or herbs -   -   -   -   -   - **Guttiferae**
      Stamens up to twice as many as the petals :
        Sepals united below into a tube :
          Petals perigynous, inserted at the mouth of the calyx-tube ; ovary with
            numerous ovules in each loculus :
            Anthers opening by apical pores ; connective often appendaged below
              the anther ; leaves with 3 or more parallel longitudinal nerves ;
              herbs or shrubs -   -   -   -   - **Melastomataceae**
            Anthers opening by longitudinal slits ; connective unappendaged ;
              leaves without parallel longitudinal nerves ; herbs or shrubs or trees
                                                          Lythraceae
          Petals hypogynous, not inserted on the calyx-tube ; ovary with 1–2 ovules
            in each loculus ; trees or shrubs :
            Stamens twice as many as the petals ; leaves with interpetiolar
              stipules ; petals fringed at the apex ; plants not spiny
                                                          Rhizophoraceae
            Stamens as many as the petals ; leaves without stipules ; petals not
              fringed ; plants spiny -   -   -   -   - **Salvadoraceae**

Sepals free or almost so :
Ovary with 1 ovule in each fertile loculus (sometimes 1 or 2 of the loculi empty) ; stamens twice as many as the petals ; fruit winged ; shrubs, often climbing -   -   -   -   -   -   **Malpighiaceae**
Ovary with 2 or more ovules in each loculus :
Leaves lobed or deeply divided ; ovary beaked ; herbs or shrublets
                                Geraniaceae
Leaves not lobed or divided ; ovary unbeaked :
Styles 3–5 ; herbs or shrublets :
Ovary with numerous ovules in each loculus ; sepals entire ; leaves with stipules -   -   -   -   -   **Elatinaceae**
Ovary with 2 ovules in each loculus ; sepals toothed or lobed at the apex ; leaves without stipules -   -   -   **Linaceae**
Style 1 ; trees or shrubs, sometimes climbing :
Stamens as many as the petals, alternating with petaloid staminodes ; leaves dotted with pellucid glands -   -   **Rutaceae**
Stamens as many as or fewer than the petals, not alternating with staminodes ; leaves without pellucid glands -   **Celastraceae**
Leaves alternate, sometimes all radical :
Ovary with 1 ovule in each fertile loculus (sometimes 1 or 2 of the loculi empty) :
Flowers unisexual or polygamous :
Leaves without stipules ; trees or shrubs without stellate hairs or scales ; flowers small, in racemes or narrow raceme-like panicles   **Sapindaceae**
Leaves with stipules ; plants often with stellate hairs or scales :
Style 1, short, with a large 2–3-lobed stigma ; fruit with 4 longitudinal wings ; stamens numerous ; trees -   -   -   **Tiliaceae**
Styles 2–5, free or united at the base, often lobed or branched ; fruit unwinged ; stamens 5 or more ; trees or shrubs or herbs **Euphorbiaceae**
Flowers bisexual :
Anthers 1-thecous ; stamens numerous, the filaments ± united into a tube ; sepals valvate, with or without an epicalyx of bracteoles ; herbs or shrublets, often with stellate hairs -   -   -   **Malvaceae**
Anthers 2-thecous :
Leaves without stipules :
Sepals 4 ; style 1 ; leaves entire or toothed or ± pinnately divided ; fruit a flattened silicule or separating into 2 cocci ; herbs
                                Cruciferae
Sepals 5 ; styles 2 ; leaves entire ; fruit separating into 2 cocci :
Shrubs, often climbing ; ovary 3-locular or 2-locular by abortion ; styles much elongated ; cocci with a long subterminal wing
                                Malpighiaceae
Herbs ; ovary 2-locular ; styles short ; cocci unwinged or winged along the back -   -   -   -   -   **Aizoaceae**
Leaves with stipules ; fruit drupaceous ; trees or shrubs :
Petals and stamens perigynous, inserted at the mouth of the calyx-tube ; ovary lateral, inserted on one side of the calyx-tube ; style arising laterally from near the base of the ovary
                                Chrysobalanaceae
Petals and stamens hypogynous ; ovary central, not inserted on the calyx ; style or styles central or terminal :
Ovary deeply lobed, with 1 style, the carpels separating in fruit ; stamens twice as many as the petals or more numerous, free
                                Ochnaceae
Ovary entire, with 1 or 3 styles, either with 2 fertile loculi or with 1 fertile loculus and 2 empty loculi, the carpels not separating in fruit ; stamens twice as many as the petals, the filaments united at the base into a cup -   **Erythroxylaceae**
Ovary with 2 or more ovules in each loculus :
Stamens as many as or fewer than the petals :
Ovules numerous in each loculus ; flowers in axillary pedunculate umbels ; trees or shrubs -   -   -   -   **Brexiaceae**
Ovules 2 in each loculus :
Ovary 5-locular :
Petals imbricate ; leaves ± lobed or divided, with stipules ; ovary and fruit beaked ; herbs or shrublets -   -   **Geraniaceae**
Petals contorted ; leaves not lobed or divided ; ovary and fruit unbeaked :
Herbs ; petals not persistent ; styles 5 -   -   -   **Linaceae**

Trees or shrubs ; petals persistent ; style 1   -   **Ixonanthaceae**
Ovary 2–4-locular ; trees or shrubs, sometimes climbing :
   Petals 2-lobed ; fruit drupaceous ; leaves entire   **Chailletiaceae**
   Petals not 2-lobed :
      Flowers unisexual ; styles 2–3, free or united at the base, often
         lobed or branched   -   -   -   -   - **Euphorbiaceae**
      Flowers bisexual ; style 1   -   -   -   -   **Celastraceae**
Stamens more numerous than the petals :
   Leaves without stipules :
      Filaments united into a tube ; petals elongated ; trees or shrubs
                                Meliaceae
      Filaments free :
         Leaves dotted with pellucid glands, sometimes deeply divided ;
           ovary with 4 or more ovules in each loculus ; trees or shrubs or
           herbs   -   -   -   -   -   -   -   -   - **Rutaceae**
         Leaves without pellucid glands ; ovary with 2 ovules in each loculus ;
           trees or shrubs, secreting resin   -   -   -   **Burseraceae**
   Leaves with stipules :
      Fruit indehiscent, surrounded by the 5 persistent accrescent wing-like
         sepals ; stamens numerous, free ; ovary 3-locular with 2 ovules in
         each loculus ; trees or shrubs   -   -   -   **Dipterocarpaceae**
      Fruit not surrounded by persistent wing-like sepals :
         Sepals imbricate ; plants without stellate hairs ; ovary 5-locular
           with 2 ovules in each loculus :
            Petals imbricate ; ovary beaked, with a 5-branched style ; herbs
              or shrublets   -   -   -   -   -   **Geraniaceae**
            Petals contorted ; ovary unbeaked, with 5 free styles ; trees or
              shrubs, sometimes climbing   -   -   -   -   **Linaceae**
         Sepals (or calyx-lobes) valvate ; plants often with stellate hairs :
           Filaments free or almost so ; trees or shrubs or herbs
                                  Tiliaceae
           Filaments ± united into a tube or into groups of 2 or 3 :
            Anthers 2-thecous ; trees or shrubs   -   -   **Sterculiaceae**
            Anthers 1-thecous ; herbs or shrubs or sometimes trees
                              Malvaceae

Group 11

Ovary 1-locular, sometimes septate towards the base :
  Flowers dioecious, the males with united petals, the females with 1–4 free petals ;
    stamens united into a synandrium ; leaf-lamina peltate or subpeltate ; woody
    climbers -   -   -   -   -   -   -   -   - **Menispermaceae**
  Flowers bisexual ; stamens free or with united filaments :
    Ovary with 1 ovule :
      Flowers zygomorphic, papilionaceous, the lowermost 2 petals united and forming
        a carina ; stamens 10, diadelphous, the uppermost 1 free, the other 9 with
        united filaments ; herbs   -   -   -   -   -   - **Leguminosae**
      Flowers actinomorphic ; stamens not diadelphous :
        Stamens twice as many as the calyx-lobes, inserted on the elongated calyx-tube ;
          corolla ring-like, inserted at the mouth of the calyx-tube ; ovule pendulous
          from the apex of the ovary ; shrubs, sometimes climbing
                                Thymelaeaceae
        Stamens as many as the calyx-lobes, not inserted on the calyx ; corolla 4–5-lobed,
          not ring-like ; ovule arising from the base of the ovary :
          Leaves alternate ; stamens opposite to the corolla-lobes ; calyx covered with
            stalked glands ; herbs or shrubs, sometimes climbing   **Plumbaginaceae**
          Leaves opposite ; stamens alternating with the corolla-lobes ; calyx without
            stalked glands ; shrubs or trees   -   -   -   -   - **Salvadoraceae**
    Ovary with 2 or more ovules :
      Stamens fewer than the corolla-lobes ; flowers ± zygomorphic ; herbs, often
        aquatic :
        Stamens 2 ; corolla strongly zygomorphic, the tube spurred ; leaves undivided
          or much divided and often bearing insectivorous bladders   **Lentibulariaceae**
        Stamens 4 ; corolla only slightly zygomorphic, the tube not spurred ; leaves
          undivided, without bladders   -   -   -   -   - **Scrophulariaceae**
      Stamens as many as the corolla-lobes ; flowers actinomorphic or almost so :
        Stamens opposite to the corolla-lobes :
          Trees or shrubs ; fruit indehiscent, usually 1-seeded ; leaves alternate
                                  Myrsinaceae

Herbs ; fruit a many-seeded circumscissile or 5-valved capsule, or sometimes 1-seeded and indehiscent ; leaves opposite or alternate   -   **Primulaceae**

Stamens alternating with the corolla-lobes ; trees or shrubs :

Leaves opposite ; corolla-lobes and stamens 4 ; ovules 4, pendulous from the apex of a 4-winged free-basal placenta ; fruit a compressed 2-valved capsule

Verbenaceae

Leaves alternate ; corolla-lobes and stamens 5 ; ovules 2, pendulous from the apex of the ovary ; fruit a drupe -   -   -   -   -   **Icacinaceae**

Ovary 2- or more-locular :

Corolla-lobes numerous (10 or more) :

Styles 5 ; fruit a 5-valved capsule ; herbs with fleshy leaves   -   -   **Aizoaceae**

Style 1 ; fruit a berry ; trees or shrubs, sometimes climbing :

Corolla-lobes in 2–3 series, imbricate ; stamens twice as many as the inner corolla-lobes, or as many as and opposite to the inner corolla-lobes and alternating with staminodes ; ovary with 1 ovule in each loculus ; leaves alternate

Sapotaceae

Corolla-lobes in 1 series ; stamens as many as or fewer than the corolla-lobes and alternating with them ; ovary with 2 or more ovules in each loculus :

Stamens 10 or more, as many as the corolla-lobes ; corolla-lobes contorted ; ovules numerous in each loculus ; leaves opposite   -   -   **Potaliaceae**

Stamens 2 ; corolla-lobes imbricate ; ovules 2–4 in each loculus ; leaves opposite or sometimes alternate -   -   -   -   -   -   **Oleaceae**

Corolla-lobes fewer than 10 :

Stamens more numerous than the corolla-lobes :

Leaves with stipules, often digitate or palmately lobed ; flowers unisexual ; trees or shrubs or herbs -   -   -   -   -   -   -   **Euphorbiaceae**

Leaves without stipules :

Flowers zygomorphic, the lowest (median) petal forming a carina ; filaments united into a sheath split on the upper side ; ovary 2-locular with 1 ovule in each loculus ; herbs or shrubs -   -   -   -   -   **Polygalaceae**

Flowers actinomorphic ; filaments not united into a sheath ; trees or shrubs :

Ovary with 1–2 ovules in each loculus ; fruit indehiscent, baccate ; flowers bisexual or unisexual   -   -   -   -   -   -   **Ebenaceae**

Ovary with several or numerous ovules in each loculus ; fruit a loculicidal capsule ; flowers bisexual :

Stamens grouped into bundles of 3 alternating with the corolla-lobes ; anthers opening by apical pores ; style 5-branched   -   -   **Theaceae**

Stamens not grouped into bundles ; anthers opening by apical pores or by longitudinal slits ; style unbranched ; plants often heath-like

Ericaceae

Stamens as many as or fewer than the corolla-lobes :

Stamens fewer than the corolla-lobes, 2–4 :

Ovary with more than 4 ovules in each loculus ; flowers zygomorphic :

Leaves pinnate, opposite or ternate ; stamens 4 ; fruit a loculicidal capsule with winged seeds ; trees or shrubs, sometimes climbing

Bignoniaceae

Leaves simple, sometimes deeply divided or reduced to scales :

Ovary completely or incompletely 4-chambered, each of the 2 original loculi becoming divided into 2 by a false septum ; stamens 4 ; herbs or shrubs -   -   -   -   -   -   -   -   **Pedaliaceae**

Ovary 2-locular, the loculi not becoming divided by false septa :

Ovules arranged in more than 2 series on each placenta ; stamens 4 or 2 ; fruit a capsule or sometimes a berry, the seeds not borne on hardened hook-like funicles ; leaves alternate or opposite or verticillate, sometimes reduced to scales ; herbs or shrubs or sometimes trees

Scrophulariaceae

Ovules arranged in 1–2 series on each placenta ; leaves opposite or sometimes alternate :

Fruit indehiscent or tardily dehiscent, armed with horns bearing recurved spines, the seeds not borne on hardened hook-like funicles ; flowers solitary, axillary ; stamens 4 ; leaves ± sinuate or lobed ; herbs   -   -   -   -   -   -   -   **Pedaliaceae**

Fruit a loculicidal capsule, without horns, the seeds often borne on hardened hook-like funicles (retinacula) ; flowers solitary or grouped in inflorescences ; stamens 4 or 2 ; leaves unlobed ; herbs or shrubs

Acanthaceae

Ovary with 1–4 ovules in each loculus :

Stamens 3 ; peduncles adnate to the petioles of the subtending leaves ; trees or shrubs with alternate leaves   -   -   -   -   -   **Chailletiaceae**

Stamens 4 or 2 ; peduncles not adnate to the leaves :
   Perianth actinomorphic ; stamens 2 ; leaves simple or compound, opposite
   or sometimes ternate or alternate ; trees or shrubs, sometimes climbing
                                                Oleaceae
   Perianth zygomorphic :
      Ovary ± deeply 4-lobed, the style gynobasic ; fruit separating into 4 nutlets
      (or fewer by abortion) ; leaves simple, opposite or verticillate or some-
      times alternate ; herbs or shrubs, often aromatic   -   - **Labiatae**
      Ovary not deeply 4-lobed, the style not gynobasic :
         Fruit a capsule, dehiscent loculicidally or sometimes failing to dehisce,
            the seeds often borne on hardened hook-like funicles ; ovary
            2-locular with 2–4 ovules in each loculus ; herbs or shrubs with
            opposite simple leaves - - - - **Acanthaceae**
         Fruit indehiscent or separating into 2 or more pyrenes or cocci, the
            seeds not borne on hardened hook-like funicles :
            Flowers solitary, axillary ; fruit indehiscent, armed with 2 or 4 spines
               or longitudinally 4-winged ; stamens 4 ; leaves opposite or
               subopposite, often toothed or pinnately lobed ; herbs
                                     Pedaliaceae
            Flowers grouped in inflorescences ; fruit without spines or wings :
               Anthers 1-thecous ; leaves mostly alternate, simple, narrow ;
               ovary 2-locular with 1 apical pendulous ovule in each loculus ;
               herbs or shrubs - - - - - - **Selaginaceae**
               Anthers 2-thecous ; leaves opposite or verticillate, simple or dig-
               itate ; ovary 2–8-locular with 1 basal or axile ovule in each
               loculus ; herbs or shrubs or trees - - **Verbenaceae**
Stamens as many as the corolla-lobes, 4 or more :
   Leaves absent or reduced to scales :
      Slender twining parasitic plants ; ovary with 2 ovules in each loculus ; fruit
      a capsule ; corolla with or without infrastaminal scales **Convolvulaceae**
      Shrubs or succulent plants, not parasitic ; ovary with numerous ovules in each
      loculus ; fruit formed of 2 separated follicular carpels (or 1 by abortion) ;
      flowers with a corona - - - - - - **Asclepiadaceae**
   Leaves present, well developed :
      Stamens opposite to the corolla-lobes ; trees or shrubs with alternate leaves :
         Leaves 2-pinnate ; corolla-lobes valvate ; stamens not accompanied by
            staminodes, the filaments united into a tube - - - **Leeaceae**
         Leaves simple, entire ; corolla-lobes imbricate ; stamens sometimes alter-
            nating with staminodes - - - - - **Sapotaceae**
      Stamens alternating with the corolla-lobes :
         Leaves opposite or verticillate, not all radical :
            Stamens hypogynous, not inserted on the corolla ; anthers opening by
            apical pore-like slits ; heath-like shrubs with small verticillate leaves
                                            Ericaceae
            Stamens inserted on the corolla-tube :
               Corolla-lobes imbricate :
                  Style 2-branched, the branches again 2-branched ; ovary 2-locular
                    with 2 ovules in each loculus ; fruit a flattened 2-lobed loculicidal
                    capsule ; shrubs - - - - - **Loganiaceae**
                  Style unbranched or with 2 short branches at the apex :
                    Ovary with numerous ovules in each loculus ; fruit a septicidal
                      capsule or sometimes baccate ; corolla-lobes and stamens 4 :
                      Trees or shrubs ; flowers in cymes or racemes or panicles
                                        Buddlejaceae
                    Herbs ; flowers solitary or paired in the axils of the leaves
                                  Scrophulariaceae
                  Ovary with 1–2 ovules in each loculus ; fruit drupaceous or
                    separating into pyrenes or nutlets :
                    Corolla-lobes and stamens 5 ; herbs, often hispid with bulbous-
                      based hairs - - - - - **Boraginaceae**
                    Corolla-lobes and stamens 4 ; herbs or shrubs or trees
                                    Verbenaceae
         Corolla-lobes contorted or valvate :
            Flowers with a corona ; pollen agglutinated into granules or waxy
            masses ; herbs or shrubs or trees, often climbing **Asclepiadaceae**
            Flowers without a corona ; pollen not agglutinated :
               Corolla-lobes valvate ; leaves often with 3 or more longitudinal
                nerves ; trees or shrubs, sometimes climbing, often with
                axillary spines or tendrils - - - - **Strychnaceae**

Corolla-lobes contorted :

Herbs ; fruit a septicidal capsule   -   -   **Gentianaceae**

Trees or shrubs, sometimes climbing ; fruit a berry or drupe or the carpels becoming separate :

Corolla-lobes and stamens 6–9 ; filaments united into a tube ; fruit a berry   -   -   -   -   **Potaliaceae**

Corolla-lobes and stamens 4–5 ; filaments free   **Apocynaceae**

Leaves (at least the lower ones) alternate, sometimes all radical :

Leaves all radical ; flowers small, in pedunculate spikes or heads ; fruit a circumscissile capsule ; herbs   -   -   -   **Plantaginaceae**

Leaves not all radical :

Ovary with more than 2 ovules in each loculus :

Corolla-lobes contorted ; fruit formed of 2 separated carpels ; shrubs or trees   -   -   -   -   -   **Apocynaceae**

Corolla-lobes not contorted ; fruit a capsule or berry :

Styles 2–3 ; corolla-lobes imbricate ; fruit a capsule ; herbs   **Hydrophyllaceae**

Style 1 :

Corolla-lobes plicate or valvate ; herbs or shrubs or trees   **Solanaceae**

Corolla-lobes imbricate :

Fruit a capsule with winged seeds ; leaves simple or 3-foliolate ; spiny shrubs   -   -   **Bignoniaceae**

Fruit a berry or capsule, the seeds unwinged ; leaves simple ; herbs or spiny shrubs   -   -   -   **Solanaceae**

Ovary with 1–2 ovules in each loculus :

Filaments united into a sheath split on the upper side ; flowers zygomorphic, the lowest (median) corolla-lobe forming a carina ; fruit drupaceous ; shrubs or trees   -   -   **Polygalaceae**

Filaments not united into a sheath :

Corolla-tube split down the front, with 4 lobes ; anthers 1-thecous ; ovary 2-locular with 1 apical pendulous ovule in each loculus ; herbs or shrublets with spicate flowers   -   **Selaginaceae**

Corolla-tube not split down the front ; anthers 2-thecous :

Fruit drupaceous ; trees or shrubs :

Style 2-branched (the branches undivided or again 2-branched) or ending in a horizontal ring bearing a 2-lobed stigma ; leaves without stipules ; flowers bisexual, in cymes or panicles   -   -   **Boraginaceae**

Style absent, the stigma sessile ; leaves with stipules ; flowers dioecious, in axillary cymes or fascicles   **Aquifoliaceae**

Fruit not drupaceous ; herbs or shrublets, sometimes climbing :

Fruit separating into 4 nutlets (or fewer by abortion) ; corolla-lobes imbricate or sometimes contorted   **Boraginaceae**

Fruit capsular or sometimes indehiscent ; corolla-lobes plicate or valvate ; plants often twining or trailing   **Convolvulaceae**

Group 12

Ovary with 2 or more ovules in each loculus :

Plants aquatic, moss-like or liverwort-like in habit, growing on rocks ; flowers small and inconspicuous ; stamens 1–2 ; ovary with numerous ovules ; fruit a capsule   **Podostemaceae**

Plants terrestrial, not moss-like or liverwort-like in habit :

Leaves reduced to scales forming toothed sheaths surrounding the nodes ; trees with unisexual flowers ; male flowers in spikes, female flowers in heads ; stamen 1 ; ovary 1-locular with 2 collateral ovules ; style with 2 elongated branches   **Casuarinaceae**

Leaves well developed, not reduced to scales :

Leaves opposite or verticillate or all radical :

Flowers unisexual ; trees or shrubs :

Calyx absent ; leaves toothed at the apex, flabellate ; flowers in catkin-like spikes ; ovules numerous in each loculus   -   **Myrothamnaceae**

Calyx present ; leaves entire, not flabellate ; flowers solitary or fasciculate in the axils of the leaves ; ovules 2 in each loculus :

Leaves with stipules ; flowers dioecious ; stamens very numerous, spirally arranged on a prolonged torus   -   -   -   -   **Euphorbiaceae**

Leaves without stipules ; flowers monoecious ; stamens 4–6   **Buxaceae**
Flowers bisexual ; herbs :
  Calyx spurred, the spur adnate to the pedicel ; ovary beaked ; leaves toothed
    or lobed   -   -   -   -   -   -   -   **Geraniaceae**
  Calyx not spurred ; ovary unbeaked ; leaves entire :
    Sepals united below ; stamens perigynous, inserted on the calyx-tube :
      Style 1 ; ovary 1–5-locular with numerous ovules in each loculus ; fruit
        a capsule, circumscissile or dehiscing irregularly or by valves
                                                  Lythraceae
      Styles 2–5, or if style 1 then ovary 1-locular with few ovules ; fruit a
        circumscissile capsule   -   -   -   -   -   -   **Aizoaceae**
    Sepals free or almost so ; stamens hypogynous :
      Ovary 3–5-locular with axile placentas   -   -   -   **Aizoaceae**
      Ovary 1-locular with a free-basal or free-central placenta
                                Caryophyllaceae
Leaves alternate, not all radical :
  Leaves pinnate ; trees with unisexual or polygamous flowers ; ovary 2-locular
    with 2 ovules in each loculus ; fruit indehiscent   -   -   **Sapindaceae**
  Leaves simple or digitate :
    Ovary 1-locular ; flowers in racemes or spikes :
      Flowers bisexual ; calyx scarious ; fruit a circumscissile capsule ; herbs
                                Amaranthaceae
      Flowers dioecious ; calyx not scarious ; fruit a drupe ; trees or shrubs,
        sometimes climbing :
        Leaves with stipules ; calyx imbricate ; stamens opposite to the sepals ;
          styles 3, usually 2-lobed   -   -   -   **Euphorbiaceae**
        Leaves without stipules ; calyx valvate ; stamens alternating with the
          sepals ; style absent, the stigma sessile and multiradiate
                                  Icacinaceae
    Ovary 2- or more-locular :
      Gynoecium composed of 3 or more loosely united carpels, in fruit separating
        into as many follicles (or fewer by abortion) ; trees or shrubs with
        unisexual or polygamous flowers ; leaves simple or digitate ; sepals
        valvate, united below ; stamens united into a column   **Sterculiaceae**
      Gynoecium composed of completely united carpels, in fruit forming a cap-
        sule or indehiscent or finally separating into winged cocci :
        Ovary borne on a distinct gynophore ; flowers bisexual ; shrubs with
          solitary pedunculate axillary flowers   -   -   -   **Capparidaceae**
        Ovary sessile or subsessile ; flowers unisexual or polygamous :
          Leaves with stipules ; fruit a capsule or indehiscent or separating into
            2 winged cocci ; trees or shrubs or herbs   -   -   **Euphorbiaceae**
          Leaves without stipules ; fruit a septicidal capsule with 2 or more
            longitudinal membranous wings ; shrubs or trees   **Sapindaceae**
Ovary with 1 ovule in each loculus :
  Ovary 2- or more-locular :
    Leaves pinnate ; trees or shrubs with polygamous or dioecious flowers ; fruit
      drupaceous or baccate   -   -   -   -   -   -   **Sapindaceae**
    Leaves simple, sometimes lobed or much divided or reduced to scales or stipular
      spines :
    Flowers unisexual or polygamous :
      Ovary 5-locular, the carpels loosely united and becoming separate in fruit ;
        calyx present, valvate, the sepals united below ; stamens 5–10, united into a
        column ; leaves alternate ; trees or shrubs   -   -   -   **Sterculiaceae**
      Ovary 2–4-locular, the carpels completely united and not becoming separate
        in fruit ; calyx present or absent, the male flowers sometimes reduced to a
        single stamen and arranged in a common involucre with a pedicellate female
        flower in their midst ; leaves alternate or opposite, sometimes reduced to
        scales or stipular spines ; trees or shrubs or herbs, sometimes climbing or
        succulent   -   -   -   -   -   -   -   -   -   **Euphorbiaceae**
    Flowers bisexual :
      Sepals united into a tube ; stamens twice as many as the calyx-lobes, perigynous,
        inserted on the calyx-tube ; ovary 2-locular ; fruit a drupe ; shrubs or trees
                                  Thymelaeaceae
      Sepals free ; stamens hypogynous or almost so :
        Leaves ± toothed or pinnately lobed or divided ; style 1 ; fruit a flattened
          silicule or separating into 2 cocci ; herbs   -   -   -   **Cruciferae**
        Leaves entire ; styles 2 or more, free or united at the base :
          Ovary 7–10-locular ; fruit a berry ; flowers in spike-like racemes ; herbs or
            shrublets   -   -   -   -   -   -   -   **Phytolaccaceae**

Ovary 2–5-locular ; fruit a loculicidal capsule or separating into 2 cocci, the cocci sometimes winged along the back ; flowers in lax or dense cymes ; herbs - - - - - - - - - **Aizoaceae**
Ovary 1-locular :
Leaves absent or reduced to scales ; flowers spicate :
Slender twining parasitic plants ; flowers not immersed in the rhachis of the spike ; stamens 6–9, accompanied by staminodes ; anthers opening by valves
Lauraceae
Succulent maritime herbs with articulated branches, not parasitic ; flowers immersed in the rhachis of the spike ; stamens 1–2 ; anthers opening by longitudinal slits - - - - - - - **Chenopodiaceae**
Leaves present, well developed, not reduced to scales :
Leaves with stipules, the stipules sometimes forming a sheath (ocrea) surrounding the stem :
Leaves 3–4-pinnate ; calyx petaloid ; fruit a stipitate achene borne on a slender pedicel ; herbs - - - - - - - - **Ranunculaceae**
Leaves simple or digitate :
Ovule pendulous from the apex or near the apex of the ovary ; flowers unisexual or polygamous :
Flowers densely spicate or capitate, or crowded in or on an open or flat receptacle or inside a hollow almost closed receptacle (fig), the female flowers sometimes immersed in the tissue of the receptacle ; calyx sometimes absent ; trees or shrubs or herbs - - - **Moraceae**
Flowers solitary or fasciculate or in cymes or racemes or panicles ; calyx sometimes reduced or absent in the female flowers :
Annual herbs ; leaves opposite or alternate, all or the lower ones palmately divided ; male flowers in elongated panicles ; female flowers spicate, enclosed by bracts ; fruit dry, indehiscent - **Cannabiaceae**
Trees or shrubs ; leaves alternate :
Stamens as many as the calyx-segments, 4–5 ; style 2-branched with simple or divided stigmas ; fruit a drupe ; trees or shrubs
Ulmaceae
Stamens more numerous than the calyx-segments ; style unbranched, sometimes very short :
Flowers solitary or paired in the axils of the leaves ; leaves 1–3-foliolate with small narrow leaflets ; heath-like shrubs
Rosaceae
Flowers in axillary racemes or panicles ; leaves simple ; trees
Euphorbiaceae
Ovule arising from the base or near the base of the ovary :
Leaves opposite, marked with cystoliths ; flowers unisexual ; herbs or shrubs - - - - - - - - - **Urticaceae**
Leaves alternate :
Calyx absent ; flowers minute, in dense spikes ; shrubs, sometimes climbing - - - - - - - - **Piperaceae**
Calyx present :
Styles 2–3, free or united below ; stamens 4–8 ; fruit a small nut ; stipules often forming an ocrea surrounding the stem ; herbs or shrubs, sometimes climbing - - - - **Polygonaceae**
Style 1 or absent ; stamens 5 or fewer :
Flowers bisexual ; sepals united below, the free portions alternating with the lobes of an epicalyx ; style arising laterally from near the base of the ovary ; herbs with palmately lobed leaves
Rosaceae
Flowers unisexual ; sepals free or united, without an epicalyx ; style or sessile stigma terminal :
Leaves palmate or palmately lobed, without cystoliths ; flowers dioecious, the male flowers crowded in branched spikes, the female flowers in heads ; trees - - - - **Moraceae**
Leaves simple, sometimes pinnately lobed, often marked with cystoliths ; flowers monoecious or dioecious, cymose, the cymes often dense or head-like ; herbs or shrubs or trees, sometimes climbing or epiphytic - - - - - **Urticaceae**
Leaves without stipules :
Submerged aquatic herbs ; leaves verticillate, deeply and bifurcately divided with linear or filiform segments ; flowers unisexual, solitary and sessile in the axils of the leaves - - - - - - **Ceratophyllaceae**
Plants not aquatic ; leaves alternate or opposite, undivided or sometimes pinnately divided :

Calyx absent ; flowers in spikes :
    Flowers unisexual ; stamens 3–12 ; styles 2, free or shortly united below ;
      leaves alternate ; trees or shrubs -   -   -   -   **Myricaceae**
    Flowers bisexual ; stamens 2 ; style absent, the stigma sessile ; leaves
      alternate or opposite or verticillate ; herbs, sometimes trailing or climbing
      or epiphytic -   -   -   -   -   -   -   **Piperaceae**
Calyx present (sometimes absent in female flowers) :
    Stamens twice as many as the calyx-segments or more numerous :
      Ovule erect from the base of the ovary ; fruit enclosed in the persistent
        longitudinally 4-winged calyx-tube ; spiny shrubs with small linear
        often fasciculate leaves; flowers polygamo-dioecious **Nyctaginaceae**
      Ovule pendulous from the apex of the ovary ; fruit not enclosed in a
        winged calyx-tube ; plants not spiny :
        Flowers bisexual ; sepals united into an elongated tube ; stamens
          inserted on the calyx-tube ; style elongated, slender ; shrubs, often
          heath-like, or sometimes trees, with alternate or opposite leaves
              Thymelaeaceae
        Flowers dioecious ; sepals shortly united at the base ; stamens not
          inserted on the calyx-tube ; style absent, the stigma sessile ; shrubs
          or trees with opposite or subopposite leaves   -   **Monimiaceae**
    Stamens fewer than twice as many as the calyx-segments, sometimes
      accompanied by staminodes :
      Anthers opening by valves ; calyx 6-lobed ; stamens 6–9, accompanied
        by staminodes ; trees -   -   -   -   -   -   **Lauraceae**
      Anthers opening by longitudinal slits ; calyx with 3–5 segments or lobes :
        Leaves opposite or subopposite :
          Sepals ± petaloid, united into a tube constricted above the ovary,
            the lower portion of the tube persistent and enclosing the fruit
            and often glandular on the outside ; stamens not accompanied
            by staminodes ; herbs or climbing shrubs -   **Nyctaginaceae**
          Sepals dry and ± scarious, free or shortly united at the base, not
            forming a tube constricted above the ovary ; stamens as many as
            and opposite to the calyx-segments, often alternating with
            staminodes ; herbs or shrubs -   -   -   **Amaranthaceae**
        Leaves alternate :
          Twining herbs ; sepals united below into a tube with 2 adnate brac-
            teoles outside; flowers in axillary pedunculate spikes **Basellaceae**
          Plants not twining ; sepals free or united, without adnate bracteoles :
            Calyx with 2–4 valvate segments ; trees or shrubs :
              Flowers dioecious, in small solitary or clustered (sometimes
                paniculate) heads ; stamens 3–5, united into a column ; fruit
                with a thick fleshy dehiscent pericarp ; seed arillate
                    Myristicaceae
             Flowers bisexual, in large bracteate heads or in elongated spikes
               or racemes ; stamens 4, free, opposite to and inserted on the
               calyx-segments ; fruit a nut ; seed not arillate   **Proteaceae**
            Calyx with 3–5 imbricate segments, or sometimes almost com-
              pletely tubular with indistinct segments ; herbs or shrubs :
              Stamens fewer than the calyx-segments and mostly alternating
                with them ; leaves small, linear ; flowers in elongated
                simple spikes -   -   -   -   **Phytolaccaceae**
              Stamens as many as or fewer than the calyx-segments and
                opposite to them :
                Sepals dry and ± scarious, free or shortly united at the base ;
                  leaves entire -   -   -   -   **Amaranthaceae**
                Sepals herbaceous, free or ± united into a tube ; leaves entire
                  or ± toothed or pinnately divided -   **Chenopodiaceae**

Group 13

Parasitic shrubs growing on other shrubs or trees ; ovule scarcely distinguishable from the
  surrounding tissue of the ovary ; calyx truncate or obsolete ; stamens as many as and
  opposite to the petals and inserted on them ; leaves simple, entire, opposite or alternate,
  sometimes reduced to scales -   -   -   -   -   -   -   **Loranthaceae**
Plants not parasitic ; ovule or ovules clearly distinguishable within the ovary :
    Ovary 1-locular :
      Trees or shrubs, sometimes climbing ; style 1, unbranched :
        Ovule 1, pendulous from the apex of the ovary ; flowers in axillary cymes ;
        stamens as many as the petals ; leaves alternate, often unequal-sided at the base
                    Alangiaceae

Ovules 2 or more :

Ovules pendulous from the apex of the ovary ; fruit indehiscent, with 2–6 longitudinal wings or angles ; flowers in racemes or spikes or heads ; stamens twice as many or sometimes as many as the petals ; leaves opposite or sometimes verticillate or alternate - - - - - - **Combretaceae**

Ovules borne on a free-basal or free-central placenta ; fruit baccate ; flowers in axillary cymes or fascicles or umbels ; stamens twice as many as the petals ; leaves opposite - - - - - - - **Melastomataceae**

Herbs ; styles or style-branches 2–6 :

Ovules numerous on a free-basal placenta ; sepals 2, often deciduous ; fruit a circumscissile capsule - - - - - - - **Portulacaceae**

Ovule or ovules pendulous from the apex of the ovary or on pendulous apical placentas ; sepals or calyx-lobes 2–5 ; fruit not circumscissile :

Ovules numerous on pendulous placentas ; fruit a capsule dehiscing at the apex ; leaves opposite ; flowers bisexual, in axillary pairs ; petals and stamens 5
Vahliaceae

Ovules 1–4, pendulous from the apex of the ovary ; fruit indehiscent ; leaves alternate or opposite, sometimes all radical ; flowers bisexual or unisexual, paniculate-spicate or in axillary clusters ; petals and stamens 2–4
Haloragaceae

Ovary 2- or more-locular :

Ovule 1 in each loculus :

Stamens twice as many as the petals ; styles 3–10 :

Leaves verticillate, pinnatisect with filiform segments ; aquatic herbs with flowers in interrupted spikes - - - - - **Haloragaceae**

Leaves alternate ; plants not aquatic :

Herbs with solitary axillary flowers ; ovary 10-locular ; fruit a prickly capsule ; leaves ± pinnately lobed - - - - - - **Rosaceae**

Trees or shrubs with spicate flowers ; ovary 3–5-locular ; fruit a drupe ; leaves entire - - - - - - - **Rhizophoraceae**

Stamens as many as the petals :

Stamens opposite to the petals ; shrubs, often climbing with tendrils ; fruit often separating into cocci, sometimes winged - - - - **Rhamnaceae**

Stamens alternating with the petals ; plants without tendrils :

Flowers solitary in the axils of the leaves ; floating aquatic herbs ; leaves alternate, rosulate, with a ± inflated petiole ; fruit large, indehiscent, with hard endocarp and armed with 2 or 4 horns - - - **Trapaceae**

Flowers grouped in inflorescences ; plants not aquatic :

Fruit separating into 2 cocci ; herbs, sometimes arborescent ; flowers in simple or compound umbels, sometimes capitate ; leaves often much divided or compound - - - - - - **Umbelliferae**

Fruit drupaceous or capsular, not separating into cocci ; trees or shrubs :

Leaves pinnate or digitate or palmately divided ; flowers in umbels or racemes or spikes ; fruit drupaceous - - - **Araliaceae**

Leaves simple, undivided :

Flowers unisexual, in heads ; petals and stamens 5 ; fruit a 2-valved capsule ; leaves entire - - - - **Hamamelidaceae**

Flowers bisexual or unisexual, in panicles or umbel-like cymes ; petals and stamens 4 ; fruit a drupe ; leaves opposite, entire or toothed
Cornaceae

Ovules 2 or more in each loculus :

Leaves alternate :

Leaves with stipules, often unequal-sided ; flowers monoecious ; stamens numerous ; ovules very numerous in each ovary-loculus ; herbs, sometimes epiphytic - - - - - - - - **Begoniaceae**

Leaves without stipules :

Styles 2 ; flowers unisexual or polygamous, in axillary panicles ; stamens as many as the petals ; ovary 2-locular ; shrubs - - **Escalloniaceae**

Style 1 with an entire or lobed stigma ; flowers bisexual or sometimes polygamous :

Stamens as many or twice as many as the petals ; fruit a capsule, dehiscing by valves or irregularly or sometimes failing to dehisce ; herbs or shrubs, sometimes aquatic - - - - - - **Onagraceae**

Stamens numerous, more than twice as many as the petals ; fruit a berry or drupe :

Flowers solitary or paired in the axils of the leaves ; ovary 2-locular ; leaves gland-dotted ; shrublets - - - - **Myrtaceae**

Flowers in terminal racemes ; ovary 4-locular ; leaves not gland-dotted ; trees - - - - - - - - - **Lecythidaceae**

Leaves opposite :

    Stamens as many as the petals ; petals alternating with incurved scales ; trees or shrubs - - - - - - - - - **Oliniaceae**

    Stamens twice as many as the petals or more numerous ; petals not alternating with scales :

        Stamens numerous, more than twice as many as the petals ; trees or shrubs with gland-dotted leaves - - - - - - **Myrtaceae**

        Stamens twice as many as the petals :

            Leaves with interpetiolar stipules ; ovules 2 in each loculus ; viviparous maritime trees or shrubs - - - - **Rhizophoraceae**

            Leaves without stipules ; ovules numerous in each loculus ; plants not viviparous :

                Anthers opening by an apical pore ; connective often appendaged below the anther ; leaves with 3 or more parallel longitudinal nerves ; seeds without a tuft of hairs ; herbs or shrubs or small trees **Melastomataceae**

                Anthers opening by longitudinal slits ; connective unappendaged ; leaves without parallel longitudinal nerves ; seeds with an apical tuft of hairs ; herbs - - - - - - - **Onagraceae**

Group 14

Parasitic shrubs growing on other shrubs or trees ; ovule scarcely distinguishable from the surrounding tissue of the ovary ; calyx truncate or shortly lobed ; stamens as many as and opposite to the corolla-lobes and inserted on them ; leaves simple, entire, opposite or alternate or sometimes ternate - - - - - - - **Loranthaceae**

Plants not parasitic ; ovule or ovules clearly distinguishable within the ovary :

Ovary 1-locular :

    Ovule 1 ; flowers actinomorphic or zygomorphic, in involucrate heads ; fruit indehiscent, often crowned by the persistent calyx forming a pappus of bristles or scales :

        Anthers united into a tube surrounding the style ; ovule erect from the base the ovary ; herbs or shrubs or trees ; corolla of the outer (ray) flowers often differing from that of the inner flowers - - - - **Compositae**

        Anthers free ; ovule pendulous from the apex of the ovary ; herbs with opposite leaves - - - - - - - - - **Dipsacaceae**

    Ovules numerous on a free-basal placenta ; flowers actinomorphic, not in involucrate heads :

        Calyx composed of 2 often deciduous sepals ; fruit a circumscissile capsule ; stamens as many as and alternating with the corolla-lobes or more numerous ; herbs - - - - - - - - - **Portulacaceae**

        Calyx 4–5-lobed ; fruit not circumscissile ; stamens as many as and opposite to the corolla-lobes :

            Trees or shrubs ; flowers in axillary panicles or racemes ; fruit a berry **Myrsinaceae**

            Herbs ; flowers in terminal racemes ; fruit a 5-valved capsule **Primulaceae**

Ovary 2- or more-locular (sometimes 3-locular with 1 fertile loculus and 2 empty loculi) :

    Petals united into a deciduous mass (calyptra) ; stamens very numerous ; trees or shrubs with gland-dotted leaves - - - - - **Myrtaceae**

    Petals more or less united into a tube, not into a deciduous mass ; stamens not more than twice as many as the corolla-lobes :

        Ovary 3-locular with 2 empty loculi, the fertile loculus with 1 apical pendulous ovule ; stamens 3 ; fruit indehiscent, often crowned by the persistent calyx forming a feathery pappus ; herbs with opposite leaves - **Valerianaceae**

        Ovary 2- or more-locular without empty loculi :

            Trailing or climbing herbs or shrubs with tendrils ; flowers unisexual ; stamens 3–5, the anthers sometimes curved or flexuous or folded ; leaves often palmately or pedately lobed or deeply divided - - - **Cucurbitaceae**

            Plants without tendrils ; flowers bisexual or sometimes unisexual :

                Leaves opposite or verticillate, with interpetiolar or intrapetiolar (sometimes leaf-like) stipules and entire margin ; trees or shrubs or herbs, sometimes climbing - - - - - - - - - **Rubiaceae**

                Leaves alternate or opposite, without stipules, sometimes reduced to scales :

                    Stamens twice as many as the corolla-lobes ; anthers opening by apical pores ; flowers in axillary racemes ; trees or shrubs with alternate leaves - - - - - - - - - - **Ericaceae**

                  Stamens as many as the corolla-lobes ; anthers opening by longitudinal slits :

                      Ovules 1 in each loculus ; fruit a drupe ; flowers zygomorphic, the corolla-tube split down the back ; anthers free ; maritime shrubs **Goodeniaceae**

Ovules numerous in each loculus ; fruit a capsule :

Corolla-lobes imbricate ; capsule circumscissile ; flowers actino-morphic, in terminal spikes ; anthers free ; annual herbs
Sphenocleaceae

Corolla-lobes valvate ; capsule dehiscing by valves ; flowers actino-morphic or zygomorphic, the corolla-tube sometimes split down the back ; anthers free or united into a tube surrounding the style ; herbs or shrubs, sometimes climbing or growing very tall
Campanulaceae

Group 15

Ovary 2- or more-locular :

Flowers unisexual ; leaves with stipules :

Flowers in heads ; stamens as many as the calyx-segments ; ovary with 1 ovule in each loculus ; trees or shrubs with entire leaves - - **Hamamelidaceae**

Flowers in cymes or panicles ; stamens numerous ; ovary with very numerous ovules in each loculus ; sepals petaloid ; herbs, sometimes epiphytic, with alternate often unequal-sided leaves - - - - - - **Begoniaceae**

Flowers bisexual ; leaves without stipules :

Leaves opposite ; flowers solitary and subsessile in the axils of the leaves ; herbs
Onagraceae

Leaves alternate :

Sepals 5, free or shortly united at the base ; stamens accompanied by numerous staminodes, the outer series of staminodes united and resembling a corolla ; anthers free from the style ; trees or shrubs - - - **Lecythidaceae**

Sepals united into an elongated tube inflated below, the mouth 3-lobed or ± pro-duced unilaterally into a single lobe ; stamens not accompanied by staminodes ; anthers adnate to the style ; herbs or shrubs, often climbing **Aristolochiaceae**

Ovary 1-locular :

Ovules 2 or more :

Plants leafless, parasitic on roots ; flowers solitary, subsessile on a rhizome ; ovules very numerous on apical pendulous placentas - - - **Hydnoraceae**

Plants with leaves, the leaves sometimes reduced and very small ; flowers solitary in the leaf-axils or grouped in inflorescences ; ovules 2–4 :

Stamens twice as many as the calyx-lobes ; flowers in racemes or spikes or heads ; fruit indehiscent, often with 2–5 longitudinal wings ; trees or shrubs
Combretaceae

Stamens as many as the calyx-lobes :

Ovules 2–4, pendulous from a free-basal placenta ; style 1 with an entire or divided stigma ; flowers solitary or in cymes or racemes, bisexual or uni-sexual ; herbs or shrubs, often parasitic on roots and sometimes with reduced leaves - - - - - - - - **Santalaceae**

Ovules 4, pendulous from the apex of the ovary ; styles 4 ; flowers clustered in the axils of the leaves, unisexual or polygamous ; herbs, not parasitic
Haloragaceae

Ovule 1 :

Fleshy herbs, parasitic on roots ; leaves reduced to scales ; flowers unisexual, in involucrate heads, or the male flowers paniculate and the female flowers united in paniculate heads - - - - - - **Balanophoraceae**

Plants not parasitic ; leaves well developed :

Flowers unisexual, densely capitate or arranged in or on an open or flat receptacle, the female flowers sometimes solitary in the receptacle ; trees or shrubs or herbs
Moraceae

Flowers cymose or paniculate-spicate :

Trees with entire or palmately 3–5-lobed leaves ; flowers cymose, unisexual or polygamous ; fruit a nut crowned by 2 elongated wings formed from per-sistent accrescent calyx-lobes - - - - - **Hernandiaceae**

Herbs with radical long-petiolate toothed reniform leaves ; flowers paniculate-spicate, bisexual or unisexual ; fruit small, unwinged, drupaceous
Haloragaceae

MONOCOTYLEDONES

Gynoecium composed of 2 or more free carpels with separate styles and stigmas
Group 1 (see p. 73)

Gynoecium composed of 1 carpel or of 2 or more united carpels with free or united styles :

Ovary superior :

Perianth present, composed of 4 or more free or united segments, not reduced to bristles :

Perianth composed of separate calyx and corolla, the calyx often herbaceous, the corolla usually petaloid or otherwise different from the calyx, the sepals and petals either free or united among themselves but never united into a single perianth-tube   -   -   -   -   -   -   - *Group 2* (see below)

Perianth composed of similar or subsimilar segments in 2 series, usually petaloid but sometimes herbaceous or dry and glumaceous, either free or united into a single perianth-tube -   -   -   -   -   - *Group 3* (see below)

Perianth absent or reduced to bristles or to 1–3 scales :

Flowers unisexual, not arranged in small spikes with scale-like bracts

Group 4 (see p. 74)

Flowers (florets) bisexual or unisexual, arranged in small spikes (spikelets) with scale-like bracts (glumes or lemmas), the spikelets sometimes 1-flowered; grasses and sedges   -   -   -   -   -   - *Group 5* (see p. 74)

Ovary ± inferior ; perianth present :

Perianth composed of separate calyx and corolla, the calyx herbaceous or otherwise different from the petaloid corolla, the sepals and petals either free or united among themselves but never united into a single perianth-tube   - *Group 6* (see p. 75)

Perianth composed of similar or subsimilar segments, usually 6 in 2 series, sometimes 3 in 1 series, usually petaloid, either free or united into a single perianth-tube

Group 7 (see p. 75)

Group 1

Leaves pinnate ; palms -   -   -   -   -   -   -   -   - **Palmae**

Leaves simple ; aquatic or marsh herbs :

Perianth absent or cupular ; stamens 1 or 2, united ; submerged marine or fresh-water plants with narrow leaves ; flowers unisexual, axillary, solitary or in cymes

Zannichelliaceae

Perianth present, composed of 1–6 free segments ; stamens 3 or more ; fresh-water or marsh plants :

Flowers with bracts, in whorls or in simple or compound umbels, sometimes spicate ; perianth composed of 3 sepals and 3 petals, or the petals sometimes absent ; plants aquatic or terrestrial   -   -   -   -   - **Alismataceae**

Flowers without bracts, in spikes ; perianth composed of 1–4 similar segments ; plants aquatic :

Leaves radical ; spikes simple or 2-branched on elongated peduncles, at first enclosed in a spathe ; perianth-segments 1–3 ; stamens usually 6, with elongated filaments ; ovules 2 or more in each carpel -   -   - **Aponogetonaceae**

Leaves borne on elongated stems ; spikes simple on axillary peduncles, without a spathe ; perianth-segments usually 4 ; stamens usually 4, the anthers sessile ; ovules 1 in each carpel   -   -   -   -   -   - **Potamogetonaceae**

Group 2

Leaves pinnate or flabellate ; palms -   -   -   -   -   -   - **Palmae**

Leaves simple, not flabellate ; herbs :

Ovary 1-locular with parietal placentas bearing numerous ovules :

Leaves filiform, borne on elongated stems ; flowers solitary, axillary on elongated peduncles ; petals free ; anthers opening by an apical pore ; style unbranched ; plants aquatic   -   -   -   -   -   -   - **Mayacaceae**

Leaves mostly radical ; flowers in spikes or heads on elongated peduncles ; petals united below into a tube ; anthers opening by longitudinal slits ; style 3-branched

Xyridaceae

Ovary 2–3-locular with axile or apical placentas bearing 1 or more ovules :

Flowers monoecious, small and actinomorphic, in bracteate heads on elongated peduncles ; corolla inconspicuous, not brightly coloured, often minute in the male flowers ; style branched ; leaves narrow, radical or crowded   - **Eriocaulaceae**

Flowers bisexual or polygamous, actinomorphic or zygomorphic, in open or congested cymes or panicles or fascicles often subtended by folded or boat-shaped bracts ; corolla conspicuous, usually brightly coloured, often blue or yellow ; style unbranched -   -   -   -   -   -   -   - **Commelinaceae**

Group 3

Perianth-segments 6, dry and glumaceous or the inner hyaline ; leaves narrow or reduced to sheaths ; rush-like herbs :

Flowers dioecious ; ovary 2-locular with 1 apical pendulous ovule in each loculus ; styles 3 -   -   -   -   -   -   -   -   -   - **Restionaceae**

Flowers bisexual ; ovary 1- or 3-locular with 3 or more ovules in each loculus ; style 1, 3-branched   -   -   -   -   -   -   -   -   - **Juncaceae**

Perianth-segments petaloid or herbaceous :
Aquatic herbs ; flowers in spikes or racemes :
Perianth-segments 6, petaloid, united below into a tube ; ovary 1–3-locular with
numerous ovules in each loculus ; fruit a capsule ; leaves with an expanded ovate
lamina, or linear and submerged   -   -   -   -   -   **Pontederiaceae**
Perianth-segments 4, herbaceous, free ; ovary 1-locular with 1 ovule ; fruit a small
drupe ; leaves narrowly linear or filiform, submerged   -   **Potamogetonaceae**
Plants terrestrial :
Inflorescence a spadix enclosed in a spathe ; flowers monoecious, very small, female
at the bottom of the spadix, male above ; ovary 1–2-locular ; leaves (or solitary
leaf) radical, 1–4-pinnate or with a sagittate or hastate lamina ; herbs   **Araceae**
Inflorescence not a spadix enclosed in a spathe ; flowers bisexual or dioecious ; ovary
3-locular :
Leaves reduced to scales or spines, their function often fulfilled by leaf-like
acicular or flattened branches (cladodes) ; herbs or shrublets, often climbing
                                                Liliaceae
Leaves (or solitary leaf) well developed, sometimes appearing after the flowers :
Flowers dioecious, small, in axillary umbels ; leaves with reticulate venation
and stipular tendrils ; fruit a berry ; climbing shrubs, often with prickly
stems   -   -   -   -   -   -   -   -   -   **Smilacaceae**
Flowers bisexual ; leaves without stipular tendrils :
Flowers in umbels subtended by 1 or more spathaceous bracts and borne on
naked peduncles ; herbs with radical leaves ; fruit a loculicidal capsule
                                             Amaryllidaceae
Flowers solitary or in racemes or spikes or panicles :
Perianth-segments united below into a tube :
Anthers opening by an apical pore ; filaments short, inserted at the
mouth of the perianth-tube ; herbs   -   -   -   **Tecophilaeaceae**
Anthers opening by longitudinal slits ; herbs or shrubs or trees, sometimes
succulent or climbing   -   -   -   -   -   -   **Liliaceae**
Perianth-segments free :
Anthers sessile ; fruit separating into 3 cocci ; perianth herbaceous ;
flowers without bracts, in racemes on elongated peduncles ; herbs
with radical linear or filiform leaves   -   -   **Juncaginaceae**
Anthers borne on ± elongated filaments ; fruit a capsule or berry or
drupe ; perianth ± petaloid :
Ovary with 1 ovule in each loculus ; fruit a drupe ; perianth sub-
petaloid ; herbaceous climbers, the leaves ending in tendrils ;
flowers small, in panicles   -   -   -   -   **Flagellariaceae**
Ovary with 2 or more ovules in each loculus ; fruit a capsule or berry ;
perianth petaloid ; herbs, sometimes climbing, the leaves sometimes
ending in tendrils   -   -   -   -   -   -   **Liliaceae**

Group 4

Leaves absent ; minute aquatic herbs, the plant-body reduced to a thallus-like " frond "
with or without 1 or more pendent rootlets   -   -   -   -   **Lemnaceae**
Leaves (or solitary leaf) present, well developed :
Submerged aquatic herbs with ± elongated stems and narrow linear leaves :
Leaves opposite or verticillate ; flowers solitary or few together in the axils of the
leaves ; fresh-water plants   -   -   -   -   -   -   **Najadaceae**
Leaves alternate ; inflorescence a spike of alternating male and female flowers,
enclosed in a spathe ; marine plants   -   -   -   -   **Zosteraceae**
Plants terrestrial or aquatic but not submerged :
Floating aquatic herbs, stemless with a rosette of sessile leaves   -   -   **Araceae**
Plants not floating :
Trees or shrubs, often with aerial roots ; leaves linear or ensiform, armed with
spiny teeth on the margin and often also on the midrib beneath   **Pandanaceae**
Herbs or woody climbers ; leaves not armed with spiny teeth :
Leaves narrow, linear ; inflorescence a dense cylindric spike, female below, male
above ; ovary 1-locular with 1 apical pendulous ovule ; swamp herbs, often
growing in water at the base   -   -   -   -   -   **Typhaceae**
Leaves (or solitary leaf) broad, often deeply divided ; inflorescence a spadix subtended by or enclosed in a spathe, female below, male above and sometimes
with a terminal sterile appendix ; ovary 1–4-locular with 1 or more ovules on
basal or axile or parietal placentas ; terrestrial herbs or climbers   **Araceae**

Group 5

Florets each enclosed by a bract (lemma) on the outside and a bracteole (palea) on the
inside, the spikelet usually having 2 empty bracts (glumes) at its base ; perianth absent

or represented by 2–3 minute scales (lodicules) ; stems usually with hollow internodes, terete or compressed ; leaf-sheaths usually open ; seed usually adnate to the pericarp of the fruit ; grasses, sometimes shrubby or arborescent (bamboos)   **Gramineae**
Florets each enclosed by a single bract (glume) on the outside, without a bracteole or female florets sometimes each surrounded by a closed bracteole (utricle) ; perianth absent or represented by bristles or scales ; stems usually solid, often triquetrous ; leaf-sheaths usually closed ; seed free from the pericarp of the fruit ; sedges
Cyperaceae

Group 6

Aquatic herbs ; flowers actinomorphic, bisexual or dioecious, solitary or 2 or more together in a tubular spathe ; spathes sessile or borne on elongated peduncles
Hydrocharitaceae
Terrestrial or epiphytic herbs ; flowers ± zygomorphic or asymmetric :
   Fertile stamens 5 ; flowers unisexual or bisexual; tall herbaceous plants with large leaves   -   -   -   -   -   -   -   -   -   - **Musaceae**
   Fertile stamen 1 ; flowers bisexual :
      Stamen not accompanied by petaloid staminodes ; pollen agglutinated into masses (pollinia) ; ovary 1-locular with numerous ovules on parietal placentas ; flowers zygomorphic, the median petal (lip) ± different from the lateral petals ; plants terrestrial or epiphytic, sometimes climbing or saprophytic   -   **Orchidaceae**
      Stamen accompanied by 1 or more petaloid staminodes ; pollen not agglutinated into masses ; plants terrestrial :
         Sepals united into a tube ; flowers zygomorphic ; anther 2-thecous ; ovary 3-locular with numerous ovules in each loculus   -   -   - **Zingiberaceae**
         Sepals free ; flowers asymmetric ; anther 1-thecous :
            Ovary 3-locular with numerous ovules in each loculus   -   -   **Cannaceae**
            Ovary 1–3-locular with 1 ovule in each loculus   -   -   - **Marantaceae**

Group 7

Aquatic herbs, sometimes marine ; flowers actinomorphic, dioecious, solitary (or the male flowers sometimes many together) in a tubular or 2-lobed spathe ; spathes sessile or borne on elongated peduncles   -   -   -   -   -   - **Hydrocharitaceae**
Plants not aquatic :
   Flowers unisexual, small ; twining plants ; leaves simple or digitate, with reticulate venation ; fruit a 3-winged or 3-angled capsule   -   -   -   - **Dioscoreaceae**
   Flowers bisexual :
      Leaves deeply divided with pinnatipartite segments, radical ; flowers in bracteate umbels, the outer bracts broad, the inner bracts long and thread-like ; herbs with a tuberous rootstock   -   -   -   -   -   -   - **Taccaceae**
      Leaves simple and undivided or sometimes absent :
         Stamen 1 ; pollen agglutinated into masses (pollinia) ; ovary 1-locular with numerous ovules on parietal placentas ; median inner segment (lip) of the perianth ± different from the other segments ; terrestrial or epiphytic herbs, sometimes climbing or saprophytic   -   -   -   -   - **Orchidaceae**
         Stamens 3 or 6 ; pollen not agglutinated into masses :
            Stamens 3 :
               Stamens opposite the inner perianth-segments ; saprophytic herbs with a few narrow basal leaves or the leaves reduced to scales ; perianth-segments united into a tube ; flowers actinomorphic, racemose or cymose
Burmanniaceae
               Stamens opposite the outer perianth-segments ; cormous or rhizomatous herbs with narrow leaves, not saprophytic ; perianth-segments free or united into a tube ; flowers actinomorphic or zygomorphic, spicate or paniculate or sometimes solitary or few together in a spathe   **Iridaceae**
            Stamens 6 :
               Saprophytic herbs, the leaves reduced to scales ; ovary 1-locular with parietal placentas ; perianth-segments united into an urceolate tube ; flowers zygomorphic   -   -   -   -   -   -   - **Burmanniaceae**
               Plants not saprophytic, the leaves well developed ; ovary 3-locular :
                  Flowers in umbels (sometimes 1-flowered) subtended by 1 or more spathaceous bracts and borne on naked peduncles ; herbs with radical leaves ; rootstock a bulb   -   -   -   -   -   **Amaryllidaceae**
                  Flowers solitary or in racemes or corymbs not subtended by spathaceous bracts ; rootstock not a bulb :
                     Shrubs or shrublets with a branched or unbranched woody stem densely clothed with persistent bases of old leaves ; flowers solitary or fasciculate   -   -   -   -   -   -   - **Velloziaceae**

Herbs with a tuberous or cormous rootstock, the stem not woody :
    Anthers opening by an apical pore ; ovary with 2 ovules in each
    loculus ; flowers in racemes or panicles ; plants glabrous
                                    Tecophilaeaceae
    Anthers opening by longitudinal slits ; ovary with several ovules in
    each loculus ; flowers solitary or in racemes or corymbs ; plants ±
    hairy - - - - - - - - - **Hypoxidaceae**

INDEX TO FAMILIES IN THE KEY

GYMNOSPERMAE

By John Lewis

1. CYCADACEAE

Dioecious trees or shrubs. Trunk simple or sparsely branched, covered with spirally arranged leaf-bases and scale-leaves, either up to 15 m. tall or very short and almost subterranean. Leaves usually long and pinnate, in close spirals, forming an often wide-spreading crown on the trunk and (except in one species of *Macrozamia*) intercalated with leathery scale-leaves. Staminate strobilus forming a terminal cone of leathery scales with numerous pollen-sacs on their abaxial surface. Female strobilus either a loose apical whorl of fronds bearing erect ovules marginally (*Cycas*) or a terminal cone of thick and often woody scales bearing a pair of inverted ovules adaxially.

Leaflets 1-nerved ; female strobilus of fronds in a loose apical whorl, with flat blades and several (rarely only 2) ovules marginally - - - - - - **1. Cycas**
Leaflets many-nerved : female strobilus a cone, the scales subpeltate with short apices and only 2 ovules on their adaxial side - - - - - **2. Encephalartos**

1. CYCAS L.

Cycas L., Sp. Pl. **2** : 1188 (1753).

Trees or shrubs, usually unbranched ; trunks bearing woody leaf-bases and crowned by a tuft of large divergent pinnately compound leaves. Leaflets entire, linear, 1-nerved. Male cones of closely-set cuneate scales often with acuminate apices ; pollen-sacs broadly ellipsoid, in groups of 3–5. Fronds of the female strobilus densely crowded and woolly at first, each developing into a thickened axis bearing one or more pairs of sessile ovules and having a narrowly elliptic to rotund blade. Seeds ovoid, obovoid or globose.

Cycas thuarsii R. Br., Prodr.: 347 (1810).—Gaudich. in Freyc., Voy. Monde, Bot. : 434 (1829).—Stapf in Kew Bull. **1916** : 1 (1916).—Prain, F.T.A. **6**, 2 : 345 (1917).—Melville, F.T.E.A. Gymnosp.: 1, t. 1 (1958). TAB. **1** fig. A. Type from Madagascar.
   Cycas madagascariensis Miq., Comm. Phyt. : 127 (1840). Type from Madagascar.
   Cycas circinalis subsp. *thuarsii* (R. Br.) Engl., Pflanzenw. Afr. **2** : 82 (1908). Type as for *C. thuarsii*.
   Cycas circinalis subsp. *madagascariensis* (Miq.) Schuster in Engl., Pflanzenr. IV, 1 : 73 (1932). Type as for *C. madagascariensis*.

Palm-like rarely branched tree, up to 9 m. tall ; trunk up to 45 cm. in diam. Scale-leaves c. 6 cm. long, 3 cm. broad at the base, brown, tomentose abaxially. Leaves 1·5–3 m. long, petiolate ; leaflets c. 30 × 1 cm., falcate, very narrowly lanceolate, in part parallel-sided but gradually attenuating to an acute spinulose apex, decurrent below, bright green with a yellower midrib ; petiole up to 60 cm. long, subterete, with short spines abruptly replacing the leaflets. Male cones very narrowly ovoid ; scales up to 5 cm. long, distal barren part deltoid with an acuminate upturned abaxially tomentose apex. Fronds of the female strobilus up to 30 cm. long ; fertile axis up to 20 cm. long bearing 4 or 5 pairs of ovules ; lamina lanceolate or narrowly ovate, margin irregularly dentate above. Seeds about 4 cm. in diam. at maturity, ovoid.

Mozambique. Z : Pebane, 1946, *Munch* 454 (SRGH) ; Quelimane, *Munch* in Herb. Christian 670. MS : Zambezi Delta, along coast between Kongone and Melambe mouths, 1858 (♀), *Kirk* (K).
Also in Tanganyika at Tanga, and in Zanzibar, Madagascar and the Comoro Is. Occurring sparsely as individuals or small groups in open bushland not far from the coast.

2. ENCEPHALARTOS Lehm.

Encephalartos Lehm., Pugill. Pl. **6** : 3 (1834).

Trees or shrubs, rarely branched ; trunk (sometimes prostrate) bearing woody leaf-bases and crowned by a tuft of large pinnately compound linear to oblong

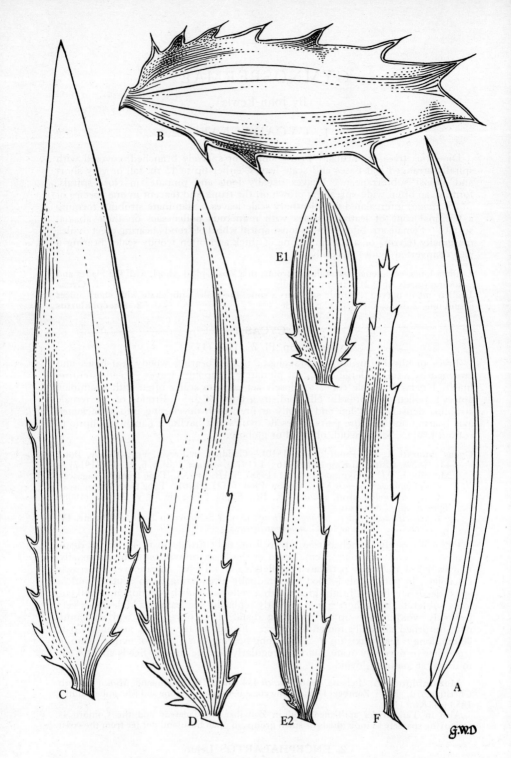

Tab. 1. Median leaflets of CYCADACEAE (all ×⅔). A.—CYCAS THUARSII. B.—ENCE-
PHALARTOS FEROX. C.—ENCEPHALARTOS GRATUS. D.—ENCEPHALARTOS HILDE-
BRANDTII. E.—ENCEPHALARTOS MANIKENSIS. E1, characteristic shape ; E2, variant
shape. F.—ENCEPHALARTOS VILLOSUS.

leaves. Leaflets linear to ovate, usually spiny, with many parallel nerves. Male cones one or more, pedunculate ; scales closely imbricate with an enlarged, sub-peltate, barren, often deflexed apex terminating a cuneate axis bearing densely packed spherical or ellipsoid pollen-sacs. Female cones sessile or shortly peduncu-late ; scales imbricate each with a much enlarged peltate apex which is more or less dorsiventrally flattened and laterally expanded and bears a pair of backwardly directed ovules. Seeds obloid to ovoid.

Median leaflets at least 4·5 cm. broad (TAB. 1 fig. B) - - - - 1. *ferox*
Median leaflets less than 4·5 cm. broad :
  Median leaflets more than 2 cm. broad :
    External facets of the cone-scales puberulous - - - - 2. *gratus*
    External facets of the cone-scales glabrous :
      Median leaflets straight, either narrowly elliptic with the marginal spines almost confined to the base or, more rarely, narrowly lanceolate with the marginal spines ± regularly spread (TAB. 1 fig. E) - - - - 3. *manikensis*
      Median leaflets falcate, lanceolate with the marginal spines more closely set towards the base (TAB. 1 fig. D) - - - - - 4. *hildebrandtii*
  Median leaflets less than 2 cm. broad (TAB. 1 fig. F) - - - - 5. *villosus*

1. **Encephalartos ferox** Bertol. f. in Mem. Accad. Sci. Bol. **3** : 264 (1851).—Prain in Kew Bull. **1916** : 180 (1916) ; F.T.A. **6**, 2 : 352 (1917).—Schuster in Engl., Pflanzenr. IV, 1 : 113 (1932).—R. A. Dyer in Journ. S. Afr. Bot. **22** : 3 (1956). TAB. 1 fig. B. Type : Mozambique, at or near Inhambane, *Fornasini* (K, painting of holotype).
  Encephalartos kosiensis Hutch. in Kew Bull. **1932** : 512 (1932) ; in Hook., Ic. Pl. : t. 3220 (1933).—Hutch. & Rattr. in A. W. Hill, F.C. **5**, 2, Suppl.: 34 (1933).—Ogilvie in Kew Bull. **1939** : 655 (1940). Type from S. Africa (Zululand).

Short-stemmed or with a trunk up to 1·3 m. tall. Leaves about 1 × 0·25 m. ; median leaflets up to 20 × 7 cm., ovate with the margin fairly regularly divided into deltoid spinescent teeth which are c. 0·5 cm. long on the lateral margins but appear longer apically ; leaflets diminishing in size towards the leaf-base, ultimately becoming bifurcated spines. Median male cone-scales ascending with the head deflexed; head triangular in outline with a hexagonal terminal facet. Female cone, 30 × 15 cm., 1–3, subsessile, bright red ; median cone-scales up to 7 cm. long, glabrous, with a dorsiventrally flattened head of almost square outline bearing two long processes that extend to the cone-axis between the seeds of adjacent scales, the umbo concave terminally and not deflexed. Seeds 5 × 1·5 cm., bright vermilion-red, becoming black, the fleshy part exceeding the stony inner part distally by about 2 cm.

  Mozambique. SS : 32 km. inland from Inhambane, Zavala, iv.1945, *Key* in Herb. Christian 762 (SRGH) and xii.1944 (♂ cone), *Mendonça* 3299 (LISC) ; Gaza, Vila João Belo, xi.1941 (♀), *Torre* 3879 (LISC). LM : Inhaca I., v.1933 (♂ cone), *Mogg* 18993 (SRGH).
  Also in Zululand. Occasional in forest margins ; also locally common behind sand dunes and elsewhere not far from the coast.

2. **Encephalartos gratus** Prain in Kew Bull. **1916** : 181 (1916) ; F.T.A. **6**, 2 : 352 (1917).—Schuster in Engl., Pflanzenr. IV, 1 : 121 (1932).—Melville in Kew Bull. **1957** : 255 (1957). TAB. 1 fig. C. Type : Nyasaland, Mt. Mlanje, *Davy* 417 (K, lectotype).

Trunk globose or (especially in male plants) up to 1·2 m. tall. Leaves up to 2 m. long, somewhat recurved ; median leaflets up to 25 cm. long, narrowly lanceolate, straight or slightly falcate, oblique at the base, margin with 1–5 short spiniferous teeth on each side which are aggregated towards the base ; leaflets reducing in size and becoming narrowly ovate towards the leaf-base, ultimately reduced to short spines ; petiole about 10 cm. long. Male cones up to 40 × 9 cm., fusiform, reddish-brown ; peduncle up to 19 cm. long ; median cone-scales ascending with deflexed heads ; head truncate-deltoid in outline with very dis-tinctly angular edges, terminal facet flat, rhomboid. Female cones up to 68 × 20 cm., broadest near the base, brown ; peduncle c. 12 cm. long ; median cone-scales straight with only the terminal facet slightly deflexed, head dorsiventrally compressed, broadly truncate-triangular in outline and with the external margin curved convexly and with a median flat, denticulate, backwardly directed extension abaxially, the external faces of the head smooth and puberulous on both surfaces. Seeds up to 4 × 2 cm. ellipsoid, red.

Nyasaland. S: Mt. Mlanje, Likabula, 29.iii.1956, *Williams* (K). **Mozambique.** Z: near Mt. Milange, 30.viii.1950, *Munch* 453 (SRGH).
Known only from this area. Commonest amidst rocks in ravines by rivers ; rarely also in forest margins and wooded grassland.
The record from Mozambique is from a sterile gathering but probably represents this species.

3. **Encephalartos manikensis** (Gilliland) Gilliland in Proc. & Trans. Rhod. Sci. Ass. **37** : 133 (1939).—Wild in Proc. & Trans. Rhod. Sci. Ass. **44** : 57 (1956).—Melville in Kew Bull. **1957** : 256 (1957). TAB. **1** fig. E. Type : S. Rhodesia, Mt. Gorongowe, *Gilliland* 2016 (BM, holotype).
 Encephalartos gratus var. *manikensis* Gilliland in Journ. S. Afr. Bot. **4** : 153 (1938). Type as above.

Trunk up to 1·5 m. tall. Leaves up to 2 × 0·35 m., straight ; median leaflets mostly 12–15 × 2·5 cm. and characteristically narrowly elliptic, sometimes up to 19 × 3·5 cm. and narrowly lanceolate, margin usually with 2–4 divergent spines on each side towards the base or rarely entire ; leaflets reducing in size and becoming ovate towards the leaf-base, ultimately reduced to short simple spines ; petiole 5–10 cm. long. Median male cone-scales at right angles to the axis, with their heads slightly deflexed ; heads bluntly triangular in outline with curving acute-angled margins, terminal facet concave. Female cones up to 53 × 18 cm., green ; median scales at right angles to axis, c. 5 cm. long with slightly deflexed heads, glabrous ; heads dorsiventrally flattened and blunt, deltoid in outline with curving acute-angled margins, terminal facet concave. Seeds about 3·5 cm. long, broadly ellipsoid, scarlet, darkening with age.

S. Rhodesia. E : Mt. Gorongowe, 1935, *Meikle* in Herb. Christian 265 (SRGH)· S : Belingwe, near Mnene Mission, 1370 m., 7.vii.1953 (♀), *Wild* 4131 (SRGH). **Mozambique.** MS : Mt. Nhangwe, 14.xii.1943, *Torre* in Herb. Christian 582 (SRGH). Known only from our area ; at forest margins and in woodland.

Closely allied to *E. gratus*, distinguishable principally by the lack of puberulence on the mature cone-scales.

4. **Encephalartos hildebrandtii** A. Braun & Bouché in Index Sem. Hort. Berol. **1874** : 18 (1875?).—Stapf in Kew Bull. **1914** : 386 (1914) ; in Curt. Bot. Mag. : t. 8592 et 8593 (1915).—Prain in F.T.A. **6**, 2 : 351 (1917) ; Kew Bull. **1918** : 127 (1918).— Schuster in Engl., Pflanzenr. IV, 1 : 118 (1932).—Melville in Kew Bull. **1957** : 246 (1957) ; F.T.E.A. Gymnosp. : 6, fig. 3 (1958). TAB. **1** fig. D. Type from Kenya.

Trunk up to 6 m. tall. Leaves up to 3 m. long, white woolly when young ; median leaflets up to 35 × 4·5 cm., lanceolate, falcate, but with the tips curved forwards above, margin with 2–6 spiniferous teeth on each side which are more closely set towards the base ; leaflets reducing in size towards the base, and becoming elliptic, being ultimately replaced by short spines which continue to the petiole-base. Male cones up to 50 × 9 cm., cylindric-fusiform ; peduncle 5–25 cm. long ; median cone-scales distinctly ascending with the heads sharply deflexed so that the terminal facet is retained in a plane parallel to the cone-axis ; heads triangular-rhomboid with a rounded adaxial margin, terminal facet rhombic and concave. Female cones up to 60 × 25 cm., cylindric, dull yellow ; peduncle 4–6 cm. long ; median cone-scales truncate-deltoid in outline with the rhomboid heads deflexed ; terminal facet hexagonal and concave. Seeds 3 cm. long, oblong-ovoid, vermilion.

Mozambique. Z : near Mocuba, Mugeba, 15.ix.1954 (♂), *Mendonça* 2082 (LISC). Well known in the coastal districts of Kenya and Tanganyika between Dar es Salaam and Mombasa (and reported beyond as far as the northern coastal boundary of Kenya), and on Zanzibar I. ; also in Toro District, Uganda.* Common in bushland and lowland forest up to 600 m. within about 80 km. of the sea ; also (in Uganda) on river banks at higher altitudes (1,200 m.).

* Less certainly, since the specimens from this area approach *E. laurentianus* De Wild. (in which species some authors have placed them) and are geographically, altitudinally and ecologically distinct from the great majority of specimens of *E. hildebrandtii* in the strict sense (see Melville in Kew Bull. **1957** : 248 (1957)).

This Mozambique record is supported only by sterile material. It is therefore impossible to be certain that this extension of the range of *E. hildebrandtii* is valid. It is not, however, inconsistent with the ecological and geographical evidence.

The acceptance of *E. hildebrandtii* as a distinct species follows current practice but the classification of *Encephalartos* is still in the early stage when all characters are given equal value ; fresh evidence of the phenotypic variability in the genus could alter the present system radically. The relationship between *E. hildebrandtii* and *E. villosus* is discussed under that species.

5. **Encephalartos villosus** (Gaertn.) Lem., Ill. Hort. **14,** misc. : 79 (1867) ; op. cit. **15** : t. 557 (1868).—Regel in Gartenfl. **24** : 41 (1875).—Dyer in Curt. Bot. Mag. : t. 6654 (1882).—Pearson in Trans. S. Afr. Phil. Soc. **16** : 345 (1906).—Marloth, Fl. S. Afr. **1** : 96, t. 15 fig. B, t. 16 fig. B (1913).—Schuster in Engl., Pflanzenr. IV, 1 : 118 (1932).—Hutch. & Rattr. in A. W. Hill, F.C. **5,** 2, Suppl. : 30 (1933).—Verdoorn in Afr. Wild Life **5** : 158 (1951). TAB. **1** fig. F. Type from S. Africa, probably Cape Province.
　　Zamia caffra [var.] *villosa* Gaertn., Fruct. **1** : 15, t. 3 (1788). Type as above.
　　Zamia villosa (Gaertn.) Willd. in Mag. Ges. Naturf. Fr. Berl. **4** : 107 (1810). Type as above.

Trunk not developed, only the crown above ground. Leaves up to 2 m. long, recurved, densely woolly when young, glabrate ; leaflets up to 20 × 1·8 cm., linear or narrowly oblong, ± straight, margin with usually 1–5 spines on either side more closely set apically, sometimes entire especially on the distal side ; leaflets reducing in size and becoming oblong towards the base, ceasing abruptly or reducing ultimately to simple spines ; petiole 20–50 cm. long. Male cones up to 60 × 10 cm., cylindric, strongly smelling ; peduncle up to 20 cm. long ; median cone-scales ascending with the head deflexed so that the terminal facet is retained in a plane parallel to the axis of the cone ; head rhombic with a rounded dentate adaxial margin, terminal facet rhombic and flat. Female cones up to 45 cm. long, oblong, ovoid, yellowing, pedunculate ; median cone-scales straight with the head much inflated ; head elliptic with somewhat blunted edges ; adaxial margin rounded and medially dentate, terminal facet deflexed. Seeds 3 cm. long, ovoid, crimson.

Mozambique. LM : Maputo Distr., Goba, 23.viii.1944 (♂), *Mendonça* 1817 (LISC) ; Libombo Mts., 20 km. from Goba towards Namaacha, vi.1945, *Sousa* 212 (SRGH).

Also in Swaziland and along the SE. coast of S. Africa (up to 80 km. inland) from East London to the border of our area. Locally common in the shadier parts of coastal woodland.

There has been in the past considerable difference of opinion over the distinctness of *E. hildebrandtii* from this species. Following Regel's suggestions (in Gartenfl. **25** : 204 (1876) ; op. cit. **26** : 215 (1876)), Hennings (in Gartenfl. **39** : 234 (1890)) made the combination *E. villosus* forma *hildebrandtii* (Gaertn.) Hennings. Eichler (in Monatsschr. Verein. z. Beförd. Gartenb. **23** : 50 (1880)) had disagreed with Regel, and Stapf (in Kew Bull. **1914** : 390 (1914)), after discussing the whole issue, decided in favour of Eichler's view that the two species should be retained. This treatment has been generally adopted.

The young leaves of *E. hildebrandtii* resemble closely the leaves of *E. villosus* and the male cones of the two species are very similar, but they differ in habit as well as in foliage and the female cones are distinguishable in the field. On the other hand, *E. hildebrandtii* var. *dentatus* Melville (in Kew Bull. **1957** : 248 (1957)), described from cultivated material, has female cone-scales which, in having dentate margins, approach those of *E. villosus* forma *intermedius* as illustrated by Hennings (loc. cit.). Further reconsideration may yet lead to the reduction in status of *E. hildebrandtii*, but scarcely to less than that of subspecies.

2. PODOCARPACEAE

Evergreen trees or shrubs, usually dioecious (always in our area). Leaves linear, lanceolate, narrowly ovate or more rarely scale-like, spirally arranged and sometimes disposed in one plane or apparently opposite. Staminate strobili terminal or axillary, forming single or fascicled usually bracteate catkin-like cones ; fertile scales subpeltate, bearing 2 pollen-sacs towards the base of the blade, pollen

Tab. 2. A.—PODOCARPUS FALCATUS. A1, foliage (×1); A2, fruit (×2) both from *Barbosa & Lemos* 7502. B.—PODOCARPUS MILANJIANUS. B1, foliage (×1) *Brass* 16183; B2, female cone (×2) *Brass* 16183; B3, fruit (×2) *Brass* 16607; B4, male cone (×1) *Gilliland* 891; B5, detail of male cone (×8) *Gilliland* 891.

grains winged. Female strobilus small with usually only 1 or 2 fertile scales. Ovule solitary, erect or inverted, soon becoming enclosed by a secondary integument variously developed from part of the strobilus.

PODOCARPUS L'Hérit. ex Pers.

Podocarpus L'Hérit. ex Pers., Syn. **2** : 580 (1807) *nom. conserv.*

Evergreen trees or shrubs, usually dioecious. Leaves linear, lanceolate or narrowly ovate, usually spirally arranged. Male strobili forming shortly stalked, short catkin-like basally bracteate cones with a pair of ± elliptic pollen-sacs on each median scale. Female strobilus of 1 or 2 fertile scales on a bracteolate axis ; ovule inverted and soon becoming enclosed in an epimatium (false aril) which arises from the scale and ultimately forms a resinous leathery integument that becomes confluent with the woody testa. Base (" receptacle ") of female strobilus sometimes swollen and fleshy at maturity.

Leaves with stomata on both surfaces ; receptacle not swollen (TAB. 2 fig. A2) 1. *falcatus*
Leaves with stomata on under surface only ; receptacle swollen at maturity (TAB. 2 fig. B3) - - - - - - - - - - - - 2. *milanjianus*

1. **Podocarpus falcatus** (Thunb.) R. Br. ex Mirb. in Mém. Mus. Hist. Nat. Par. **13** : 75 (1825).—W. Scott in Cape For. Rep. **1895** : 183 (1896).—Pilg. in Engl., Pflanzenr. IV, 5 : 72 (1903).—Sim, For. Fl. Port. E. Afr. : 108, t. 97A (1909).—Stapf in A. W. Hill, F.C. **5**, 2, Suppl. : 10, fig. 1 (1933).—Robyns in Bull. Inst. Roy. Col. Belg. **6** : 237 (1935).—Melville in Kew Bull. **1954** : 568 (1955). TAB. 2 fig. A. Type from S. Africa (Cape Province).
   Taxus falcata Thunb., Prodr. Pl. Cap. **2** : 117 (1800). Type as above.

Tree up to 30 m. tall, with thin greyish-brown bark. Leaves 2–4 cm. × c. 3 mm. and closely set on older plants, longer and somewhat spreading on juveniles, shortly petiolate, strap-shaped, long-attenuate in the upper half, stomata on both surfaces. Male cones 1 cm. long, solitary or several, sessile, bracteate. Female strobili solitary or several ; fertile scales solitary ; seed at maturity (" fruit ") about 1·5 cm. long, subglobose, drupe-like, green, surmounting a terete pedicel (the receptacle not swollen) ; testa very hard, crustaceous, tubercled, enclosed in a very resinous somewhat fleshy integument.

Mozambique. LM : R. Maputo, near Quinta da Pedra, fr. 20.ii.1948, *Gomes e Sousa* 3684 (COI) ; Bela Vista, fr. 17.i.1957, *Barbosa & Lemos* 7502 (BM ; COI ; LISC ; LMJ).
Also widespread and moderately common in montane forest, in eastern and central S. Africa.

A closely related East African species, *P. gracilior* Pilg. occurs in our area in cultivation at Manica, Mozambique (*Pedro* 4932 (LMJ)).

2. **Podocarpus milanjianus** Rendle in Trans. Linn. Soc., Bot., Ser. 2, **4** : 61 (1894).— Pilg. in Engl. Pflanzenr. IV, 5 : 92 (1903).—Stapf in Prain, F.T.A. **6**, 2 : 340 (1917).—Brenan in Mem. N.Y. Bot. Gard. **8**, 3 : 210 (1953).—Keay, F.W.T.A. ed. 2, **1**, 1 : 42 (1954).—Melville, F.T.E.A. Gymnosp. : 11 (1958). TAB. 2 fig. B. Syntypes : Nyasaland, Mt. Mlanje, *Whyte* 34 & 39 (BM).

Tree up to 35 m. tall with thin flaking reddish-brown bark. Leaves 2–15 cm. × 5–12 mm., shortly petiolate, spreading, strap-shaped, margin long-attenuate in the upper half, stomata on the under surface only ; juvenile leaves similar but often longer and slightly broader (see note p. 86). Male cones solitary or paired, up to 5 cm. long, pinkish. Female strobili solitary ; fertile scales 1 or 2 ; seed at maturity (" fruit ") c. 1 cm. long, subglobose, drupe-like, green to purple, surmounting a fleshy, red to purple, glaucous receptacle, testa thin and brittle, enclosed in a thin, very resinous integument.

N. Rhodesia. W : Mwinilunga Distr., xi.1953, *Fanshawe* 1144 (SRGH). **S. Rhodesia.** W : Gwaai (♂), *Eyles* 7445 (SRGH). E : Nyumkombe valley, x.1934(♂), *Gilliland* 891 (BM) & (♀), *Gilliland* 909 (BM). **Nyasaland.** N : Nyika Plateau, 2300 m., 13.viii.1946 (♀), *Brass* 17208 (SRGH). S : Mt. Mlanje, Luchenya Plateau, 1890 m., 2.vii.1946 (♀), *Brass* 16607 (SRGH). **Mozambique.** Z : Gúruè, 1600 m., st. 20.ix.1944, *Mendonça* 2176D (LISC). MS : Gorongosa range, Mt. Gogogo, 1600 m., 26.ix.1943 (♀), *Torre* 5958 (LISC).

Also in Angola, Belgian Congo, Cameroons, Sudan, and widespread in E. Africa. Often dominant in montane forest ; 900–3150 m. generally, 1300–2300 m. in our area. A valuable timber tree.

The lowest extreme of leaf-size is due to the occurrence, amidst rocks in exposed parts of the Chimanimani mountains of a stunted form of this species. These small trees or shrubs, represented by *Wild* 2912, *Phipps* 438, *Goodier* 168, and *Finlay* 12, have all their leaves less than 4 cm. long which gives the specimens a quite unusual appearance.

Very closely related to *P. latifolius* (Thunb.) Mirb. Trees of this S. African species bear parallel-sided leaves with deltoid apices during all phases of their life-history although in addition a few leaves on juvenile individuals may be long-attenuate ; in our species leaves attenuating from the middle or below are the rule except on senescent individuals where some may be parallel-sided although even then their apices are scarcely deltoid. Mozambique specimens especially show this tendency to have leaves not so noticeably attenuate. These two species are related in the same way as are some of the small-leaved Podocarps and a study like that of Melville of *P. usambarensis* and *P. gracilior* (in Kew Bull. **1954** : 568 et seq.) is needed to establish their differences statistically.

3. CUPRESSACEAE

Trees or shrubs, monoecious or dioecious. Leaves on adult plants scale-like, appressed and apparently decussate ; on juveniles subulate, spreading and spirally arranged or irregularly disposed. Staminate strobili in small cones, terminal or on short lateral shoots ; scales few, subpeltate, bearing 2–many pollen sacs. Female strobili of 1–12 scales ; scales woody or fleshy. Ovules 1–many per scale, erect. Mature female strobilus usually a cone with woody peltate persistent opposite scales, sometimes a berry-like fruit with fleshy confluent scarcely distinguishable scales.

Mature female strobilus a woody dehiscent cone ; leaves at maturity 2 mm. or more long, tightly clasping and giving the young shoots a smooth silhouette (TAB. **3** fig. B2) ; bark grey-brown - - - - - - - - - - 1. **Widdringtonia**
Mature female strobilus a berry-like indehiscent fruit ; leaves at maturity less than 2 mm. long, tightly clasping only at the base and giving the young shoots an irregular silhouette (TAB. **3** fig. A2) ; bark reddish-brown - - - - - - 2. **Juniperus**

1. WIDDRINGTONIA Endl.

Widdringtonia Endl., Gen. Pl., Suppl. 2 : 25 (1842).

Evergreen trees, monoecious (always ?). Leaves apparently decussate, appressed, imbricated and very small on the ultimate branches of mature plants but spirally arranged and cultrate on juveniles and sometimes on parts of adults. Male cones terminal with few or several decussate scales ; pollen-sacs 2–4. Female cones variously arranged, of two pairs of opposite scales ; ovules 3 or more at the base of each scale. Mature cones ovoid or globose, opening into 4 thick woody valves. Seeds winged, testa hard.

Widdingtonia whytei Rendle in Trans. Linn. Soc., Bot., Ser. 2, **4** : 60, t. 9 fig. 6–11 (1894).—Sim, For. Fl. Port. E. Afr.: 109 (1909).—Stapf in Prain, F.T.A. **6**, 2 : 334 (1917) ; in A. W. Hill, F.C. **5**, 2, Suppl. : 17 (1933).—Gilliland, in Journ. S. Afr. Bot. **4**: 155 (1938).—Brenan in Mem. N.Y. Bot. Gard. **8**, 3 : 210 (1953). TAB. **3** fig. B. Type : Nyasaland, Mt. Mlanje, 1891, *Whyte* (BM, holotype).

Trees, up to 40 m. tall ; bark grey-brown. Leaves of adults c. 2 mm. long, rhombic, tightly appressed. Male cones broadly elliptic with about 8 pairs of subpeltate scales ; pollen-sacs 2 or 4. Female cones in short bracteolate spikes up to 1 cm. long ; scales ovate, with up to 5 ovules each. Mature cones up to 2 cm. long, ovoid or globose, dark brown, woody and resinous, dehiscent ; valves abaxially umbonate. Seeds 2 or 3 per scale ; body c. 6 mm. long, beaked, very dark brown or black ; wing scarious to reddish-brown.

S. Rhodesia. E : Umtali Distr., Engwa, 1.ii.1955 (♀), *E.M. & W.* 78 (BM; SRGH). **Nyasaland.** S : Mt. Mlanje, 1950 m., 7.ix.1956 (♀ cone), *Newman & Whitmore* 692 (SRGH). **Mozambique.** MS : Manica Distr., Mavita, Mt. Xiroso, 26.x.1944 (♀ cone), *Mendonça* 2667 (LISC).

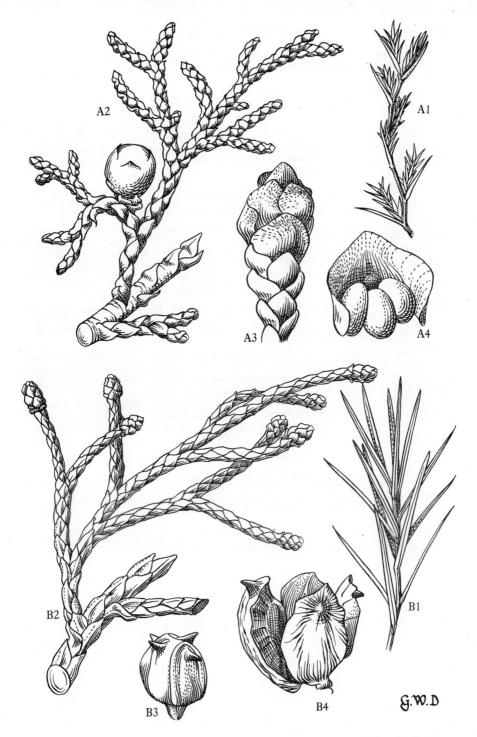

Tab. 3. A.—JUNIPERUS PROCERA. A1, juvenile foliage (×2) *Buchwald* 450 ; A2, fruiting branch (×2) *Brass* 17159 ; A3, male cone (×6) *Cons. of Forests* 77 ; A4, male cone-scale viewed adaxially (×12) *Cons. of Forests* 77. B.—WIDDRINGTONIA WHYTEI. B1, juvenile foliage (×2) *Chase* 2872 ; B2, male coning branch (×2) *Chase* 2872; B3, female cone (×2) *Swynnerton* 1963 ; B4, dehisced female cone (×2) *Swynnerton* 1963.

Also in the Transvaal. Characteristically in montane forest; rarely, stunted, in *Brachystegia spiciformis* woodland at the lower altitude limit of its range (*E.M. & W.* 78); 1680–2400 m.

This, the Mlanje Cedar, is common and flourishes on the mountains of Nyasaland where it is one of the most important timber trees. It does not as a rule grow so well in S. Rhodesia and has not been very successfully cultivated elsewhere.

Attention has been drawn by Mr. G. Jackson to a variant of this species in which the spreading cultrate juvenile foliage is not produced (the young plant bearing only the small appressed scale-leaves usual in adults) and which cones precociously. Evidence indicates that this is due to cytological rather than ecological causes and if this be confirmed it would warrant distinction as a variety.

2. JUNIPERUS L.

Juniperus L., Sp. Pl. **2** : 1038 (1753) ; Gen. Pl. ed. 5 : 461 (1754).

Evergreen trees or shrubs, monoecious or dioecious. Leaves usually decussate, elliptic, appressed, imbricate and small on mature plants but subulate and spreading on juveniles. Male cones of several rounded decussate scales bearing 2–6 pollen-sacs. Female strobili on short lateral shoots, of 2–8 usually decussate fleshy scales each bearing one ovule ; mature strobilus a berry-like fruit, the scales swollen, fleshy and confluent. Seeds 1 or few ; testa woody.

Juniperus procera Hochst. ex Endl., Syn. Conif. : 26 (1847).—A. Rich., Tent. Fl. Abyss. **2** : 278 (1851).—Stapf in Prain, F.T.A. : **6**, 2 : 336 (1917).—Brenan in Mem. N.Y. Bot. Gard. **8**, 3 : 211 (1953).—Melville, F.T.E.A. Gymnosp. : 16 (1958). TAB. 3 fig. A. Type from Ethiopia.

Straight evergreen trees up to 40 m. tall ; bark papery, reddish-brown. Leaves on young plants in threes, subulate, spine-tipped, with an elongate gland abaxially ; mature foliage of decussate scale-like leaves, varying in size as the individual ages down to 1 mm. long, abaxial gland elliptic. Male cones c. 3 mm. long, elliptic; scales c. 10, each bearing 2–3 pollen-sacs. Female strobili of 6–8 scales. Ripe fruit c. 6 mm. in diam., globose, blue-black, with the tips of the scales just distinguishable. Mature seeds 5 mm. long, 1 or 2, brown ; aborted seeds 1–3, smaller.

Nyasaland. N: Nyika Plateau, 2250 m., fr. 11.viii.1946, *Brass* 17159 (BM; K; SRGH).

Occurring at comparable altitudes on the mountains of eastern Africa. From western Arabia, Eritrea and Somaliland to our area. In evergreen forests between 1500 m. and 3000 m., except where the rainfall is more than 125 cm. annually; often locally dominant. Especially important in Kenya where it is a valuable source of easily worked durable timber.

Differing only in the generally smaller size of the fruits from the European *J. excelsa* Bieb., of which it has sometimes been regarded as a variety.

ANGIOSPERMAE

4. RANUNCULACEAE

By A. W. Exell and E. Milne-Redhead

Annual or perennial herbs with radical and alternate or spiral leaves, or shrublets or woody climbers with opposite or verticillate leaves. Leaves simple or compound, entire, lobed or dissected, usually with sheathing bases. Inflorescences generally terminal and many-flowered or more rarely with few or solitary flowers. Flowers actinomorphic or zygomorphic, hypogynous, bisexual or sometimes polygamous or dioecious. Sepals free, often petaloid. Petals free, sometimes nectariferous, often absent. Stamens indefinite in number ; anthers attached at the base, dehiscing longitudinally. Carpels indefinite in number, few or solitary, free or more or less united ; ovules 1-several. Fruit usually of achenes, drupelets or follicles, rarely a capsule or berry. Seeds with endosperm.

Flowers actinomorphic :
 Sepals more or less petaloid, no true petals present :
  Fruit of achenes (sometimes solitary) with persistent styles :
   Sepals valvate ; climbing or trailing woody plants - - - **1. Clematis**
   Sepals more or less imbricate ; herbs, often woody at the base, or erect shrublets :
    Styles plumose, many times longer than the achenes ; leaves opposite or verticillate - - - - - - - **2. Clematopsis**
    Styles not plumose, equalling or shorter than the achenes ; leaves spirally arranged - - - - - - - **3. Thalictrum**
   Fruit of fleshy drupelets with deciduous styles ; leaves all basal - **4. Knowltonia**
 Sepals not petaloid ; petals present and clearly distinguishable from the sepals, nectariferous - - - - - - - - **5. Ranunculus**
Flowers zygomorphic ; sepals more or less petaloid, dorsal one produced into a long spur ; fruit of follicles - - - - - - - **6. Delphinium**

1. CLEMATIS L.

Clematis L., Sp. Pl. **1** : 543 (1753) ; Gen. Pl. ed. 5 : 242 (1754).

Mostly woody climbers or shrubby trailing plants. Leaves opposite, usually pinnately or ternately compound with petiole and rhachis capable of twining. Flowers actinomorphic in many- to few-flowered panicles or solitary. Petals and nectaries absent but staminodes transitional between sepals and stamens sometimes present. Carpels indefinite in number, free, normally 1-ovulate by abortion. Achenes capitate with a persistent, usually elongated, plumose or naked style.

Plants woody climbers on trees or bushes ; leaflets entire or toothed, very variable in shape but usually ovate-lanceolate, ovate or suborbicular in outline :
 Leaves once to twice divided :
  Leaflets ovate to ovate-lanceolate, entire or regularly dentate, except near the base, rarely lobed, apex usually somewhat prolonged ; plant usually a strong climber along edges of forests - - - - - - - *1. simensis*
  Leaflets very variable in shape, suborbicular or ovate-oblong, irregularly and often coarsely dentate and at least some usually lobed ; plants climbing in woodland, bush or edges of riverine forests :
   Flower-buds and back of sepals with yellow indumentum when dried *2. iringaensis*
   Flower-buds and back of sepals with white, grey or cream indumentum :
    Sepals very thin, semi-transparent ; leaves usually twice divided, leaflets membranous at time of flowering ; achenes (including persistent style) up to 7 cm. long - - - - - - - *3. viridiflora*
    Sepals thicker, not transparent ; leaves usually once divided, leaflets usually chartaceous at time of flowering ; anthers usually 1·5–2 mm. long ; achenes (including persistent style) up to 4 cm. long - - - *4. brachiata*
 Leaves at least twice divided and the ultimate divisions usually deeply incised ; anthers usually 1–1·5 mm. long ; achenes (including persistent style) 1–2 cm. long 5. *oweniae*
Plants prostrate, usually trailing among grass or over rocks ; leaflets narrowly elliptic to elliptic-oblong :
 Sepals 1·3–2·5 cm. long :
  Sepals densely hairy all over on the outside - - - - *2. iringaensis*

Sepals usually only sparsely hairy outside except towards the edges   6. *welwitschii*
Sepals 2·5–3·5 cm. long   -   -   -   -   -   -   -   -   7. *thalictrifolia*

1. **Clematis simensis** Fresen. in Mus. Senckenb. **2** : 267 (1837).—Eyles in Trans. Roy.
    Soc. S. Afr. **5** : 352 (1916).—Exell & Mendonça, C.F.A. **1, 1** : 2 (1937).—Staner &
    Léonard, F.C.B. **2** : 191 (1951).—Milne-Redh. & Turrill, F.T.E.A. Ranunc. : 2
    (1952).—Milne-Redh. in Mem. N.Y. Bot. Gard. **8,** 3 : 211 (1953). Type from
    Ethiopia (Semen).

Tall woody climber up to 20 m. or more, often behaving as a strong liane ;
younger stems more or less hairy but usually becoming glabrous or nearly so,
longitudinally ribbed and furrowed. Leaves imparipinnate with 5 leaflets, but
frequently reduced in association with the inflorescence ; leaflets ovate to ovate
lanceolate, acuminate or shortly acuminate, rounded to cordate at the base, entire
to regularly dentate but nearly always entire towards the base, usually not lobed
but leaflets of the upper leaves occasionally with 1–2 lobes, nearly glabrous to
densely pubescent on lower surface, a few scattered hairs on upper surface.
Inflorescence many-flowered ; pedicels 1–3·5 cm. long ; flower-buds ellipsoid.
Sepals 7–18 mm. long, cream or white.

S. Rhodesia. E : Mt. Nuza, 1830 m., fl. & fr. 19.vi.1934, *Gilliland* 392 (BM ; FHO ;
SRGH). **Nyasaland.** N : NW. Nyasaland, 1896, *Whyte* (K). S : Mt. Mlanje, 1650
m., fr. vii.1946, *Brass* 16860 (K ; SRGH). **Mozambique.** N : Maniamba, Serra Gessi,
fl. 29.v.1948, *Pedro & Pedrógão* 4090 (LMJ).
Widespread on the mountains of E. Africa and in the Cameroons, Fernando Po and
Angola. Forest edges from about 1100–1860 m.

All the specimens from our area that can be confidently named *C. simensis* come from a
belt, narrow from east to west but considerably extended from north to south, along the
escarpments of the Great Rift and the eastern escarpment mountains of the S. Rhodesian
Plateau—a typical distribution of a rain-forest species. We could not accept as pure *C.
simensis* any of the N. Rhodesian specimens that had been so named in the past but some
of them show tendencies towards this species, which may have hybridized at times with
C. brachiata. Similar intermediates occur in Mozambique.
Martineau's reference (Rhod. Wild Fl. : 28 (1954)) probably refers partly to this species
and partly to *C. brachiata* Thunb.

2. **Clematis iringaensis** Engl., Bot. Jahrb. **28** : 388 (1900). Type from Tanganyika
    (Iringa).
        Clematis commutata sensu Staner & Léonard, F.C.B. **2** : 187 (1951).—Milne-
    Redh. & Turrill, F.T.E.A. Ranunc. : 5 (1952).

Woody climber or occasionally described as prostrate ; the younger stems pilose
becoming more or less glabrous with age, longitudinally ridged and furrowed.
Leaves imparipinnate with usually 5 leaflets ; leaflets ovate to lanceolate-ovate,
often strongly asymmetric, curved in upper part (especially the terminal leaflet),
sometimes lobed, irregularly and unequally dentate to almost entire, glabrous on
upper surface or with scattered pubescence on the veins, shortly pilose or pubescent
on the veins beneath. Inflorescences 1- to many-flowered ; pedicels 1–2 cm. long ;
buds broadly ovoid to spherical, with yellow indumentum. Sepals 1·7–2 cm. long,
cream, pilose to sparsely pilose outside, more densely so along the edges and inside.

N. Rhodesia. N : Abercorn, Lake Chila, fl. ii.1954, *Nash* 48 (BM). W : Ndola,
fl. 18.ii.1956, *Fanshawe* 2781 (K). **Nyasaland.** C : Dedza Distr., Dzenza Forest
Reserve, 1220 m., fl. 21.iii.1955, *E.M. & W.* 1101 (BM ; LISC ; SRGH).
Also in Uganda, Tanganyika and the Belgian Congo. Wooded grassland and bush at
1400–1500 m.

Miss A. H. Gamwell notes that it " crawls over rocks or small bushes " but it may
possibly have been confused with the trailing *C. welwitschii* which grows in the same
neighbourhood. The flowers are considerably smaller than those of *C. commutata*
Kuntze, a southern Angolan species.

3. **Clematis viridiflora** Bertol., Miscell. Bot. **19** : 7,˙t. 3 (1858).—Milne-Redh. & Turrill,
    F.T.E.A. Ranunc. : 5 (1952). Type : Mozambique, Inhambane, *Fornasini* (BO,
    holotype, †).

Tall woody climber ; younger stems pubescent, becoming glabrous, longitudinally ribbed and furrowed. Leaves pinnate or pinnate-ternate ; leaflets
5–12 × 4–10 cm., ovate to broadly ovate in general outline, often obliquely asym-

metric with slightly acuminate lobes and with few, large, often asymmetric teeth usually mucronate at the apex, with minute scattered appressed hairs sparse on the upper surface, denser below. Inflorescences few- to many-flowered; pedicels 1–5 cm. long; flower-buds broadly ovoid. Sepals 1·2–2 cm. long, pale yellow or greenish, very thin in texture and slightly transparent, pubescent especially along the margins. Anthers 1·5–2 mm. long. Achenes (including persistent style) up to 7 cm. long.

Nyasaland. S: Fort Johnston, Citimera, fl. 1.vi.1955, *Jackson* 1663 D (BM; LISC; SRGH). **Mozambique.** Z: between Marral and Mopeia, fr. 27.vi.1942, *Torre* 4430 (BM; LISC). T: between Tete and Zobue, fl. 16.vi.1941, *Torre* 2852 (LISC). MS: between Amatongas and Gondola, fr. 10.vii.1948, *Barbosa* 1720 (BM; LISC). SS: between João Belo and Chibuto, fr. 27.vii.1944, *Torre* 6788 (BM). LM: Maputo, Salamanga, fl. 3.vii.1948, *Mendonça* (BM; LISC).
Also in Zanzibar. Forest margins and secondary woodlands, usually at low altitudes.

Flowers and leaves rubbed together smell of formaldehyde and are used as an inhalant in Nyasaland.
Jackson 1663 C (BM) from Kalembo, near Fort Johnston, Nyasaland, seems intermediate between this species and *C. brachiata* Thunb. and may possibly be a hybrid.

4. **Clematis brachiata** Thunb., Prodr. Pl. Cap. **2**: 94 (1800).—Harv. & Sond., F.C. **1**: 2 (1860).—Weim. in Bot. Notis. **1936**: 30 (1936).—M. Henderson in Fl. Pl. Afr. **30**: t. 1197 (1955). Type from S. Africa.
 Clematis petersiana Klotzsch in Peters, Reise Mossamb. Bot. **1**: 170 (1861). Type: Mozambique, Tete, *Peters* (B, holotype †).
 Clematis thunbergii sensu Eyles in Trans. Roy. Soc. S. Afr. **5**: 352 (1916).— Garcia in Bol. Soc. Brot., Sér. 2, **19**: 509 (1945).
 Clematis wightiana sensu Eyles, tom. cit.: 353 (1916).
 Clematis viorna sensu Eyles, loc. cit. (err. " virona ").
 Clematis brachiata var. *burkei* Burtt Davy, F.P.F.T. **1**: 111 (1926). Type from S. Africa (Orange Free State).
 Clematis inciso-dentata sensu Garcia, loc. cit.—Martineau, Rhod. Wild Fl.: 28 (1954).
 Clematis orientalis sensu Merxm. in Proc. & Trans. Rhod. Sci. Ass. **43**: 12 (1951).
 Clematis hirsuta sensu Milne-Redh. in Mem. N.Y. Bot. Gard. **8**, 3: 211 (1953).
 Clematis simensis sensu Wild, Guide Fl. Vict. Falls: 143 (1953).

Woody climber up to about 4 m. tall; younger stems more or less softly hairy, longitudinally ribbed and furrowed. Leaves pinnate with 5–7 leaflets or pinnate-ternate; leaflets suborbicular to ovate in outline, shortly acuminate, acute or more rarely subobtuse, cordate to rounded or obtuse at the base, often a longer central lobe with a shorter lateral one on each side but sometimes asymmetric, margins crenate-dentate, from glabrous to densely sericeous-pubescent or tomentose on the lower surface and from glabrous to sparsely appressed-pubescent on the upper surface. Inflorescences generally many-flowered; flowers sweet-scented; pedicels 0·5–3 cm. long; flower-buds spherical to ellipsoid, rounded to acuminate. Sepals 0·8–1·5 cm. long; cream or white. Anthers usually 1·5–2 mm. long. Achenes (including persistent style) up to 4 cm. long.

Bechuanaland Prot. N: Ngamiland, Toakhe R., Gomane, fr. 19.vi.1937, *Evans* 252 (PRE). SE: Mochudi, fr. viii.1914, *Harbor* (SRGH). **N. Rhodesia.** W: Solwezi, fr. 12.ix.1952, *White* 2333 (FHO). C: Chalimbana, fl. 31.v.1937, *Robinson* 2215 (K). E: Mvuvye, fr. 16.viii.1955, *Lees* 32 (K). S: Namwala, on Kalahari sand, fr. 24.vi.1952, *White* 2986 (FHO). **S. Rhodesia.** N: Mazoe, 1310 m., fl. iv.1906, *Eyles* 326 (BM). W: Matopos, near World's View, fl. 13.iv.1955, *E.M. & W.* 1488 (BM; LISC; SRGH). C: Marandellas, fr. 21.vii.1947, *Newton* 94 (SRGH). E: Inyanga, 1550 m., fr. 24.xi.1930, *F.N. & W.* 3195 (LD); Melsetter, 1460 m., fl. 30.iii.1950, *Williams* 87 (BM; SRGH). S: SE. Ndanga, fr. 14.vii.1955, *Mowbray* 49 (SRGH). **Nyasaland.** N: Nyika Plateau, Rumpi, 1520 m., fl. ix.1902, *McClounie* 129 (K). C: Nchisi Mt., 1350 m., fl. 3.viii.1946, *Brass* 17112 (BM; K; PRE; SRGH). S: Fort Johnston, Kalembo, fl. 31.v.1955, *Jackson* 1663 A (BM; LISC; SRGH). **Mozambique.** N: Vila Cabral, 500 m., fl. 3.viii.1946, *Torre* 151 (BM; COI). Z: Montes do Ile, fr. 26.vi.1943, *Torre* 5584 (BM; LISC). MS: Chimoio, Serra de Garuso, fl. 5.iv.1948, *Garcia* 896 (BM; LISC). SS: between Macia and Monianga, fr. 9.vii.1947, *Pedro & Pedrógão* 1385 (LMJ; PRE; SRGH).
Widespread in tropical and S. Africa. In woodlands and wooded grassland from 500–1550 m.

In Southern Rhodesia and Mozambique this species is certainly not separable from the South African *C. brachiata* Thunb., which is the earliest name available. In the northwest and northeast of the region our material is equally inseparable from an Angolan and East African " species " which has been called *C. hirsuta* Perr. & Guill. in the recent floras of those countries. Although we suspect that much of the material at present referred to *C. hirsuta* in various parts of tropical Africa will eventually have to be transferred to *C. brachiata* we have not formally included *C. hirsuta* in our synonymy because: (*a*) this latter species needs further collecting from the type locality (Cape Verde Peninsula) ; and (*b*) the problem of the " species " in East Africa is very complex (see Milne-Redh. & Turrill, F.T.E.A. Ranunc.: 6 (1952)) and does not directly concern us here. In our region the species is much less variable and identification is only difficult in a few cases where there has perhaps been hybridization with *C. simensis* in the wetter regions, with *C. oweniae*, more particularly in S. Africa, or with *C. viridiflora* in Nyasaland and Mozambique.

5. **Clematis oweniae** Harv., Thes. Cap. **1** : 6, t. 9 (1859) ; in Harv. & Sond., F.C. **1** : 2 (1860).—Burtt Davy, F.P.F.T. **1** : 111 (1926). Type from Natal (Port Natal).
    Clematis brachiata Thunb. forma.—Bremek. & Oberm. in Ann. Transv. Mus. **16** : 413 (1935).

Woody climber very similar to *C. brachiata* but with leaves twice divided and the ultimate segments deeply incised, sepals not exceeding 1·2 cm. in length, anthers usually about 1 mm. long, sometimes up to 1·5 mm. and achenes (including persistent style) 1–2 cm. long.

Bechuanaland Prot. SE : Metsimotlaba, near Gaberones, fl. 15.iii.1930, *van Son* in Herb. Transv. Mus. 28990 (BM ; PRE). **S. Rhodesia.** S : Victoria Distr., fl. 1909, *Monro*, 1002 (BM ; SRGH). **Mozambique.** LM ? unlocalized, *Almeida* in Herb. Pret. 12402 (PRE).
    Also in the Transvaal, Swaziland and Natal. Probably in woodland and wooded grassland, up to about 1200 m.

This species, which has a " Limpopo " type of distribution is only known, in our area, from a few gatherings. In S. Africa it is said to hybridize with *C. brachiata* Thunb. and some specimens with anthers 1·5 mm. or so in length, which are difficult to classify, may be of hybrid origin. M. Henderson (in Fl. Pl. Afr.: sub. t. 1197 (1955)) says " The two species are closely allied and *C. brachiata* is distinguished by its usually larger less hairy and darker green leaves, longer anthers and the more pleasant scent of the flowers. The young achenes of *C. brachiata* are usually greenish in colour, while those of *C. oweniae* are pink or brown ". It should be possible to find this species again in the Victoria District, where Monro collected it in 1909.

6. **Clematis welwitschii** Hiern ex Kuntze in Verh. Bot. Verein. Brand. **26** : 171 (1885).—Exell & Mendonça, C.F.A. **1**, 1 : 3 (1937).—Staner & Léonard, F.C.B. **2** : 188, t. 15 (1951).—Milne-Redh. & Turrill, F.T.E.A. Ranunc. : 3 (1952). Type from Angola (Pungo Andongo).
    Clematis prostrata Hutch., Botanist in S. Afr. : 484 (1946) *nom. nud.*
    Clematis probably n. sp.—Eyles in Trans. Roy. Soc. S. Afr. **5** : 353 (1916).

Woody trailing or climbing plant ; younger stems pilose, glabrescent with age, longitudinally ribbed and furrowed. Leaves bipinnate, pinnate-ternate or pinnate with the lateral leaflets more or less deeply trilobed ; leaflets typically oblong with few rather irregular teeth, glabrous or glabrescent on upper surface, pubescent to pilose, especially on the veins, or nearly glabrous on lower surface. Inflorescence loose, 3–7 or more-flowered or flowers occasionally solitary ; pedicels 1·5–6·5 cm. long ; buds ellipsoid, rounded or acute. Sepals 1·2–2·5 cm. long, outer surface pale pink or purple with scattered hairs or glabrescent outside except at the margin which is densely hairy ; inner surface densely lanate.

N. Rhodesia. N : Abercorn Distr., Lucheche R., 1500 m., fl. 24.iv.1936, *Burtt* 6262 (BM ; K). W : Kitwe, fl. 10.iv.1957, *Fanshawe* 3155 (K). C : 10 km. south of Kapiri Mposhi, fl. 27.iii.1955, *E.M. & W.* 1218 (BM ; LISC ; SRGH). S : Mazabuka, fl. 27.iii.1952, *White* 2340 (FHO ; K). **S. Rhodesia.** N : Mazoe Distr., 1460–1500 m., fl. iii.1906, *Eyles* 281 (BM ; K : SRGH). C : Salisbury, the Kopje, fl. iii.1909, *Rand* 1401 (BM). S : Lundi R., fr. 30.vi.1930, *Hutchinson & Gillett* 3255 (K). **Nyasaland.** N : Fort Hill, 1100 m., *Whyte* (K). C : Kasungu, fl. 4.iv.1955, *Jackson* 1569 (ZOM). S : *Buchanan* 369 (BM).
    Also in Angola, southern Belgian Congo and Tanganyika. Upland grassland, wooded grassland, *Brachystegia* woodland and abandoned cultivation, 1000–1500 m.

7. **Clematis thalictrifolia** Engl., Bot. Jahrb. **45** : 270 (1910).—Staner & Léonard, F.C.B. **2** : 187 (1951). Type from the Belgian Congo (Katanga, Kundelungu).

Woody trailing plant, younger stems longitudinally ribbed and furrowed, rather sparsely pubescent. Leaves bipinnate or pinnate-ternate ; leaflets and segments of deeply cut leaflets narrowly elliptic, glabrous or nearly so on the upper surface, sparsely pilose beneath. Flowers solitary (always ?), axillary, with peduncles up to 7·5 cm. long and pedicels up to 11 cm. long. Sepals 4, 2·5–3·5 cm. long, white, purplish outside, pubescent outside, more densely so on the margins and tomentose within. Anthers 3·5 mm. long.

N. Rhodesia. N : Abercorn, on road to Kalambo Falls, fl. 30.iii.1952, *Richards* 1247 (K).
Belgian Congo (Upper Katanga), Rhodesia and Tanganyika. In *Brachystegia* woodland.

This little-known species appears to be related to *C. welwitschii* Hiern ex Kuntze, from which it can be distinguished by the larger, solitary flowers.

2. CLEMATOPSIS Boj. ex Hutch.

Clematopsis Boj. ex Hutch. in Kew Bull. **1920** : 12 (1920).

Perennial herbs with erect stems occasionally somewhat woody towards the base. Leaves opposite, simple or pinnately compound. Flowers regular, solitary or several at the ends of stems or branches. Sepals 4 (rarely more) more or less imbricate, petaloid. Petals and nectaries absent. Carpels indefinite in number, with one fertile ovule. Achenes capitate, with a persistent, elongated, plumose style.

A genus confined to tropical and subtropical Africa and Madagascar. Probably derived from *Clematis* as an adaptation to savanna conditions and only somewhat doubtfully separable from that genus.

Leaves pinnate, bipinnate or pinnate-trifoliolate (sometimes with some simple leaves near base or apex of stem, if so usually with petioles at least 1 cm. long) :
    Anthers up to 5 mm. long ; stems usually several-flowered - - 1. *scabiosifolia*
    Anthers 5–9 mm. long ; stems usually 1-flowered - - - - 2. *homblei*
Leaves all simple, petioles very short ; stems usually 1-flowered - 3. *uhehensis*

1. **Clematopsis scabiosifolia** (DC.) Hutch. in Kew Bull. **1920** : 20 (1920).—Exell & Mendonça, C.F.A. **1**, 1 : 5 (1937).—Garcia in Bol. Soc. Brot., Sér. 2, **19** : 509 (1945).— Exell, Léonard & Milne-Redh. in Bull. Soc. Roy. Bot. Belg. **83** : 407 (1951).—Milne-Redh. & Turrill, F.T.E.A. Ranunc. : 7 (1952). Type from Angola.
Clematis scabiosifolia DC., Syst. Nat. **1** : 154 (1818). Type as above.

Stems 0·7–1·5 m. tall, longitudinally ribbed and furrowed, indumentum very variable, from silky-tomentose to nearly glabrous. Leaves pinnate, bipinnate or pinnate-trifoliolate, occasionally simple near the base or apex of the stem. Flowers 3·5–7 cm. in diam., usually several (occasionally solitary) at the ends of the stems or main branches. Sepals softly hairy on both surfaces, white, cream, pink, mauve or lilac. Anthers up to 5 mm. long. Achenes in heads up to 10 cm. in diam. (including the persistent styles).

Throughout the area and widespread in tropical and S. Africa. *Brachystegia* woodland and upland grasslands from about 1000 m. upwards.

This beautiful species with its drooping, often sweetly scented, white, cream or pale pink flowers is common throughout the region at altitudes above about 1000 m. It is equally conspicuous when flowering in the wet season and by its heads of bearded fruits in the dry season.

F. White has noted that in N. Rhodesia the acid juice from the leaves is used to cause the cicatrices of tribal markings.

Insect visitors recorded : *Sturmia inconspicua* Mg., *Rhynchomyia pruinosa* Vill., *Rhinia apicalis* Wied.

The taxonomic problem of this very variable aggregate species has been discussed by Exell, Léonard and Milne-Redhead (loc. cit.) who recognized seven main groups, which they designated A–G, and various intermediates. As the Flora Zambesiaca region is the meeting-place of the areas of distribution of most of these groups, the populations met with are mainly of intermediates which often defy analysis. The name *C. scabiosifolia* will suffice for most purposes but we have tried below to give an account of some of the forms more commonly met with.

Key to the Groups of Clematopsis scabiosifolia

Leaves bipinnate ; ultimate segments linear to narrowly-elliptic, 5–6 times longer than
wide - - - - - - - - - - - - Group A
Leaves pinnate ; ultimate segments not linear and rarely more than 3 times longer than
wide (except in Group E, confined to Angola) :
  Stems 1–3-flowered, flowers usually solitary ; internodes generally longer than the
  leaves (except in transition forms) - - - - - - Group D
  Stems usually many-flowered ; at least the lower leaves longer than the internodes :
    Sepals markedly acuminate - - - - - - - Group G
    Sepals not markedly acuminate :
      Leaves sericeous-tomentose, the indumentum usually concealing the tertiary
      nervation :
        Pinnae narrowly elliptic, up to 5 times longer than broad, tending to become
        entire or 1–5-dentate in the upper half (Angola only) - - Group E
        Pinnae usually suborbicular in outline, never more than twice as long as broad
        Group F
      Leaves pubescent or sparsely pubescent, pubescence mainly on the nerves,
      tertiary nervation visible :
        Pinnae suborbicular in outline, crenate-serrate, usually rounded at the base
        Group C
        Pinnae ovate to elliptic, irregularly serrate or incised, tending to be cuneate at
        the base - - - - - - - - - - - Group B

Group A

Clematis stanleyi Hook., Ic. Pl. **6** : t. 589 (1843). Type : from the Transvaal
(Magalisberg).

Clematopsis stanleyi (Hook.) Hutch. in Kew Bull. **1920** : 21 (1920).—Pole Evans,
Fl. Pl. S. Afr. : t. 81 (1923). Type as above.

Bechuanaland Prot. N : Leshumo R., fl. i.1876, *Holub* (K). **N. Rhodesia.** B :
Sesheke, on Kalahari Sand in open *Baphia-Terminalia* scrub, fl. 20.xii.1952, *Angus* 981
(FHO ; K). **S. Rhodesia.** W : Wankie, fl. 19.ii.1956, *Wild* 4744 (BM ; LISC ; SRGH).

Southern Angola, N. Rhodesia (Barotseland), S. Rhodesia, Bechuanaland Prot. and
Transvaal. Abundant in the Transvaal but apparently rare elsewhere.

On the hypothesis that *Clematopsis* has evolved from *Clematis* and that various species
of the latter genus may have contributed elements to the polymorphic *Clematopsis scabiosifolia*, it is possible that in Group A we may have a considerable inheritance from *Clematis
oweniae*, which has a fairly similar distribution.

Group A transitions

In our area there are many transitional series connecting Group A with other groups,
more especially with Groups B and F. The type of the epithet " scabiosifolia " (from
Angola) comes in the A–F series. Most of the material from Southern Rhodesia belongs
here and is therefore " typical " in the sense that it agrees fairly closely with the nomenclatural type. For anyone who wishes to construct a more conventional nomenclature
than our own, the specimens cited below, under Series A–F, will indicate our idea of
" typical " *C. scabiosifolia.*

Transition A–F

Clematis pulchra Weim. in Bot. Notis. **1936** : 27, fig. 8 (1936). Type : S. Rhodesia,
Inyanga, *F. N. & W.* 3711 (LD, holotype).

N. Rhodesia. C : Mufulira, 6.iii.1949, *Cruse* 497 (K). S : Mazabuka, Siambambo
Forest Reserve, fl. 15.i.1952, *White* 1918 (FHO ; K). **S. Rhodesia.** W : Rhodes Dam,
between Bulawayo and Matopos, fl. ii.1903, *Eyles* 1172 (BM ; SRGH). C : Salisbury,
fl. xii.1897, *Rand* 1 (BM). E : Inyanga, 1520 m., fl. i.1919, *Nobbs in Eyles* 1461 (SRGH).
S : Victoria Distr., fl. 1909–1912, *Monro* (BM).

Group B

N. Rhodesia. N : Abercorn, 1770 m., fr. 20.iv.1936, *Burtt* 6379 (BM ; K). **Mozambique.** " Zambeziland," *Kirk* (K).

Also in the Belgian Congo.

Group B transitions

Clematis villosa subsp. *stanleyi* var. *pubescens* Kuntze in Verh. Bot. Verein. Brand.
26 : 174 (1885). Type : Nyasaland, *Simons* (BM, holotype).

Clematis stanleyi var. *pubescens* (Kuntze) Dur. & Schinz, Consp. Fl. Afr. **1** : 7
(1898). Type as above.

Transitions occur between Group B and Groups A, C and D mainly in the eastern
half of the region.

Group C

Clematis kirkii Oliv., F.T.A. **1** : 5 (1868). Type : Nyasaland, Manganja Hills,
Kirk (K, holotype).

Clematis villosa subsp. *normalis* var. *kirkii* (Oliv.) Kuntze in Verh. Bot. Verein. Brand. **26** : 173 (1885). Type as above.
Clematopsis kirkii (Oliv.) Hutch. in Kew Bull. **1920** : 17 (1920). Type as above.

Nyasaland. N : 50 km. north of Rukuru R., fl. 4.v.1944, *Pole Evans & Erens* 694 (PRE). C : Mwera Hill, 1460 m., fl. 13.iii.1950, *Foster* 2 (K). S : Ncheu, 1180 m., fl. 17.iii.1949, *Wiehe* N/23 (K). **Mozambique.** N : Massangulo, fl. iii.1933, *Gomes e Sousa* 1314 (COI ; K). T : Maravia, Vila Vasco da Gama, 1200 m., fr. 12.vii.1941, *Torre* 3256 (BM ; LISC).
Tanganyika, Nyasaland and Mozambique.

Transition C–D
Clematopsis kirkii sensu Weim. in Bot. Notis. **1936** : 26 (1936).
Clematopsis costata Weim., tom. cit. : 28, fig. 9 (1936). Type : S. Rhodesia, Inyanga, *F.N. & W.* 3482 b (BM, isotype ; LD, holotype).

N. Rhodesia. E : Lundazi Distr., Kangapanda Mt., Nyika Plateau, 2130 m., fr. 7.v.1952, *White* 2750 (FHO). **S. Rhodesia.** N : Sinoia Cave, fr. 21.iv.1948, *Rodin* 4368 (PRE ; SRGH). E : Inyanga, 1900 m., fl. 29.i.1931, *Norlindh & Weimarck* 4657 (BM ; L). **Nyasaland.** S : Limbe, 1220 m., fl. 30.i.1948, *Goodwin* 52 (BM). **Mozambique.** Z : between Mugeba and Ile, fl. 1.iv.1943, *Torre* 5020 (BM ; LISC).

Transition C–F—Milne-Redh. in Mem. N.Y. Bot. Gard. **8**, 3 : 212 (1953).

Nyasaland. C : Kasungu, 1000 m., fl. 27.viii.1946, *Brass* 17435 (BM ; K ; SRGH). **Mozambique.** T : Angonia, between Melengo Balame and Vila Coutinho, fr. 11.v.1948, *Mendonça* 4161 (BM ; LISC).

Group D
Nyasaland. N : between Kondowe and Karonga, *Whyte* 351 (K).
Also in Nigeria, Cameroons, Sudan and E. tropical Africa.

In some localities, especially in Tanganyika, there are quite distinct and homogeneous populations of Group D.

Group D transitions
Transitions occur between Group D and all the other groups found in the region.

Group F
N. Rhodesia. W : Mwinilunga, Matonchi, 14.ii.1938, *Milne-Redhead* 4562 (K).
Also in Angola and the Belgian Congo.

Group G
Nyasaland. N : Fort Hill, *Whyte* (K).

2. **Clematopsis homblei** (De Wild.) Staner & Léonard, F.C.B. **2** : 196 (1951).—Exell, Léonard & Milne-Redh. in Bull. Soc. Roy. Bot. Belg. **83** : 423 (1951). Type from the Belgian Congo (Upper Katanga).
Clematis homblei De Wild. in Fedde, Repert. **13** : 200 (1914). Type as above.

Stems up to about 1 m. tall, strongly longitudinally striate, pilose to hirsute or only sparsely hairy. Leaves pinnate or pinnate-ternate, tending to become simple near the base and apex of the stem. Leaflets elliptic, cuneate at the base, coarsely toothed and often deeply lobed, sericeous-pilose especially on the nerves beneath. Flowers 4·5–8 cm. in diam., pink or white, usually solitary. Anthers 5–9 mm. long.

N. Rhodesia. W : Ndola, fr. 29.v.1953, *Fanshawe* 40 (K) ; Mwinilunga Distr., south-east of Dobeka Bridge, fl. 7.xii.1937, *Milne-Redhead* 3538 (K).
Belgian Congo, Angola and N. Rhodesia. In *Cryptosepalum* woodland and in bush at 1300–1500 m.

3. **Clematopsis uhehensis** (Engl.) Hutch. ex Staner & Léonard in Bull. Soc. Roy. Bot. Belg. **82** : 342 (1950).—Exell, Léonard & Milne-Redh. in Bull. Soc. Roy. Bot. Belg. **83** : 423 (1951).—Milne-Redh. & Turrill, F.T.E.A. Ranunc. : 7 (1952). Type from Tanganyika (Iringa).
Clematis uhehensis Engl., Bot. Jahrb. **28** : 287 (1900). Type as above.
Clematopsis simplicifolia Hutch. & Summerh. in Kew Bull. **1925** : 361 (1925). Type from Tanganyika (Rungwe Mts.).

Stems 0·4–0·9 m. tall, strongly longitudinally striate with spreading, villous hairs. Leaves up to 9 × 5·5 cm., simple, ovate, sessile or very shortly petiolate, coarsely and somewhat irregularly dentate, sometimes obscurely lobed, with scattered hairs. Flowers 6·5–12 cm. in diam., usually solitary at the ends of the stems. Sepals spreading.

N. Rhodesia. E : Nyika Plateau, 2130 m., 27.xi.1955, *Lees* 104 (K). **Mozambique.** N : east of Lake Nyasa, *Johnson* (K).

Also in Tanganyika.

Specimens cited by Garcia (in Bol. Soc. Brot., Sér. 2, **19** : 509 (1945)) collected in Mozambique by Gomes e Sousa (Niassa : S. Antonio de Mecango, *Gomes e Sousa* 1641 (COI) and Torre (Niassa : Lichinga Plateau, Vila Cabral, *Torre* 34 (BM ; COI)) seem to form a series intermediate between this species and *C. scabiosifolia*. The flowers, which are usually in threes, are much smaller than in *C. uhehensis* and the leaves, of which the upper pairs are usually simple, are distinctly petiolate.

One of the three specimens constituting *Lees* 104 (K) has, in addition to simple leaves, two 3-foliolate leaves and longer petioles than in the other two specimens. This would seem to have some genes from *C. scabiosifolia* group D.

3. THALICTRUM L.

Thalictrum L., Sp. Pl. **1** : 545 (1753) ; Gen. Pl. ed. 5 : 242 (1754).

Herbs with compound spirally arranged leaves with sheathing bases. Flowers actinomorphic, usually rather small in terminal panicles, bisexual or some male only, sepals 3–5, imbricate in bud, green or petaloid. Petals absent. Stamens 3 to numerous, often with conspicuous anthers. Carpels 1 to numerous, 1-ovulate. Achenes sessile or stipitate, stigma deciduous or persistent forming a long beak.

Carpels 1–4 ; sepals 1·5–4 mm. long ; some flowers male only   1. *rhynchocarpum*
Carpels 10–20 ; sepals 8 mm. long ; all flowers bisexual -   -   -   -   2. *zernyi*

1. **Thalictrum rhynchocarpum** Dill. & Rich. in Ann. Sci. Nat., Sér. 2, Bot. **14** : 262 (1840).—Oliv., F.T.A. **1** : 8 (1868).—R.E.Fr., Schwed. Rhod.-Kongo-Exped. **1** : 43 (1914).—Eyles in Trans. Roy. Soc. S. Afr. **5** : 353 (1916).—Weim. in Bot. Notis. **1936** : 32 (1936).—Staner & Léonard, F.C.B. **1** : 181 (1951).—Milne-Redh. & Turrill, F.T.E.A. Ranunc. : 11 (1952).—Milne-Redh. in Mem. N.Y. Bot. Gard. **8**, 3 : 213 (1953). Type from Ethiopia (Tigre).

Thalictrum innitens B. Boiv. in Rhodora **46** : 394 (1944). Type from S. Africa (Cape Province).

Perennial herb up to 4 m. tall, sometimes semi-scandent. Leaves up to 40 cm. or more in length (including petiole), tripinnate to quadripinnate ; leaflets elliptic to broadly ovate or oblate, often somewhat trilobed or with coarse rounded teeth, apex usually apiculate, base rounded to cordate. Inflorescence lax, many-flowered, some flowers male only. Sepals 1·5–3 mm. long, green to purplish. Stamens 3–13 ; anther-thecae 0·8–1·7 mm. long, prolongation of connective 0·05–0·5 mm. long. Carpels 1–4. Stigma linear, 2–7 mm. long at anthesis, covered with stigmatic papillae. Achenes up to 8 × 2 mm., asymmetric, flattened-fusiform, longitudinally striate, glabrous, tapering to the persistent stigma ; pedicels elongating to 6–11 cm., hair-like.

S. Rhodesia. C : Salisbury 1490 m., fl. 19.ii. fr. 19.iii.1919, *Eyles* 1512 (SRGH) ; Inyanga, 1900 m., fl. 29.i.1931, *Norlindh & Weimarck* 4653 (BM ; E ; SRGH) ; Chipinga Distr., Chipete Forest, 1220 m., 1906, *Swynnerton* 352 (BM ; K). **Nyasaland.** N : Nyika Plateau, 2350 m., fr. 17.viii.1946, *Brass* 17281 (K ; SRGH) ; Nkata Bay Distr., Rumpi, *Chapman* 164 (BM). C : Dedza Mt. (recorded but not collected, *E.M. & W.*). S : Mt. Mlanje, 1830 m., *Whyte* 159 (BM). **Mozambique.** N : Maniamba, Serra Gessi, fl. 29.v.1948, *Pedro & Pedrogão* 4094 (LMJ). MS : Tsetsera, 1850 m., fl. 9.ii.1955, *E.M. & W.* 333 (BM ; LISC ; SRGH).

Widespread on uplands and mountains of tropical Africa and in S. Africa. In undergrowth and along margins of montane evergreen forest, 1100–2350 m.

According to Swynnerton, forest birds use the slender fruiting pedicels to build their nests.

" False Maidenhair."

2. **Thalictrum zernyi** Ulbr. in Notizbl. Bot. Gart.Berl. **15** : 715 (1942).—Milne-Redh. & Turrill, F.T.E.A. Ranunc.: 9 (1952). Type from Tanganyika (Matengo Highlands).

Perennial herb up to 50 cm. tall, with erect, glabrous stems. Leaves 1·5–6 cm. in length (including petiole), bipinnate ; leaflets elliptic to oblate, often trilobed, apices of lobes rounded and slightly apiculate, base cuneate to truncate. Inflorescence few-flowered, pedicels not greatly elongated in infructescence. Sepals 8 mm. long, white inside, pink outside. Stamens about 15, anther-thecae 1·7–2·3 mm.

long, prolongation of connective up to 0·2 mm. long. Carpels 10–20, stigma 1·5 mm. long at anthesis, curled. Achenes sessile, 2·5 mm. long, with 3–4 longitudinal ribs.

Nyasaland. N : Nyika Plateau, fl. & fr. 17.ii.1956, *Chapman* 387 (BM ; SRGH). Also in southern Tanganyika. In submontane grassland.

4. KNOWLTONIA Salisb.

Knowltonia Salisb., Prodr. : 372 (1796).

Perennial herbs with 3-foliolate to biternate leaves and many-flowered cymes or compoundly umbellate inflorescences on long scapes with an involucre of reduced leaves at the branching of the inflorescence. Flowers actinomorphic. Tepals more or less petaloid, about 15 but variable in number, the outer often somewhat smaller and slightly sepaloid, without nectariferous pits. Stamens indefinite in number. Carpels indefinite in number, 1-ovulate, with glabrous deciduous styles. Fruit of fleshy drupelets.

Knowltonia transvaalensis Szyszyl., Polypet. Thalam. Rehm. : 99 (1887).—Burtt Davy, F.P.F.T. **1** : 110 (1926).—Milne-Redh. & Turrill, F.T.E.A. Ranunc. : 12 (1952).—Milne-Redh. in Mem. N.Y. Bot. Gard. **8**, 3 : 212 (1953). TAB. **4**. Type from the Transvaal.

Anemone whyteana Bak. f. in Trans. Linn. Soc., Bot., Ser. 2, **4** : 4 (1894). Type : Nyasaland, Mt. Mlanje, *Whyte* 100 (BM, holotype ; K).

Anemone peneensis Bak. f. in Journ. Linn. Soc., Bot. **40** : 16 (1911).—Eyles in Trans. Roy. Soc. S. Afr. **5** : 352 (1916). Type : S. Rhodesia, Mt. Pene, *Swynnerton* 783 (BM holotype).

Anemone transvaalensis (Szyszyl.) Burtt Davy in Ann. Transv. Mus. **3** : 121 (1912). Type as for *Knowltonia transvaalensis* Szyszyl.

Knowltonia whytei Engl., Pflanzenw. Afr. **3**, 1 : 170 (1915) in obs. : sphalm. pro *K. whyteana*.

Rhizome short and stout with long silky hairs on the sheathing leaf-bases. Leaves with petioles 7–22 cm. long, densely lanate towards the base, pilose or rarely glabrous above ; basal leaves biternate to trifoliolate, the segments varying in size and lobing, lobes 2–10 × 0·8–5·5 cm. Flowering stems 30–80 cm. tall, lanate towards the base. Flowers 2–3 cm. in diam. in simple or compound umbels, 2–15, sweet scented. Tepals white, pink or tinged with purple. Carpels glabrous.

S. Rhodesia. E : Inyanga, 2070 m., fl. 23.x.1955, *Chase* 5833 (BM ; SRGH) ; Chimanimani Mts., 1520 m., fl. 9.vi.1949, *Wild* 2974 (SRGH). **Nyasaland.** S : Mlanje, Luchenya Plateau, 2000 m., fl. 18.vii.1946, *Brass* 16870 (K ; SRGH). **Mozambique.** MS : Tsetsera, 1850 m., fl. 9.ii.1955, *E.M. & W.* 327 (BM ; LISC ; SRGH).

Also in southern Tanganyika and the Transvaal. Damp places in submontane grassland, often among bracken, 1500–2100 m.

The freshly crushed leaves contain anemonol, a strong irritant.

5. RANUNCULUS L.

Ranunculus L., Sp. Pl. **1** : 548 (1753) ; Gen. Pl. ed. 5 : 243 (1754).

Perennial herbs with simple or compound petiolate radical leaves and spirally arranged cauline leaves with sheathing bases. Flowers actinomorphic in 1–many-flowered inflorescences. Sepals spreading or reflexed in anthesis. Petals 5–8 or more, usually yellow or white with a basal nectariferous pit with or without a scale. Stamens indefinite in number. Carpels indefinite in number, 1-ovulate. Fruit of numerous achenes with usually glabrous styles.

Basal leaves dissected or deeply lobed :
    Basal leaves bi- or tripinnatisect, pubescent ; segments more or less tripartite, coarsely and irregularly toothed ; stigma a short persistent curved beak ; achenes about 40–60 in a spherical to slightly elongated head - - - - 1. *multifidus*
    Basal leaves 3–5-palmatisect or deeply lobed with a few coarse teeth; stigma filiform, often wholly or partly deciduous in fruit ; achenes about 12–25 in a spherical head
                                      2. *raeae*
Basal leaves suborbicular, cordate with crenate margins - - - - 3. *meyeri*

1. **Ranunculus multifidus** Forsk., Fl. Aegypt.-Arab. : CXIV, 102, t. 11 fig. 4 (1775).—Staner & Léonard, F.C.B. **2** : 175, fig. 1 (5) (1951).—Exell & Mendonça, C.F.A.,

Tab. 4. KNOWLTONIA TRANSVAALENSIS. 1, rhizome and basal leaves (× ⅔); 2, inflorescence (× ⅔); 3, longitudinal section of flower (× 2); 4, stamen (× 6); 5, carpel
(× 6); 6, fruit (× 2). Rhizome and leaves from *Schlieben* 1242 and *Whyte* 100
respectively; flowers and details from *Shinn* 5A.

1, 2 : 353 (1951).—Milne-Redh. & Turrill, F.T.E.A. Ranunc. : 19, fig. 4 (1) (1952).
—Milne-Redh. in Mem. N.Y. Bot. Gard. **8,** 3 : 213 (1953). Type from Arabia.
Ranunculus pubescens Thunb., Prodr. Pl. Cap. **2** : 94 (1800).—Burtt Davy,
F.P.F.T. **1** : 107 (1926).—Weim. in Bot. Notis. **1936** : 32 (1936).—Exell & Men-
donça, C.F.A. **1,** 1 : 7 (1937). Type from S. Africa (Cape Province).
Ranunculus forskoehlii DC., Syst. Nat. **1** : 303 (1817) *nom. illegit.* Type as for
R. multifidus Forsk.
Ranunculus membranaceus Fresen. in Mus. Senckenb. **2** : 270 (1837).—R.E.Fr.
in Schwed. Rhod.-Kongo-Exped. : 43 (1914). Type from Ethiopia.
Ranunculus pinnatus sensu Oliv., F.T.A. **1** : 9 (1868) excl. var. *extensa.*—Eyles in
Trans. Roy. Soc. S. Afr. **5** : 353 (1916).
Ranunculus plebeius sensu Bak. f. in Journ. Linn. Soc., Bot. **15** : 17 (1911).—Eyles,
loc. cit.

Perennial herb up to about 1 m. tall, appressed-pubescent with hairs directed
upwards to patent-pilose ; stems much branched in the upper part with numerous
yellow flowers about 12–16 mm. in diam. Leaves bi- or tripinnatisect, the final
segments coarsely and irregularly toothed, strongly to weakly hirsute or pilose.
Sepals reflexed. Petals 5, yellow, 3–7 mm. long. Achenes in a spherical to slightly
elongated head, numerous, smooth or with small scattered tubercles, with bordered
margin and with short or well-developed beak.

N. Rhodesia. N : Abercorn Distr., 1580 m., fl. 26.i.1952, *Richards* 562 (K). C :
Chilanga, 1070 m., fl. & fr. xi.1909, *Rogers* 1851 (K ; SRGH). W : Mwinilunga, fl. &
fr. 1.xii.1937, *Milne-Redhead* 3463 (BM ; K). **S. Rhodesia.** N : Mazoe, 1370 m.,
fl. & fr. xi.1906, *Eyles* 450 (BM ; SRGH). W : Bulawayo, fl. v.1898, *Rand* 287 (BM).
C : Salisbury, fl. & fr. vii.1898, *Rand* 435 (BM). E : Inyanga, fl. 19.x.1946, *Wild* 1395
(K ; LD ; SRGH) ; Chipinga Distr., Chirinda Forest, 1100–1200 m., fl. & fr., 28.x.1906,
Swynnerton 345 (BM ; K ; LISC ; SRGH). **Nyasaland.** N : Nyika Plateau, *Whyte*
(K) ; C : Mwera Hill Station, Dowa, *Jackson* 221 (K ; ZO). S : Blantyre, Shire High-
lands, *Adamson* 56 (BM ; K). **Mozambique.** N : between Unango and Metonia,
Johnson (K). MS : Tsetsera, 1900–2100 m., fl. & fr. 10.ii.1955, *E.M. & W.* 345 (BM ;
LISC ; SRGH). SS : Morrumbene, fl. & fr. 27.x.1947, *Barbosa* 555 (LM). LM :
Maputo, fl. & fr. 20.xi.1944, Catuane, fl. & fr. 20.xi.1944, *Mendonça* 2978 (BM ; LISC).
Widespread in tropical and in S. Africa, Madagascar and Arabia. Wet places in
submontane grassland and along river-banks from near sea-level to 2000 m. (reaching
3600 m. in Kenya).

Insect visitors recorded : *Cuphocera haemorrhoidalis* Macquart and *Tachina duplaria*
Vill.

2. **Ranunculus raeae** Exell in Journ. of Bot. **73** : 262 (1935).—Staner & Léonard,
F.C.B. **2** : 177, fig. 1 (7) (1951).—Milne-Redh. & Turrill, F.T.E.A. Ranunc. :
16, fig. 4 (4) (1952). TAB. **5.** Type from Tanganyika (Livingstone Mts.).

Tuberous-rooted perennial herb up to 60 cm. tall with sparsely pilose stems and
rather small, yellow flowers 15–20 mm. in diam. Basal leaves 3–5-palmatisect or
palmatilobed, with a few coarse teeth, appressed-pilose later glabrescent. Sepals
spreading, pilose on the outside, soon caducous. Petals 5, yellow, 8–10 mm. long.
Achenes smooth with narrowly bordered margin and straight or slightly curved beak.

N. Rhodesia. N : Kawimbe (Fwambo), *Carson* 55 (K). **Nyasaland.** N : Vipya
Plateau, 1830–2140 m., fl. i-iii.1948, *Benson* 1471 (BM). S : Zomba Mt., fl. & fr.
2.xi.1958, *Jackson* 2249 (BM). **Mozambique.** N : Vila Cabral, *Gomes e Sousa* 1052
(COI).
Also in the Belgian Congo (Katanga) and Tanganyika. In wet places in submontane
grassland up to 2140 m. and along river-banks at lower altitudes.

3. **Ranunculus meyeri** Harv. in Harv. & Sond., F.C. **1** : 7 (1860).—Burtt Davy,
F.P.F.T. **1** : 109 (1926). Type from S. Africa (Cape).
Ranunculus meyeri var. *transvaalensis* Szyszyl., Polypet. Thalam. Rehm. : 102
(1887). Type from the Transvaal.

Perennial herb rooting at the nodes, 3–16 cm. tall. Leaves suborbicular, 6–25
mm. in diam., crenate at the margins, cordate or subcordate at the base ; petioles
up to 16 cm. long, appressed-pilose with the hairs directed upwards or glabrous.
Flowers 10–14 mm. in diam., yellow, solitary on peduncles 3–16 cm. long. Sepals
3·5–4 mm. long, spreading, glabrous. Petals narrowly oblong to ligulate, glabrous.
Achenes in a subspherical head, broadly ellipsoid, shortly beaked, with narrowly
bordered margin, glabrous.

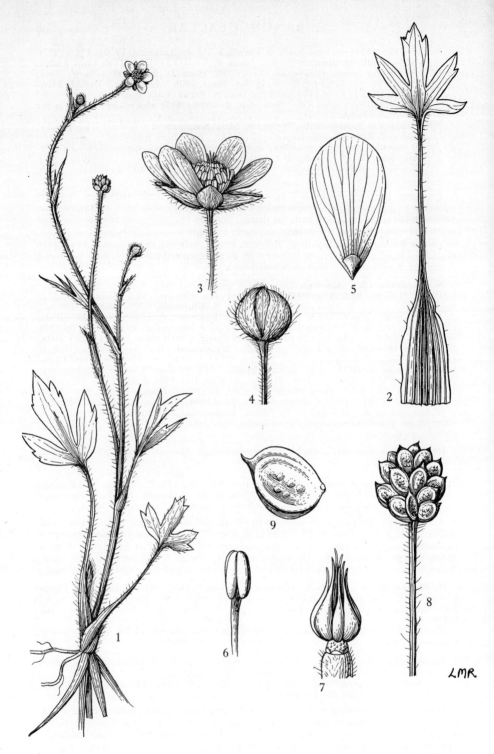

Tab. 5. RANUNCULUS RAEAE. 1, plant in flower and fruit (×⅔) ; 2, leaf (×1) ; 3, flower (×3) ; 4, flower-bud (×4) ; 5, petal (×7) ; 6, stamen (×12) ; 7, carpels (×6) ; 8, fruit (×4) ; 9, achene (×14). No. 8 from *Zimmer* 251, rest from *Rae* A6.

S. Rhodesia. E : 10 km. N. of Troutbeck, Inyanga, fl. & fr. 16.xi.1956, *Robinson* 1912 (K). **Mozambique.** MS : Tsetsera, 1800 m., fl. i.1956, *Drewe* (SRGH). Also in S. Africa. In swamps and shallow ponds at 1500–2000 m.

Our specimen has glabrous petioles, as in *R. meyeri* var. *transvaalensis* Szyszyl., but the amount of hairiness in the Transvaal specimens is very variable and the variety is scarcely worth maintaining.

6. DELPHINIUM L.

Delphinium L., Sp. Pl. **1** : 530 (1753) ; Gen. Pl. ed. 5 : 236 (1754).

Annual or perennial herbs with radical and spirally arranged palmatinerved, palmatilobed or palmatisect petiolate leaves and terminal racemose or paniculate inflorescences. Flowers zygomorphic. Sepals 5, petaloid, the posterior one produced into a spur. Petals 2–5, the 2 posterior ones with spurs more or less joined and fitting into the calyx-spur. Stamens indefinite in number, often with flattened filaments. Carpels 1–5 with numerous ovules, sessile, ripening to follicles.

In addition to the two indigenous species, horticultural forms of *D. elatum* L. are grown in the cooler parts of the region and an annual species, recorded as *D. ajacis* L., is grown in gardens in Lourenço Marques and probably elsewhere.

Spur 0·5–0·8 cm. long ; sepals usually blue, occasionally pinkish or white   1. *dasycaulon*
Spur 3–4 cm. long ; sepals usually white, occasionally pale mauve or purplish-white
     2. *leroyi*

1. **Delphinium dasycaulon** Fresen. in Mus. Senckenb. **2** : 272 (1837).—Oliv., F.T.A, **1** : 11 (1868).—Staner & Léonard, F.C.B. **2** : 168 (1951).—Milne-Redh. & Turrill, F.T.E.A., Ranunc. : 22 (1952).—Milne-Redh. in Mem. N.Y. Bot. Gard. **8**, 3 : 213, fig. 2 (1953). TAB. **6**. Type from Ethiopia (Semen).

Erect perennial herb up to about 1 m. high with yellowish-pubescent stems and inflorescences. Leaves mostly arising from the base of the stem, often persisting after the inflorescences have died, lamina palmatisect, usually nearly to the base, segments acute, pubescent especially on the nerves ; petioles of the lower leaves up to 15 cm. or more long. Flowers 2–2·5 cm. across. Sepals usually deep blue, more rarely violet, pinkish or white, pubescent ; spur 0·5–0·8 cm. long, stout, almost straight. Follicles 3, shortly and densely pubescent.

N. Rhodesia. N : Abercorn, near Malombe, 1650 m., fl. & fr. 2.vi.1936, *Burtt* 6114 (BM ; K). W : Mwinilunga, Matonchi, 1370 m., fl. 12.ii.1938, *Milne-Redhead* 4552 (K). E : Lundazi Distr., Kangampande Mt., 2130 m., fl. 7.v.1952, *White* 2742 (FHO ; K). **Nyasaland.** N : Nyika, Nchena-chena, 1700 m., fl. 10.viii.1946, *Brass* 17150 (K ; SRGH). C : Kota Kota Distr., Nchisi, 1400 m., fl. 27.vii.1946, *Brass* 16990 (NY). S : Zomba Plateau (recorded in flowerless state by *E.M. & W.*, 9.iii.1955). **Mozambique.** N : between Unango and Lake Chilwa, *Johnson* 19 (K) ; Maniamba, Serra Gessi, Malulo, 1360 m., fl. 27.v.1948, *Pedro & Pedrogão* 3969 (LMJ). T : Angónia, Kirk Range, fl. 17.iii.1955, *E.M. & W.* 994 (BM ; LISC ; SRGH).

Sudan, Eritrea, Ethiopia, Cameroons, Belgian Congo, Tanganyika, N. Rhodesia, Nyasaland and Mozambique. Submontane grasslands and *Brachystegia* woodland up to 2150 m.

2. **Delphinium leroyi** Franch. ex Huth in Engl., Bot. Jahrb. **20** : 474, t. 6 fig. 5 (1895). —Staner & Léonard, F.C.B. **2** : 169 (1951).—Milne-Redh. & Turrill, F.T.E.A. Ranunc. : 20, fig. 5 (1952).—Milne-Redh. in Mem. N.Y. Bot. Gard. **8**, 3 : 213 (1953). Type from Kilimanjaro.

Erect perennial 0·5–1·5 m. high, stems pubescent to sparsely pubescent often nearly glabrous towards the base ; inflorescences with dense yellowish pubescence. Leaves mostly arising from near the base of the stem, lamina palmatisect often almost to the base, rather sparsely pubescent, sometimes nearly glabrous ; petioles of the basal leaves up to 18 cm. or more long. Flowers 3·5–6 cm. across, sweetly scented. Sepals white, sometimes with a green spot, occasionally pale mauve or purplish-white ; spur 3–4 cm. long. Anthers black or dark purple. Follicles 3, pubescent, 2·5 cm. long.

N. Rhodesia. E : Lundazi Distr., Kangampande Mt., 2130 m., fl. 7.v.1952, *White* 2741 (FHO ; K). **Nyasaland.** N : Nyika, Kassaramba, above Nchena-chena, 2300 m., fl. iv.1953, *Chapman* 136 (BM).

Sudan (Imatong Mts.), Uganda, Kenya, Tanganyika, Belgian Congo, N. Rhodesia and Nyasaland. Abundant in submontane grasslands up to 2440 m. (higher in Tanganyika and the Sudan).

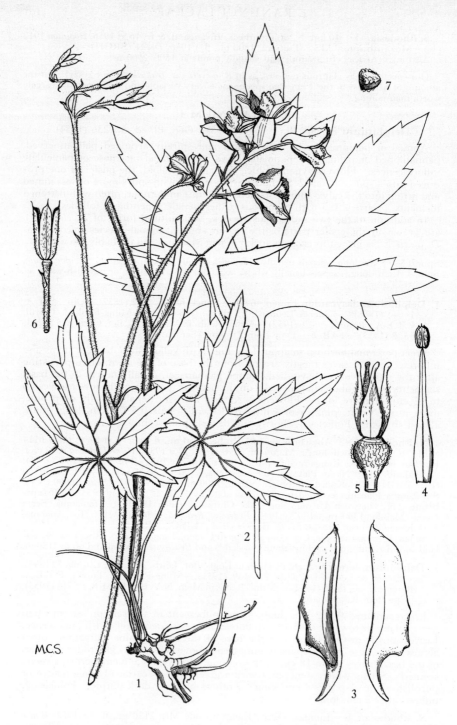

Tab. 6. DELPHINIUM DASYCAULON. 1, rootstock, basal leaves, flowering and fruiting stems
($\times \frac{2}{3}$) ; 2, leaf ($\times \frac{2}{3}$) ; 3, posterior petals ($\times 3$) ; 4, stamen ($\times 6$) ; 5, carpels ($\times 4$) ; 6,
fruit ($\times \frac{2}{3}$) ; 7, seed ($\times 6$). Composite from *Chapman* 321 and *Jackson* 1860.

5. DILLENIACEAE
By H. Wild

Shrubs, trees or lianes, rarely herbs.   Leaves alternate, entire or variously toothed (rarely pinnatifid or 3-lobed but not in our area), with prominent, parallel lateral nerves.   Stipules absent or represented by a narrow wing adnate to the petiole.   Flowers bisexual (in all our species) or unisexual.   Sepals 5, imbricate. Petals 5 or fewer, imbricate, often crumpled in bud.   Stamens numerous, rarely definite (never in our area), hypogynous, free or variously united into bundles at the base, often persistent ;   anthers opening lengthwise or by terminal pores. Carpels free, rarely one ;   ovules single or more, erect from the base or on the ventral suture of the carpel ;   styles free.   Ripe carpels dehiscent or baccate.   Seeds usually with a crested or laciniate aril ;   endosperm copious, fleshy ;   embryo minute.

TETRACERA L.

Tetracera L., Sp. Pl. **1** : 533 (1753) ; Gen. Pl. ed. 5 : 237 (1754).

Shrubs, small trees or lianes with entire or denticulate often scabrous leaves ; stipules absent or represented by a narrow wing adnate to the petiole.   Flowers moderate or large in terminal cymes, racemes or panicles.   Sepals 5, convex, imbricate, persistent and becoming leathery in fruit.   Petals as many as the sepals or fewer, often emarginate.   Stamens very numerous, dilated towards the apex, anther-thecae borne on the margins of the dilated connective parallel or divergent below.   Carpels 3–5, free, each with a simple style ;   ovules few to many borne on the ventral suture of the carpel.   Fruiting carpels brittle and woody or leathery when ripe, dehiscing along the ventral or along both sutures.   Seeds few with a laciniate aril.

Sepals densely silky-pilose outside ; low shrubs or small trees :
   Carpels densely hirsute when young, setulose-pilose when ripe -   1. *boiviniana*
   Carpels glabrous in all stages - - - - - - -   2. *masuiana*
Sepals glabrous or very sparsely pubescent outside ; a tall liane - -   3. *alnifolia*

1. **Tetracera boiviniana** Baill., Adansonia **7** :   300, t. 7 (1867).—Oliv., F.T.A. **1** :   13 (1868) pro parte excl. specim. Welwitsch.—Engl., Pflanzenw. Afr. **3, 2** : 476, fig. 220 (1921).—Gilg & Werderm. in Engl. & Prantl, Nat. Pflanzenfam. ed. 2, **21** : 17, fig. 7 (1925).—Brenan, T.T.C.L. : 182 (1949).   TAB. 7 fig. B.   Syntypes from Zanzibar.

Shrub or small tree up to 4·5 m. tall ; young branchlets tomentose, soon glabrescent.   Leaf-lamina up to 13 ×6 cm., elliptic to obovate, apex acute or obtuse, mucronulate, base narrowly rounded or broadly cuneate, margin entire or denticulate with gland-tipped teeth, rugose above with impressed nervation and venation, rather scabrous with scattered hairs, greyish, tomentose below with prominent midrib, nerves and raised reticulate venation, petioles up to 1·5 cm. long, tomentose or glabrescent.   Flowers pinkish-white in few-flowered terminal cymes up to c. 5 cm. long ;   peduncles and pedicels tomentose.   Sepals c. 1·2 ×1 cm., concave, rotund, obtuse, densely pilose outside, sparsely so or glabrescent within.   Petals c. 2 ×1·5 cm., broadly obovate, emarginate.   Stamens very numerous, 3–4 mm. long ;   anther-thecae subparallel or divergent below.   Carpels 3–5, narrowly obovoid, densely hirsute ; styles divergent, c. 5 mm. long.   Fruiting carpels 3–5, c. 1·8 ×1 cm., brittle, woody, obovoid, apiculate, densely or sparsely setulose-pilose, surrounded by the persistent leathery calyx.   Seeds 1–5 per carpel, c. 5 mm. in diam., shining black, somewhat compressed and subglobose, very minutely reticulate, with a yellowish deeply fimbriate aril c. twice the length of the seed.

Mozambique.   N :   Nampula, fl. 21.xi.1936, *Torre* 1077 (COI).   Z :   Inhamacurra, fr. 28.viii.1949, *Barbosa & Carvalho* 3875 (LM ; SRGH).
Also in Zanzibar and the coastal districts of Tanganyika and Kenya.   In coastal woodland.

2. **Tetracera masuiana** De Wild. & Th. Dur. in Ann. Mus. Cong., Bot., Sér. 1, **1** : 61, t. 31 (1899).—Gilg in Engl., Bot. Jahrb. **33** : 196 (1902).—Engl., Pflanzenw. Afr. **3**, 2 : 476 (1921).—Gilg & Werderm. in Engl. & Prantl, Nat. Pflanzenfam. ed. 2, **21** : 17 (1925).—Exell & Mendonça, C.F.A. **1**, 1 : 8 (1937).—Staner in Bull. Jard. Bot. Brux. **15** : 297 (1939). TAB. **7** fig. A.   Type from the Belgian Congo.
  Tetracera boiviniana sensu Oliv., F.T.A. **1** : 13 (1868) quoad specim. Welwitsch.
  Tetracera strigillosa Gilg in Engl., Bot. Jahrb. **33** : 196 (1902).—Hutch., Botanist in S. Afr. : 514 (1946).   Syntypes from Sudan.

Shrublet c. 1 m. tall very like *T. boiviniana* but the leaves more consistently narrowly obovate and more coarsely dentate-serrate with gland-tipped teeth, the flowers white and the carpels and ripe fruit quite glabrous.   Ripe fruits red.

N. Rhodesia.   N : Mporokoso, fl. 16.x.1947, *Brenan & Greenway* 8122 (FHO ; K). W : Mwinilunga, fl. 14.xi.1955, *Holmes* 1328 (K).
  Also in the Sudan, Belgian Congo, Cameroons, Angola and Tanganyika.   In open *Brachystegia* or mixed woodland.

3. **Tetracera alnifolia** Willd. in L., Sp. Pl. ed. 4, **2**, 2 : 1243 (1800).—Oliv., F.T.A. **1** : 12 (1868).—Engl., Pflanzenw. Afr., **3**, 2 : 477 (1921).—Gilg & Werderm. in Engl. & Prantl, Nat. Pflanzenfam. ed. 2, **21** : 18 (1925).—Exell & Mendonça, C.F.A. **1**, 1 : 9 (1937).—Staner in Bull. Jard. Bot. Brux. **15** : 302 (1939). TAB. **7** fig. C.   Type from the Guinea Coast.

Woody liane up to 15 m. or more tall, main stems up to 10 cm. in diam., young stems sparsely appressed-hairy or glabrescent.   Leaf-lamina up to 15 × 8 cm., oblong, oblong-elliptic or broadly elliptic, apex rounded, subacute or emarginate, margin entire or sinuate-dentate, often recurved, very minutely tuberculate-scabrid on both sides or almost smooth, very sparsely appressed-hairy beneath or glabrous, petiole up to 2·5 cm. long, appressed-pubescent or glabrescent, often very narrowly winged.   Flowers white in ample, terminal panicles c. 20 × 10 cm. ; branches of inflorescence pubescent or glabrescent.   Sepals c. 7 × 5 mm., rotund-concave, obtuse, sparsely pubescent or glabrescent outside, glabrous within. Petals c. 8 × 6 mm., broadly obovate, truncate or emarginate.   Stamens c. 5 mm. long.   Carpels 3, glabrous or nearly so ; styles divergent, protruding a short way beyond the anthers.   Fruiting carpels c. 1 × 0·75 cm., dull red, woody, ovoid, apiculate.   Seeds few, c. 5 mm. in diam., shining black, very minutely reticulate, compressed globose; aril orange, enclosing the seed, fimbriate.

N. Rhodesia.   N : Kawambwa, Lumangwe Falls, fr. 14.xi.1957, *Fanshawe* 4018 (FHO ; K).   W : Mwinilunga, Zambezi R., fr. 20.ix.1952, *White* 3310 (FHO ; K).
  Widely distributed through tropical W. Africa and in the islands of the Gulf of Guinea, the Cameroons and the Belgian Congo.   In evergreen forest and riverine forest.

6. ANNONACEAE

By N. K. B. Robson

Trees, shrubs, or lianes (rarely rhizomatous shrublets), glabrous or with an indumentum of simple, stellate or lepidote hairs ; bark usually smooth and entire, pale grey or buff to brown, the branches often reddish to purple-black with lozenge-shaped striations, pubescent or tomentose (rarely glabrous) when young ; pith septate, oil cells present.   Leaves alternate in two rows, entire, penninerved, membranous to coriaceous, exstipulate.   Flowers morphologically terminal (i.e. terminal, supra-axillary, extra-axillary or leaf-opposed) or axillary, solitary or paired to fasciculate or cymose, on the young or old wood, sessile or pedicellate, more rarely pedunculate, actinomorphic, bisexual or more rarely unisexual, often fragrant.   Sepals 3 (2), usually valvate in bud, free or ± united.   Petals 6 (4) in two equal or ± unequal whorls (more rarely one whorl of 6, 4 or 3), imbricate or valvate (rarely open) in bud, free or ± united at the base, usually alternating with the sepals.   Stamens ∞, spiral, or 6–12 and whorled (sometimes staminodial) with anthers linear to semi-orbicular, lateral or extrorse (rarely apical) ; connective usually prolonged beyond the thecae, with the apex truncate, oblique, capitate, convex, conical or acute ; filaments usually very short or absent, free, rarely more

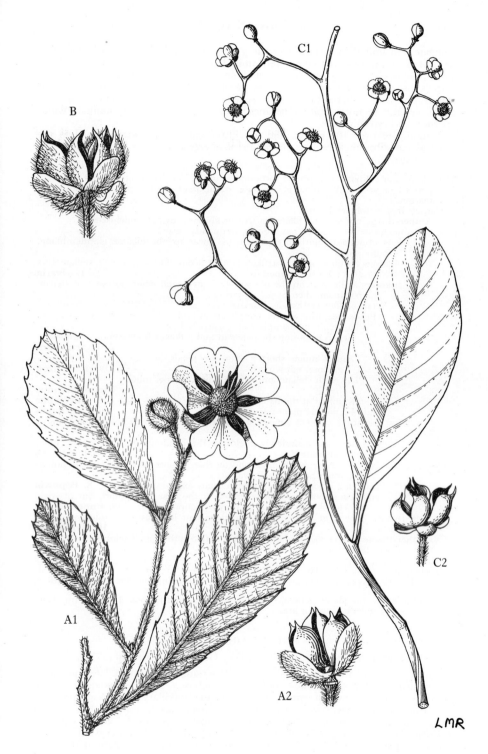

Tab. 7. A.—TETRACERA MASUIANA. A1, flowering branch (×⅔) *Fanshawe* 3957 ; A2, fruit (×1) *Bullock* 2911. B.—TETRACERA BOIVINIANA, fruit (×1) *Faulkner* 888. C.—TETRACERA ALNIFOLIA. C1, flowering branch (×⅔) *Corbisier* 937 ; C2, fruit (×⅘) *Léonard* 189.

6. ANNONACEAE

elongate and united in a cone over the gynoecium (*Xylopia*). Carpels ∞-1 free or united at the base, or completely united to form a 1-locular ovary (*Monodora*) ; ovules ∞-1 : styles free or united, or absent. Fruit apocarpous with ripe carpels baccate (fleshy or lignified) or rarely dehiscent, stipitate or sessile, or syncarpous with either aggregated 1-seeded carpels or 1-locular and ∞-seeded, Seeds vertical to horizontal, sometimes arillate, with abundant ruminate endosperm ; embryo minute.

A family of over 100 genera and over 1,000 species, almost confined to the tropical and subtropical regions of both hemispheres apart from *Asimina* in eastern N. America.

Carpels free or solitary, or if united forming a multilocular syncarp ; stigmas ± erect ;
  petals free or slightly united at the base, equal or unequal, rarely with undulating
  margins :
Carpels free or solitary in fruit :
  Anther-thecae not septate, filaments free or absent ; carpels indehiscent :
    Peduncles neither hooked nor zig-zag, frequently absent :
      Seeds oblique to horizontal ; stamens linear to obconic, or elliptic-oblong,
      numerous :
        Flowers terminal or extra-axillary ; outer petal-whorl imbricate ; stellate hairs
        present ; carpels numerous - - - - - - **1. Uvaria**
        Flowers axillary or on the old wood ; outer petal-whorl valvate ; if stellate
        hairs present then carpel solitary :
        Petals 6 ; flowers usually bisexual :
          Petals free, in 2 whorls, not plicate in bud :
            Plant with only simple hairs ; carpels 3-∞ :
              Sepals enclosing the petals in bud ; flower-buds conic, ± winged
                      2. Uvariastrum
              Sepals much shorter than the petals in bud ; buds depressed-
              globose, not winged - - - - **3. Uvariodendron**
            Plant with stellate and simple hairs ; carpel solitary
                    4. Dielsiothamnus
          Petals united at the base, plicate in bud - - - **5. Hexalobus**
          Petals 4 ; flowers monoecious - - - - **6. Uvariopsis**
      Seeds vertical ; stamens cuneate-quadrate or clavate, rarely linear, numerous or
      few :
        Calyx entire and completely enclosing the petals in bud, glabrous ; inner
        petals imbricate - - - - - - **7. Cleistochlamys**
        Calyx of 3 valvate sepals much shorter than the petals in bud, usually pubes-
        cent ; inner petals usually valvate :
          Petals in 2 whorls, free ; flowers usually extra-axillary - **8. Popowia**
          Petals contiguous or shortly united at the base in 1 whorl, but biseriate
          above ; flowers usually axillary or supra-axillary **9. Enneastemon**
      Peduncles hooked or zig-zag owing to persistent leaf-bases, ± thickened, leaf-
      opposed - - - - - - - - - **10. Artabotrys**
  Anther-thecae transversely septate, filaments united at the base, enclosing the ovaries ;
  carpels dehiscent - - - - - - - **11. Xylopia**
Carpels united in fruit to form a fleshy syncarp - - - **12. Annona**
Carpels united into a 1-locular ovary with parietal placentas : stigmas radiating ; petals
united in the lower part, very unequal, often with undulating margins **13. Monodora**

In addition to the above genera, *Cananga odorata* (Lam.) Hook. f. & Thoms. from tropical Asia is cultivated in our area.

1. UVARIA L.

Uvaria L., Sp. Pl. **1** : 536 (1753) ; Gen. Pl. ed. 5 : 240 (1754).

Shrubs or small trees, usually climbing or scrambling, with stellate and some-times also simple hairs. Buds conic or ± globose. Flowers usually bisexual, solitary or in few-flowered cymes, terminal or extra-axillary (often leaf-opposed), rarely on old branches, sessile or pedicellate. Bracteoles 2, respectively basal and median or at the base of the calyx, often caducous. Sepals 3, valvate, often ± connate. Petals 6, in two whorls, both imbricate, equal or the outer whorl slightly longer, expanding and spreading at anthesis, free or rarely connate at the base. Stamens ∞, the outer sometimes sterile and foliar, linear or obconic, with thecae lateral or extrorse and connective prolongations thin, tapering, rounded or trun-cate, rarely capitate ; filaments very short or absent. Carpels ∞, free, linear or cylindric, with numerous ovules in 2 rows, rarely with 1–3 ovules ; style very short or absent ; stigma horse-shoe-shaped or funnel-shaped with a slit down the inner side. Ripe carpels indehiscent, dry or succulent, often sweet and edible, globular to cylindric, usually on median or lateral stipes, many- to few-seeded. Seeds horizontal or oblique ; aril small, carunculoid.

A palaeotropical genus of over 100 species which ranges from Africa and Madagascar to S. China, New Guinea, New Caledonia and Oceania.

Calyx cupular in bud, splitting ± irregularly into 3 lobes at anthesis, or remaining
    scarcely lobed ; prolongation of anther-connective thin or tapering :
    Leaves with lateral nerves impressed and venation not or scarcely prominent above,
        densely or sparsely stellate-pubescent below when mature :
        Fruiting carpels cylindric, dark brown, usually velutinous, smooth or rarely slightly
            rugose, with stipes 1–2 cm. long ; petals 12–20 mm. long, tomentellous above,
            yellow - - - - - - - - - - - 1. *angolensis*
        Fruiting carpels ovoid to cylindric, orange, tomentellous, rugose, sessile or with stipes
            up to 0·35 cm. long ; petals 4–6 mm. long, glabrous above (except near the
            margin) - - - - - - - - - - - 2. *edulis*
    Leaves with lateral nerves and venation usually ± prominent above, glabrous or
        sparsely stellate-pubescent below when mature ; fruiting carpels dark- or orange-
        brown, tomentellous, ± smooth, with stipes 0·4–1·1 cm. long ; petals 7–13 (–20) mm.
        long, glabrous above (except near the margin), green - - - - 3. *virens*
Calyx with lobes distinct in bud, separating into regular sepals at anthesis, free to the
    middle or to the base ; prolongation of anther-connective truncate or rounded :
    Fruiting carpels globose or broadly ovoid, with stipes 3–5 cm. long ; sepals 8–13 mm.
        long - - - - - - - - - - - 5. *welwitschii*
    Fruiting carpels globose to cylindric, with stipes up to 0·8 cm. long ; sepals 3–7 mm.
        long :
        Petals 15–40 mm. long ; fruiting carpels cylindric, (1) 1·7–2·2 cm. long, tomentellous
            4. *kirkii*
        Petals 6–13 mm. long ; fruiting carpels globular or ovoid, rarely shortly cylindric,
            0·8–1·5 cm. long, glabrous to tomentellous :
            Leaves cuneate to rounded at the base, glabrous or subglabrous below ; fruiting
                carpels ovoid or shortly cylindric, glabrous or sparsely rusty-pubescent, with
                stipes 0·3–0·8 cm. long :
                Leaves attenuate at the base ; petiole (2) 3–6 mm. long ; flowering pedicels
                    1–2 cm. long, stout, thickened upwards, densely pubescent - 6. *caffra*
                Leaves cuneate to rounded at the base ; petiole 1–3 mm. long ; flowering
                    pedicels c. 2·2–3 cm. long, slender, not thickened upwards, ± sparsely
                    pubescent - - - - - - - - - 7. *gracilipes*
        Leaves cordate (rarely rounded) at the base, stellate-pubescent below ; fruiting
            carpels globose, tomentellous, with stipes 0·2–0·3 cm. long - 8. *acuminata*

1. **Uvaria angolensis** Welw. ex Oliv., F.T.A. **1** : 23 (1868).—Engl. & Diels in Engl.,
    Mon. Afr. Pflanz. **6** : 17 (1901).—Exell & Mendonça, C.F.A. **1**, 1 : 13 (1937) ;
    1, 2 : 354 (1951) ; **2** : XIII (1954).—Boutique, F.C.B. **2** : 295 (1951). Type from
    Angola (Pungo Andongo).
    Uvaria bukobensis Engl., Pflanzenw. Ost-Afr. **C** : 178 (1895).—Engl. & Diels, op.
    cit. : 17, t. 3D (1901).—Robyns, Fl. Parc Nat. Alb. **1** : 186 (1948).—Brenan,
    T.T.C.L. : 45 (1949). Type from Tanganyika (Bukoba).
    Uvaria nyassensis sensu R.E.Fr., Wiss. Ergebn. Schwed. Rhod.-Kongo-Exped. **1** :
    44 (1914).
    Uvaria variabilis De Wild., Pl. Bequaert. **1** : 461 (1922). Type from the Belgian
    Congo.

Subsp. **angolensis**

Liane, scrambling shrub or small tree, 2–9 m. high. Branches rusty-pubescent at first, eventually glabrous. Leaves petiolate ; lamina 4–13·5 (16) × (1·5) 2–5·6 (8) cm., oblong to obovate, or oblanceolate, obtuse or shortly acuminate at the apex, broadly cuneate or rounded to shallowly cordate at the base, subcoriaceous, greyish-green and puberulous or glabrous (except along the veins) above, bright green and sparsely to rather densely stellate-pubescent (especially on the nerves) below, with nerves impressed above and prominent below, reticulate venation not usually prominent above ; petiole 3–8 mm. long, rusty-pubescent. Flowers solitary or in 2–3-flowered condensed cymes, terminal and extra-axillary ; pedicels absent or up to 8 mm. long ; bracteoles ovate-orbicular, tomentellous. Sepals c. 4–8 mm. long, united to form a ± unlobed cupule in bud, splitting ± irregularly

into 3 parts at anthesis, rusty-pubescent on both surfaces.   Petals yellow or
yellowish-green, fleshy, subequal, 12–15 (20) mm. long, ovate to obovate, obtuse
to rounded, tomentellous on both surfaces.   Stamens 3–5 mm. long, linear, brown ;
connective-prolongation ovate to orbicular, acute to rounded, thin, flattened.
Carpels pubescent.   Fruit on a pedicel c. 10 mm. long ; ripe carpels numerous,
1- to several-seeded, 1·3–3·8 × 0·8–2 cm., cylindric (or ± globose when few-
seeded), apiculate, not or only slightly constricted between the seeds, dark-brown-
tomentellous or -velutinous to smooth or sightly rugose, with stipes 1–1·5 (2)
cm. long.   Seeds c. 9 mm. long, oblong.

N. Rhodesia.   N : Abercorn Distr., Sausia Falls, Kalambo R., fl. 1.i.1957, *Richards*
7430 (K).   W : Mwinilunga Distr., 6·4 km. N. of Mayowa Plains, fr. 4.x.1952, *White* 3456
(FHO ; K).
In Angola, N. Rhodesia, Belgian Congo, Tanganyika, Uganda and the Sudan.   Subsp.
guineensis Keay occurs in W. Africa from Sierra Leone to Fr. Cameroons.   Forest margins,
thickets and fringing forest.

U. angolensis subsp. *guineensis* Keay has a longer indumentum on stem, petioles etc.
than the typical subspecies.   No constant difference in leaf-shape exists between *U.
angolensis* and *U. bukobensis* but specimens from the eastern part of the range of the species
tend to have narrower, subglabrous leaves and sometimes smaller flowers.   In these
characters, therefore, they approach *U. virens*.

2. **Uvaria edulis** N. Robson in Bol. Soc. Brot., Sér. 2, **32** : 151 (1958). Type : N.
Rhodesia, Mwinilunga, *Holmes* 1176 (K, holotype).

Liane or scrambling shrub up to 18 m. high.   Branches rusty-tomentose at first,
eventually subglabrous.   Leaves as in *U. angolensis* but sometimes remaining more
densely stellate-pubescent below, especially on the nerves.   Flowers solitary or in
2–3-flowered condensed cymes, extra-axillary, ± sessile.   Sepals forming a cupular
rusty-pubescent calyx, as in *U. angolensis*.   Petals brown, fleshy, subequal, 4–6
mm. long, ovate, obtuse to rounded, glabrescent above, tomentellous towards the
base.   Stamens c. 2 mm. long, linear, orange-brown ; connective-prolongation
orbicular-oblate, rounded, tapering.   Carpels pubescent.   Fruit on a pedicel 3–4
mm. long ; ripe carpels 5–10, (1) 2–5 × 1–2 cm., 1- to several-seeded, ovoid to
cylindric (or ± globose when few-seeded), not or scarcely apiculate, not constricted
between seeds, orange-tomentellous, rugose, sessile or with stipes not exceeding
0·35 cm. long.   Seeds c. 1 cm. long.

N. Rhodesia.   W : Mwinilunga Distr., near Zambezi R., 6·4 km. N. of Kalene Hill
Mission, fr. 24.ix.1952, *White* 3384 (FHO ; K).
Known only from Mwinilunga.   " In seasonally flooded forest, near Zambezi R."
(*White* 3384).

U. edulis is very similar vegetatively to *U. angolensis* although it appears to grow taller ;
but the smaller flowers and subsessile fruits are more like those of *U. pulchra* Louis ex
Boutique, from the Belgian Congo.   In both these species, however, the fruits are
velutinous, not markedly rugose.

3. **Uvaria virens** N.E.Br. in Kew Bull. **1896** : 16 (1896). TAB. **8** fig. A.   Type :
Mozambique, Delagoa Bay (K, holotype) ; cult. in Hort. Kew., 1895, from seed coll.
Mrs. Monteiro, 1886.
Uvaria nyassensis Engl. & Diels in Engl., Mon. Afr. Pflanz. **6** : 17 (1901).—Burtt
Davy & Hoyle, N.C.L. : 30 (1936). Syntypes : Nyasaland, *Buchanan* 20 (K), 899
(K), 1129 (K).
Uvaria gazensis Bak. f. in Journ. Linn. Soc., Bot. **40** : 17 (1911).—Eyles in Trans.
Roy. Soc. S. Afr. **5** : 354 (1916).—Steedman, Trees etc. S. Rhod. : 8 (1933). Type :
S. Rhodesia, Chirinda outskirts, *Swynnerton* 1326 (BM, holotype ; K).

Liane or shrub, 1·2–7 (9) m. high.   Branches shortly rusty-pubescent at first,
eventually glabrous.   Leaves petiolate ; lamina 5–14·3 × (2)2·5–6 cm., elliptic-
oblong to obovate or oblanceolate, obtuse or shortly acuminate at the apex,
narrowly cuneate to rounded at the base (rarely subcordate), ± coriaceous, greyish-
green and glabrous (except sometimes on the nerves) above, brighter green and
glabrous or sparsely pubescent below, with nerves (and often also reticulation)
± prominent on both surfaces ; petiole 3–8 mm. long, rusty-pubescent or glabrous,
often transversely ribbed.   Flowers solitary or in 2–3-flowered condensed cymes,
terminal or extra-axillary ; pedicel (2) 4–10 mm. long, rusty- or fawn-tomentellous;
bracteoles ovate to orbicular, sometimes very small.   Sepals 4–8 mm. long,

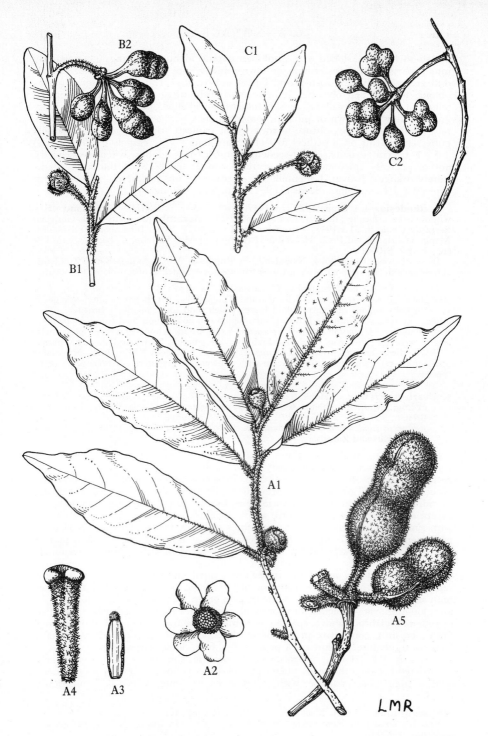

LMR

Tab. 8. A.—UVARIA VIRENS. A1, branch with flower-buds (×⅔) ; A2, flower (×1) ; A3, stamen (×6) ; A4, carpel (×6) ; A5, fruit (×⅔). All cult. ex *Monteiro* except fruit, *Hack* 56/50. B.—UVARIA CAFFRA. B1, leaves and flower-bud (×⅔) *Schlechter* 11997 ; B2, fruit (×⅔) *Wood* 1327. C.—UVARIA GRACILIPES. C1, leaves and bud (×⅔) *Phelps* 207 ; C2, fruit (×⅔) *E.M. & W.* 642.

united to form a 3-lobed or ± unlobed cupule in bud, splitting ± irregularly into 3 parts at anthesis, rusty- or fawn-tomentellous on both surfaces. Petals dull green, rarely yellowish-green, fleshy, subequal, 7–13 (20) mm. long, ovate-oblong to orbicular-obovate, sometimes slightly unguiculate, sparsely tomentellous or subglabrous above, densely tomentellous below. Stamens 2–4 mm. long, linear, yellowish-brown or pinkish, tomentellous ; connective-prolongation ovate to orbicular or oblong, obtuse to rounded or truncate, thick, tapering. Carpels pubescent. Fruit on a pedicel 5–9 (11) mm. long ; ripe carpels c. 10–20, (1·3) 2–5 × 1–1·8 cm., usually several-seeded, cylindric (or ± globose when few-seeded), not or scarcely apiculate, not or scarcely constricted between the seeds, dark-(rarely orange-) brown, tomentellous, ± smooth (not or scarcely rugose), with stipes 0·3–1·1 cm. long. Seeds (7) 9–11 mm. long, oblong.

S. Rhodesia. E : Chipete, fl. 22.x.1947, *Wild* 2135 (K ; SRGH). **Nyasaland.** S? : without precise locality, fl. 1891, *Buchanan* 899 (K). **Mozambique.** MS : Vila Gouveia, Serra de Chôa, fl. 31.x.1943, *Torre* 6106 (BM ; LISC). SS : Inhambane, Vilanculos, Mapinhane, fr. 31.viii.1942, *Mendonça* 52 (LISC). LM : Maputo, Goba, fr., *Borle* 159 (K ; PRE ; SRGH).
In south-west Tanganyika, Nyasaland, S. Rhodesia, Mozambique, Natal (Zululand) and Swaziland (Libombos). Evergreen or fringing forest, thickets or scrubland, 0–1500 m.

U. virens is very closely allied to *U. lucida* Boj. ex Benth. (=*U. fruticosa* Engl.) from Zanzibar and eastern regions of Kenya and Tanganyika, which has smaller, orange or fawn, usually rugose ripe carpels and leaves which tend to be relatively broader. Two other close relatives from the Lower Congo region can be distinguished from *U. virens* by fruit characters, *U. smithii* Engl. (*U. pecoensis* Exell) by its yellow verrucose ripe carpels, and *U. versicolor* Pierre ex Engl. in which they are smooth, brown and shortly stipitate.
Where the areas of *U. virens* and *U. angolensis* meet, in eastern N. Rhodesia, Nyasaland and south-west Tanganyika, forms occur with some characters intermediate between these species.

4. **Uvaria kirkii** Oliv. ex Hook. f. in Curt. Bot. Mag. **98** : t. 6006 (1872).—Engl. & Diels in Engl., Mon. Afr. Pflanz. **6** : 19 (1901) pro parte excl. *U. schelei* et *U. fruticosa*.— Brenan, T.T.C.L. **2** : 45 (1949). Type from Tanganyika.
     Uvaria stuhlmannii Engl., Pflanzenw. Ost-Afr. **C** : 178 (1895). Syntypes from Tanganyika and Zanzibar.

Shrub 1–2 m. high, sometimes scrambling or climbing, when it may reach 7–8 m. Branches ± sparsely rusty-pilose at first, eventually glabrous. Leaves petiolate ; lamina 4·7–10·7 (12) ×(2·5) 3–5 (6) cm., elliptic to oblong or obovate, obtuse to rounded or slightly emarginate at the apex, not or scarcely acuminate, rounded or cordate at the base, coriaceous, bluish- or greyish-green and glabrous except sometimes along the midrib above, brighter green and sparsely pubescent with simple and stellate hairs or sometimes glabrous below, with nerves slightly prominent below and slightly impressed above or both surfaces plane ; petiole 2–4 mm. long, sparsely pilose, usually transversely ribbed. Flowers terminal, solitary ; pedicels absent or up to 3 mm. long, rusty-pilose ; bracteoles lanceolate. Sepals 6–7 mm. long, united only in the lower third, covering the petals in bud and separating at anthesis, ovate, acute or obtuse, fawn- or greyish-tomentellous on both surfaces. Petals cream to white or pale green, thin, not fleshy, equal, 15–40 mm. long, obovate-orbicular or rhomboid, sometimes slightly unguiculate, obtuse or rounded, glabrous above, sparsely tomentellous below. Stamens 1–2 mm. long, linear, cream ; connective-prolongation thick, truncate or rounded, pubescent. Carpels tomentose. Fruit on a pedicel 4–8 mm. long ; ripe carpels c. 5–10, (1) 1·7–2·2 ×0·8–1·1 cm., several-seeded, cylindric, sometimes apiculate, slightly constricted between the seeds when ripe and dried, yellowish-fawn or grey, tomentellous, rugose or verrucose, with stipes 6–8 mm. long. Seeds c. 7 mm. long, ovoid.

Mozambique. N : R. Msalo, fl. 3.xii.1911, *Allen* 85 (K).
Near the coast from Kenya to northern Mozambique. In open woodland, thickets and grassland, 0–450 m.

5. **Uvaria welwitschii** (Hiern) Engl. & Diels in Engl., Mon. Afr. Pflanz. **6** : 18 (1901).— Exell & Mendonça, C.F.A. **1**, 1 : 12 (1937).—Brenan, T.T.C.L. : 46 (1949).— Boutique, F.C.B. **2** : 289 (1951). Type from Angola (Pungo Andongo).
     Uvaria sp. nova ? Oliv., F.T.A. **1** : 23 (1868).

Oxymitra? Welwitschii Hiern, Cat. Afr. Pl. Welw. **1** : 10 (1896).   Type as above.
Uvaria valvata De Wild., Pl. Bequaert. **1** : 460 (1922).   Type from the Belgian
Congo (Angi).

Shrub, tree or liane, 2–4 m. high (or higher ?).   Branches rusty-pubescent or
rusty-puberulent at first, eventually glabrous.   Leaves petiolate ; lamina 7–13
(16) × 3–5·4 cm., elliptic to oblong or oblanceolate, obtuse to acute or shortly
acuminate at the apex, cuneate to rounded at the base, coriaceous, bright green,
concolorous, glabrous on both sides except sometimes along the midrib, with
lateral nerves and densely reticulate venation prominent on both surfaces ; petiole
2·5–4 mm. long, rusty-pubescent or almost glabrous.   Flowers terminal or extra-
axillary, solitary or rarely in cymose pairs ; pedicels 1–2·8 cm. long, rusty-tomentellous ; bracteoles soon caducous.   Sepals 8–13 mm. long, ± free to the base,
covering the petals in bud and separating at anthesis, ovate, often cucullate, acute
or obtuse, rusty-tomentellous on both surfaces.   Petals yellow or brownish-yellow,
sometimes pinkish at the base, somewhat fleshy, subequal, 12–22 mm. long, ovate
or broadly ovate, glabrous above, tomentellous below, ciliate round the margin.
Stamens c. 1 mm. long, linear or obconic, glabrous or pubescent along the midline ;
connective-prolongation broadened, truncate or rarely rounded.   Carpels tomentose.   Fruit on a pedicel 20–28 mm. long ; ripe carpels numerous (c. 20–80),
1–1·5 × 0·8–1·2 cm., 2-seeded, globose or broadly ovoid, not apiculate, not constricted between the seeds, orange, rusty-tomentellous, finely rugose, with stipes
30–50 mm. long.   Seeds c. 6–8 mm. long, plano-ovoid.

N. Rhodesia. N : Abercorn Distr., Mbashi, fl. 15.xi.1956, *Richards* 6991 (K).
From Angola (Cuanza N. & S.) to N. Rhodesia, eastern Belgian Congo, Tanganyika
(Bukoba) and western Uganda.   In fringing forest, forest margins or thickets, c. 1000–
1500 m.

6. **Uvaria caffra** E. Mey. ex Sond. in Harv. & Sond., F.C. **1** : 8 (1860).—Engl. & Diels
     in Engl., Mon. Afr. Pflanz. **6** : 20, t. 6E (1901).—Wood, Ill. Pl. Natal, **3**, 2 : 18,
     t. 241 (1901).—Sim, For. Fl. Port. E. Afr. : 8 (1909).—Hutch., Botanist in S. Afr. :
     264 (1949).   TAB. **8** fig. B.   Type from Natal.

Shrub, small tree, or climber, c. 1–2 m. high or higher, often scrambling.
Branches sparsely rusty-puberulent at first, soon glabrous.   Leaves petiolate ;
lamina 5–13·2 × 1·8–4 cm., oblong or elliptic-oblong, acute or shortly acuminate to
rounded or slightly emarginate at the apex, cuneate and ± decurrent at the base,
subcoriaceous, bright or bluish green, concolorous, glabrous on both sides when
fully grown except sometimes in the region of the base of the midrib below, with
lateral nerves and densely reticulate venation prominent on both surfaces but
especially below ; petiole (2) 3–6 mm. long, rusty-pubescent.   Flowers terminal or
extra-axillary, usually solitary ; pedicels 1– c. 2 cm. long, densely fawn- or rusty-
pubescent, stout, ± thickened upwards ; bracteoles soon caducous.   Sepals 3–5
mm. long, free to the base, covering the petals in bud and separating at anthesis,
broadly ovate, ± obtuse, rugose, rusty-pubescent on the outside.   Petals greenish,
not fleshy, subequal or the outer rather longer, 7–13 mm. long, ovate to suborbicular, cucullate, often ± clawed, rugose, glabrous or tomentellous above,
tomentellous below.   Stamens 2–3 mm. long, linear, glabrous ; connective-prolongation broadened, truncate.   Carpels pubescent.   Fruit on a pedicel 18–25 (34)
mm. long ; ripe carpels few (c. 5–10), 1–3-seeded, 0·8–1·5 × 0·6–1 cm., ovoid or
shortly cylindric, often apiculate, slightly constricted between the seeds when
dried, yellow or orange, glabrous or sparsely rusty-pubescent, finely rugose, with
stipes 3–8 mm. long.   Seeds c. 7–8·5 mm. long, ovoid or plano-ovoid, horizontal.

Mozambique. LM : Delagoa Bay, fl. 5.i.1898, *Schlechter* 11997 (BM ; COI ; E ; K).
In southern Mozambique, Natal and Cape Province (Pondoland).   Tropical or subtropical forest, frequently at the margin, up to 600 m.

U. caffra has been confused with *U. virens*, but can easily be distinguished from the
latter species by the free sepals and the quite different fruit.   It is allied to *U. scheffleri*
Diels from E. Africa, which has larger, yellow flowers and longer fruits.

7. **Uvaria gracilipes** N. Robson in Bol. Soc. Brot., Sér. 2, **32** : 152 (1958).   TAB. **8** fig. C.
     Type : S. Rhodesia, Ndanga Distr., Lundi R., *Phelps* 207 (K, holotype ; SRGH).

Closely related to *U. caffra*, but differs in the following characters :
Leaf-lamina 5–7·2 × 1·5–2·5 cm., ovate-lanceolate to elliptic or elliptic-oblong,

acute or subacute at the apex, cuneate to rounded but not decurrent at the base, very sparsely stellate-pubescent below or almost glabrous when mature, with nerves often dark above ; petiole 1–3 mm. long. Flowers extra-axillary ; pedicels c. 2·2–3 cm. long, slender, not thickened upwards (except in fruit), ± sparsely fawn-pubescent ; bracteoles eventually caducous. Petals yellow, 6–10 mm. long. Stamens c. 1·5 mm. long. Fruiting carpels 1–4-seeded, blackish-purple when dried. Seeds oblique, clustered.

S. Rhodesia. S : Ndanga Distr., Lundi R. bank, fl. 23.xi.1957, *Phelps* 207 (K ; SRGH). **Mozambique.** SS : Inhambane, Massinga, Inhachengo, fr. 26.ii.1955, *E.M. & W.* 642 (BM ; LISC ; SRGH).

Known only from the above regions. Semi-evergreen forest and river banks, sea-level to c. 300 m.

8. **Uvaria acuminata** Oliv., F.T.A. **1** : 21 (1868).—Engl. & Diels in Engl., Mon. Afr.
   Pflanz. **6** : 25 (1901).—Sim, For. Fl. Port. E. Afr.: 8 (1909).—Brenan, T.T.C.L.
   2 : 45 (1949). Type : Mozambique (or Tanganyika), R. Rovuma, *Kirk* (K, holotype).
   Uvaria holstii Engl., Pflanzenw. Ost-Afr. **C** : 178 (1895). Type from Tanganyika.
   Uvaria leptocladon var. *holstii* (Engl.) Engl. & Diels, loc. cit.: t. 4C (1901). Type
   as for *U. holstii*.

Shrub, small tree or liane, 2–6 m. high, much-branched. Branches rusty-pilose at first, eventually shortly pubescent or glabrous. Leaves petiolate ; lamina (2) 3–8 (11·5) × 1·3–3·9 cm., oblanceolate or obovate to oblong, obtuse or usually with an acute acumen up to 2 cm. long at the apex, cordate (rarely rounded) at the base, subcoriaceous, deep glossy green or greyish-green and sparsely pubescent with simple (rarely stellate) hairs or glabrous (except along the midrib) above, paler green and sparsely or ± densely stellate-pubescent below (pilose with simple hairs along the nerves), with lateral nerves prominent below and densely reticulate venation prominent on both surfaces ; petiole 1–2·5 (4) mm. long, rusty-pilose. Flowers terminal or extra-axillary, solitary or in cymose pairs, pendulous ; pedicels 0·5–2 cm. long, rusty-pilose ; bracteoles lanceolate, caducous. Sepals 4–6 mm. long, shortly united at the base, covering the petals in bud and separating at anthesis, ovate or ovate-lanceolate, obtuse to acute, fawn- or rusty-pubescent on the outside. Petals pale yellow, thin or slightly fleshy, subequal, 6–10 mm. long, oblong or oblong-lanceolate, acute to obtuse or rounded, puberulous on both surfaces. Stamens c. 1 mm. long, linear or obconic, yellowish ; connective-pro-longation broadened, truncate, puberulous. Carpels pubescent ; stigma orange. Fruit on a pedicel 10–20 mm. long ; ripe carpels few (c. 5–15), 1–2 (3)-seeded, c. 0·9–1·2 × 0·9–1·2 cm., globose, not apiculate, not constricted between the seeds, rusty- or fawn-tomentellous, rugose or verrucose, with stipes 2–3 mm. long. Seeds c. 6–7 mm. long.

Mozambique. N : Fernão Velosa, fl. 15.viii.1948, *Pedro & Pedrógão* 4805 (SRGH). Z : Between Inhamacurra and Maganja da Costa, fl. 25.iii.1943, *Torre* 4986 (BM ; LISC).

Coastal regions from Kenya to Mozambique ; Madagascar. In coastal or dry forest or scrub, thickets, forest margins and clearings, 0–750 m.

U. acuminata can be distinguished from the E. African *U. leptocladon* Oliv. by the absence of short densely tomentose hairs on the underside of the leaves. A plant from that region with whitish-pilose sepals and white flowers is probably a variety of *U. acuminata.*

2. UVARIASTRUM Engl.

Uvariastrum Engl. apud Engl. & Diels in Engl., Mon. Afr. Pflanz. **6** :
31 (1901).

Shrubs or trees, not climbing, glabrous or with simple hairs (and also stellate hairs in *U. modestum* Diels). Buds conic. Flowers bisexual (rarely unisexual), solitary or paired, axillary, sometimes on the old wood. Bracteole 1, usually caducous. Sepals 3, valvate, free, enclosing the petals and with the margins folded to form three ± prominent longitudinal wings in bud, often densely brown-pubescent. Petals 6, in two whorls, both valvate (or the inner one imbricate in *U. hexaloboides* (R.E.Fr.) R.E.Fr.), expanding and spreading at anthesis, free, subequal (the outer whorl usually slightly longer). Stamens ∞, linear, with thecae extrorse

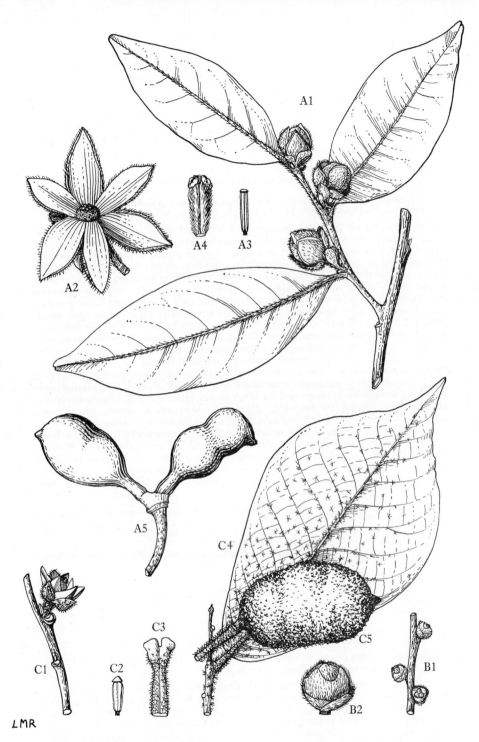

Tab. 9. A.—UVARIASTRUM HEXALOBOIDES. A1, branch with flower-buds (× ⅔) *Milne-Redhead* 3413 ; A2, flower (× ⅔) ; A3, stamen (× 2) ; A4, carpel (× 2), all from *Angus* 919 ; A5, fruit (× ⅔) *Duff & Miller* 300/35. B.—UVARIODENDRON SP. B1, part of branch with flower-buds (× ⅔) ; B2, bud (× 2) ; both from *Mendonça* 2558A. C.—DIELSIOTHAMNUS DIVARICATUS. C1, part of branch with flower (× ⅔) ; C2, stamen (× 2) ; C3, carpel (× 2), all from *Anderson* 798 ; C4, leaf, lower surface (× ⅔) ; C5, fruiting carpel (× ⅔) both from *Kirk* s.n.

and connective-prolongations ± capitate. Carpels 3–10 or sometimes more numerous, free, cylindric or obconic, sometimes ± angular, with numerous ovules in two rows ; style very short or absent ;. stigma ± expanded, bilobed or irregular. Ripe carpels indehiscent, succulent (? or sometimes dry), ellipsoid to cylindric, shortly stipitate, many- to few-seeded. Seeds ± horizontal ; aril absent.

A tropical African genus of 8 species which can be distinguished from *Uvaria* by the axillary flowers and usually also by the valvate corolla and simple indumentum.

Uvariastrum hexaloboides (R.E.Fr.) R.E.Fr. in Ark. Bot., Ser. 2, **3** : 42 (1953). TAB. **9** fig. A. Syntypes : N. Rhodesia, between Abercorn and Kalambo, *R.E.Fries* 1260 (UPS) and 1260a (UPS).
    Uvaria hexaloboides R.E.Fr., Wiss. Ergebn. Schwed. Rhod.-Kongo-Exped. **1** : 44 (1914).—Boutique, F.C.B. **2** : 297 (1951). Syntypes as above.

Small tree, much branched, evergreen, 3·5–13 m. high ; bark flaking. Branches pliant, pendulous, densely-rusty- or fawn-pubescent at first, eventually glabrous, with fluted bark and prominent lenticels. Leaves petiolate ; lamina 5–14 × 2–4 (5) cm., elliptic to oblong or oblanceolate, acuminate (more rarely obtuse) and emarginate at the apex, broadly to narrowly cuneate at the base, subcoriaceous, bright glossy green and glabrous above except for the rusty- or fawn-pubescent midrib, ± matte and sparsely subappressed-pubescent below but densely pubescent along the midrib, usually with slightly prominent venation on both sides ; petiole 2·5–6 mm. long, ± densely rusty- or fawn-pubescent. Flowers solitary or paired, axillary, on both young and old branches ; pedicels almost absent or up to 15 mm. long, appressed-pubescent ; bracteoles ovate-orbicular, cucullate, brown-sericeous outside, glabrous within, enclosing the bud, caducous. Sepals 8–15 mm. long, ovate-triangular or oblong, acute or obtuse, densely brown-sericeous on both sides. Petals yellow or cream, sometimes greenish, 15–30 mm. long, fleshy, subequal, the outer whorl valvate and the inner one imbricate in bud, elliptic or oblong-oblanceolate, obtuse to rounded, grey-silky-tomentellous above, grey-brown-sericeous below. Stamens 1·5–3 mm. long, linear, glabrous ; connective-prolongation capitate. Carpels densely pubescent. Fruit on a pedicel c. 10–20 mm. long ; fruiting carpels 1–2 (8–several-seeded), (2) 3–5 × 2–2·5 cm., cylindric to obovoid, straight or curved, apiculate, not or only slightly constricted between the seeds, glabrous or sparsely pubescent, finely rugose, yellow, becoming blackish-purple and woody when dry, edible, with stipes 0·3–0·8 cm. long. Seeds c. 1·5 cm. long, discoid.

N. Rhodesia. B : Lialui Distr., between Lukulu R. and Kabompo R., fr. 30.xii.1938, *Martin* 924/38 (FHO). N : Between Shiwa Ngandu and Chinsali, fr. 24.ix.1938, *Greenway* 5771 (FHO ; K). W : Mufulira, fr. 6.vi.1934, *Eyles* 8178 (BM ; K ; SRGH). S ? : Luampa-Kafue traverse, st. 16.x.38, *Martin* 880 (FHO).
In N. Rhodesia and the Belgian Congo (Upper Katanga). In *Brachystegia* and mixed woodland, 1200–1450 m.

U. hexaloboides is a typical *Uvariastrum* except for the imbricate inner whorl of petals. Since it cannot be included in *Uvaria*, which always has stellate hairs and terminal or extra-axillary flowers, the most satisfactory solution is to widen the circumscription of *Uvariastrum* to include species in which the inner whorl of petals is imbricate in bud.

3. UVARIODENDRON (Engl. & Diels) R.E.Fr.

Uvariodendron (Engl. & Diels) R.E.Fr. in Act. Hort. Berg. **10** : 51 (1931).
Uvaria sect. *Uvariodendron* Engl. & Diels in Engl., Mon. Afr. Pflanz. **6** : 8 (1901).

Shrubs or trees, not climbing, with simple hairs. Buds depressed-globose. Flowers bisexual, solitary or paired, axillary, sometimes on the old wood. Bracteoles 2–6, usually biseriate, forming an involucre in sessile flowers. Sepals 3, valvate, free or united at the base, not enclosing the petals in bud, with plane margins, densely brown- or golden-sericeous outside, glabrous within. Petals 6, in two whorls, the outer valvate, the inner valvate above and open below, expanding and spreading at anthesis or the inner whorl remaining connivent at the apex, free, subequal, thick. Stamens very numerous, linear, with thecae lateral or extrorse and connective-prolongations capitate. Carpels ∞, cylindric, with numerous ovules in two rows ; style very short ; stigma truncate, horse-shoe-shaped. Ripe carpels indehiscent, ovoid or ellipsoid to cylindric, straight, subsessile, many-seeded. Seeds ± horizontal ; aril absent.

A tropical African genus of about 12 species, differing from *Uvariastrum* in the shape of the buds and in the sepals, which do not enclose the petals in bud. It is closely related to *Polyceratocarpus* Engl. & Diels in which the inner petals are wholly contiguous and the fruits are usually curved.

Uvariodendron sp. TAB. 9 fig. B.

Shrub or small tree ? Branches appressed-golden-pubescent at first, soon glabrous. Leaves petiolate ; lamina 8–13·5 × (2) 3–5 cm., oblong-elliptic, acuminate at the apex, cuneate at the base, undulate along the margin (when dried), coriaceous, bluish-green and glabrous above, paler green and appressed-pilose along the midrib or wholly glabrous below, with c. 12–16 pairs of primary nerves and densely reticulate venation prominent on both sides ; petiole 3–5 mm. long, glabrous. Flowers solitary, axillary, on leafless shoots, sessile or subsessile ; bracteoles 4–5, imbricate, suborbicular or semi-lunar, often slightly keeled, pale golden-sericeous, with marginal fringe. Sepals c. 4 mm. long, united in the lower half in bud to form a 3-lobed cupule, the lobes acute or obtuse, golden-sericeous outside, glabrous within. Petals in bud golden-sericeous outside (or the inner whorl glabrous except along the midrib), glabrous within. Stamens numerous, linear-obconic. Carpels (young) 5, glabrous except for the pilose stigma.

Mozambique. MS : Quedas do Revuè, bud ix.1944, *Mendonça* 2558A (LISC). Margin of riverine forest.

The only record. This species appears to be related to *U. angustifolium* (Engl. & Diels) R.E.Fr. from Ghana and British Cameroons, but the leaves are shorter and broader, the venation more densely reticulate, and the nerves form a wider angle with the midrib. These last two characters also separate it from an undescribed species of coastal districts in Kenya, N. Tanganyika and Zanzibar which, in addition, has pedicellate flowers. It would be unwise to name our species until further material with mature flowers and fruit is available.

4. DIELSIOTHAMNUS R.E.Fr.

Dielsiothamnus R.E.Fr. in Ark. Bot., Ser. 2, **3** : 35 (1953).

Shrub or small tree, sometimes climbing, with stellate and sparse simple hairs. Buds globose. Flowers bisexual (rarely male), solitary, axillary, shortly pedicellate. Bracteoles 2, enclosing the bud. Sepals 3, valvate, enclosing the petals in bud. Petals 6, in 2 whorls, both valvate, free, the outer whorl slightly larger. Stamens ∞, linear-obconic with thecae lateral and connective-prolongations obliquely capitate, apiculate ; filaments short. Carpel solitary, cylindric, with numerous ovules in two rows ; style absent ; stigma bilobed. Ripe carpel indehiscent, sessile or subsessile, many-seeded. Seeds horizontal ; aril absent.

A monotypic genus having affinities with *Uvaria* and *Hexalobus*, and apparently restricted to Tanganyika, Nyasaland and Mozambique.

Dielsiothamnus divaricatus (Diels) R.E.Fr. in Ark. Bot., Ser. 2, **3** : 36 (1953). TAB. 9 fig. C. Type from Tanganyika (Lindi).
    Uvaria divaricata Diels in Notizbl. Bot. Gart. Berl. **13** : 265 (1936).—Brenan, T.T.C.L. : 45 (1949). Type as above.

Shrub or small tree, sometimes scrambling or climbing, up to 4 m. high or higher ; bark flaking. Branches spreading, rather sparsely stellate-tomentose at first, soon glabrous, with prominent and persistent leaf-bases. Leaves petiolate ; lamina 8–17 (24) × 4–9·7 cm., obovate, acuminate at the apex, cordate or rounded at the base, membranous, concolorous, sparsely stellate-pubescent above (at least along the nerves) or sometimes almost glabrous, more densely stellate-pilose below, with venation prominent below but not above ; petiole 3–7 mm. long, stellate-tomentose. Flowers solitary (? or in few-flowered groups), axillary, on leafless shoots ; pedicels c. 5 mm. long, rusty-stellate-pubescent ; bracteoles ovate-orbicular, cucullate, densely brown-pubescent outside, almost glabrous within, enclosing the bud, caducous. Sepals 5–7 mm. long, triangular or orbicular-ovate, acute or apiculate, brown, stellate-velutinous outside, greyish-tomentose within. Petals brownish-yellow, thick, rather fleshy, subequal (6) 8–10 (12) mm. long, triangular-ovate, obtuse or apiculate, shortly grey-stellate-pubescent outside, glabrous within (except sometimes round the margin). Stamens 1·5–2 mm. long, linear or obconic, glabrous. Carpel densely pubescent. Fruit on a pedicel c. 15 mm. long ; fruiting carpel several-seeded, 5 × 3–4 cm. when ripe, cylindric, not constricted between the seeds, not usually apiculate, with scattered stellate hairs, rugose, oily, green when immature, probably ripening to yellowish, sessile or sub-sessile. Seeds c. 2 × 1 cm., bean-like.

Nyasaland. C ; Dedza, Bembeke-Mua Escarpment, fr. 20.i.1959, *Robson* 1275 (K ; LISC ; SRGH). **Mozambique.** N : between Nampula and Lumbo, fl. 30.x.1948, *Barbosa* 1148 (BM ; LISC). Z : Morrumbala, fr. 18.i.1863, *Kirk* (K).
Confined to southern Tanganyika and northern Mozambique. Open woods and rocky places, 200–540 m.

The leaves and fruits of *D. divaricata* have been confused with those of *Monodora grandidieri* Baill., but the stellate indumentum at once distinguishes *Dielsiothamnus* from all other genera in our area except *Uvaria*.

5. HEXALOBUS A. DC.

Hexalobus A. DC. in Mém. Soc. Phys. Hist. Nat. Genève, **5** : 212 (1832).

Shrubs or trees, not climbing, with simple hairs. Buds ovoid-conic. Flowers bisexual, solitary, axillary, sessile or shortly pedicellate. Bracteoles 2 to several, biseriate, caducous. Sepals 3, valvate, free, enclosing the petals in bud, with margins not or scarcely bent outwards, densely brown- or fawn-pubescent. Petals 6, in two whorls, both valvate, transversely plicate in bud, expanding and spreading at anthesis, connate at the base, equal or subequal. Stamens ∞, linear or obconic, with thecae extrorse and connective-prolongation capitate ; filaments short. Carpels (1?) 3–12, free, cylindric-ellipsoid, with numerous biseriate ovules ; style very short ; stigma bilobed with shortly clavate lobes. Ripe carpels indehiscent, succulent, cylindric-ellipsoid to obovoid, usually not or scarcely constricted, sub-sessile, several-seeded. Seeds horizontal ; aril absent.
An African genus of 5–6 species which occurs from Senegal to the Sudan and south to Bechuanaland Protectorate and the Transvaal. Its gamopetalous corolla with lobes plicate in bud provides a good " spot " character.

Leaves obtuse to rounded at the apex, nerves prominent below or on both sides, reticulate
　venation usually prominent above ; bracteoles and sepals dark brown　　1. *monopetalus*
Leaves acute to obtusely acuminate at the apex, nerves not prominent on either side,
　reticulate venation not prominent above ; bracteoles and sepals fawn
　　　　　　　　　　　　　　　　　　　　　　　　　　　　　　　2. *mossambicensis*

1. **Hexalobus monopetalus** (A. Rich.) Engl. & Diels in Engl., Mon. Afr. Pflanz. **6** : 56, t. 20B (1901).—R.E.Fr. in Act. Hort. Berg. **10** : 66 (1931).—Boutique, F.C.B. **2** : 370 (1951). Type from Senegambia.
　　Uvaria monopetala A. Rich. in Guill., Perr. & Rich., Fl. Senegamb. Tent. : 8, t. 2 (1831). Type from Senegambia.
　　Hexalobus senegalensis A. DC. in Mém. Soc. Phys. Hist. Nat. Genève, **5** : 213 (1832).—Benth. in Trans. Linn. Soc. **23** : 468 (1862).—Oliv., F.T.A. **1** : 27 (1868). —Gibbs in Journ. Linn. Soc., Bot. **37** : 428 (1906).—Eyles in Trans. Roy. Soc. S. Afr. **5** : 354 (1916). Type from Senegambia.
　　Uvaria huillensis Engl. & Diels in Notizbl. Bot. Gart. Berl. **2** : 296 (1899). Type from Angola (Mossâmedes).
　　Hexalobus huillensis (Engl. & Diels) Engl. & Diels in Engl., Mon. Afr. Pflanz. **6** : 56 (1901).—Exell & Mendonça, C.F.A. **1**, 1 : 16 (1937). Type as for *Uvaria huillensis*.
　　Hexalobus monopetalus var. *parvifolius* Bak. f., Cat. Talb. Pl. : 5 (1913).—Keay, F.W.T.A. ed. 2, **1** : 48 (1954).—Heine in Mitt. Bot. Staatss. München **18** : 350 (1957). Type from Ubangi.
　　Hexalobus glabrescens Hutch. & Dalz. ex Burtt Davy, F.P.F.T. **1** : 103 (1926).— Hutch. & Dalz. in Kew Bull. **1927** : 152 (1927) ; F.W.T.A. **1** : 52 (1927).— Verdoorn, Edible Wild Fr. Transv. : 15, t. 3 (1938).—Hutch., Botanist in S. Afr. : 317, 455 (1946).—Meeuse in Fl. Pl. Afr. **30** : t. 1195 (1955). Syntypes from the Transvaal.

Shrub or tree, sometimes much branched but not climbing, 2–7 (9) m. high. Branches spreading, ± densely brownish-pubescent or -tomentellous at first, eventually glabrous, with ± prominent and persistent petiole-bases. Leaves

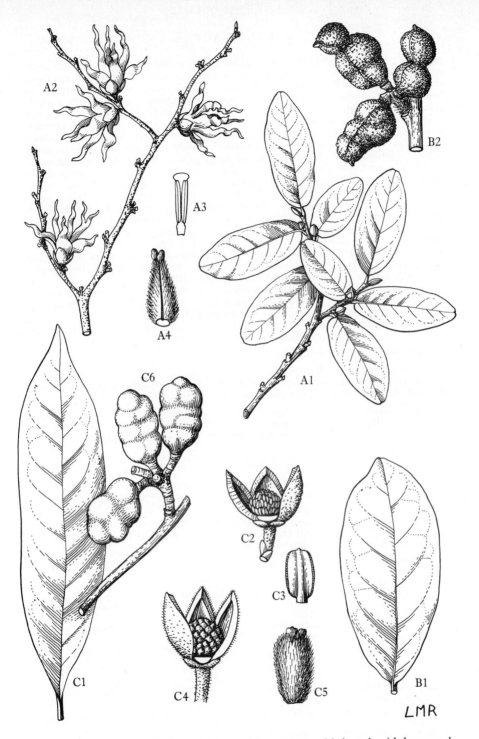

Tab. 10. A.—HEXALOBUS MONOPETALUS VAR. MONOPETALUS. A1, branch with leaves and flower-buds (× ⅔) *Davies* 1248 ; A2, branch with flowers (× ⅔) ; A3, stamen (× 8) ; A4, carpel (× 8). A2–A4 from *Miller* 1989. B.—HEXALOBUS MONOPETALUS VAR. OBOVATUS. B1, leaf (× ⅔) *Milne-Redhead* 557 ; B2, fruit (× ⅔) *Milne-Redhead* 4526. C.—UVARIOPSIS CONGENSIS. C1, leaf (× ⅔) *Dawe* 484 ; C2, male flower, one petal removed (× 2) *Eggeling* 2291 ; C3, stamen (× 14) *Eggeling* 2291 ; C4, female flower, one petal removed (× 2) *Eggeling* 3154 ; C5, carpel (× 10) *Eggeling* 3154 ; C6, fruit (× ⅔) *Dawe* 484.

petiolate ; lamina 3–10 (15) × 1·5–5 (6) cm., elliptic-oblong to obovate, obtuse to rounded or emarginate at the apex, cuneate to rounded or subcordate at the base, chartaceous or subcoriaceous, concolorous, ± glabrescent above and often also below, but persistently appressed-pubescent or tomentellous along the midrib below, with arcuate-ascending nerves slightly prominent on both sides or only below and reticulate venation usually prominent only above ; petiole 1–4 (8) mm. long, pubescent. Flowers solitary, axillary, subsessile, opening after leaf-fall ; bracteoles 2–3, elliptic-orbicular, cucullate, densely brown-appressed-pubescent outside, glabrous and rugose within, the inner pair enclosing the bud, caducous. Sepals 4–6 mm. long, ovate-elliptic, acute or obtuse, cucullate, brown-sericeo-pilose outside, glabrous and rugose within, spreading widely at anthesis. Petals yellow or cream, thin, subequal, the inner whorl narrower, 10–20 mm. long, narrowly lanceolate, corrugated, obtuse, brown-sericeous outside near the base, appressed-puberulous towards the apex and within. Stamens 1–1·5 mm. long, linear or obconic, glabrous ; connective-appendage capitate. Carpels 4–6, densely pubescent. Fruit subsessile ; fruiting carpels 1–3 (4), several-seeded, 2–3·5 × 1–1·5 cm., cylindric-ellipsoid to obovoid, not apiculate, sometimes constricted between the seeds, sparsely brown-puberulous, finely rugose, scarlet, sessile. Seeds 1·2–1·5 × 0·6–0·7 cm., semi-lunar, triquetrous.

Throughout tropical Africa from Senegal to the Sudan and Uganda, and south-ward to SW. Africa, Bechuanaland Protectorate and the Transvaal. Fringing forest, woodland or bush, often among rocks, 480–1650 m.

A variable species which is distinguished from all other species of *Hexalobus*, except *H. mossambicensis*, by the subsessile flowers. Two varieties occur in our area.

Var. monopetalus. TAB. **10** fig. A.

Leaves oblong to elliptic, rarely oblanceolate, usually relatively narrow, glabres-cent or persistently appressed-pubescent below or sometimes tomentellous along the midrib, sometimes without prominent nerves and venation above.

Bechuanaland Prot. SE : Palapye Distr., Moremi, bud 6.iii.1957, *de Beer* T24 (K ; SRGH). **S. Rhodesia.** W : Matopo National Park, bud 14.vii.1949, *West* 2943 (K ; SRGH). S : Gwanda Distr., Diti Special Native Area, bud v.1955, *Davies* 1229 (K ; SRGH).

From Senegal to Fr. Cameroons and Uganda, and also in Angola, south and west S. Rhodesia, SW. Africa, Bechuanaland Protectorate and the Transvaal.

Var. obovatus Brenan in Mem. N.Y. Bot. Gard. **8**, 3 : 214 (1953). TAB. **10** fig. B.
Type : N. Rhodesia, Mwinilunga Distr., E. of Matonchi Farm, *Milne-Redhead* 4536 (BM ; K, holotype).

Leaves obovate to oblong-obovate, usually relatively broad, glabrescent or per-sistently appressed-pubescent (rarely tomentellous) near the midrib only below, always with prominent nerves and venation above.

N. Rhodesia. W : Ndola, Dola Hill, bud 13.viii.1952, *Angus* 197 (FHO ; K). C : Broken Hill, fl. 13.ix.1947, *Brenan & Greenway* 7856 (BM ; FHO). E : 46·4 km. from Fort Jameson towards Petauke, bud 24.v.1952, *White* 2872 (FHO ; K). S : Choma to Pemba, st. 17.viii.1929, *Burtt Davy* 20721 (FHO). **S. Rhodesia.** N : Sebungwe, Kariangwe, fr. 20.xi.1951, *Lovemore* 197 (BM ; FHO ; SRGH). C : Hartley, near Gambiza, fr. vi.1930, *Eyles* 6398 (K ; SRGH). E : Nyumquarara Valley, fr. ii.1935, *Gilliland* 1592 (BM ; FHO). S : Ndanga, bud 25.vii.1955, *Mowbray* 54 (SRGH). **Nyasaland.** N : N. Nyasa Distr., near Ntalire village, fr. 1940, *Lewis* A8/40 (FHO). C : Kota Kota Distr., Chia area, fl. 1.ix.1946, *Brass* 17478 (BM ; K ; SRGH). **Mozambique.** N : Cuamba, st. 13.v.1948, *Pedro & Pedrógão* 3383 (SRGH). T : 54·7 km. from Chicoa towards Fingué, bud 26.vi.1949, *Barbosa & Carvalho* 3302 (SRGH). MS : Chimoio, between Vila Pery and Garuzo, bud 19.x.1944, *Mendonça* 2519 (LISC).

In western Tanganyika, Belgian Congo (Katanga), N. Rhodesia, S. Rhodesia, Nyasa-land, and western Mozambique.

Intermediates between these two varieties occur in regions where their distributions are adjacent or overlap (Angola, S. Rhodesia, Transvaal). Subdivision of *H. monopetalus* according to indumentum characters proves to be quite impracticable.

In the sterile state *H. monopetalus* has frequently been mistaken for a species of *Diospyros*.

2. **Hexalobus mossambicensis** N. Robson in Bol. Soc. Brot., Sér. 2, **32** : 153 (1958).
Type : Mozambique, Nacala, between Fernão Velosa and Quissangulo, *Barbosa* 2428 (BM ; LISC, holotype).

Shrub or tree, 4–5 m. high, much branched. Branches spreading, appressed-pubescent at first, glabrescent, with persistent petiole-bases. Leaves petiolate; lamina 4–7 × 1·6–3·1 cm., elliptic to oblong-elliptic, acute or obtusely acuminate at the apex, ± narrowly cuneate at the base, subcoriaceous, concolorous, appressed-pubescent at first, soon completely glabrous, with midrib impressed above and prominent below, nerves not prominent on either side, with densely reticulate venation scarcely visible below; petiole 2–5 mm. long, pubescent. Flowers solitary, axillary, subsessile; bracteoles densely fawn- or greyish-sericeous outside. Sepals 6–8 mm. long, fawn- or greyish-sericeous outside. Petals yellow. Stamens 2 mm. long. Carpel 1 (? or more), densely pubescent. Fruit subsessile; fruiting carpels unknown. Otherwise similar to H. monopetalus.

Mozambique. N: between Quiterajo and R. Msalo, bud 12.ix.1948, *Pedro & Pedrógão* 5189 (SRGH).
Known from the eastern part of Niassa Province only.

H. mossambicensis differs from *H. monopetalus* by its acuminate or subacuminate leaves, paler buds, somewhat larger sepals and usually more " twiggy " habit. *H. salicifolius* Engl. from French Cameroons has somewhat larger acuminate leaves, but these have prominent nerves beneath and the buds are chocolate-brown and pedicellate, not subsessile.

6. UVARIOPSIS Engl. & Diels

Uvariopsis Engl. & Diels in Notizbl. Bot. Gart. Berl. **2**: 298 (1899), emend. Robyns & Ghesq. in Ann. Soc. Sci. Brux., Sér. B, **53**: 314 (1933).
Tetrastemma Diels apud Winkler in Engl., Bot. Jahrb. **38**: 241 (1906).

Shrubs or trees, not climbing, with simple hairs. Flowers monoecious or dioecious, solitary or in fascicles, axillary, sometimes on the old wood, pedicellate, female larger than male. Bracteoles 2, respectively basal and median, ± persistent. Sepals 2, connate to form a bilobed or patelliform calyx, much shorter than the petals in bud. Petals 4, in one whorl, valvate, expanding and spreading at anthesis, free or ± connate, subequal, thick. Stamens very numerous, oblong to oblong-obovoid, with thecae extrorse; connective-prolongation absent or very short; filaments absent or very short. Carpels numerous, free, obovoid, pubescent, with numerous biseriate ovules; stigma sessile, cylindric or depressed-obconic. Ripe carpels indehiscent, succulent, cylindric or ellipsoid to obovoid, frequently constricted, stipitate, several-seeded. Seeds horizontal; aril absent.

A genus of 11 species in tropical Africa from Sierra Leone to Kenya and south to Angola, Belgian Congo and N. Rhodesia. The tetramerous perianth distinguishes it from the other genera in our area.

Uvariopsis congensis Robyns & Ghesq. in Ann. Soc. Sci. Brux., Sér. B, **53**: 317 (1933).—Boutique, F.C.B. **2**: 381 (1951). TAB. **10** fig. C. Type from the Belgian Congo.

Shrub or small tree, 4–7 (9) m. high. Branches spreading, glabrous even when young. Leaves petiolate; lamina 7–15 × 2–4·5 (6) cm., narrowly elliptic to oblong or oblanceolate, obtuse or acute to acuminate at the apex, cuneate or rarely rounded at the base, membranous, bluish-green, concolorous, glabrous, with relatively loosely reticulate venation prominent on both sides; petiole 2–4 mm. long, glabrous or rarely sparsely pubescent. Flowers monoecious, solitary or rarely paired, axillary; bracteoles 2, semi-orbicular, appressed-pubescent, ± persistent, near base of pedicel. Male flowers on leafy shoots; pedicels 4–5 mm. long, appressed-pubescent; calyx bilobed, sepals 1–1·5 mm. long, semi-orbicular, appressed-pubescent outside; petals yellow or pale orange, free, ± fleshy, subequal, 5–6 mm. long, elliptic, acute, appressed-pubescent outside, glabrous within; stamens 0·5 mm. long, oblong, glabrous, connective not produced. Female flowers on leafy or defoliated shoots or on the old wood; pedicels 6–11 mm. long; sepals c. 2 mm. long, subacute; petals 5–7 (9) mm. long, ovate; carpels numerous, densely pubescent, stigma black when dry, glabrous; otherwise similar to male flowers. Fruit on a pedicel 1–1·5 mm. long; fruiting carpels usually 4–6, several-seeded (1·2) 2·5–3·5 (4·8) × 1·2–1·8 cm., cylindric, not usually apiculate, ± deeply constricted between the seeds, sparsely pubescent or glabrescent, finely rugose, red, black when dry, with stipes 3–6 mm. long. Seeds 4–6, 10–13 mm. long, biseriate.

N. Rhodesia. W : Mwinilunga Distr., W. of Kalene Mission, fr. 22.ix.1952, *White* 3334 (FNO ; K).

In N. Rhodesia, Belgian Congo, Uganda and Kenya. An understorey tree of fringing forest and forest margins.

7. CLEISTOCHLAMYS Oliv.

Cleistochlamys Oliv. in Journ. Linn. Soc., Bot. **9** : 175 (1865).

Shrub or small tree, erect or straggling, with simple hairs. Buds globose. Flowers bisexual, solitary, axillary, sessile. Bracteoles 4–5, biseriate, persistent. Calyx entire and completely enclosing the petals in bud, splitting into 3 (rarely 2) $\pm$ equal sepals, glabrous. Petals 6, in two whorls the outer valvate, the inner imbricate and somewhat shorter, expanding and spreading at anthesis, free. Stamens c. 30–40, cuneate-quadrate, with thecae extrorse, divergent and connective-prolongation broadly and obliquely capitate ; filaments absent or almost so. Carpels 6–10, free, ovoid, with a single basal ovule ; style cylindric, oblique, almost as long as the ovary ; stigma capitate. Ripe carpels indehiscent, succulent, cylindric, stipitate, 1-seeded. Seed vertical ; aril absent.

A monotypic genus near *Popowia* and *Ophrypetalum*, but distinguished from both by the entire calyx. It is confined to Mozambique and the adjacent regions of Tanganyika, N. and S. Rhodesia and Nyasaland.

Cleistochlamys kirkii (Benth.) Oliv. in Journ. Linn. Soc., Bot. **9** : 175 (1865) ; F.T.A. **1** : 24 (1868).—Engl. & Diels in Engl., Mon. Afr. Pflanz. **6** : 36, t. 13 B (1901).— Sim, For. Fl. Port. E. Afr. : 8 (1909).—Burtt Davy & Hoyle, N.C.L. : 30 (1936).— Brenan, T.T.C.L. : 41 (1949). TAB. **11** fig. A. Syntypes : Mozambique, opposite Sena, *Kirk* (K) ; foot of Morrumbala, *Kirk* (K).

Popowia ? kirkii Benth. in Trans. Linn. Soc. **23** : 470 (1862). Syntypes as above.

Shrub or small tree (2·4) 3–7·5 (9) m. high, much-branched, sometimes straggling ; bark pale, flaking. Branches spreading, glabrous, pale at first, becoming dark brown with prominent lenticels. Leaves petiolate ; lamina (4·5) 6–11 (13) × (1) 2–3·5 (6·5) cm., narrowly oblong to oblanceolate or obovate, obtuse to rounded or emarginate at the apex, cuneate to rounded at the base, membranous, bright green, concolorous or slightly paler below, glabrous above, very sparsely appressed-pubescent below, with pinnate nervation and reticulate venation prominent on both sides ; petiole 2–5 mm. long, glabrous. Flowers solitary, axillary, sessile, opening after leaf-fall ; bracteoles 4–5, biseriate, imbricate, cucullate, increasing in size towards the flower, reddish-brown, glabrous or fringed with short hairs, persistent. Calyx scarious-coriaceous, completely investing the petals in bud, splitting at anthesis into 3 (rarely 2) $\pm$ equal cucullate, broadly triangular sepals, c. 3 mm. long, glabrous, yellow- or reddish-brown, finely rugose. Petals cream or white, coriaceous, the outer whorl 5–6 mm. long, the inner one somewhat shorter, obovate to oblong, rounded, densely white-sericeo-pilose outside, glabrous within or sericeous-pilose near the apex. Stamens c. 0·75 mm. long, cuneate-quadrate, glabrous ; connective-prolongation broadly and obliquely capitate. Carpels 1·5 mm. long, glabrous or subglabrous. Fruit subsessile ; fruiting carpels (1) 3–10, 1·3–2·3 × 0·7–1 cm., cylindric, often apiculate, glabrous, finely rugose, purplish-black when ripe, edible, with stipes 4–5·5 mm. long. Seeds c. 1–2 cm. long, cylindric.

N. Rhodesia. C : Feira Distr., near Feira Boma, st. 29.v.1952, *White* 2898 (FHO ; K). E : Petauke Distr., Luangwa R. above Beit Bridge, st. 5.ix.1947, *Brenan* 7802 (BM ; FHO ; K). **S. Rhodesia.** N : Urungwe, Kariba Gorge, fl. 15.x.1954, *Lovemore* 405 (K ; SRGH). E : Chipinga, fr. 19.i.1957, *Phipps* 27 (BM ; K ; SRGH). S : Sabi-Lundi Junction, Chitsa's Kraal, st. 5.vi.1950, *Chase* 2309 (BM ; K ; SRGH). **Nyasaland.** S : Fort Johnston Distr., Namingundi R., st. 21.vi.1954, *Jackson* 1321 (BM ; FHO ; K); **Mozambique.** N : Nampula, R. Mutivasi, fr. 3.xi.1942, *Mendonça* 1216 (BM ; LISC). Z : Namagoa, Mocuba, fl. ix.1944, *Faulkner* 225 (K ; SRGH). T : Baroma Prov., Sisitso, R. Zambezi, st. 16.vii.1950, *Chase* 2744 (BM ; SRGH). MS : Mossurize, R. Save, Massangena, st. 11.vi.1942, *Torre* 4306 (BM ; LISC). SS : Macovane, st. 1.vi.1947, *Hornby* 2719 (K ; PRE ; SRGH).

Also in south-east Tanganyika. Bush and thickets in dry valleys, often on alluvium, 60–900 m.

C. kirkii is easily distinguished by the solitary, globose flower-buds completely covered by the calyx and sessile in the leaf axils ; and by the flowers on leafless branches, with petals entirely different from those of *Hexalobus* species.

Tab. 11. A.—CLEISTOCHLAMYS KIRKII. A1, branch with leaves and flower-buds (× ⅔)
White 2898 ; A2, branch with flowers (× ⅔) ; A3, flower, opened out (× 2) ; A4,
stamen side and front views (× 8) ; A5, carpel (× 8), all from *Faulkner* 225 ; A6,
fruit (× ⅔) *Phipps* 27. B.—ENNEASTEMON SCHWEINFURTHII. B1, branch with leaves,
flower-buds and flower (× ⅔) ; B2, flower opened out (× 2) ; B3, stamen front and side
views (× 8) ; B4, carpel (× 8), all from *Milne-Redhead* 3044 ;B5, fruit (× ⅔) *Fanshawe*
1474.

8. POPOWIA Endl.

Popowia Endl. in Walp., Repert. **1** : 74 (1842).

Shrubs or small trees, often climbing or scrambling, with simple (very rarely also forked or stellate) hairs. Leaves rarely bearing 2 small basal glands. Flower-buds broadly ovoid or depressed-globose. Flowers bisexual or more rarely unisexual, solitary or in 2- to many-flowered cymes, extra-axillary or occasionally axillary, sometimes on leafless branches, pedicellate. Bracteole 1, persistent, on the lower half of the pedicel, sometimes foliaceous. Sepals 3, valvate, free or ± united, much shorter than the petals in bud, usually pubescent. Petals 6, in two ± unequal whorls, both valvate or rarely the inner imbricate, thick, free, the outer ± spreading at anthesis, the inner smaller, concave, often somewhat connivent and erect, usually shortly clawed. Stamens 8 to few (6-8), cuneate-quadrate (rarely linear) or flattened and staminodial, with thecae extrorse, ± parallel, and connective broadened, with or without an obliquely capitate prolongation ; filaments usually present, sometimes longer than the anthers. Carpels rather numerous to few (6), free ; ovary clyindric or ellipsoid to obconic, with 1-4 (5) uniseriate ovules ; style as long as the ovary or shorter, cylindric or obconic, oblique ; stigma sub-capitate or tapering, often bifurcating. Ripe carpels indehiscent, succulent, cylindric-torulose, stipitate, 1-4-seeded. Seeds vertical, serial ; aril absent.

A palaeotropical genus of c. 60 species, most frequent in Africa and Madagascar but also ranging from S. India and Hainan to New Guinea and Queensland.

Inflorescence terminal or axillary ; anther-connective narrow, thecae elongate ; bracteole small, c. 0·5 mm. long ; fruiting carpels with segments 1-2 cm. long, glabrous
                                                                                              1. *gracilis*
Inflorescence extra-axillary (rarely terminal in *P. obovata*) ; anther-connective broad, thecae short :
  Bracteole foliaceous, (5) 10-25 mm. long ; fruiting carpels with segments 1·3-2 cm. long, puberulous - - - - - - - - - - - 2. *obovata*
  Bracteole small, 0·5-2 mm. long ; fruiting carpels with segments 0·5-1·1 cm. long, glabrous or pubescent :
    Stems red to black when mature ; leaves mostly obovate or oblanceolate, with reticulation prominent on both sides ; fruiting carpels 1-4-seeded :
      Fruiting carpels glabrous, straight ; young shoots hirsute or ± appressed-pubescent :
        Leaves 7-18·5 cm. long and (3) 3·5-7 cm. wide ; young shoots ± appressed-pubescent ; outer petals 8-9 mm. long - - - - - 3. *chasei*
        Leaves 2·5-7·5 cm. long, if longer then usually not wider than 3 cm. ; young shoots hirsute or ± appressed-pubescent ; outer petals 2·5-8 mm. long :
          Leaves cuneate to rounded (rarely subcordate) at the base, obtuse to acute or subacuminate (rarely rounded or emarginate) at the apex ; fruiting carpels 1 (2)-seeded ; young shoots often ± appressed-pubescent and glabrescent, occasionally ± zig-zag - - - - - - - 4. *caffra*
          Leaves usually cordate at the base, usually ± emarginate at the apex ; fruiting carpels 1-4-seeded ; young shoots always ± persistently and densely hirsute, usually markedly zig-zag - - - - - 5. *oliverana*
      Fruiting carpels hispid-pilose, often curved ; young shoots densely hirsute
                                                                                              6. *trichocarpa*
    Stems fawn (rarely reddish-brown) when mature ; leaves mostly oblong or elliptic, with reticulation ± prominent above only ; fruiting carpels 1 (2)-seeded
                                                                                              7. *buchananii*

1. **Popowia gracilis** Oliv. ex Engl. & Diels in Engl., Mon. Afr. Pflanz. **6** : 48 (1901).— Brenan, T.T.C.L.: 44 (1949).—N. Robson in Bol. Soc. Brot., Sér. 2, **32**: 154 (1958). Type from Tanganyika (Dar es Salaam).

Shrub, small tree or liane, 3-12 m. high. Branches yellowish-brown, turning blackish, glabrous or glabrescent, with spreading or ascending axillary or supra-axillary shoots and ± prominent petiole-bases. Leaves petiolate ; lamina (3) 5-12 × (1·3) 2-5 (5·8) cm., oblong or elliptic to ovate or obovate, obtuse to shortly acuminate (more rarely rounded or acute) at the apex, cuneate at the base, coria-ceous, dark- or bluish-green and glabrous above, glaucous and glabrous or sparsely appressed-pubescent below, with nerves and loosely reticulate venation ± prom-inent on both sides or the venation prominent above only, glandular ; petiole 3-7 (8) mm. long, glabrous. Flowers bisexual, solitary or in cymose pairs, terminal or axillary, pendulous ; pedicel 0·4-1·5 cm. long, broadening upward, glabrous or ±

densely appressed-pubescent; bracteole small, c. 0·5 mm. long, triangular or semi-orbicular, rusty-pubescent or glabrous, below the middle of the pedicel. Sepals 1·5–2·5 mm. long, ovate-triangular, obtuse to rounded or apiculate, free or ± connate at the base, appressed-pubescent or glabrous. Petals greenish- to dark-yellow, thick, the outer ones valvate, 5–14 mm. long, ovate or elliptic, acute or obtuse, ± densely sericeous-puberulous except near the base within, the inner ones imbricate (? or sometimes valvate), about half as long as the outer, obovate, obtuse, appressed-pubescent outside, glabrous within. Stamens ∞, 0·75–1·5 mm. long, linear, glabrous; connective prolongation obliquely capitate; filament not distinct. Carpels c. 18–30, c. 2 mm. long, curved-cylindric; ovary appressed-pubescent or glabrous, 1–2-ovulate; style equal to the ovary or sometimes shorter, cylindric or flattened, usually bifurcating, red, glabrous. Fruit pendulous on a pedicel 0·8–1·5 cm. long; fruiting carpels 2–19 (23), 1–2-seeded, segments (1·1) 1·3–2 × 0·6–1 cm., ellipsoid or cylindric, sometimes apiculate, glabrous, finely rugose, crimson, glaucous, with stipes 5–14 mm. long. Seeds c. 1–1·8 cm. long.

From Angola to Tanganyika and Mozambique. Dry evergreen forest, open woodland and bush, often on sand, 0–c. 1300 m.

P. gracilis and *P. englerana* Exell & Mendonça are very closely related, differing mainly in number and size of parts and in degree of indumentum. Although these taxa have distinct geographical distributions, it is preferable to treat them as subspecies because some Tanganyika material has intermediate characters.

Subsp. gracilis

Young shoots, pedicels, sepals, ovaries and sometimes the undersides of the leaves appressed-pubescent. Leaves oblong or elliptic to obovate, usually without prominent reticulate venation beneath. Pedicels ± slender.

Mozambique. N : Macomia, between Ingoane and Quiterajo, fl. 19.ix.1948, *Barbosa* 2083 (BM ; LISC). Z : Maganja da Costa, Régulo Muza, fl. 29.ix.1949, *Barbosa & Carvalho* 4248 (SRGH). MS : Beira, near Dondo, fl. 22.ix.1943, *Torre* 5917 (BM ; LISC). SS : Macia, Licilo, st. 8.vii.1947, *Pedro & Pedrógão* 1348 (K ; PRE ; SRGH). LM : Lourenço Marques fr. 1908, *Sim* 21115 (PRE).
From eastern Tanganyika to Lourenço Marques.

Subsp. englerana (Exell & Mendonça) N. Robson in Bol. Soc. Brot., Sér. 2, **32** : 154 (1958). Type from Angola (Pungo Andongo).
   Unona? sp.—Oliv., F.T.A. **1** : 36 (1868).
   Popowia macrocarpa Engl. ex. Engl. & Diels in Engl., Mon. Afr. Pflanz. **6** : 48 (1901) non Baill. (1868). Type as for *P. gracilis* subsp. *englerana*.
   Popowia englerana Exell & Mendonça, C.F.A. **1**, 1 : 23 (1937) ; op. cit. **2**, 1 : XIV (1954).—Boutique, F.C.B. **2** : 350 (1951). Type as for *P. gracilis* subsp. *englerana*.

Young shoots, pedicels, ovaries, the undersides of the leaves, and frequently the sepals, glabrous. Leaves oblong or elliptic to ovate, rarely obovate, with prominent reticulate venation on both sides. Pedicels stouter and longer, stamens and carpels more numerous, and leaves, floral parts and fruits tending to be larger than in subsp. *gracilis*.

N. Rhodesia. N : Fort Rosebery, Samfya Mission, L. Bangweulu, fl. 21.viii.1952, *White* 3108 (FHO ; K). W : Ndola, fr. 21.xi.1954, *Fanshawe* 1674 (K ; SRGH).
In Angola (Pungo Andongo), N. Rhodesia and the Belgian Congo (Katanga).

2. **Popowia obovata** (Benth.) Engl. & Diels in Engl., Mon. Afr. Pflanz. **6** : 44, t. 17B (1901).—Bak. f. in Journ. Linn. Soc., Bot. **40** : 18 (1911).—R.E.Fr., Wiss. Ergebn. Schwed. Rhod.-Kongo-Exped. **1** : 45 (1914).—Eyles in Trans. Roy. Soc. S.Afr. **5** : 354 (1916).—Burtt Davy & Hoyle, N.C.L. : 30 (1937).—Exell & Mendonça, C.F.A. : **1**, 1 : 24 (1937).—Hutch., Botanist in S. Afr. : 461 (1946).—Brenan, T.T.C.L. **2** : 44 (1949).—Boutique, F.C.B. **2** : 345 (1951).—O. B. Mill. in Journ. S. Afr. Bot. **18** : 14 (1952).—Pardy in Rhod. Agr. Journ. **53** : 434, cum tab. (1956). TAB. 12 fig. A. Type : Mozambique, foot of Morrumbala, *Kirk* (K).
   Unona obovata Benth. in Trans. Linn. Soc. **23** : 469 (1862).—Oliv. F.T.A. **1** : 35 (1868).—Sim., For. Fl. Port. E. Afr. : 8 (1909). Type as for *Popowia obovata*.
   Popowia stormsii De Wild. in Ann. Mus. Cong., Bot., Sér. 5, **1** : 242 (1906).—Diels in Engl., Bot. Jahrb. **39** : 478 (1907). Type from Tanganyika (Karema).

Shrub or small tree 1–5 m. high, sometimes scrambling or subscandent. Branches spreading or drooping, grey-brown or yellowish, yellowish-tomentellous at first, eventually glabrous.   Leaves petiolate ;   lamina (4·7) 6–14 (20) × 3·4–9·5 (11) cm., obovate to obovate-oblong, broadly obtuse to rounded or truncate-emarginate at the apex, rounded to cordate (more rarely broadly cuneate) at the base, chartaceous, bright green above, paler and glaucous below, sparsely crisped-pubescent on both sides or glabrescent above, hairs simple and (occasionally) forked or stellate, with nerves and loosely reticulate venation prominent on both sides and a gland frequently terminating each basal lobe ; petioles 3–9 mm. long. Flowers bisexual, solitary, terminal or usually extra-axillary ; pedicels (1) 2·5–5 (6) cm. long, ± slender, tomentellous ; bracteole large (0·5) 1–2·5 cm. long, foliaceous, orbi-cular to broadly ovate, obtuse or rounded at the apex, cordate-amplexicaul at the base, puberulous, near the base of the pedicel. Sepals 4–6 mm. long, broadly triangular to orbicular or rarely oblong, puberulous to tomentellous. Petals yellow or greenish-yellow, thick, (4) 6–12 mm. long, the outer ones broadly ovate or orbi-cular to reniform, puberulous-tomentellous but glabrous near the base within, the inner ones rather shorter, thicker, obovate, obtuse to rounded, unguiculate or attenuate, connivent at the apex at first, remaining curved over the stamens, tomentellous except near the base within. Stamens ∞, 0·75–1 mm. long, oblong or cuneate, glabrous ; connective prolongation broadly and obliquely capitate-truncate ; filament not distinct. Carpels ∞, c. 2 mm. long, ± cylindric ; ovary pubescent, 4-ovulate ; style about a third as long at the ovary, elliptic or obconic, oblique. Fruit pendulous, on a pedicel 3–6 cm. long ; fruiting carpels 3–9 (13), 1–3 (4)-seeded segments 1·3–2 × 0·7–1 cm., ellipsoid to cylindric, often apiculate, ± sparsely puberulous, finely rugose, red, edible, with stipes 0·7–1·5 cm. long. Seeds (1) 1·2–1·8 cm. long, cylindric.

Bechuanaland Prot. N : Kasane to Chobe R., fr. vii.1930, *van Son* in Herb. Mus. Transv. 28766 (BM ; K ; PRE). **N. Rhodesia.** B : Sesheke, Lonzi Forest, fl. 19.xii.1952, *Angus* 945 (FHO ; K). N : L. Mweru, fl. 13.xi.1957, *Fanshawe* 3949 (K). W : Kitwe, fr. 14.ii.1954, *Fanshawe* 815 (FHO ; K ; SRGH). C : 25 km. SW. of Lusaka towards Kafue Flats, fr. 17.vi.1956, *Angus* 1342 (FHO). E? : between Lunsemfwa R. and Nyimba R., st. 25.viii.1929, *Burtt Davy* 20917 (FHO). S : Livingstone Distr., near Knife Edge, Victoria Falls, fr. 28.viii.1947, *Brenan* 7771 (FHO ; K). **S. Rhodesia.** N : Lomagundi, fr. vii.1921, *Eyles* 3134 (K ; SRGH). W : Victoria Falls, fl. 9.ii.1912, *Rogers* 5618 (BM ; K ; PRE ; SRGH). C. Hartley, Umfuli R. at Poole, fr. 25.vii.1943, *Hornby* 2254 (K ; PRE). E : Melsetter, Umvumvumvu Valley, fl. 18.xii.1948, *Chase* 1252 (BM ; K ; SRGH). S : Nuanetsi, Rhino Hotel, Lundi R., fl. xii.1955, *Davies* 1743 (K ; SRGH). **Nyasaland.** N : Nyika Plateau, Mwanemba, fl. & fr. ii–iii.1903, *McClounie* 148 (K). C : Dedza Distr., Mua-Livulezi Forest Reserve, Nam-kokwe R., fr. 19.iii.1955, *E.M. & W.* 1058 (BM ; SRGH). S : Fort Johnston Distr., fr. 20.v.1954, *Jackson* 1315 (FHO ; K). **Mozambique.** N : between Meconta and Nam-pula, fl. 22.ix.1936, *Torre* 1119 (COI ; LISC). Z : Hot Spring, foot of Morrumbala, fl. 31.xii.1858, *Kirk* (K). T : between Chiôco and Tete, st. 27.ix.1942, *Mendonça* 464 (LISC). MS : Chimoio, Bandula, near Chibata, fl. 8.i.1948, *Barbosa* 823 (BM ; LISC).
     Also in Tanganyika, Belgian Congo (Upper Katanga) and Angola (Cubango). Open woodland, thickets, kopjes and near rivers, often on granitic soils, (120) 420–1200 m.

P. obovata is a species characteristic of our area, easily recognized by the foliar bracteole.

3. **Popowia chasei** N. Robson in Bol. Soc. Brot., Sér. 2, **32**: 155 (1958). Type: S. Rhodesia, Umtali, SE. Commonage, *Chase* 5375 (BM ; K, holotype ; SRGH).

Shrub or liane, 3–6 m. high (or higher?). Branches sarmentose, blackish, rusty-or yellowish-subappressed-pubescent at first, eventually glabrous. Leaves petio-late ; lamina 7–18·5 × (3) 3·5–7 cm., obovate or oblanceolate, obtuse or shortly acuminate at the apex, cordate or more rarely rounded at the base, chartaceous, dark green and glabrescent above, glaucous and sparsely subappressed-pubescent below especially on the golden or reddish nerves with simple hairs, and venation prominently reticulate on both sides, eglandular ; petiole 3–7 mm. long, appressed-pubescent. Flowers bisexual, solitary or in 2–5-flowered cymes, extra-axillary ; pedicels 1·5–2·5 (3·5) cm. long, slender, subappressed-pubescent ; bracteole small, 1–2 mm. long, ovate-triangular, subappressed-pubescent, ± median or near the base of the pedicel. Sepals c. 3 mm. long, broadly ovate-triangular, obtuse, ± connate at the base, golden-tomentellous. Petals yellowish, thick, the outer ones c. 8–9 mm. long, ovate-orbicular, obtuse or shortly acuminate, densely golden-tomentellous but glabrous near the base within, the inner ones shorter, narrower, oblong to obovate, obtuse or subacute, golden-tomentellous except at the base within. Stamens and staminodes c. 20–30, 1–1·5 mm. long, glabrous ; connective-prolongation very short or absent ; filament ± equal to the anther. Carpels 15–18, c. 2 mm. long, cylindric-obconic ; ovary glabrous, 4-ovulate ; style c. ⅓ as long as the ovary, cylindric. Fruit on a pedicel 1–3 cm. long ; fruiting carpels 2–7, 1–4-seeded, segments 0·7–1 × 0·5–0·6 cm., orbicular to ellipsoid, apiculate, glabrous, finely rugose, scarlet, with stipes 2–5 mm. long. Seeds c. 6–9 mm. long.

S. Rhodesia. E : Umtali, SE. Commonage, fl. 26.xii.1954, *Chase* 5375 (BM ; K ; SRGH). S : Nyoni Hills, S. of Chibi, fr. 13.iii.1956, *Furness* 30/56 (K ; SRGH). **Nyasaland.** S : Mwanza, W. of Shire, st., *Topham* 705 (FHO). **Mozambique.** MS : Chimoio, Gondola, Serra Nharo-Nharo, fr. 13.ii.1948, *Garcia* 191 (BM ; LISC).
In S. Rhodesia, Nyasaland and Mozambique. Evergreen forest, c. 900 m.

P. chasei is closely related to *P. discolor* Diels (S. Tanganyika) and *P. ferruginea* (Oliv.) Engl. & Diels (Angola, Belgian Congo and Uganda), but differs in indumentum characters and in the frequently cymose inflorescence. It is further distinguished from *P. discolor* by the glabrous ovaries and larger flowers with more numerous stamens and carpels, and from *P. ferruginea* by the flower colour and the markedly unequal petals.

4. **Popowia caffra** (Sond.) Benth. in Trans. Linn. Soc. **23** : 470 (1862).—Baill., Adansonia, **8** : 319 (1868) ; Hist. Pl. **1** : 219, fig. 251 (1868).—Engl. & Diels in Engl., Mon. Afr. Pflanz. **6** : 46 (1901).—Sim, For. Fl. Port. E. Afr. : 8 (1909). TAB. **12** fig. B. Syntypes from Natal.
Guatteria caffra Sond. in Harv. & Sond., F. C. **1** : 9 (1860). Syntypes as for *Popowia caffra*.
Popowia buchananii (Engl.) Engl. & Diels, tom. cit. : 46 (1901) pro parte quoad specim. mossamb. et t. 18 fig. B. excl. fr.

Shrub or liane. Branches sarmentose, red or blackish with prominent lenticels, rusty-subappressed-pubescent at first, glabrescent, rarely zig-zag, with axillary shoots usually spreading or ascending. Leaves petiolate ; lamina 3·5–8 (10) × (1·2) 1·5–3·4 cm., oblanceolate to oblong or obovate, acute to obtuse or rounded (rarely shortly acuminate or emarginate) at the apex, broadly cuneate to subcordate at the base, chartaceous, dark green and glabrescent above, glaucous and ± sparsely subappressed-pubescent especially on the nerves (more rarely glabrescent) below, with venation prominently reticulate on both sides, eglandular ; petiole 1–6 mm. long, rusty-pilose or glabrescent. Flowers bisexual, solitary or in 2–4-flowered cymes, extra-axillary ; peduncle c. 1–2 mm. long ; pedicels 10–20 mm. long, slender, pubescent or almost glabrous ; bracteole small, c. 1 mm. long, ovate-oblong or squamiform, pubescent. Sepals 1–2 mm. long, broadly ovate, obtuse or apiculate, ± connate at the base, rusty-pubescent. Petals cream or pale yellow, rather thick, the outer ones (2·5) 4–6·5 mm. long, ovate, obtuse or shortly apiculate to rounded, golden-tomentellous or -puberulous except near the base within, the inner ones shorter and narrower, obovate or unguiculate, obtuse, golden-tomentellous except near the base within. Stamens and staminodes c. 20, 1–1·5 mm. long, glabrous ; connective-prolongation absent ; filament about as long as the anther. Carpels c. 14–18, 2–3 mm. long, 1 (2)-ovulate, glabrous except for a tuft of hairs at the base, ovoid-conic ; style ± equal to the ovary, cylindric. Fruit on a pedicel 1–2 cm. long ; fruiting carpels 3–8 (15), 1 (2)-seeded, segments 0·6–0·8 × 0·5–0·6 cm., suborbicular or ellipsoid, usually apiculate, glabrous, finely rugose, scarlet, with stipes 2–5 mm. long. Seeds c. 5–7 mm. long.

Mozambique. SS : between Vila de João Belo and Lumane, fr. 6.iii.1941, *Torre* 2635 (BM). LM : Delagoa Bay, fl. 6.i.1898, *Schlechter* 12006 (BM ; COI ; E ; K).
In Mozambique, E. Transvaal, Natal, Basutoland and E. Cape Province. Evergreen forest and bush, 0–600 m.

Mozambique specimens of *P. caffra* tend to have smaller leaves and flowers and a sparser indumentum than those from further south.
Hooker f. and Thomson (Fl. Ind. **1** : 105 (1855)) referred to this species as probably belonging to *Popowia* ; but they neither described it nor gave it a specific name, and therefore the binomial *Popowia caffra* cannot be attributed to them.

5. **Popowia oliverana** Exell & Mendonça, C.F.A. **1**, 1 : 24 (1937); op. cit. **2**, 1 : XIV (1954).—Boutique, F.C.B. **2** : 353 (1951). Type from Angola (Golungo Alto).
Unona parvifolia Oliv., F.T.A. **1** : 36 (1868). Type as for *P. oliverana*.
Popowia parvifolia (Oliv.) Engl. & Diels in Engl. Mon. Afr. Pflanz. **6** : 46 (1901) non Kurz (1875) nec Scheff. (1881).—R.E.Fr., Wiss. Ergebn. Schwed. Rhod.-Kongo-Exped. **1** : 45 (1914). Type as for *P. oliverana*.

Shrub or liane, 3–6 m. high. Branches sarmentose, red or blackish, rusty-pubescent at first, eventually glabrous, frequently ± zig-zag, with axillary shoots reflexed. Leaves petiolate ; lamina 2·5–7·5 × 1·5–3·5 (4·2) cm., oblanceolate or oblong to obovate or elliptic, rounded or usually emarginate at the apex, cordate or rounded at the base, chartaceous, dark green and glabrescent above, ± glaucous and pubescent especially on the nerves below, with venation prominently reticulate on both sides, eglandular ; petiole 2–3 (5) mm. long, rusty-pubescent. Flowers bisexual, solitary or rarely paired, extra-axillary ; pedicels 0·3–1·2 cm. long, slender, rusty-pubescent ; bracteole small, 1–2 mm. long, linear-lanceolate, pubescent, near the base of the pedicel. Sepals 2–3 (4) mm. long, broadly ovate, obtuse or rounded, ± connate at the base, rusty appressed-pubescent. Petals yellow, rather thick, the outer ones (4) 6–8 mm. long, broadly ovate, obtuse, golden-tomentellous except near the base within, the inner ones shorter and narrower, elliptic, obtuse, golden-tomentellous except near the base within. Stamens and staminodes 22–24, 1–1·5 mm. long, glabrous ; connective-prolongation broadly and obliquely trun-cate ; about half as long as the anther. Carpels 11–17 (26), 1·5–2 mm. long, cylindric ; ovary glabrous 1–3 (5)-ovulate ; style about a third as long as the ovary, cylindric. Fruit on a pedicel 1–1·5 cm. long ; fruiting carpels 3–5 (11), 1–3 (5)-seeded, segments 0·5–0·8 (1) × 0·4–0·5 cm., ellipsoid, apiculate or obtuse, glabrous, finely rugose, orange-red, with stipes 2–5 mm. long. Seeds c. 4–8 mm. long.

N. Rhodesia. N : Luapula R., st. 6.ix.1911, *Fries* 558 (K). W : Mwinilunga, Kalene Hill Mission, fl. 27.ix.1952, *White* 3391 (FHO ; K).
In Angola, N. Rhodesia, Belgian Congo and Tanganyika (Bukoba). Margins of wood-land and fringing forest, 40–1500 m.

P. oliverana is closely related to *P. caffra*, but can usually be recognized by its cordate leaf-bases.

6. **Popowia trichocarpa** Engl. & Diels in Engl., Mon. Afr. Pflanz. **6** : 47, t. 18 A (1901).
　—Brenan, T.T.C.L. : 44 (1949). Type from Tanganyika (Uzaramo).
　　Unona ferruginea forma.—Engl., Pflanzenw. Ost-Afr. **C** : 178 (1895). Type as for *P. trichocarpa*.
　　Popowia ferruginea sensu Bak. f. in Journ. Linn. Soc., Bot. **40** : 18 (1911).

Shrub (climbing or straggling) or liane. Branches red to purplish-black, densely rusty-hirsute, not zig-zag, with axillary shoots spreading or ascending. Leaves petiolate ; lamina (5) 6–14·2 (19·3) × 2·2–6 (6·9) cm., obovate to oblanceolate-oblong, acute to rounded or emarginate (sometimes shortly mucronate) at the apex, cordate at the base, chartaceous or subcoriaceous, dark green and glabrescent above, glaucous and sparsely pilose below especially on the nerves, with venation prom-inently reticulate on both sides, eglandular ; petiole 2–5 mm. long, rusty-hirsute. Flowers bisexual, solitary or rarely in cymose pairs, extra-axillary ; pedicels (0·6) 1–2·5 cm. long, very slender, hirsute ; bracteole small, 1–2 mm. long, ovate-oblong, hirsute, near the base of the pedicel. Sepals 1·5–3 mm. long, broadly ovate, rounded to obtuse or acute, free or ± connate at the base, rusty-hirsute. Petals yellow or greenish, rather thick, the outer ones 6–12 mm. long, ovate, obtuse to rounded or apiculate, rusty- or golden-tomentellous except near the base within, the inner ones shorter and narrower, obovate to elliptic, obtuse, golden-tomen-tellous except near the base within. Stamens (and staminodes?) c. 18–36, c. 1 mm. long, glabrous ; connective-prolongation broadly and ± obliquely truncate ; fila-ment ± as long as the anther. Carpels 6–12, c. 2 mm. long, ovoid-cylindric ; ovary densely rusty-pilose, 1–4-ovulate ; style ± equal to the ovary, cylindric-obconic, capitate. Fruit on a pedicel 1·8–3 cm. long ; fruiting carpels 2–8 (12), 1–4-seeded, curved, segments 0·9–1·1 × 0·5–0·6 cm., ovoid or cylindric, apiculate, rusty-hirsute, finely rugose, pinkish to orange-yellow, with stipes 3–5 mm. long. Seeds 8–9 mm. long.

Mozambique. Z : Namagoa, Mocuba, fl. x–xi.1944, *Faulkner* 80 (BM ; K ; PRE). MS : Chimoio, Gondola, R. Nhamissanguere, fr. 19.ii.1948, *Garcia* 295 (LISC).
In Kenya, Tanganyika, Zanzibar, Mozambique and (?) Nyasaland. Evergreen or fringing forest, 0–870 m.

The reference to *P. ferruginea* (Oliv.) Engl. & Diels in Burtt Davy & Hoyle, N.C.L. : 30 (1936), probably refers to *P. trichocarpa*, but I have seen no specimen from Nyasaland.
　When not in fruit, *P. trichocarpa* can usually be distinguished from *P. ferruginea* by the smaller bracteoles (1–2 mm. as opposed to 2·5–9 mm. long).

LMR

Tab. 12. A.—POPOWIA OBOVATA. A1, flowering branch (× ⅔) ; A2, flower (× 4/3) ; A3, stamen (× 8) ; A4, carpel (× 8), all from *Martin* 157/31 ; A5, fruit (× ⅔) *Macaulay* 386. B.—POPOWIA CAFFRA, leaf and fruit (× ⅔) *Hornby* 2613. C.—POPOWIA BUCHANANII, leaves and fruit (× ⅔) *Richards* 680.

7. **Popowia buchananii** (Engl.) Engl. & Diels in Engl., Mon. Afr. Pflanz. **6** : 47,
t. 18 fig. B quoad fr. (1901) excl. specim. ex Delagoa Bay.—Burtt Davy & Hoyle,
N.C.L. : 30 (1936).—Brenan, T.T.C.L. **2** : 43 (1949). TAB. **12** fig. C. Syntypes
from Tanganyika (Uzaramo) and Nyasaland (sine loc.), *Buchanan* 1152 (B, holotype
† ; BM ; K).
    Unona buchananii Engl., Pflanzenw. Ost-Afr. **C** : 179 (1895). Syntypes as above.
    Popowia djurensis Schweinf. ex Engl. & Diels, tom. cit. : 49, t. 19 A (1901).
Syntypes from the Sudan.

Shrub, tree or liane, 1·5–6 m. high (or higher?) ; bark smooth. Branches
slender, fawn, appressed-pubescent at first, glabrescent, not zig-zag, with spreading
or ascending axillary shoots and prominent leaf-bases. Leaves petiolate ; lamina
(4) 4·8–9·6 × 1·4–3·1 (3·5) cm., narrowly oblong to oblanceolate or elliptic, obtuse
(rarely acute) to rounded or emarginate at the apex, cuneate to rounded or shallowly
cordate at the base, membranous, bright- or yellowish-green and glabrescent above,
glaucous and appressed-pubescent below, with nerves slightly prominent on both
sides but reticulate venation scarcely prominent above only, eglandular (or with
two basal glands when cordate) ; petiole 2–5 mm. long, appressed-pubescent or
glabrescent. Flowers bisexual, solitary or in 2–4-flowered shortly pedunculate
cymes, extra-axillary ; pedicels 1–2 (3) cm. long, very slender, appressed-pubescent;
bracteole small, 0·5–1 mm. long, triangular-ovate, appressed-pubescent, near the
base of the pedicel. Sepals c. 1–2 mm. long, ovate-triangular, obtuse to rounded,
free or ± connate at the base, appressed-pubescent. Petals lemon or darker
yellow, rather thick, brittle, the outer ones 4–5 mm. long, broadly ovate, obtuse to
rounded or apiculate, silver- or rusty-appressed-pubescent outside, puberulous
within except near the base, the inner ones shorter and narrower, oblong to
oblanceolate, acute or obtuse, puberulous except near the base within. Stamens
c. 12–18, c. 1 mm. long, glabrous ; connective broad, truncate, prolongation
absent ; filament ± as long as the anther. Carpels c. (6) 9–20, c. 2 mm. long,
ovoid-ellipsoid ; ovary densely appressed-pilose, 1–2-ovulate ; style ± equal to
the ovary, cylindric, entire or bifurcating. Fruit on a pedicel 1·3–3 cm. long ;
fruiting carpels 2–11, 1 (2)-seeded, segments 0·5–0·9 × 0·4–0·5 cm. ovoid or cylindric, apiculate, glabrescent, finely rugose, orange or red, with stipes 3–8 mm. long.
Seeds 4–8 mm. long.

N. Rhodesia. N : Abercorn Distr., Lunzua Valley, fr. 6.ii.1952, *Richards* 680 (K).
Nyasaland. C : Dedza Distr., Mua-Livulezi Forest Reserve, Namkokwe R., fr.
19.iii.1955, *E.M. & W.* 1056 (BM ; SRGH). S : Nyambi, Fort Johnston, fr. 21.iv.1955,
Jackson 1630 (FHO ; K). **Mozambique.** N : Marrupa, Maúa, Serra de Mecopo, st.
15.x.1942, *Mendonça* 845 (LISC). T : Ngami Valley, st. 6.i.1942, *Hornby* 4481 (PRE).
Z : Maganja da Costa, fr. 20.iv.1943, *Torre* 5201 A (BM ; LISC). MS : Cheringoma
Durundi, fr. 26.v.1948, *Barbosa* 1672 (BM ; LISC).
    In East Africa from eastern Chari, the Sudan and Kenya to Mozambique and Nyasa-
land. Thickets and undergrowth of evergreen forest, often on well-drained soil, 100–
1290 m.

P. buchananii cannot be separated from *P. djurensis* by the characters used by Engl. &
Diels. The stamen number varies continuously between them, and the style can bifurcate
in both species.
P. buchananii is closely related to *P. boivinii* Baill. from Madagascar, which differs in
having more ovate-elliptic leaves with reticulate venation prominent on both sides,
purplish-red branches, and fruiting carpels with stipes longer on the average.

9. ENNEASTEMON Exell

Enneastemon Exell in Journ. of Bot. **70**, Suppl. Polypet. : 209 (1932).
? *Clathrospermum* Planch. ex Benth. in Benth. & Hook., Gen. Pl. **1** : 29 (1861).

Shrubs or lianes with simple appressed ferrugineous hairs. Leaves frequently
bearing 2 basal glands. Buds broadly ovoid or depressed-subglobose. Flowers
bisexual, solitary or in 2- to several-flowered cymes or fascicles, axillary or supra-
axillary (more rarely extra-axillary), pedicellate. Bracteole single, persistent, near
the base of the pedicel, small. Sepals 3, valvate, fused at the base, much shorter
than the petals in bud. Petals 6, slightly united at the base to form a single whorl
or almost free, but forming two whorls above in bud, both valvate, thick, ±
spreading at anthesis, usually concave, not or scarcely connivent, the inner smaller,

sometimes shortly clawed. Stamens 8–11 in a single whorl, obconic-clavate or quadrate, with thecae extrorse, parallel and connective frequently prolonged backwards over the ovaries, not broadened between the thecae ; filaments usually present, longer than the anthers. Carpels 6–11, free ; ovary ovoid or ellipsoid, with 1–5 uniseriate ovules ; style ± cylindric, oblique, shorter than the ovary ; stigma subcapitate or tapering, usually bifurcating. Ripe carpels indehiscent, succulent, cylindric-torulose, stipitate, 1–5-seeded. Seeds vertical, serial ; aril absent.

A tropical African genus of c. 10 species, distinguished from *Popowia* by the arrangement of the petals.

Enneastemon schweinfurthii (Engl. & Diels) Robyns & Ghesq. in Ann. Soc. Sci. Brux., Sér. B, **53** : 165 (1933).—Boutique, F.C.B. **2** : 380 (1951). TAB. **11** fig. B. Type from the Sudan.

Popowia schweinfurthii Engl. & Diels in Engl., Mon. Afr. Pflanz. **6** : 51, t. 19 E (1901). Type as above.

Popowia ochroleuca Diels in Engl., Bot. Jahrb. **53** : 441 (1915).—Brenan, T.T.C.L. : 44 (1949). Type from Tanganyika (Kyimbila).

Enneastemon affinis Robyns & Ghesq., tom. cit. : 163 (1933). Type from the Belgian Congo (Katanga).

Enneastemon ochroleucus (Diels) R.E.Fr. in Ark. Bot., Ser. 2, **3** : 41 (1953). Type as for *Popowia ochroleuca*.

Shrub or liane, 2–7 m. high, with metallic-ferrugineous indumentum. Branches densely velutinous at first, eventually glabrous. Leaves petiolate ; lamina 6–16 (23) × 3–7 (9) cm., obovate to oblanceolate or more rarely elliptic-oblong, obtuse (rarely acute) or with an acumen up to 2 cm. long at the apex, cuneate to rounded or truncate at the base, chartaceous or subcoriaceous, pale to bright green and glabrescent above, ± glaucous and appressed-pubescent below, with nerves and reticulate venation ± prominent on both sides, and usually with two ± conspicuous black glands at the base ; petiole 4–10 mm. long, appressed-pubescent. Flowers solitary or in pairs (more rarely in fascicles of 3–7), axillary or supraaxillary ; pedicels (0·8) 1–2 (2·5) cm. long, usually slender, appressed-pubescent ; bracteole small, 0·5–1 mm. long. Sepals 1–2 mm. long, ovate-triangular, acute to rounded, connate at the base or forming a trilobed calyx, appressed-pubescent. Petals cream or yellow, thick, slightly united at the base, the outer ones 2·5–4 (6) mm. long, obovate or suborbicular to oblong or rhomboid, obtuse to rounded or apiculate, appressed-pubescent outside, tomentellous within, the inner ones shorter and narrower, obovate or rhomboid, obtuse, slightly clawed, tomentellous. Stamens 8–9, 1–1·5 mm. long, glabrous ; connective truncate, ± prolonged inwards behind the thecae ; filament usually longer than the anther. Carpels 6–9, c. 1·5 mm. long, ovoid ; ovary densely appressed-pilose, 1–5-ovulate ; style about a third as long as the ovary, ± cylindric, bifurcating. Fruit on a pedicel 0·8–3 cm., long ; fruiting carpels 1–5 (9), 1–5-seeded, segments 0·7–1·2 (1·6) × 0·6–1 cm., globose to elliptic or cylindric, apiculate, ± densely appressed-pubescent, orange, with stipes 1–4 mm. long. Seeds 6–15 mm. long.

N. Rhodesia. N : Mukungwa Valley, near Mpika, fr. 31.viii.1938, *Greenway* 5647 (FHO ; K). W : Mwinilunga, between Matonchi R. and Kaoomba R., fl. 1.xi.1937, *Milne-Redhead* 3044 (BM ; K).

From the Sudan and Chari to Uganda, west Kenya, south-west Tanganyika, N. Rhodesia, east and south Belgian Congo and Angola. Undergrowth of evergreen and fringing forest, 1300–2000 m.

The anther form in this species is very variable and cannot be used to separate *E. affinis* from *E. schweinfurthii* and *E. ochroleucus*. *E. seretii* (De Wild.) Robyns & Ghesq., may not be specifically distinct from *E. schweinfurthii*. It is a rain forest species from Angola and the Belgian Congo with leaves rounded to cordate at the base (usually eglandular), 2–7-fascicled flowers and anthers with long connective-prolongations.

10. ARTABOTRYS R. Br.

Artabotrys R. Br. in Bot. Reg. : t. 423 (1820).

Shrubs or rarely small trees, usually climbing or scrambling, with simple hairs. Buds ovoid. Flowers bisexual, solitary or in condensed cymes, extra-axillary (rarely terminal), pedicellate, on usually hooked and flattened peduncles. Bracts and bracteoles small, often caducous. Sepals 3, valvate, equal to the petals or

shorter. Petals 6, free, in two equal or unequal whorls, both valvate, usually thick, erect or spreading, flat or ± subulate, usually ± concave at the base and enclosing the reproductive parts. Stamens ∞, oblong or cuneiform, with thecae extrorse and connective-prolongation broadened and truncate, the outer ones sometimes staminodial ; filaments very short or absent. Carpels few (6) to numerous, free ; ovary ovoid or ellipsoid, with 2 basal ovules ; style as long as the ovary or shorter, cylindric or ± flattened, oblique ; stigma small. Ripe carpels indehiscent, succulent, cylindric or ellipsoid, 1–2-seeded. Seeds vertical, collateral ; aril absent.
A genus of over 100 species in tropical Africa and eastern Asia.

Inner petals broadly ovate, not concave at the base ; bracts deciduous, leaving prominent
   biseriate bases ; pedicels 1 cm. long or longer ; ripe carpels bluish-black, stipitate
                                                                               1. *brachypetalus*
Inner petals narrowly elliptic to subulate, concave at the base ; bracts persistent ; pedicels
   1–6 mm. long ; ripe carpels red or purple, sessile :
Outer petals ovate-lanceolate ; ripe carpels rostrate, scarlet   -   -   - 2. *collinus*
Outer petals ± linear or subulate ; ripe carpels not rostrate, reddish or purple :
   Petals 4–12 mm. long, subulate or narrowly linear ; pedicels 1–3 mm. long ; leaves
      shortly acuminate to rounded -   -   -   -   -   -   - 3. *monteiroae*
   Petals 11–18 mm. long, narrowly spathulate or linear-spathulate ; pedicels 3–6 mm.
      long ; leaves acuminate or attenuate   -   -   -   -   -   - 4. *stolzii*

1. **Artabotrys brachypetalus** Benth. in Trans. Linn. Soc. **23** : 467 (1862).—Oliv., F.T.A. **1** : 28 (1868).—Engl. & Diels in Engl., Mon. Afr. Pflanz. **6** : 71, t. 26 A (1901).—Gibbs in Journ. Linn. Soc., Bot. **37** : 428 (1906).—Sim, For. Fl. Port. E. Afr. : 8 (1909).—Bak. f. in Journ. Linn. Soc., Bot. **40** : 19 (1911).—R.E.Fr., Wiss. Ergebn. Schwed. Rhod.-Kongo-Exped. **1** : 45 (1914) ; in Bot. Notis. **1934** : 92 (1934).—Eyles in Trans. Roy. Soc. S. Afr. **5** : 354 (1916).—Pardy in Rhod. Agr. Journ. **53** : 617, cum tab. (1956). TAB. **13** fig. A. Type : Mozambique, Tete, *Kirk* (K).
   Artabotrys cf. brachypetalus.—O. B. Mill. in Journ. S. Afr. Bot. **18** : 14 (1952).

Shrub, small tree or liane, 2–10·5 m. high. Branches reddish-brown, rusty-tom-entose at first, eventually glabrous. Leaves petiolate ; lamina (2) 4·5–11 (15) × (1·2) 2·2–6·6 (7·5) cm., obovate to oblanceolate or elliptic-oblong, obtuse or rounded to apiculate or shortly acuminate at the apex, cuneate to rounded at the base, coriaceous, bluish- or bright-green, glossy and glabrous or sparsely pubescent above, ± matte and glabrescent or ± densely pubescent below, equally densely reticulate on both sides ; petiole 3–5 mm. long, shortly and ± densely tomentose. Flowers solitary in the axils of biseriate tomentose caducous bracts on short con-densed zig-zag secondary peduncles, rarely on the primary peduncle ; primary peduncle usually uncinate ; pedicels 1–3·5 (4) cm. long, broadening above, sub-appressed-pubescent. Sepals 6–11 mm. long, ovate to lanceolate-elliptic, acute or obtuse, free, subappressed-pubescent outside, subglabrous within. Petals yellow or cream to greenish, ± coriaceous, subequal, 6–15 mm. long, broadly ovate or elliptic, not concave at the base, obtuse or apiculate, sparsely appressed-pubescent or strigose outside, glabrous and rugose within. Stamens 2 mm. long, linear-cuneiform ; connective broadened above, truncate. Carpels 6–15, c. 2 mm. long, narrowly ovoid, glabrous, 2-ovulate ; style cylindric, pubescent. Fruit on a pedicel 1·5–4 cm. long ; fruiting carpels 1–8, 1·5–2·2 × 0·8–1·5 cm., 1–2-seeded, oblong-cylindric to obovoid, apiculate or not, glabrous, smooth, blackish-purple, with stipes 5–9 mm. long. Seeds c. 1·2–1·7 cm. long, plano-convex or cylindric.

Bechuanaland Prot. N : Chobe, Kasane Rapids, st. 2.viii.1950, *Robertson & Elffers* 95 (K ; PRE). **N. Rhodesia.** C : Chingombe, fl. 27.ix.1957, *Fanshawe* 3746 (K). E : Luangwa Valley, fr. 25.iii.1955, *E.M. & W.* 1195 (BM ; SRGH). S : Victoria Falls, fl. 20.xi.1949, *Wild* 3115 (SRGH). **S. Rhodesia.** N : Sebungwe, Zambezi R. above Binja's camp, st. ix.1955, *Davies* 1429 (K; SRGH). W : South Matopos, near Siloz-wane, fr. 28.ii.1954, *Plowes* 1688 (K ; SRGH). C : Umsweswe R., fr. xii.1948, *Hodgson* 32/48 (SRGH). E : Umtali Distr., Impodsi R., fr. 2.ii.1947, *Chase* 282 (BM ; K ; SRGH). S : Victoria, Zimbabwe Ruins, fr. 4.ii.1951, *Mullin* 29/51 (FHO ; SRGH). **Nyasaland.** S? : L. Nyasa shore, st. 10.i.1944, *Hornby* 4799 (K). **Mozambique.** N : Nacala, near Fernão Velosa, fl. 14.x.1948, *Barbosa* 2413 (BM ; LISC). Z : Lugela, Mocuba, Namagoa, fl. x.1945, *Faulkner* P 39 (BM ; COI ; K ; PRE). T : Tete, fl. & fr. xi.1858, *Kirk* (K). MS : Chimoio, between Vila Pery and Garuso, fl. 19.x.1944, *Men-donça* 2514 (BM ; LISC). SS : Massinga, Inhambane, fl. x.1937, *Gomes e Sousa* 1899 (COI ; K ; LISC). LM : Magude, between Mapulanguene and Macaêna, fl. 1.xii.1944, *Mendonça* 3212 (BM ; LISC).

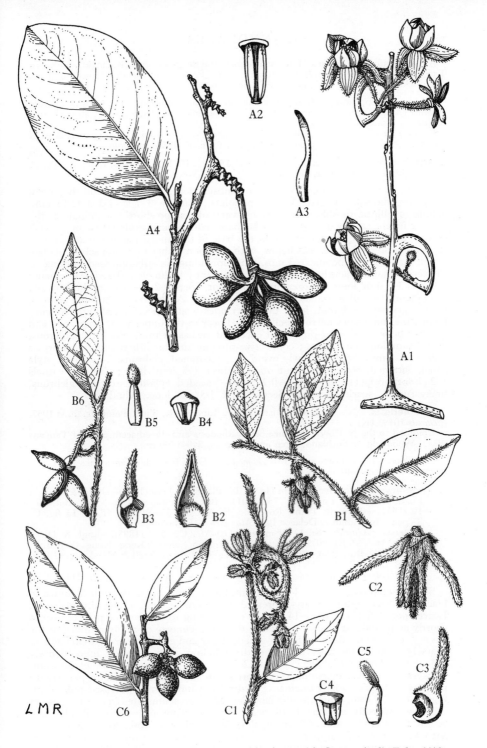

LMR

Tab. 13. A.—ARTABOTRYS BRACHYPETALUS. A1, shoot with flowers (×⅔) *Eyles* 6412 ;
A2, stamen (×8) *Faulkner* P.39 ; A3, carpel (×8) *Faulkner* P.39 ; A4, shoot with
leaf and fruit (×⅔) *Flemming* 3876. B.—ARTABOTRYS COLLINUS. B1, shoot with
leaves and flower (×⅔) ; B2, outer petal (×2) ; B3, inner petal (×2) ; B4, stamen
(×8) ; B5, carpel (×8), all from *Richards* 5190 ; B6, shoot with leaf and fruit (×⅔)
Richards 10173. C.—ARTABOTRYS MONTEIROAE. C1, flowering shoot (×⅔) ; C2, flower
(×2) ; C3, inner petal (×2) ; C4, stamen (×8) ; C5, carpel (×8), all from *Milne-
Redhead* 4630 ; C6, part of shoot with fruit (×⅔) *Fanshawe* 1513.

Almost confined to our area, but also occurs in the Belgian Congo (Katanga), south-east Tanganyika, and the Transvaal (Soutpansberg). Fringing forest, deciduous woodland and thickets, frequently on sand or river alluvium, 0–1140 m.

The zig-zag appearance of the persistent biseriate bract-bases is a means of identifying *A. brachypetalus* even if the hooked tendrils characteristic of *Artabotrys* are absent. It is frequently attacked by a gall midge.

Popowia dicranantha Diels, from the Lindi district of Tanganyika, is in fact a species of *Artabotrys* and can probably be regarded as a form of *A. brachypetalus*.

2. **Artabotrys collinus** Hutch. in Kew Bull. **1931** : 245 (1931). TAB. **13** fig. B. Type: N. Rhodesia, Kaloswe, *Hutchinson & Gillett* 3770 (K, holotype).

Shrub, c. 1–2 m. high, straggling or subscandent. Branches reddish-brown, rusty-spreading-pubescent. Leaves petiolate ; lamina 3–7·5 ×1·3–3 (3·8) cm., oblong or elliptic to oblanceolate or obovate, obtuse or shortly acuminate at the apex, cuneate (rarely rounded) at the base, coriaceous, apple-green or greyish green, glossy and glabrous (except at the base of the midrib) above, matte and ± sparsely appressed-pubescent below, with venation rather more densely reticulate below than above ; petiole 1–3 mm. long, spreading- or subappressed-pubescent. Flowers solitary, bracts linear-subulate, persistent ; peduncle reflexed or uncinate ; pedicels 3–5 mm. long, ± spreading-pubescent. Sepals 2·5–4 (6) mm. long, ovate, acuminate, free, pilose outside, pubescent within. Petals green (? turning yellow), thick, unequal, 9–11 mm. long, the outer ones ovate-lanceolate, slightly concave below, the inner ones narrowly elliptic or linear in the upper part, broadened and concave below, sericeous-pubescent except at the base within. Stamens numerous, 0·75–1 mm. long, cuneiform ; connective-prolongation broadly triangular. Carpels c. 6–9, c. 1 mm. long, narrowly ovoid or ± flattened, glabrous, 2-ovulate ; style small, flattened, glabrous. Fruit on a pedicel 3–5 mm. long ; fruiting carpels 1–2 (? or more), (1) 1·3–1·9 ×0·5–0·7 cm., 1–2-seeded, cylindric, rostrate, glabrous, finely rugose, scarlet, sessile. Seeds c. 1·1–1·3 cm. long, cylindric.

N. Rhodesia. N : Abercorn Distr., above Lunzua R., Tiger Point, fr. 26.vi.1957, *Richards* 10193 (K).

Confined to the Northern Province of N. Rhodesia and the adjacent parts of Tanganyika. Woodland or bushland, often among rocks, 840–1500 m.

A. collinus is most closely related to *A. modestus* Diels from the Tanganyika coast, and may not be specifically distinct.

3. **Artabotrys monteiroae** Oliv. in Hook., Ic. Pl. **18**: t. 1796 (1888).—Engl. & Diels in Engl., Mon. Afr. Pflanz. **6** : 75 (1901).—Bak. f. in Journ. Linn. Soc., Bot. **40** : 19 (1911).—Eyles in Trans. Roy. Soc. S. Afr. **5** : 354 (1916). TAB. **13** fig. C. Type : Mozambique, Delagoa Bay, *Monteiro* 4 (K, holotype).
    Artabotrys nitidus Engl., Pflanzenw. Ost-Afr. **C** : 179 (1895).—Engl. & Diels, loc. cit. st. t. 26 E (1901).—R.E.Fr.,Wiss. Ergebn. Schwed. Rhod.-Kongo-Exped. **1** : 46 (1914).—Burtt Davy & Hoyle, N.C.L. : 30 (1936).—Exell & Mendonça, C.F.A. **1, 1** : 22 (1937), **2** : 355 (1951). Syntypes from Tanganyika.
    Artabotrys cf. monteiroae.—Brenan in Mem. N.Y. Bot. Gard. **8**, 3 : 214 (1953).

Shrub, small tree or liane, 1·2–6 (10) m. high. Branches dark reddish-brown to blackish when mature, rusty-appressed-pubescent at first, eventually glabrous. Leaves petiolate ; lamina 3–14 ×1·5–6 cm., oblong to ovate or elliptic, obtuse to shortly acuminate or rarely rounded at the apex, cuneate or rarely rounded at the base, ± coriaceous, dark- or bluish-green, glossy and glabrous above, matte and ± sparsely strigose (at least on the midrib) below, with venation more densely reticulate below than above ; petiole 2–5 (7) mm. long, strigose or glabrescent. Flowers 1–9 (rarely more) on very short secondary peduncles, bracts persistent ; primary peduncle usually uncinate ; pedicel 1–3 mm. long, appressed-pubescent. Sepals 1·5–2·5 mm. long, ovate-triangular, acute or acuminate, free or connate at the base, sparsely appressed-pubescent outside, glabrous within. Petals deep yellow to cream or greenish, thick, flexible, equal, 4–12 mm. long, subulate or narrowly linear in the upper part, broadened and concave below, erect or arcuate-ascending, obtuse, sericeo-pubescent except at the base within, the bases of the inner ones enclosing the reproductive parts. Stamens c. 1 mm. long, cuneiform ; connective-prolongation capitate, truncate. Carpels c. 8–11, c. 2 mm. long, narrowly ovoid or ± flattened, glabrous, 2-ovulate ; style spathulate, papillose. Fruit on a pedicel 2–3 mm. long ; fruiting carpels 1–4, 0·9–1·5 ×0·5–0·9 cm., (1) 2-seeded, ellipsoid or cylindric, not apiculate, glabrous, glossy, finely rugose, red, sessile. Seeds c. 0·8–1·4 cm. long, plano-convex.

N. Rhodesia. N : Fort Rosebery, L. Bangweulu, N. of Samfya Mission, fl. 6.x.1947, *Brenan* 8039 (BM ; FHO ; K). W : Ndola, fr. 2.ix.1954, *Fanshawe* 1513 (FHO ; K ; SRGH). C : Kapiri Mposhi, fr. 22.i.1955, *Fanshawe* 1821 (K). E : Petauke, Ft. Jameson-Lusaka road, km. 135, fl. & fr. 24.v.1952, *White* 2876 (FHO ; K). **S. Rhodesia.** E : Umtali, waterfall below Drumfad, fl. 2.xi.1952, *Chase* 4684 (BM ; K ; SRGH). **Nyasaland.** N : Matipa Forest Reserve, fr. viii.1954, *Chapman* 221 (FHO). C. Kota Kota Distr., Nchisi Mt., fr. 24.vii.1946, *Brass* 16897 (K ; SRGH). S : Zomba, Likangala stream, fr. iii.1934, *Clements* 432 (FHO). **Mozambique.** N : Cabo Delgado, Macondes Plateau, Muêda, st. 19.x.1942, *Mendonça* 953 (LISC). T : Angónia, Vila Mouzinho, fl. 15.x.1943, *Torre* 6034 (LISC). MS : Madanda Forest, fl. 5.xii.1906, *Swynnerton* 1183 (BM ; K). SS : between Vila Gomes da Costa and Chibuto, fr. 4.xii.1944, *Mendonça* 3257 (BM ; LISC). LM : Maputo, between Chiquiche and Santaca, fr. 31.vii.1948, *Gomes e Sousa* 3767 (COI ; K).

Also occurs in Uganda, Tanganyika, Belgian Congo, Angola, Natal and the Transvaal (Soutpansberg). Fringing and evergreen forest, *Brachystegia* and *Combretum* woodland and thickets, frequently among granite or sandstone rocks or on termite mounds, 45–1950 m.

Although the leaves and flowers of southern representatives of this species tend to be smaller than those of specimens from northern Mozambique, N. Rhodesia and the Congo, it does not seem possible to recognize *A. nitidus* as a separate species.

A. mabifolius Diels from Madagascar should probably be regarded as a synonym of *A. monteiroae*.

4. **Artabotrys stolzii** Diels in Engl., Bot. Jahrb. **53** : 446 (1915).—Burtt Davy & Hoyle, N.C.L. : 30 (1936).—Brenan, T.T.C.L. : 41 (1949). Type from Tanganyika (Kyimbila).

Closely related to *A. monteiroae*, but differs in the following characters :
Large woody climber, up to 15 m. high. Leaf lamina 8·5–14 × 3·3–5 cm., oblong to elliptic or oblanceolate, acuminate or attenuate at the apex, membranous or subcoriaceous. Flowers 1–7 ; pedicels 3–6 mm. long. Sepals 2–4 mm. long. Petals 11–18 mm. long, narrowly spathulate or linear-spathulate.

N. Rhodesia. E : Nyika Plateau, fl. 24.xi.1955, *Lees* 66 (K). **Nyasaland.** N? : no locality, fr. 1.xi.1937, *Townsend* 211 (FHO).

In southern Tanganyika (Kyimbila) and the Nyika plateau. In mist-forest and evergreen forest relics, (700) 1500–2100 m.

11. XYLOPIA L.

Xylopia L., Syst. Nat. ed. 10 : 1250 (1759) *nom. conserv.*

Trees, shrubs or shrublets, not climbing, with simple or rarely forked hairs. Buds triquetrous or conic. Flowers bisexual, solitary or in cymes or fascicles, axillary, pedicellate or sessile, often (? always) fragrant. Bracteoles (1) 2–5, caducous or persistent. Sepals 3, valvate, much shorter than the petals, ± united. Petals 6, free, in two subequal or unequal whorls, both valvate, thick, erect or spreading, flat or triquetrous and often linear above, broad and ± concave at the base, the inner ones sometimes with a fleshy pad covering the reproductive parts. Stamens ∞, linear, with thecae extrorse, septate and connective-prolongation, capitate, the outer and inner ones often flattened and staminodial ; filaments articulated, the bases sometimes united, enclosing the gynoecium. Carpels ∞ to few (rarely solitary), free, ovoid-cylindric or ± flattened, with several 1-seriate or biseriate ovules ; styles cylindric or narrowly clavate, aggregated. Ripe carpels dehiscent, ± succulent, cylindric or obovoid, 1–8-seeded. Seeds vertical to horizontal, collateral or biseriate ; aril present (sometimes inconspicuous).

A pantropical genus of over 100 species.

Inner petals 4–6 mm. long, rhomboid, much shorter than the linear outer petals ; ripe carpel torulose, stipitate ; seeds vertical, uniseriate with a filamentous aril   1. *rubescens*
Inner petals at least ¾ as long as the outer petals ; ripe carpels not or scarcely torulose, stipitate or subsessile ; seed characters various :
   Seeds vertical, uniseriate ; aril basal, papyraceous, yellow ; ripe carpels usually 16 or
      more, subsessile, narrowly cylindric, straight   -   -   -   -   2. *aethiopica*
   Seeds oblique or horizontal, uniseriate or biseriate ; aril investing the seed or absent ;

ripe carpels 10 or fewer, usually stipitate, cylindric or obovoid, ± curved if
    cylindric :
Leaves glabrous or pubescent, not pilose-fringed, cuneate to shallowly cordate ;
    young shoots pubescent or glabrescent :
Seeds uniseriate, oblique ; flowers nearly always solitary ; outer petals 2–5 cm.
    long, magenta within in the upper part   -   -   -   -   3. *acutiflora*
Seeds biseriate, oblique or ± horizontal ; flowers varying in number ; outer petals
    up to 4·5 cm. long, yellow to white or brownish in the upper part :
Leaves ovate or lanceolate to oblong, acuminate or attenuate, glabrous and
    densely reticulate above ; flowers solitary on pedicels at least 3 mm. long,
    or 2–12 in loose cymes ; seeds almost horizontal   -   -   4. *katangensis*
Leaves elliptic to oblong, or if lanceolate then flowers subsessile, acuminate to
    emarginate, glabrous or pubescent above, with reticulation varying in
    density ; flowers 1–3 or if more numerous (–10) then in subsessile clusters ;
    seeds oblique :
Plant a tree 6–24 m. high, with clean bole and sparingly branched crown ;
    leaves narrowly oblong to elliptic, usually acute or acuminate ; flowers
    remaining erect or becoming reflexed, frequently pedunculate 5. *holtzii*
Plant a shrub or small tree up to 6 (rarely to 9) m. high, much branched ;
    leaves oblong or elliptic to lanceolate or suborbicular, subacute to emar-
    ginate ; flowers eventually spreading or reflexed, rarely shortly
    pedunculate :
Pedicels slender, 5–6 mm. long ; leaves glabrous above except at the base
    of the midrib ; petioles 2–3 mm. long   -   -   -   -   6. *torrei*
Pedicels stout (up to 7 mm. long) or absent ; leaves ± sparsely pubescent
    on the lamina above or, if glabrous except for the midrib, then flowers
    sessile or subsessile :
Fruits glabrous, often reticulate-striate when dry ; flowers 1–3, pedi-
    cellate, or rarely subsessile ; plant a shrub or small tree c. 2–6 m. high
                                                           7. *odoratissima*
Fruits ± densely brown-pubescent, not reticulate-striate when dry ;
    flowers 1–10, sessile or subsessile ; plant a shrub or rhizomatous
    shrublet 0·4–3 (4·5) m. high -   -   -   -   -   8. *tomentosa*
Leaves pilose-fringed, ± cordate ; young shoots pilose ; flowers solitary, subsessile
                                                           9. *collina*

1. **Xylopia rubescens** Oliv., F.T.A. **1** : 30 (1868).—Engl. & Diels in Engl., Mon. Afr.
   Pflanz. **6** : 60 (1901).—Pellegrin in Mém. Soc. Bot. Fr. **31** : 70 (1950).—Boutique,
   F.C.B. **2** : 322 (1951).—Keay, F.W.T.A. ed. 2, **1**, 1 : 41 (1954). TAB. **14** fig. B.
   Type from Nigeria.
       Xylopia humilis Engl. & Diels, tom. cit. : t. 21 B (1901). Type from Liberia.
       Xylopia butayei De Wild. in Ann. Mus. Cong., Bot., Sér. 4, **1** : 33 (1902). Type
   from the Belgian Congo.
       Xylopia zenkeri Engl. & Diels in Engl., Bot. Jahrb. **39** : 480 (1907). Type from
   Fr. Cameroons.
       Xylopia gossweileri Exell in Journ. of Bot. **64**, Suppl. Polypet. : 6 (1926) ; op. cit.
   70, Suppl. Polypet. : 212 (1932).—Exell & Mendonça, C.F.A. **1**, 1 : 20 (1937).
   Type from Angola (Cabinda).

   Tree 4·5–20 (30) m. high, erect, or rarely ± straggling, with smooth pale grey
bark and numerous stilt roots.   Branches greyish-fawn to red-brown or dark-
brown, markedly rugose, soon glabrous ; lenticels few, not conspicuous.   Leaves
petiolate ; lamina (7) 9–21·5 × (3·5) 4–8·4 cm., oblong or elliptic to oblanceolate
or ovate-lanceolate, obtuse or ± abruptly acuminate at the apex, cuneate to
rounded and decurrent at the base, coriaceous, bluish-green and glabrous above,
orange or brick-coloured and shortly appressed-pubescent or glabrescent below,
with densely reticulate venation prominent on both sides ; petiole 5–18 mm. long,
often blackish, appressed-pubescent or glabrescent.   Flowers solitary or in 2–5-
flowered fasciculate cymes ; pedicels 4–6 mm. long, appressed-rusty-pubescent ;
bracteoles 2–5, ± oblong.   Sepals 2–4 mm. long, ovate-triangular, acute or
apiculate, appressed-pubescent outside, glabrous and crimson within.   Petals very
unequal, the outer ones 2·6–3·7 × 0·3–0·4 cm., orange-yellow when mature, crimson
at the base within, linear, triquetrous above, ± concave at the base, sericeo-
pubescent outside, puberulous above and glabrous below within, the inner ones
0·4–0·6 cm. long, crimson, rhomboid to oblanceolate, apiculate, concave, carinate,
sericeo-pubescent above outside, otherwise glabrous.   Stamens c. 2 mm. long,
linear ; connective-prolongation obliquely capitate, papillose.   Carpels 8–11, c. 4
mm. long ; ovary cylindric, appressed-pilose with simple or forked hairs, 9–10-

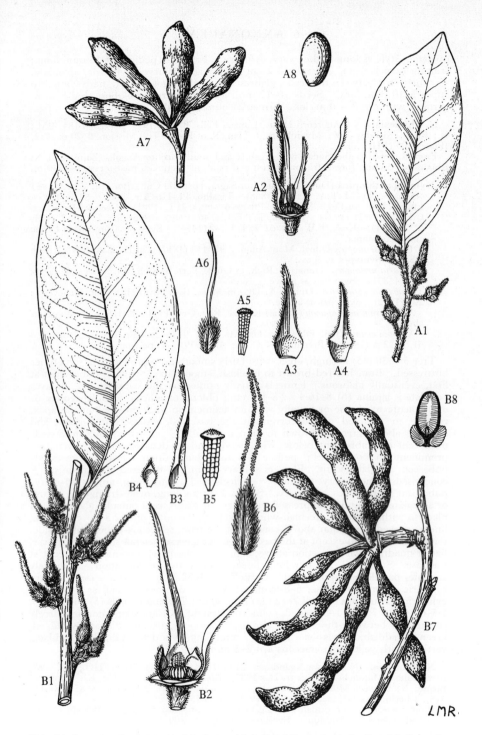

Tab. 14. A.—XYLOPIA HOLTZII. A1, shoot with leaf and flower-buds (× ⅔) ; A2, flower in section (× ⅔) ; A3, outer petal (× 2) ; A4, inner petal (× 2) ; A5, stamen (× 8) ; A6, carpel (× 8), all from *Hutchinson* 2240 ; A7, fruit (× ⅔) *S. African Forest Dept.* ; A8, seed (× ⅘) *Hutchinson* 2240.   B.—XYLOPIA RUBESCENS. B1, shoot with leaf and flower-buds (× ⅔) ; B2, flower in section (× ⅘) ; B3, outer petal (× 1) ; B4, inner petal (× 1) ; B5, stamen (× 8) ; B6, carpel (× 8) ; B7, fruit (× ⅔), all from *Brenan & Greenway* 8022 ; B8, seed in section (× 1) from Engl. & Diels in Engl., Mon. Afr. Pflanz. **6** : t. 2 fig. Be (1901).

ovulate ; style as long as the ovary, cylindric. Fruit on a pedicel 10–15 mm. long ; fruiting carpels 1–11, 2–7 ×0·8–1·2 cm., 1–8-seeded, curved-cylindric, apiculate, ± constricted between the seeds, glabrescent, finely rugose, green, narrowing into stipes 5–10 mm. long. Seeds 10–12 mm. long, cylindric or globose, red-brown, vertical, with aril 2–4 mm. long formed of numerous filaments.

N. Rhodesia. N : Abercorn Distr., Lunzua Falls, fl. 26.x.1952, *Robertson* 184 (BM ; K ; PRE ; SRGH). W : Mwinilunga, 6·7 km. N. of Kalene Hill Mission, fl. 20.ix.1952, *White* 3309 (FHO ; K).

From Liberia to the Sudan and Uganda and southward to Angola (Zaire) and N. Rhodesia. Swamp forest, fringing forest and relics of evergreen mist-forest, 0–1530 m.

2. **Xylopia aethiopica** (Dunal) A. Rich. in Sagra, Hist. Ile Cub., Bot., Pl. Vasc. **1** : 53 (1845).—Oliv., F.T.A. **1** : 30 (1868).—Ficalho, Pl. Ut. Afr. Port. : 83 (1884).— Engl. & Diels in Engl., Mon. Afr. Pflanz. **6** : 60, t. 21 C (1901).—Exell & Mendonça, C.F.A. **1**, 1 : 21 (1937).—Brenan, T.T.C.L. : 46 (1949).—Boutique, F.C.B. **2** : 324, t. 32 (1951).—Keay, F.W.T.A. ed. 2, **1**, 1 : 42 (1954).—Syntypes from Sierra Leone and " Aethiopia ".

　　Unona aethiopica Dunal, Mon. Anon. : 97, 113 (1817) ; in DC., Syst. Nat. : 496 (1817). Syntypes as above.

　　Uvaria aethiopica (Dunal) A. Rich. in Guill., Perr. & Rich., Fl. Senegamb. Tent. **1** : 9 (1831). Syntypes as above.

　　Habzelia aethiopica (Dunal) A. DC. in Mém. Soc. Phys. Hist. Nat. Genève, **5** : 207 (1832). Syntypes as above.

　　Xylopicrum aethiopicum (Dunal) Kuntze, Rev. Gen. Pl. **1** : 8 (1891). Syntypes as above.

　　Xylopia eminii Engl., Pflanzenw. Ost-Afr. **C** : 179 (1895).—Engl. & Diels, op. cit. : 61, t. 22 A (1901).—Brenan, T.T.C.L. : 47 (1949). Type from Uganda.

Tree 5·4–30 (45) m. high, erect, with fairly smooth grey-brown bark, sometimes buttressed. Branches red-brown to blackish, smooth or ± rugose, puberulent at first, eventually glabrous ; lenticels usually numerous and conspicuous. Leaves petiolate ; lamina (6) 8–16·4 ×2·8–6·5 cm., oblong or elliptic to oblanceolate or ovate, acute to obtuse or usually with an acumen up to 2 cm. long at the apex, cuneate to rounded and decurrent at the base, coriaceous, bluish-green and glabrous above, greenish-brown to orange (usually ± glaucous) and appressed-pubescent or glabrescent below, with nerves and ± densely reticulate venation prominent on both sides ; petiole 3–6 mm. long, often blackish, appressed-pubescent or glabrescent. Flowers solitary or in 2–5-flowered fasciculate or ± condensed cymes, very fragrant ; pedicels 4–11 mm. long, appressed-rusty-pubescent ; bracteoles 2, reduced. Sepals 3–5 mm. long, ovate-triangular, obtuse or apiculate to rounded, appressed-pubescent or glabrescent outside, glabrous and blackish within. Petals greenish-white to yellow, the outer ones 2·5–5 ×c. 0.5 cm., linear, curved in section above, concave at the base, sericeous-pubescent outside, tomentellous within except at the base, the inner ones somewhat shorter, narrower, flat or quadrangular in section above, concave at the base, tomentellous except at the base. Stamens 1–1·5 mm. long, linear ; connective-prolongation ± obliquely capitate, papillose or pubescent. Carpels c. 24–32 (42), c. 4 mm. long ; ovary cylindric, appressed-pilose, 6–8-ovulate ; style c. 3 times as long as the ovary, cylindric. Fruit on a pedicel 7–12 (22) mm. long ; fruiting carpels (5) 16–24 (42), 1·5–6 ×0·5–0·7 cm., 1–8-seeded, cylindric, straight, not apiculate, not or scarcely constricted between the seeds, glabrous, ± smooth, usually diagonally ridged, green or reddish, subsessile. Seeds 5–7 mm. long, cylindric, orange-red to black, vertical, with yellow papyraceous aril 2–3 mm. long.

N. Rhodesia. N : Shiwa Ngandu, fr. 21.vii.1938, *Greenway* 5452 (FHO ; K). W : Mwinilunga, Matonchi R., fl. & fr. 21.x.1937, *Milne-Redhead* 2883 (BM ; K). **Mozambique.** N : between Muêda and Chomba, st. 20.ix.1948, *Pedro & Pedrógão* 5285 (SRGH). MS : Cheringoma, Macuácua, fl. 22.xi.1944, *Simão* 256 (LISC).

From Senegal to the Sudan and Uganda and southward to Angola, Belgian Congo, N. Rhodesia and Mozambique. Evergreen rain forest, 800–1620 m.

The population from W. Africa and the Congo basin tends to differ from that in Uganda, Tanganyika, Mozambique and N. Rhodesia by the leaves in which the reticulation is usually lax on both sides. If dense reticulation is visible at all it is less prominent than the nerves and usually apparent only below. However, the zone of overlap between these two populations (in the Belgian Congo and Uganda) is so wide that it does not seem possible to recognize the latter(*X. eminii* Engl.) even as a subspecies.

3. **Xylopia acutiflora** (Dunal) A. Rich. in Sagra, Hist. Ile Cub., Bot., Pl. Vasc. **1** : 55 (1845).—Oliv., F.T.A. **1** : 32 (1868).—Engl. & Diels in Engl., Mon. Afr. Pflanz. **6** : 63, t. 22 C (1901).—Exell & Mendonça, C.F.A. **1**, 1 : 20 (1937), 355 (1951).— Boutique, F.C.B. **2** : 326 (1951).—Keay, F.W.T.A. ed. 2, **1** : 42 (1954). Type from Sierra Leone (" America meridionale " =lapsus calami !).

Unona acutiflora Dunal, Mon. Anon. : 98, 116, t. 22 (1817). Type as above.
Unona oxypetala DC., Syst. Veg. **1** : 496 (1817). Type from Sierra Leone.
Coelocline ? *oxypetala* (DC.) A. DC. in Mém. Soc. Phys. Hist. Nat. Genève, **5** : 209, t. 5 b (1832). Type as above.
Coelocline acutiflora (Dunal) A. DC., loc. cit.: t. 5 c (1832). Type as for *Xylopia acutiflora*.
Xylopia oxypetala (DC.) Oliv. ex Engl. & Diels, tom. cit.: 63, t. 22 fig. E (1901). Type as for *Unona oxypetala*.

Shrub or tree, 2–20 (30) m. high, sometimes scrambling. Branches orange-red to purple-brown, smooth, spreading-pubescent at first, eventually glabrous ; lenticels usually numerous ; bark often exfoliating. Leaves petiolate ; lamina 5·5–11·5 × 2–4·6 cm., oblong to elliptic or ovate-elliptic, acute or usually acuminate at the apex, cuneate to rounded at the base, membranous, bluish-green or greyish and glabrous (except along the midrib) above, green to orange-brown (not glaucous) and ± densely sericeous- or appressed-pubescent below, with densely reticulate venation usually more prominent below than above ; petiole 1·5–6 mm. long, blackish, appressed-pubescent. Flowers solitary (very rarely 2–3), fragrant ; pedicels absent or up to 6 mm. long, appressed-rusty-pubescent ; bracteoles 2–5, oblong to orbicular. Sepals 2–3·5 mm. long, ovate to semi-orbicular, obtuse or apiculate to rounded, appressed-pubescent outside, glabrous and magenta within. Petals white or pale yellow outside, magenta within, the outer ones 2–5 × 0·3–0·4 cm., linear, curved in section above, broadened and concave at the base, sericeous-pubescent outside, puberulous or glabrous within ; the inner ones somewhat shorter and narrower, otherwise similar. Stamens 1·5–2 mm. long, linear or cuneate ; connective-prolongation ± obliquely capitate, papillose. Carpels 5–10, 3·5–4·5 mm. long ; ovary ovoid-cylindric, appressed-pilose ; style c. 1·5–2 times as long as the ovary, cylindric, with a terminal tuft of hairs. Fruit on a pedicel 4–7 mm. long ; fruiting carpels 3–9, 1–4·8 × 0·8–1·1 cm., 1–8-seeded, cylindric or obovoid, curved, not apiculate, glabrescent, finely rugose, scarcely vertically ridged, scarlet outside, pink within, on stipes 4–10 mm. long. Seeds c. 1 cm. long, ellipsoid, reddish-brown, oblique, uniseriate ; aril inconspicuous.

N. Rhodesia. W : Mwinilunga, fl. 13.x.1955, *Holmes* 1273 (K).
From Sierra Leone to the Sudan and southwards to the Belgian Congo, N. Rhodesia and Angola. Fringing and evergreen forest, 15–1050 m.

4. **Xylopia katangensis** De Wild. in Ann. Mus. Cong., Bot., Sér. 4, **1** : 32 (1902) ; Contrib. Fl. Kat.: 62 (1921).—Diels in Engl., Bot. Jahrb. **39** : 481 (1907).—Th. & H. Dur., Syll. Fl. Cong.: 21 (1909).—Boutique, F.C.B. **2** : 329 (1951). Type from the Belgian Congo (Upper Katanga).

Tree 10–25 m. high. Branches purple-brown, shortly pubescent at first, soon glabrous ; lenticels numerous. Leaves petiolate, lamina 5–10·6 × 2–3·7 cm., lanceolate to elliptic-oblong or ovate, acute or shortly acuminate at the apex, broadly cuneate to rounded at the base, membranous to subcoriaceous, bluish- or sage-green and glabrous (rarely pubescent at the base) above, glaucous or pale green and appressed-pubescent below, with densely reticulate venation equally prominent on both sides ; petiole 5–9 (11) mm. long, orange-brown or blackish, pubescent or glabrescent. Flowers solitary or in c. 2–12-flowered cymes ; primary peduncle up to 3 mm. long ; pedicels 3–7 (8) mm. long, slender, appressed-pubescent ; bracteole 1, oblong to elliptic. Sepals 1–5·2 mm. long, triangular-ovate, acute to obtuse, appressed-pubescent outside, glabrous and green within. Petals cream to greenish-yellow, the outer ones 2–4·5 × 0·5 cm. narrowly linear, ± flat in section above, broadened but scarcely concave at the base, appressed-pubescent outside, puberulous within, the inner ones purple and thickened at the base, somewhat shorter and narrower, otherwise similar. Stamens c. 1 mm. long, linear, purplish ; connective-prolongation capitate, rugulose. Carpels 4–6, c. 4 mm. long ; ovary ovoid, appressed-pilose ; style 3 times as long as the ovary, cylindric. Fruit on a pedicel 10–15 mm. long ; fruiting carpels 3–6, (1·7) 2·3–3·8 × 1·3–2 cm. ; 3–7-seeded, cylindric or obovoid-ellipsoid, straight, rounded at the apex, glabrous, smooth or finely rugose, with or without a vertical ridge, with stipes 2–4 mm. long. Seeds c. 1 cm. long, ellipsoid, brown, almost horizontal, biseriate, entirely covered by an aril when fresh.

N. Rhodesia. N : Kawambwa Distr., edge of L. Mweru near Kafulwe Mission, fl. 4.xi.1952, *White* 3610 (FHO ; K). S : Mazabuka, fr. 1931, *Stevenson* 265/31 (K).

Confined to the Belgian Congo (Katanga and Yangambi) and N. Rhodesia. Fringing forest, lake margins and islands, c. 700–1500 m.

X. katangensis is closely related to *X. parviflora* (A. Rich.) Benth. (*X. vallotii* Chipp ex Hutch. & Dalz.), a W. African and Congo rain-forest species in which the petioles are shorter (3–5 mm.), the midrib of the lamina ± pubescent above, the pedicels longer (usually 10–15 mm.), the flowers usually solitary or paired (rarely cymose), the styles longer (5–6 mm.) and the fruiting carpels cylindric with 1–4 vertical ridges.

5. **Xylopia holtzii** Engl., Bot. Jahrb. **34** : 159 (1904).—Brenan, T.T.C.L. : 47 (1949). TAB. **14** fig. A. Type from Tanganyika (Pugu Hills.).
  Xylopia antunesii var. *shirensis* Engl. & Diels in Engl., Mon. Afr. Pflanz. **6** : 66 (1901).—Burtt Davy & Hoyle, N.C.L. : 30 (1936).—Hutch., Botanist in S. Afr. : 338 (1946). Type : Nyasaland, Shire Highlands, *Buchanan* 237 (K).

Tree 6–24 m. high, erect, with clean bole and sparingly branched crown. Branches red to purplish-black, usually rather sparsely yellowish-fawn-pubescent at first, soon glabrous. Leaves petiolate ; lamina 4·5–11 × 1·7–4·7 cm., narrowly oblong to elliptic, acute or more rarely obtuse or rounded to ± bluntly acuminate at the apex, narrowly cuneate to rounded or truncate at the base, membranous, dull greyish- or ± glossy green and glabrous or sparsely appressed-pubescent above, dull pale green or greyish and more densely appressed-pubescent below, with reticulate venation varying in density and prominence ; petiole 4–10 mm. long, reddish to purple-black, ± pubescent or glabrous. Flowers solitary or 2 (3–4), fasciculate or with peduncle up to 3 mm. long, usually remaining erect ; pedicels 4–8 mm. long, pubescent or glabrescent, ± slender ; bracteoles 2, broadly ovate, near the flower, deciduous. Sepals 1·5–2 mm. long, triangular-ovate, obtuse or apiculate, pubescent and greenish outside, glabrous and reddish within. Petals yellow or greenish, the outer ones 1–2 (2·5) cm. long, narrowly linear or subulate, concave and c. 2·5 mm. wide at the base, golden-brown-sericeo-pubescent outside, puberulous within except at the base, the inner ones somewhat shorter and slightly narrower, shortly unguiculate, puberulous on both sides except at the base. Stamens 1 mm. long, linear, with connective prolongation capitate, rugulose. Carpels 6–8, 3 mm. long ; ovary ovoid-cylindric or flattened, appressed-pilose ; style twice as long as the ovary, cylindric, pilose. Fruit on a pedicel 5–8 mm. long ; fruiting carpels 1–8, 1–3·7 × 0·7–1·1 (1·5) cm., 1–6-seeded, obovoid or cylindric, ± curved, apiculate or rounded at the apex, sometimes slightly constricted between the seeds when dry, glabrous, rugose, red, reticulate-striate or not, with stipes 2–9 mm. long. Seeds c. 0·8–1·1 cm. long, ellipsoid or ± flattened, orange, oblique, biseriate; aril inconspicuous.

S. Rhodesia. E : Lower Haroni R., Melsetter, st. xi.1952, *Ball* 32 (K ; SRGH). **Nyasaland.** N : Deep Bay, fl., *Lewis* 95 (FHO). S : Zomba, fl. xi.1915, *Purves* 257 (K). **Mozambique.** Z : Bajone, between Namuera and Murroa, fr. 2.x.1949. *Barbosa & Carvalho* 4284 (LISC ; SRGH). MS : Mossurize, Espungabera, fl. & fr. 14.x.1943, *Torre* 6192 (LISC).

From the Sudan, Uganda and Kenya southward to S. Rhodesia, Mozambique and the Transvaal. Fringing and mist-forest, and sometimes in secondary forest, c. 50–1200 m.

Specimens from Nyasaland approach *X. odoratissima* by their relatively broad-based obtuse leaves.

6. **Xylopia torrei** N. Robson in Bol. Soc. Brot., Sér. 2, **32** : 157 (1958). Type : Mozambique, Gaza, Chibuto, *Torre* 2350 (LISC, holotype).
  Closely related to *X. holtzii*, but differs in the following characters :

Shrub, up to 2 m. high. Branches red-brown, turning greyish, eventually glabrous. Leaves with lamina 2–4 × 1·2–1·9 cm., elliptic or oblong to suborbicular, obtuse or rounded to shallowly emarginate at the apex, cuneate to rounded at the base ; petiole 2 mm. long. Flowers solitary or more rarely paired, spreading or reflexed ; pedicels 4–5 mm. long. Petals (outer) 1–1·5 cm. long. Carpels 4–7. Fruits on pedicel c. 8 mm. long ; fruiting carpels c. 4–5, 0·7 × 0·5 cm. (immature), 1-seeded, subglobular, with stipes 1–2 mm. long.

Mozambique. SS : Panda, fl. 25.ii.1955, *E.M. & W.* 598 (BM ; LISC).
At present known from only three collections made at Chibuto and Panda. Dry forest
and forest margins, c. 100–150 m.

7. **Xylopia odoratissima** Welw. ex Oliv., F.T.A. **1** : 31 (1868) pro parte quoad specim.
Welw.—Welw. in Trans. Linn. Soc. **27** : 12 (1869).—Engl. & Diels in Engl., Mon.
Afr. Pflanz. **6** : 66 (1901).—R.E.Fr., Wiss. Ergebn. Schwed. Rhod.-Kongo-Exped.
1 : 45 (1914).—Exell & Mendonça, C.F.A. **1, 1** : 19 (1937) excl. specim. ex Zaire,
2 : 355 (1951).—Boutique, F.C.B. **2** : 330 (1951).—Pardy in Rhod. Agr. Journ.
53 : 634, cum tab. (1956). Type from Angola (Huila).
Xylopicrum odoratissimum (Welw. ex Oliv.) Kuntze, Rev. Gen. Pl. **1** : 8 (1891).
Type as above.
Xylopia antunesii Engl. & Diels in Notizbl. Bot. Gart. Berl. **2** : 299 (1899) ; in
Engl., tom. cit. : 66 (1901).—Brenan, T.T.C.L. : 46 (1949).—O. B. Mill. in Journ.
S. Afr. Bot. **18** : 14 (1952). Type from Angola (Huila).

Shrub or small tree (1) 2–6 (9) m. high, erect or ± spreading, much branched.
Branches reddish-brown, ± densely yellowish-fawn-pubescent at first, eventually
glabrous. Leaves petiolate ; lamina 3·5–11·5 (13·5) × 1·6–5·7 cm., oblong to
elliptic or rarely lanceolate, obtuse or rarely subacute to rounded or emarginate at
the apex, broadly cuneate to truncate or cordate at the base, membranous, dull
greyish- or ± glossy green and ± sparsely appressed-pubescent (rarely glabrous)
above, dull pale green or greyish and more densely appressed-pubescent below,
with reticulate venation varying in density and prominence ; petiole 3–10 (17) mm.
long, red- or yellow-brown, pubescent or glabrous. Flowers solitary or 2–3-
fasciculate (rarely shortly pedunculate), eventually reflexed ; pedicels absent or up
to 7 mm. long, pubescent and ± stout ; bracteoles 2, broadly ovate, near the
flower, deciduous. Sepals 1·5–3 mm. long, triangular-ovate, obtuse or apiculate,
pubescent and green outside, glabrous and reddish within. Petals greenish-white
to yellow, the outer ones 1·8–3 cm. long, narrowly linear, concave and c. 3 mm.
wide at the base, golden-sericeo-pubescent outside, puberulous within except at the
base, the inner ones somewhat shorter and slightly narrower, shortly unguiculate,
puberulous on both sides except at the base. Stamens 1 mm. long, linear, with
connective prolongation capitate, rugulose. Carpels 7–14, 3 mm. long ; ovary
ovoid-cylindric or flattened, appressed-pilose ; style twice as long as the ovary,
cylindric, pilose. Fruit on a pedicel 4–7 mm. long ; fruiting carpels 1–8, 1–3 × 0·8–
1·1 cm., 1–5-seeded, obovoid or cylindric, ± curved, rounded at the apex, constricted between the seeds when dry, glabrous, finely rugose, red, reticulate-striate
or not, with stipes 3–4 mm. long. Seeds c. 1·2 cm. long, ellipsoid or ± flattened,
yellow-brown, oblique, biseriate ; aril inconspicuous.

Bechuanaland Prot. N : Kazungula, fl. iv.1936, *Miller* 312 (BM ; FHO). **N.
Rhodesia.** B : Near Kalabo Boma, fl. 13.xi.1952, *White* 2062 (FHO ; K). N : Abercorn Distr., Sunzu Hill, fl. 18.xi.1952, *White* 3711 (FHO ; K). W : Bwana Mkubwa,
fr. vii.1909, *Rogers* 8383 (K ; SRGH). S : Livingstone, fl. 12.ii.1956, *Gilges* 586 (SRGH).
S. Rhodesia. W : Nyamandhlovu Distr., fl. i.1931, *Pardy* in GHS 4497 (BM ; FHO ;
K ; SRGH).
In Angola, Bechuanaland Prot., S. Rhodesia, N. Rhodesia, Belgian Congo (Katanga)
and south-west Tanganyika. Open *Brachystegia* or *Baikiaea* woodland, usually on sand
or rocky ground, 900–1680 m.

In the eastern part of its range *X. odoratissima* may not always be easily distinguishable
from *X. holtzii* morphologically, but these two species appear to have different ecological
requirements. *X. odoratissima* is very variable, and several recognizable local races occur.
These are specially noticeable in the north-west of N. Rhodesia, where the plants tend to
have less densely pubescent stems and leaves than usual. In one gathering from Mwinilunga (*Holmes* 887) the leaf lamina is apiculate at the apex and decurrent down the petiole ;
while in another, from the shore of Lake Bangweulu (*Brenan* 8041), the flowers are smaller
than usual and the leaves are mostly lanceolate rather than elliptic.

8. **Xylopia tomentosa** Exell in Journ. of Bot. **64**, Suppl. Polypet. : 7 (1926) ; op. cit.
70, Suppl. Polypet. : 212 (1932) ; op. cit. **73**, Suppl. Polypet. Addend. : 5 (1935).—
Exell & Mendonça, C.F.A. **1, 1** : 17 (1937), **2** : 355 (1951).—Boutique, F.C.B. **2** :
334 (1951). Type from Angola (Bié).
Xylopia mendoncae Exell in Journ. of Bot. **72** : 280 (1934).—Exell & Mendonça
C.F.A. **1, 1** : 18 (1937), **2** : 355 (1951). Type from Angola (Lunda).

Shrub or rhizomatous shrublet, 0·4–3 (4·5) m. high, much branched. Branches
reddish-brown, ± densely yellowish- or brown-pubescent at first, eventually

glabrous. Leaves petiolate ; lamina 2·4–6 (8·6) × 1·2–3·5 cm., oblong-ovate or suborbicular to lanceolate or elliptic ; obtuse to rounded or emarginate at the apex, cuneate to truncate or subcordate at the base, subcoriaceous, dull or ± glossy greyish- or bluish-green and glabrous or glabrescent above, dull pale green and appressed-pubescent or ± densely tomentose below, with densely reticulate venation prominent on both sides (sometimes hidden by the indument below) ; petiole 2–7 mm. long, brown or blackish, pubescent. Flowers solitary or in clusters of 2–10 ; pedicels absent or up to 2 mm. long ; bracteoles 2, ovate-orbicular, often forming an involucre, persistent. Sepals 2–3 (4) mm. long, ovate or triangular-ovate, obtuse or rounded to apiculate or acute, densely brown-pubescent outside, glabrous and red-brown within. Petals greenish-white to yellow, the outer ones 0·5–1·3 cm. long, triangular-lanceolate to linear, 2·5–4·5 mm. wide at the base, golden- to dark-brown-sericeous-tomentose on both sides or greyish-puberulous within except at the base, the inner ones shorter and narrower, ± unguiculate, brown-tomentose to greyish-puberulous on both sides except at the base. Stamens 1 mm. long, linear ; connective-prolongation rugulose. Carpels 5–11, c. 3 mm. long ; ovary ovoid-cylindric or flattened, appressed-pilose ; style twice as long as the ovary, cylindric, pilose. Fruit on a pedicel 4–6 mm. long ; fruiting carpels 1–7, 1–3·2 × 0·7–0·9 cm., 1–6-seeded, obovoid or cylindric, ± curved, rounded at the apex, ± constricted between the seeds when dry, yellow-green and ± densely brown-pubescent outside, cerise within, with stipes 2–4 mm. long. Seeds 0·6–1 cm. long, ovoid-ellipsoid, orange-red, oblique, 2-seriate, entirely covered by an orange aril when fresh.

N. Rhodesia. B : Balovale, fr. vii.1952, *Gilges* 122 (K; PRE; SRGH). W : just N. of Mwinilunga, fl. 2.x.1937, *Milne-Redhead* 2528 (K).
In Angola, N. Rhodesia and the Belgian Congo (Kasai, Upper Katanga). Dry woodland or scrub, often on Kalahari sand, 1000–1700 m.

Although the plants from Mwinilunga and the Belgian Congo tend to be low rhizomatous shrubs with leaves larger and pubescent below, while those from further south are often taller with leaves smaller and tomentose below, it does not seem possible to recognize *X. mendoncae* as a separate species. Indeed the distinction between *X. tomentosa* and *X. odoratissima* is not always clear-cut ; but the sessile or subsessile flowers and pubescent fruits are fairly good diagnostic characters for *X. tomentosa*. *X. odoratissima* var. *minor* Engl. (from Angola) has pubescent fruits and therefore can probably be included in *X. tomentosa*. The inflorescences of *X. tomentosa* are frequently galled.

9. **Xylopia collina** Diels in Notizbl. Bot. Gart. Berl. **13** : 271 (1936).—Brenan, T.T.C.L : 46 (1949). Type from Tanganyika (Lindi).

Shrub or small tree, 1–2·4 m. high, erect, much branched. Branches reddish-brown, slender, spreading-fawn-pilose at first, eventually glabrous. Leaves petiolate ; lamina (2) 3–6·6 × (1·4) 2–3·5 cm., oblong to elliptic-lanceolate or obovate, acute or obtuse to rounded or apiculate at the apex, cordate (rarely rounded) at the base, membranous, bright green above, paler below, appressed-pilose on both surfaces, spreading-pilose along the midrib below and round the margin, with densely reticulate venation not or scarcely prominent on both sides ; petiole 2–3 (5) mm. long, brown or blackish, ± spreading-pilose. Flowers solitary ; pedicel absent or up to 2·5 mm. long, pilose ; bracteoles 1–2, ovate. Sepals 2 mm. long, triangular-ovate, obtuse or subacute, pilose or pubescent and green outside, glabrous and reddish within. Petals carmine, the outer ones 1·3–1·6 cm. long, linear-lanceolate, concave and 3–3·5 mm. wide at the base, sericeous-pubescent outside, puberulous within, glabrous at the base, the inner ones shorter and much narrower, puberulous. Stamens 1 mm. long, linear ; connective-prolongation capitate, rugulose. Carpels c. 7–10, c. 2·5 mm. long ; ovary cylindric or ± flattened, appressed-pilose ; style about equal to the ovary, cylindric, pilose. Fruit on a pedicel 6 mm. long ; fruiting carpels 3–10, 1–2·5 × 0·6–1·3 cm., obovoid or cylindric, slightly curved, rounded at the apex, glabrous finely rugose, green and pruinose outside, scarlet within, vertically striate, with stipes 2–2·5 mm. long. Seeds c. 1 cm. long, ellipsoid, yellow-brown, oblique, biseriate, without aril ?

Mozambique. N : Between Muêda and Chomba, st. 20.ix.1948, *Pedro & Pedrógão* 5279 (LMJ ; SRGH).
Confined to south-east Tanganyika (Lindi Distr.) and north Mozambique. Open woodland or thickets on sand or termitaria, 200–810 m.

A specimen in Herb. Kew. from the Rondo Plateau (*Milne-Redhead & Taylor* 7617) has longer sepals (c. 4 mm.) and much broader petals (c. 8 mm.) than the other Tanganyika material examined. If this belongs to a distinct species or variety it will not be possible to allocate the Mozambique material to the correct taxon until flowers are available.

12. ANNONA L.

Annona L., Sp. Pl. **1** : 536 (1753) ; Gen. Pl. ed. 5 : 241 (1754).

Trees, shrubs or shrublets, not climbing, glabrous or with simple or stellate hairs. Buds globose to conic or triquetrous. Flowers usually bisexual, solitary or in few-flowered cymes or fascicles, terminal or extra-axillary or sometimes on the old wood, pedicellate. Bracteoles 0–2, persistent. Sepals 3, valvate, much shorter than the petals, free. Petals 6, free or connate at the base, in two equal or ± unequal whorls, or the inner whorl sometimes absent, both valvate or the inner whorl imbricate, thick, coriaceous, connivent or suberect, concave at the base or throughout. Stamens ∞, linear or linear-clavate, with thecae extrorse and often unequal at the base, and connective-prolongation obliquely capitate or apiculate, sometimes ± papillose ; filaments short. Carpels numerous, free at first or united from the beginning, ± cylindric, with a single basal ovule ; style clavate ; stigma muricate. Fruit a fleshy syncarp, ovoid-globose to cylindric, many-seeded. Seeds irregularly arranged in the syncarp ; aril carunculoid.

A genus of over 100 species, mostly tropical American, not native in Asia but cultivated widely throughout the tropics.

Inner petals present :
    Fruit globose to ovoid, obtusely squamose or almost smooth ; leaves usually ± pubescent or tomentose ; outer petals obtuse or rounded :
        Plant a shrub or small tree at least 1·5 m. high ; leaves mostly ovate or elliptic
                1. *senegalensis*
        Plant a low shrub or rhizomatous shrublet up to 1 m. high, sometimes with annual stems ; leaves mostly oblong to obovate or oblanceolate - - 2. *stenophylla*
    Fruit ovoid to cylindric, curved-spinose ; leaves glabrous ; outer petals acuminate
                3. *muricata*
Inner petals minute or absent :
    Leaves glabrous (or rarely sparsely pubescent) below ; young shoots appressed-pubescent :
        Fruit obtusely squamose ; leaves obtuse to rounded at the apex   4. *squamosa*
        Fruit almost smooth ; leaves acuminate at the apex - - - 5. *reticulata*
    Leaves tomentose below ; young shoots spreading-pubescent - 6. *cherimolia*

1. **Annona senegalensis** Pers., Syn. Pl. **2** : 95 (1807).—Harv. in Harv. & Sond., F.C. **2** : 563 (1862).—Oliv., F.T.A. **1** : 16 (1868) pro parte excl. syn. *A. arenaria* Thonn.— Engl. & Diels in Engl., Mon. Afr. Pflanz. **6** : 78 (1901) pro parte excl. syn. *A. arenaria* Thonn.—Gibbs in Journ. Linn. Soc., Bot. **37** : 429 (1906).—Sim, For. Fl. Port. E. Afr. : 7, t. 1 (1909).—Bak. f. in Journ. Linn. Soc., Bot. **40** : 18 (1911).— R.E.Fr., Wiss. Ergebn. Schwed. Rhod.-Kongo-Exped. **1** : 46 (1914) ; in Bot. Notis. **1934** : 93 (1934).—Eyles in Trans. Roy. Soc. S. Afr. **5** : 354 (1916).—Burtt Davy, F.P.F.T. **1** : 102 (1926).—Steedman, Trees, etc. S. Rhod. : 8 (1933).—Robyns & Ghesq. in Bull. Soc. Roy. Bot. Belg. **67** : 35, fig. 4 et t. 3 (1934). Type from Senegal.
    Annona chrysophylla Boj. in Ann. Sci. Nat., Sér. 2, Bot. **20** : 53 (1843).—Robyns & Ghesq., tom. cit. : 28, fig. 3 et t. II (1934).—Brenan, T.T.C.L. : 39 (1949).— Boutique, F.C.B. **2** : 272 (1951). Type from Anjouan Island.
    Annona senegalensis var. *latifolia* Oliv., F.T.A. **1** : 17 (1868).—Engl. & Diels, tom. cit. : 79 (1901).—R.E.Fr., Wiss. Ergebn. Schwed. Rhod.-Kong. Exped. **1** : 47 (1914). Type from Uganda.
    Annona porpetac Boiv. ex Baill. in Bull. Soc. Linn. Par., **1** : 341 (1882). Type from Madagascar.
    Annona senegalensis var. *porpetac* (Boiv. ex Baill.) Diels in Notizbl. Bot. Gart. Berl. **9** : 356 (1925).—Burtt Davy, F.P.F.T. **1** : 103, fig. 6 (1926). Type as above.
    Annona chrysophylla var. *porpetac* (Boiv. ex Baill.) Robyns & Ghesq., tom. cit. : 32, t. 2 (1934).—Hutch., Botanist in S. Afr. : 336 (1946).—Brenan, T.T.C.L. : 40 (1949).—Boutique, F.C.B. **2** : 273 (1951). Type as above.
    Annona senegalensis var. *chrysophylla* (Boj.) Sillans in Bull. Mus. Hist. Nat. Par., Sér. 2, **24** : 581 (1953). Type as for *A. chrysophylla*.

Shrub or small tree, 1·5–8 (10) m. high. Branches cylindric or with opposite pairs of raised lines decurrent from the leaf-bases, ± densely brown- to yellow-

or greyish-tomentose at first, eventually glabrous. Leaves petiolate ; lamina 6–18·5 (30) × 3–11·5 (21) cm., oblong to ovate or elliptic, obtuse or apiculate to rounded or slightly emarginate at the apex, cordate to truncate or cuneate at the base, chartaceous to coriaceous (more rarely membranous), yellowish- or glaucous-green and ± sparsely puberulous or more rarely glabrous above, paler or glaucous and densely sericeous-tomentose to glabrescent below, with green to reddish-purple nerves and densely reticulate venation prominent below ; petiole (5) 10–20 (25) mm. long, densely tomentose to glabrous. Flowers solitary or rarely 2–4-fascicu-late, extra-axillary, erect or ± deflexed ; pedicels 10–20 mm. long, shortly tomentose ; bracteoles 0–1 (2), small. Sepals 3–4 mm. long, ovate-triangular, obtuse to acute, pubescent or tomentose outside, glabrous within. Petals greenish outside, yellow to cream within, fleshy, the outer ones 8–12 (15) mm. long, broadly ovate, concave, obtuse, shortly and densely greyish- or yellowish-appressed-pubescent outside, glabrous, or minutely papillose within, the inner ones somewhat shorter, narrowly oblong, almost triquetrous. Stamens 1·75–2·5 mm. long, linear, with thecae equal or unequal at the base ; connective-prolongation obliquely capitate and minutely papillose ; filament ± cuneate. Carpels c. 1–1·5 mm. long, cylin-dric, glabrescent ; stigmas subclavate. Fruit on a pedicel 15–30 mm. long, erect or spreading ; syncarp 2·5–5 × 2·5–3 cm., ovoid or globose, obtusely squamose, glabrescent, orange or yellow. Seeds numerous, c. 10 mm. long, cylindric or ± flattened, orange-brown ; aril pectinate.

N. Rhodesia. N : Abercorn Distr., near Mpulungu, fl. & fr. 10.xi.1953, *Brenan* 8165 (FHO ; K). W : Ndola Golf Course, fr. 31.xii.1951, *White* 1837 (FHO ; K). C : Balengwe camp, E. of Kafue, fr. 2.xii.1919, *Shantz* 462 (K). S. Rhodesia. N : Loma-gundi, fl. x.1910, *Bell* in GHS 1030 (K ; SRGH). E : Umtali, Commonage, fl. 15.xi.1955, *Chase* 5864 (BM ; K ; SRGH). S : near Ndanga, fl. 21.x.1930, *F.N. & W.* 2178 (BM ; SRGH). Nyasaland. N : Between Katowo and Tayale Kanjere, fl. 10.ix.1952, *Chap-man* 37 (FHO ; K). C : Dedza Escarpment, st. 13.ix.1929, *Burtt Davy* 21635 (FHO). S : Mandala, Shire Highlands, fr. 1893–4, *Scott Elliot* 8440 (BM). Mozambique. N : Metangula, margin of L. Nyasa, fl. 10.x.1942, *Mendonça* 747 (BM ; LISC). Z : Milange, Sobado de Mapinga, fr. 13.xi.1942, *Mendonça* 1433 (BM ; LISC). T : Monte Zóbuè, fl. 3.x.1942, *Mendonça* 629 (BM ; LISC). MS : Chibuli, Bandula, fl. xi.1923, *Honey* 750 (K ; PRE). SS : Macia, S. Martinho, st. 16.vii.1947, *Pedro & Pedrógão* 1461 (K ; SRGH). LM : Lourenço Marques, Palmar de Polane, fr. 21.xii.1935, *Pomba Guerra* 309 (COI).

From Senegal to Fr. Cameroons, Belgian Congo, the Sudan and Kenya, and southward to S. Rhodesia, the Transvaal, Swaziland and Natal. Also in Madagascar, the Comoro Is. and the Cape Verde Is. Dry open woodland, bush and grassland, 0–1500 m. (to 2400 m. in E. Africa).

There is a tendency for the specimens from W. Africa to be more nearly glabrous than those from the Sudan southwards, and to have reddish-purple venation and a larger average number of flowers ; but too many intermediate conditions occur (especially in the Sudan region) to allow the southern group specific status as *A. chrysophylla*.

2. **Annona stenophylla** Engl. & Diels in Engl., Mon. Afr. Pflanz. **6** : 78 (1901).— R.E.Fr., Wiss. Ergebn. Schwed. Rhod.-Kongo-Exped. **1** : 46 (1914).—Robyns & Ghesq. in Bull. Soc. Roy. Bot. Belg. **67** : 35, fig. 8 et t. 4 (1934). Type : N. Rhodesia, Stevenson Road, *Scott Elliot* 8287 (K).

  Annona stenophylla var. *nana* R.E.Fr., loc. cit. Type : N. Rhodesia, Abercorn, near Katwe, *Fries* 1208 (UPS).

  Annona friesii Robyns & Ghesq., tom. cit. : 47, fig. 9 (1934). Type as for *A. steno-phylla* var. *nana*.

  Annona friesii var. *elongata* Robyns & Ghesq., tom. cit. : 48 (1934). Type : N. Rhodesia, Abercorn, *Fries* 1208 b (B† ; UPS).

Shrub or rhizomatous shrublet, up to 1 m. high, with stems simple or ± branched, usually annual. Branches cylindric or with opposite pairs of raised lines decurrent from the leaf-bases, red, fawn-tomentose at first, eventually glab-rous or persistently tomentose. Leaves petiolate ; lamina 4–18 (24) × 1–6 (12·6) cm., narrowly oblong or oblanceolate to ovate or obovate, rounded or slightly emarginate to acute or apiculate at the apex, narrowly cuneate to rounded at the base, membranous, bright or bluish-green and glabrous or sparsely pubescent above, usually paler or glaucous and ± densely tomentose (more rarely glabrous) below, with green to orange or reddish-purple nerves and densely reticulate vena-tion usually ± prominent above as well as below ; petiole 2–10 mm. long, densely

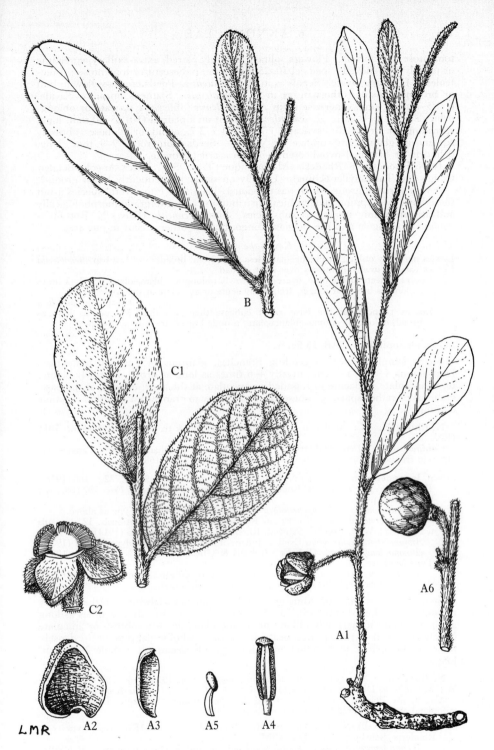

LMR

Tab. 15. ANNONA STENOPHYLLA. A.—SUBSP. STENOPHYLLA. A1, shoot and flower (× ⅔) ; A2, outer petal (× 4/3) ; A3, inner petal (× 4/3) ; A4, stamen, dehisced (× 8) ; A5, carpel (× 8) ; A6, fruit (× ⅔), all from *Richards* 7011. B.—SUBSP. LONGEPETIOLATA, leaves (× 2⅔) *White* 3397. C.—SUBSP. NANA. C1, leaves (× ⅔) *Hislop* Z.277 ; C2, flower, petals and part of receptacle removed (× 2) *Sandwith* 47.

tomentose to glabrous. Flowers solitary or rarely paired, extra-axillary, spreading or usually ± deflexed ; pedicel 10–25 mm. long, pubescent or tomentose ; bracteoles 0–1, oblong-ovate. Sepals as in *A. senegalensis*. Petals cream-yellow to buff or brownish inside, otherwise as in *A. senegalensis*. Stamens linear to clavate, c. 1·5–2 mm. long, otherwise as in *A. senegalensis* ; filament cuneate or oblong-orbicular. Carpels as in *A. senegalensis*. Fruit on a pedicel (10) 15–40 mm. long, spreading or deflexed ; syncarp 1·7–2·7 × 1·7–2·7, ovoid or globose, obtusely squamose, glabrescent, orange or yellow. Seeds numerous, 7–9 mm. long, cylindric or flattened-ovoid, orange-brown ; aril pectinate.

In N. Rhodesia, S. Rhodesia, Mozambique (Tete), SW. Africa, Angola, Belgian Congo and Tanganyika (Ufipa). Usually on sandy or frequently burnt ground.

A very variable group which has previously been split into several species ; but these have overlapping geographical distributions and intergrade morphologically and are preferably treated as subspecies. Subsp. *cuneata* (Oliv.) N. Robson is confined to Angola and the Belgian Congo, but the others occur in our area.

Key to the subspecies

Leaves narrowly oblong or oblanceolate, less than 2 cm. broad   -   subsp. *stenophylla*
Leaves varying in shape, mostly over 2 cm. broad :
    Leaves cuneate at the base, ovate to narrowly oblong or oblanceolate, mostly 3 times
    as long as broad or longer, frequently acute or apiculate at the apex
                                         subsp. *longepetiolata*
    Leaves rounded at the base, or if cuneate then less than 3 times as long as
    broad, oblong or oblong-oblanceolate, rounded or emarginate at the apex
                                               subsp. *nana*

Subsp. **stenophylla**. TAB. 15 fig. A.

Stems simple or with ascending branches, glabrous or sparsely tomentose. Leaf-lamina 4–8·3 × 1–2 cm., mostly 4–6 times as long as broad, narrowly oblong to oblanceolate, subacute to rounded or apiculate at the apex, cuneate at the base, glabrous or rather sparsely pubescent, with yellow to orange-red venation ; petiole c. 2 mm. long.

N. Rhodesia. N : Abercorn Distr., Lumi Marsh, fl. & fr. 17.xi.1956, *Richards* 7011 (K).
Confined to the Abercorn district of N. Rhodesia and the adjacent part of Tanganyika. Often in frequently burnt grassland and open woodland, 1590–1680 m.

Subsp. **longepetiolata** (R.E.Fr.) N. Robson in Bol. Soc. Brot., Sér. 2, **32** : 160 (1958).
    TAB. 15 fig. B. Type : N. Rhodesia, Bangweulu, Kawendimusi, *Fries* 782 (B† ; K ; UPS, holotype).
    Annona cuneata var. *longepetiolata* R.E.Fr., loc. cit. (1914). Type as above.
    Annona cuneata sensu R.E.Fr. in Karsten & Schenk, Vegetationsb. **12**, 1 : sub t. 2 (1914) ; Wiss. Ergebn. Schwed. Rhod.-Kongo-Exped. **1** : 46 (1914).
    Annona stenophylla sensu Exell in Journ. of Bot. **70**, Suppl. Polypet. : 211 (1932).
    Annona longepetiolata (R.E.Fr.) Robyns & Ghesq., tom. cit. : 42, fig. 6 (1934). Type as above.
    Annona longepetiolata var. *precaria* Robyns & Ghesq., tom. cit. : 46 (1934). (" longepetiola " sphalm.). Type from Angola (Malange).

Stems simple or with spreading or ascending branches, glabrous to rather densely tomentellous. Leaf-lamina (5) 7·5–15 × 2–4·7 (5·5) cm., mostly 3–5 times as long as broad, ovate to narrowly oblong or oblanceolate, acute to obtuse or apiculate (rarely rounded) at the apex, cuneate at the base, wholly glabrous or ± densely tomentellous below, with yellow to reddish-purple venation ; petiole 2–13 mm. long.

N. Rhodesia. N : Bangweulu, Kawendimusi, fl. 25.ix.1911, *Fries* 782 (K ; UPS)·
W : Mwinilunga Distr., near source of Matonchi R., fl. 7.x.1937, *Milne-Redhead* 2630 (K).
Also in the Belgian Congo (Kasai) and Angola (Lunda, Malange). Dry deciduous woodland and burnt grassland, often on sand, 1000–1540 m.

Subsp. **nana** (Exell) N. Robson, tom. cit. : 162 (1958) . TAB. 15 fig. C. Type : S. Rhodesia, Salisbury, *Rand* 1342 (BM).
    Annona senegalensis var. *subsessilifolia* Engl. in Engl. & Diels, tom. cit. : 80 (1901). Type from Angola (Bié).
    Annona senegalensis var. *rhodesiaca* Engl. & Diels in Engl., Bot. Jahrb. **39** : 484 (1907).—Eyles in Trans. Roy. Soc. S. Afr. **5** : 354 (1916). Syntypes : S. Rhodesia, near Salisbury, *Engler* 3080 (B†) ; Matopos, *Marloth* 3376 (B†).

Annona senegalensis var. *cuneata* sensu Bak. f. in Journ. Linn. Soc., Bot. **40** : 18 (1911).
Annona nana Exell in Journ. of Bot. **64,** Suppl. Polypet. : 5 (1926).—Robyns & Ghesq., tom. cit.: 43, fig. 7 (1934).—Exell & Mendonça, C.F.A. **1,** 1 : 29 (1937).— Boutique, F.C.B. 2 : 276 (1951). Type as for *A. stenophylla* subsp. *nana*.
Annona nana var. *sessilifolia* Exell, tom. cit. : 6 (1926). Type from Angola. (Bié).
Annona nana var. *oblonga* Robyns & Ghesq., tom. cit. : 46 (1934).—Exell & Mendonça, C.F.A. **1,** 1 : 29 (1937). Type from Angola (Huila).
Annona nana var. *katangensis* Robyns & Ghesq., tom. cit. : 46 (1934).—Boutique, F.C.B. **2** : 276 (1951). Type from the Belgian Congo (Katanga).
Annona cuneata var. *subsessilifolia* (Engl.) R.E.Fr. in Bot. Notis. **1934** : 93 (1934) (" subsessiliflora " sphalm.). Type as for *A. senegalensis* var. *subsessilifolia*.
Annona cuneata var. *rhodesiaca* (Engl. & Diels) R.E.Fr. in Bot. Notis. **1934** : 93 (1934).—Suesseng. in Proc. & Trans. Rhod. Sci. Ass. **43** : 86 (1951). Type as for *A. senegalensis* var. *rhodesiaca*.
Annona nana var. *subsessilifolia* (Engl.) Exell & Mendonça, C.F.A. **1,** 1 : 29 (1937). Type as for *A. senegalensis* var. *subsessilifolia*.

Stems simple or more rarely with a few ascending branches, densely pubescent-tomentose to sparsely puberulous. Leaf-lamina 5·8–12 (15·2) × (2·2) 2·6–6·7 cm., mostly 2–3 times as long as broad, oblong to obovate or elliptic-oblong, rounded to emarginate at the apex, rounded or more rarely cuneate at the base, sparsely to rather densely pubescent above, sparsely pubescent to densely lanuginous below (very rarely wholly glabrous), with yellow to reddish-purple venation ; petiole 1–7 (10) mm. long.

N. Rhodesia. B : Sesheke Distr., fl. & fr., *Macaulay* 110 (K). N : Mpundu dambo, fr. 23.x.1949, *Bullock* 1355 (K). W : Muzera R., 16 km. W. of Kakoma, fl. 28.ix.1952, *White* 3397 (FHO ; K). C : Mkushi, fl. 24.ix.1957, *Fanshawe* 3714 (K). S : Livingstone, Dambwa Forest Reserve, st. 12.i.1952, *White* 3235 (FHO ; K). **S. Rhodesia.** N : Miami Settlement Area, fl. & fr. 22.xi.1944, *Hopkins* in GHS 13057 (K ; SRGH). W : Nyamandhlovu, Umgusa R., fl. x.1929, *Pardy* 4612 (BM ; FHO ; K ; SRGH). C : Marandellas, Grasslands Expt. Station, fl. 21.x.1945, *Rattray* 740 (K ; SRGH). E : Inyanga, fl. 29.x.1930, *F.N. & W.* 2385 (BM ; PRE ; SRGH). S : Victoria Distr., st. 1909, *Monro* 305 (BM ; SRGH). **Mozambique.** T : Angónia, Vila Mousinho, fl. 15.x.1943, *Torre* 6045 (BM ; LISC).

In N. Rhodesia, S. Rhodesia, Mozambique, Belgian Congo (Katanga), Angola and SW. Africa (? including Caprivi). Often in frequently burnt grassland, swamp margins and cleared woodland, usually on sand, c. 1000–1500 m.

R.E. Fries (tom. cit. : 47 (1914)) records this subspecies from the Caprivi Strip, but so far no specimens from that region have been seen.

Introduced Species of Annona

3. **Annona muricata** L., Sp. Pl. **1** : 536 (1753).
Native of Central and tropical South America and of the West Indies.

4. **Annona squamosa** L., Sp. Pl. **1** : 537 (1753).
Native of the West Indies.

5. **Annona reticulata** L., Sp. Pl. **1** : 537 (1753).
Native of the West Indies.

6. **Annona cherimolia** Mill., Gard. Dict. ed. 8 : no. 5 (1768).
Native of the northern Andean region.

Specimens of three of these species have been collected in our area. *A. reticulata* is cultivated widely throughout Africa and probably occurs there also.

13. MONODORA Dunal

Monodora Dunal, Mon. Anon. : 34, 79 (1817).

Trees or shrubs, sometimes climbing or scrambling, glabrous or with simple hairs, with young shoots which frequently blacken and become brittle on drying. Buds ovoid. Flowers bisexual, solitary or more rarely paired or in few-flowered cymes, terminal or extra-axillary or in the axils of fallen leaves, pedicellate. Bracteole solitary, usually amplexicaul, persistent, often foliaceous. Sepals 3, valvate, much shorter than the petals, free. Petals 6, in two unequal whorls, both valvate, all united at the base, often with undulate margins, the outer ones broad at the base, usually spreading or reflexed, the inner ones ± unguiculate, usually connivent at the apices or erect. Stamens ∞, oblong or cuneiform to $\pm$ orbicular, with thecae extrorse, and connective-prolongation thickened ; filaments absent. Carpels several, united to form a 1-locular globose or conic ovary with parietal placentation and numerous ovules ; stigmas sessile, radiating. Fruit a coriaceous or ligneous berry, globose to ellipsoid or obovoid, many-seeded. Seeds irregularly arranged, shining ; aril absent.

A tropical African genus of c. 15 species.

Leaves and young stems $\pm$ pubescent ; petals narrowed in the middle ; fruit elongate
  (obovoid or $\pm$ ellipsoid), smooth not ridged :
Outer petals undulating, 5–26 mm. broad in the upper part ; leaves cordate at the base
                                                                                1. *grandidieri*
  Outer petals plane, 2–4 mm. broad in the upper part ; leaves cuneate at the base
                                                                                2. *stenopetala*
Leaves and young stems glabrous ; petals not or only slightly narrowed in the middle ;
  fruit globose or ovoid-conic, ridged or wrinkled :
Outer petals undulating, yellow-green and white with purplish spots ; fruit ovoid-conic
  or globose, prominently and $\pm$ reticulately ridged   -   -   -   3. *angolensis*
Outer petals plane, turning purple-brown, not spotted ; fruit globose, smooth or
  wrinkled, sometimes slightly vertically ridged -   -   -   -   -   4. *junodii*

1. **Monodora grandidieri** Baill. in Adansonia, **8**: 301 (1868).—Engl. & Diels in Engl.,
    Mon. Afr. Pflanz. **6** : 85, t. 28 F (1901).—Diels in Engl., Bot. Jahrb. **39** : 485
    (1907).—Brenan, T.T.C.L. : 42 (1949). TAB. **16** fig. B. Type from Zanzibar.
      Monodora veithii Engl. & Diels in Engl., Bot. Jahrb. **39** : 485 (1907).—Brenan,
    tom. cit. : 43 (1949). Type from Tanganyika (Usambara).
      Monodora stocksii Sprague in Kew Bull. **1916** : 38 (1916). Type : Mozambique,
    Moçimbua, *Stocks* 96 (K).

Shrub or small tree (1·5) 2·5–12 m. high, much-branched, sometimes sub-scandent.   Branches spreading or pendulous, dark purple, buff-pubescent at first, eventually glabrous. Leaves petiolate ; lamina 7·5–20 (24) × 3·5–8·2 (8·7), obovate to oblanceolate (or oblong when young), obtuse or rounded to apiculate or shortly acuminate at the apex, cordate at the base, membranous or subcoriaceous, bluish-green and concolorous or glaucous below, $\pm$ densely and softly pubescent on both sides or almost glabrous, usually with ciliate margin, with nerves and reticulate venation $\pm$ prominent on both sides ; petiole 2–3 (5) mm. long, often blackish, softly pubescent. Flowers solitary, terminal or extra-axillary appearing with the young leaves ; pedicel 2–5·7 (7·3) cm. long, slender, pubescent ; bracteole ovate, attenuate, $\pm$ amplexicaul, glabrous above, appressed-pubescent below. Sepals c. 1–2 cm. long, lanceolate or narrowly triangular, acute or attenuate, recurved with undulate margins, $\pm$ densely sericeous-pubescent below, almost glabrous above. Petals yellow to greenish-white, variously flushed with pink or red to dark brown, or the inner ones wholly reddish, the outer ones (3) 3·7–6·5 cm. long, 0·5–1·8 (2·6) cm. broad in the upper part, oblanceolate or spathulate to linear with a broadened base, acute to acuminate or cuspidate, recurved with undulate margins, more densely pubescent below than above, the inner ones 0·7–0·9 (1·3) × 0·7–1·2 (1·4) cm., sagittate to triangular or rhomboid, acute or acuminate, with a claw 0·7–0·9 (1·3) cm. long, erect, $\pm$ connivent, setose-pilose towards the apex, pubescent towards the base. Stamens 1 mm. long, oblong-cuneiform. Ovary c. 1 mm. long, globose or conic, glabrous except the stigmas. Fruit 5–7·5 × 4–4·5 cm., obovoid or ellipsoid, finely rugose, not ridged, mottled green and white, glabrous (? or shortly pubescent), pendulous, on an elongate pedicel 6–8 cm. long. Seeds 2–2·5 cm. long, bean-shaped, smooth, reddish-brown.

Nyasaland. C : Dedza, Mua-Livulezi Reserve, fl. 9.xii.1953, *Adlard* 12 (K ; SRGH). **Mozambique.** N : Erati, between Namapa and Posto de Lurio, fl. 11.x.1948, *Barbosa* 2369 (BM ; LISC).

In south-east Kenya, eastern Tanganyika, Zanzibar, Nyasaland and Mozambique. Evergreen rain-forest, bamboo thickets and streamsides, 0–600 m.

M. grandidieri is a very polymorphic species which varies in leaf-shape, petal-shape and degree of pubescence, and correlation of these variations is rather difficult on account of the rarity of mature leaves on flowering shoots. None of the alleged differences in these characters appears to be constant, so that it is not possible to maintain *M. veithii* or *M. stocksii* as separate species.

Two specimens (*Torre* 1324 from Nampula, Niassa and *Migeod* 699 from Tendaguru,

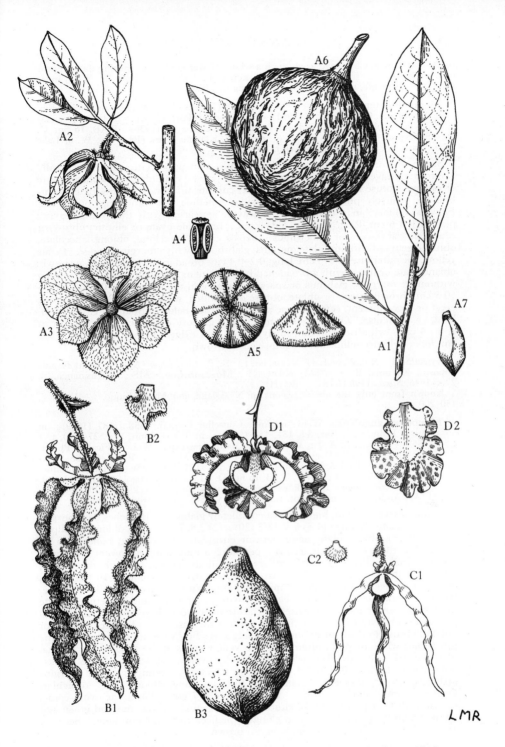

Tab. 16. A.—MONODORA JUNODII. A1, leaves (× ⅔) *Chase* 5504 ; A2, shoot with flower (× ⅔) *Lovemore* 450 ; A3, flower (× ⅔) ; A4, stamen (×10) ; A5, ovary, 2 views (×10) ; A6, fruit (× ⅔) ; A7, seed (× ⅔), A3–7 from *Chase* 5504. B.—MONODORA GRANDIDIERI. B1, flower (× ⅔) ; B2, inner petal (× ⅔), both from *Burtt* 5044 ; B3, fruit (× ⅔) *Greenway* 1207. C.—MONODORA STENOPETALA. C1, flower (× ⅔) ; C2, inner petal (× ⅔), both from *Kirk* s.n. D.—MONODORA ANGOLEṆSIS. D1, flower (× ⅔) ; D2, outer petal (× ⅔), both from *Fanshawe* 3499.

south-east Tanganyika) have leaves with a whitish lower epidermis, making them appear tomentose, and the fruits of the latter are said to be " slightly velvety ". When more material is available, these may prove to belong to a distinct species.

2. **Monodora stenopetala** Oliv., F.T.A. **1** : 39 (1868).—Engl. & Diels in Engl., Mon. Afr. Pflanz. **6** : 85, t. 28 E (1901).—Sim, For. Fl. Port. E. Afr. : 8 (1909) (" steno-phylla " sphalm.).—Burtt Davy & Hoyle, N.C.L. : 30 (1936). TAB. **16** fig. C. Syntypes : Nyasaland, Rapids of the Shire, Kavuma, Kirk (B† ; K) ; W. of Lake Nyasa, lat. 12° S, *Livingstone* (K).

Shrub or small tree. Branches dark purple, buff-pubescent at first, eventually glabrous. Leaves petiolate ; lamina c. 8·5–11 × 3–4·2 cm., obovate to oblanceolate or oblong, obtuse or rounded at the apex, cuneate to rounded at the base, mem-branous, sparsely pubescent on both sides ; petiole 3–4 mm. long, pubescent. Flowers solitary, in the axils of the leaves, appearing with the young leaves ; pedicel 1–1·6 cm. long, slender, pubescent ; bracteole ovate or elliptic, obtuse or rounded, glabrescent or puberulous. Sepals c. 4–5 mm. long, oblong-lanceolate, obtuse, recurved with undulate margins, glabrous or with a marginal fringe. Petals yellow, the outer ones 3·5–5 cm. long, 0·2–0·4 cm. broad in the upper part, narrowly oblanceolate or linear with a broadened base, acute or attenuate, widely spreading with entire margins, glabrous or sparsely pubescent, the inner ones 0·2–0·4 × 0·35 –0·5 cm., broadly ovate or oblate, obtuse or rounded, with a claw 0·3–0·7 cm. long, erect or ± connivent, setose-pilose above, glabrous below. Stamens c. 0·35 mm. long, ± orbicular. Ovary c. 0·75 mm. long, globose or conic, glabrous. Fruit c. 6 × 4 cm., cylindric-ellipsoid, finely rugose, not ridged, glabrous. Seeds 1·5–2 cm. long, cylindric-ellipsoid, smooth, yellow-brown.

Nyasaland. N : W. of Lake Nyasa, lat. 12° S., fl., *Livingstone* (K). S : Shire Rapids opposite Kavuma, fl. 4.xi.1861, *Kirk* (K). **Mozambique. MS :** Inhamitanga, fl. 27.xi.1946, *Simão* 1196 (LISC ; SRGH).

Known from only the above regions of Nyasaland and Mozambique. " Woods ", 150–200 m.

3. **Monodora angolensis** Welw. in Ann. Conselho Ultram. **1858** : 587 (1859) ; in Journ. Linn. Soc., Bot. **3** : 151, 154 (1859) ; in Trans. Linn. Soc. **27** : 10, t. 1 (1869).—Oliv., F.T.A. **1** : 38 (1868) pro parte quoad spec. Welw.—Ficalho, Pl. Ut. Afr. Port. : 86 (1884).—Engl. & Diels in Engl., Mon. Afr. Planz. **6** : 88, t. 29 fig. C (1901).—Exell & Mendonça, C.F.A. **1,** 1 : 31 (1937), 2, 356 (1951).—Boutique, F.C.B. **2** : 265, t. 24 (1951). TAB. **16** fig. D. Type from Angola (Cuanza Norte). *Monodora gibsonii* Bullock ex Burtt Davy, Uganda Checklist : 20 (1935).— F. W. Andr., Fl. Anglo-Egypt. Sudan, **1** : 6 (1950). Type from Uganda.

Shrub or tree, 3–20 m. high, almost completely glabrous. Branches dark purple. Leaves petiolate ; lamina (4·5) 6·5–18·5 (20) × (2) 2·8–7 (8·5) cm., oblong or elliptic to obovate or oblanceolate, acuminate or cuspidate at the apex, cuneate to ± rounded at the base, subcoriaceous ; petiole 2–8 mm. long. Flowers solitary or rarely paired, axillary or extra-axillary, pendulous, appearing with the mature leaves, scented ; pedicel 1·5–6 (8) cm. long, slender ; bracteole 5–10 mm. long, orbicular or ovate, acuminate or apiculate, amplexicaul. Sepals 5–10 mm. long, triangular-lanceolate, subacute, recurved, with ± undulate and sometimes ciliate margins, green or purplish-spotted. Petals (outer) greenish-yellow to orange with purplish spots, whitish or pink-flushed towards the base, 3–6 × 1·5–3 cm., ovate or oblong, obtuse to acute or attenuate, widely spreading with undulate margins, the inner ones white or pink, often with two yellow spots at the base of the lamina, (0·4) 0·6–1·1 × (0·5) 0·9–1·5 (1·9) cm., broadly ovate or sagittate, obtuse or cus-pidate, with a claw (0·3) 0·7–1·0 cm. long, erect or ± connivent, with margins glabrous or ± pubescent. Stamens c. 0·5 mm. long, oblong or ± orbicular. Ovary c. 1 mm. long, ± conic. Fruit 5–9 × 4·5–7 cm., globose or ovoid-conic, rugose with very prominent longitudinal and reticulate ridges, mottled green and white, glabrous, pendulous, on an elongate pedicel up to c. 8 cm. long. Seeds c. 1·3 cm. long, ± flattened-ovoid, smooth, brown.

N. Rhodesia. N : Kawambwa, fl. 22.viii.1957, *Fanshawe* 3499 (K).

In Angola, Cabinda, Belgian Congo, Gaboon, Fr. Cameroons, N. Rhodesia, Uganda, and the Sudan. Evergreen rain-forest, fringing forest and evergreen thickets, 30–1050 m.

4. **Monodora junodii** Engl. & Diels in Notizbl. Bot. Gart. Berl. **2** : 301 (1899) ; in Engl., Mon. Afr. Pflanz. **6** : 86, t. 28 fig. D (1901).—Bak. f. in Journ. Linn. Soc.,

Bot. **40** : 18 (1911).—Brenan, T.T.C.L. : 42 (1949). TAB. **16** fig. A. Type : Mozambique, Delagoa Bay, *Junod* 411 (B† ; Z).
Monodora sp.—Oliv., F.T.A. **1** : 39, fruct. 2 (1868).

Shrub or small tree, (1·2) 2–7 m. high, erect, sometimes virgate. Branches blackish-purple with pale lenticels at first, turning silvery-grey, glabrous. Leaves petiolate ; lamina 6·5–16·5 (17·5) × 3–5·5 (7) cm., narrowly oblanceolate to obovate or elliptic-oblong, rounded or obtuse to acuminate at the apex, cuneate to rounded at the base, usually ± membranous, glabrous, often with a reddish midrib ; petiole 1–6 mm. long, blackish, glabrous. Flowers solitary, extra-axillary or in the axils of fallen leaves, pendulous, appearing with the young leaves, scentless ; pedicels 0·8–2·0 cm. long (or rarely to 4 cm. long), sparsely pubescent or glabrescent ; bracteole 3–9 mm. long, ovate to orbicular, rounded, amplexicaul, glabrescent except for a marginal fringe. Sepals c. 5–10 (14) mm. long, elliptic to ovate or orbicular, obtuse or rounded, sparsely pubescent, spreading, with plane margins, green. Petals yellow or greenish at first, the inner ones with a reddish base, all turning purple-brown, the outer ones 2–3·5 (4·5) × 1·6–2·7 cm., obovate to broadly elliptic or suborbicular, rounded to obtuse or bluntly acuminate, widely spreading with plane margins, sparsely pubescent or puberulous, the inner ones (0·6) 1–1·6 × (0·9) 1·4–2·1 cm., broadly ovate to subhastate, obtuse or rounded to cuspidate, with a claw (0·5) 0·7–1·0 cm. long, erect or ± spreading, pubescent except for the glossy upper side of the claw. Stamens c. 0·5 mm. long, oblong. Ovary c. 0·75 mm. long, globose, sparsely pubescent. Fruit c. 4–5 cm. in diam., ± globose, wrinkled or lightly vertically ridged, mottled green-grey and brown, glabrous, reflexed, on a pedicel c. 1–3 cm. long. Seeds c. 1·5–2 cm. long, ± flattened-ovoid, smooth, yellow-brown.

S. Rhodesia. N : Mtoko Distr., Mkota Reserve, fl. 20.ix.1951, *Whellan* 571 (K ; SRGH). E : 27 km. S. of Umtali, 1·6 km. E. of Melsetter road, fl. 30.xi.1952, *Chase* 4732 (BM ; K ; SRGH). S : Nuanetsi Distr., rapids on Lundi R., fl. xi.1956, *Davies* 2199 (K ; SRGH). **Nyasaland.** S : Ft. Johnston, Nankumba, fl. 19.xi.1954, *Jackson* 1388 (BM ; K). **Mozambique.** N : Lurio, Chamba road, fl. 19.x.1948, *Andrada* 1422 (COI ; LISC). T? : R. Shire valley, fl., *Waller* (K). MS : Madanda forest, fl. 5.xii.1906, *Swynnerton* 1765 (K). SS : outskirts of Chibuto, st. 12.ii.1942, *Torre* 3971 (BM : LISC). LM : Lourenço Marques, fl. 6.xii.1897, *Schlechter* 11630 (BM ; COI ; E ; K).
In Tanganyika, Mozambique, Nyasaland, S. Rhodesia and the Transvaal. Dry open woodland or scrub, 50–900 m.

Throughout the eastern part of the range of *M. junodii* occasional specimens occur with elliptic-oblong rather than oblanceolate leaves, and relatively long and glabrous pedicels and outer petals. These might be worthy of recognition as a variety, but it has not been possible to make a decision with the material available.

GENUS UNKNOWN

Slender liane up to 12 m. (at least). Leaf-lamina 11·5–13·5 × 4·8–5 cm., obovate, shortly and bluntly acuminate at the apex, rounded at the base, subcoriaceous, glaucous-grey, glabrous, with orange nerves and reticulate venation prominent below ; petiole 8 mm. long, stout. Flowers solitary, axillary ? ; pedicels 3·8 cm. long, very slender, not broadened upwards. Sepals 2 mm. long, broadly ovate, obtuse, united at the base. Petals 6, in 2 whorls, free, unequal, the outer ones (damaged) much larger, the inner ones, c. 3 mm. long, oblong, obtuse. Stamens and carpels rather numerous. Fruiting carpels 1-seeded, elliptic, sparsely appressed-pubescent, yellow-brown. Seed erect.

N. Rhodesia. W : Mwinilunga, fr. 13.x.1955, *Holmes* 1272 (K).

The stout petioles of this specimen suggest an affinity with the genus *Neostenanthera* Exell ; but from the very slender pedicels (which are not broadened upwards), and the relatively large sepals it seems more likely to belong to *Oxymitra* (Bl.) Hook. f. & Thoms. Unfortunately it is not possible to discern whether or not the anthers are septate.

7. MENISPERMACEAE
By G. Troupin

Twining or rarely erect shrubs or small trees, dioecious. Wood in cross-section showing broad medullary rays. Leaves alternate, petiolate, exstipulate, sometimes peltate, without stipules, usually simple, entire or lobed. Inflorescence various, many-flowered, the flowers rarely solitary or geminate, axillary or borne on the leafless wood. Flowers small, actinomorphic, rarely slightly irregular. Male flowers : sepals 3–12 or more, rarely 1, free or slightly connate, imbricate or valvate ; petals 1–6 or absent, free or connate, usually imbricate ; stamens 3–6 or indefinite, rarely 2, free or variously united. Female flowers : sepals and petals generally as in male flowers, sometimes not so numerous ; staminodes present or absent ; carpels 3–6 or more, rarely 1, free ; ovules 2, soon reduced to 1 by abortion, attached to the ventral suture. Fruiting carpels drupaceous, with the scar of the style subterminal or near the base by excentric growth ; exocarp membranous or subcoriaceous, mesocarp more or less pulpy, endocarp often chartaceous or bony, rugose, tuberculate or ribbed and with the septum of the condyle,* if any, perforated or not. Seeds often curved and horseshoe-shaped, with uniform or ruminate endosperm, or without endosperm.

Key to Plants with Female Flowers or Fruits

Leaves not peltate :
　Sepals 3–40 :
　　Inner sepals connate and coriaceous ; flowers axillary, usually solitary ; carpels 4–12
　　　　　　　　　　　　　　　　　　　　　　　　　　　　　　　　1. Epinetrum
　　Inner sepals free :
　　　Carpels 8–40 ; sepals thick :
　　　　Surface of fruit glabrescent, smooth or rugose ; sepals glabrous or nearly so ;
　　　　　petals present - - - - - - - - - **3. Tiliacora**
　　　　Surface of fruit velutinous or puberulous ; sepals hairy ; petals minute or absent
　　　　　　　　　　　　　　　　　　　　　　　　　　　　　　　　4. Triclisia
　　　Carpels 3–6 ; sepals thin :
　　　　Endocarp of fruit prickly or fibrillous :
　　　　　Petals 6 ; surface of fruit bristly - - - - - **5. Jateorhiza**
　　　　　Petals 0 ; surface of fruit smooth - - - **6. Dioscoreophyllum**
　　　　Endocarp of fruit smooth, rugose or tuberculate :
　　　　　Flowers numerous, in more or less elongate and compound inflorescences :
　　　　　　Leaves nearly rounded or subangular, densely hairy, especially when young ;
　　　　　　　petals with a rib on the inner side - - - **7. Chasmanthera**
　　　　　　Leaves entire, not angular, glabrous ; petals without a rib **8. Tinospora**
　　　　　Flowers solitary or in short condensed inflorescences :
　　　　　　Flowers, in few-flowered corymbose panicles ; endocarp of fruit without
　　　　　　　prominent dorsal ridge ; seeds without endosperm - **2. Anisocycla**
　　　　　　Flowers solitary or in fascicule cymules ; endocarp of fruit with prominent
　　　　　　　dorsal ridge ; seeds with little endosperm - - - **9. Cocculus**
　Sepals 1, rarely 2 ; bracts enlarged in fruit - - - - **10. Cissampelos**
Leaves peltate ; carpel 1 :
　Bracts minute, not enlarged in fruit ; sepals 3–6 ; petals 2–4, free - **11. Stephania**
　Bracts conspicuous, enlarged in fruit, membranous or leafy ; sepals 1 ; petals 1, rarely
　　more - - - - - - - - - - - **10. Cissampelos**

Key to Plants with Male Flowers

Leaves not peltate :
　Inner sepals connate ; stamens 15–30, united into an elongated synandrium ; flowers
　　in shortly pedunculate cymules or glomerules - - - - **1. Epinetrum**
　Inner sepals free :
　　Petals 0 :
　　　Stamens free ; inflorescences not elongated - - - - **4. Triclisia**
　　　Stamens united into a subglobose synandrium ; inflorescences elongated
　　　　　　　　　　　　　　　　　　　　　　　　　　　6. Dioscoreophyllum

---

* The condyle is a prominent enlargement of the placenta and forms a hollow chamber within the cavity of the loculus round which the seed is moulded.

Petals 3–6 :
    Stamens 9–18, united into a conical synandrium ; inflorescences of corymbose
        panicles or cymules, densely flowered - - - - - **2. Anisocycla**
    Stamens 6–9, free or united :
        Petals more or less united into a cup ; stamens united into a flattened stipitate
          synandrium ; anthers opening by a transverse slit - **10. Cissampelos**
        Petals not united into a cup :
          Inner sepals valvate :
            Petals developed ; inner sepals glabrous or nearly so - **3. Tiliacora**
            Petals minute - - - - - - - - **4. Triclisia**
          Inner sepals imbricate :
            Anthers opening by a transverse slit :
              Inflorescences in long panicles of 3–7-flowered glomerules ; leaves large,
                deeply lobed, strigose - - - - - - **5. Jateorhiza**
              Inflorescences in short cymules of axillary flowers ; leaves small, generally
                entire, not strigose - - - - - - **9. Cocculus**
            Anthers opening by a longitudinal slit :
              Leaves nearly rounded or subangular, densely hairy, especially when
                young ; petals with a rib on the inner side - **7. Chasmanthera**
              Leaves entire, not angular, glabrous ; petals without a rib **8. Tinospora**
Leaves peltate ; stamens always united :
    Sepals 6–8 ; petals 3–4, free - - - - - - - - **11. Stephania**
    Sepals 4 ; petals united into a cup - - - - - - **10. Cissampelos**

1. EPINETRUM Hiern

Epinetrum Hiern, Cat. Afr. Pl. Welw. **1** : 21 (1896).

Shrubs, scandent shrubs or lianes. Leaves simple, petiolate. Male inflorescences of small 1–4-flowered condensed cymules, solitary or two together, subsessile or pedunculate, axillary. Male flowers with 6–12 sepals, the 3–9 outer ones
imbricate, bract-like, the 3 inner ones much larger than the outer ones, coriaceous,
valvate, united to form a false corolla ; petals 6 or less, truncate-reniform, fleshy ;
synandrium 15–30-locular, conical, stipitate ; anther-thecae with transverse dehiscence. Female flowers axillary, usually solitary ; sepals and petals more or less
similar to those of the male flowers ; carpels 4–12, hairy. Drupelets ovoid-ellipsoid,
exocarp echinulate, endocarp coriaceous and smooth. Seeds without endosperm.

Leaves with 1 or 2 pairs of basal nerves, the stronger pair reaching the upper quarter of the
    lamina ; sepals 6 - - - - - - - - - - 1. *delagoense*
Leaves without prominent basal nerves, lateral nerves in c. 5 pairs, ± parallel ; sepals 9
                                           2. *exellianum*

1. **Epinetrum delagoense** (N.E.Br.) Diels in Engl., Pflanzenr. IV, 94 : 95 (1910).
    TAB. **17**. Type : Mozambique, Delagoa Bay, *Bolus* 7632 (K, holotype).
    Synclisia delagoensis N.E.Br. in Kew Bull. **1892** : 196 (1892). Type as above.
    Synclisia zambesiaca N.E.Br., loc. cit., pro parte quoad fl. masc.
    Junodia triplinervia Pax in Engl., Bot. Jahrb. **28** : 22 (1899). Type : Mozam-
bique, Delagoa Bay, *Junod* 464 (B†).
    Anisocycla triplinervia (Pax) Diels in Engl., Pflanzenr. IV, 94 : 93 (1910). Type
as above.

Rhizomatous shrublet or liane with greyish hairs. Leaf-lamina 4–9 × 2–5 cm.,
oblong-elliptic to broadly elliptic, narrowed at the base, mucronulate at the apex,
subcoriaceous, glabrous except the nerves, pubescent on the upper face, puberulous
on the lower face, discolorous, nerves 4 (6), the basal ones reaching the upper
quarter of the lamina, petiole 1–1·5 cm. long. Male inflorescences in axillary
cymules, 1–3-flowered, pedicels 2–3·5 mm. long ; bracts 1·5 mm. long, linear-
lanceolate. Male flowers with 6 sepals, the 3 outer 1·5–2·5 × 0·5 mm., narrowly
lanceolate or ovate, the 3 inner 4·5–6 × 2·5–3·5 mm. ; petals 6, 0·5–0·9 × 0·5–0·9
mm., cordate-reniform, inflexed at the base ; synandrium 15–20-locular, 3–5·5
mm. long. Female flowers solitary ; sepals and petals similar to those of the male
flowers ; carpels 6. Drupelets ellipsoid, 2·2 × 1·5 cm., ferruginous-tomentellous.
Seeds 1·4 × 0·8–1 mm.

Mozambique. MS : near Dondo, fr. xii.1899, *Cecil* 261 (K). SS : Inhambane,
Homoine, fl. 29.viii.1942, *Mendonça* 18 (LISC). LM : Maputo, between Zitunde and
Ponta do Ouro, fr. 17.xi.1944, *Mendonça* 2899 (BM ; LISC).
    Also in Natal. On littoral sand.

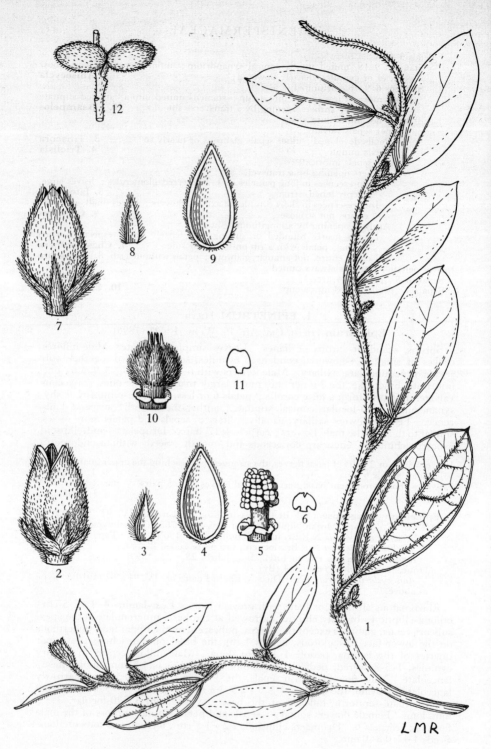

Tab. 17. EPINETRUM DELAGOENSE. 1, flowering shoot (× ⅔) *Bolus* 7632 ; 2–6, male flower,
(all × 6) *Bolus* 7632 (2, flower, 3, outer sepal, 4, inner sepal, 5, corolla and synandrium, 6, petal) ; 7–11, female flower (all × 6) *Torre* 2112 (7, flower, 8, outer sepal, 9,
inner sepal, 10, corolla and gynoecium, 11, petal) ; 12, fruit (× ⅔) *Torre* 2112.

2. **Epinetrum exellianum** Troupin in Bull. Jard. Bot. Brux. **25** : 132 (1955) ; F.T.E.A. Menisperm. : 4 (1956). Type from Ruanda-Urundi.

Liane with young branchlets tomentellous, tawny yellow, mature branchlets sparsely puberulous or glabrescent. Leaf-lamina 3·5–7 × 2·5–4·5 cm., elliptic, rounded at the base, ± rounded, obtuse or bluntly acuminate at the apex, sparsely puberulous on both sides except for the densely puberulous or tomentellous nerves or glabrescent, petiole 0·8–1·5 cm. long, yellowish tomentellous. Male inflorescences of 2–4-flowered condensed cymules, solitary or paired ; peduncles 0·5–3 mm. long or less ; pedicels absent. Male flowers with 9 sepals, the 3 outer 1·5 × 1 mm., ovate-triangular, the 3 median 2–2·5 × 1 mm., suboblong, all rusty puberulous, the 3 inner 4·5 × 1 mm., ovate-elliptic, glabrescent. Petals 6, 0·7 × 0·7–1 mm. ; synandrium 25–30-locular, 3·5–4 mm. long, on a stipe 1·5 mm. long. Female flowers solitary ; pedicels 0·5–2·5 mm. long ; carpels 3–6, 1·5–2·5 mm. long, densely yellowish puberulous. Drupelets unknown.

N. Rhodesia. N : Lake Mweru, fl. 12.xi.1957, *Fanshawe* 3934 (K). Also in Uganda and Ruanda-Urundi.

In lowland rain-forest in Uganda and lake-shore forest in N. Rhodesia.

2. ANISOCYCLA Baill.

Anisocycla Baill. in Bull. Soc. Linn. Par. **2** : 1078 (1893).

Liane or small erect shrub. Leaves simple. Male inflorescences of corymbose panicles or condensed cymes. Male flowers sessile or shortly pedicellate ; sepals 9–24, pubescent to hairy on the outer side, the 3–12 outer bract-like, sublinear to subovate, the 3–6 median subspathulate or similar to the outer ones, the inner ovate-elliptic, much larger than the others ; petals 3–6, inserted at the base of the androecium, small and glabrous ; stamens 9–18 with filaments connate into a synandrium, anthers subsessile, thecae with transverse dehiscence. Female inflorescences similar to the male. Female flowers with petals similar to those of the male flowers : petals 3 ; staminodes 3 or absent ; carpels 3–6, subovoid ; ovary pubescent ; style cylindric and glabrous. Drupe ovoid or subglobular, endocarp woody. Seeds without endosperm.

Anisocycla blepharosepala Diels in Engl., Pflanzenr. IV, 94 : 93 (1910). TAB. **18.** Type : Mozambique, Boroma, *Menyhart* 780 (BM ; G ; K ; UPS ; Z, holotype). *Synclisia zambesiaca* N.E.Br. in Kew Bull. **1892** : 196 (1892) pro parte quoad fr.

Liane with yellowish hairs. Leaf-lamina 3–7 × 2–4 cm., elliptic or ovate-elliptic, obtuse at the base, obtuse and sometimes mucronulate at the apex, papyraceous to subcoriaceous, discolorous, nerves 2–4 on each side, petiole 0·5–1·5 cm. long, whitish or yellowish puberulous. Male inflorescences of 4–6-flowered corymbs ; peduncle 1–1·5 cm. long. Male flowers with 18–24 sepals, margins ciliate with white hairs, otherwise glabrous, blackish when dry, the outer very small, less than 0·5 mm. in diam., the median larger, the inner ovate, 1·7–2·2 × 1·2–1·4 mm. ; petals 6, 0·8–1·2 × 0·8–1·2 mm., broadly ovate-elliptic, more or less fleshy ; synandrium 1·3–1·6 mm. long ; anthers 9–10. Female inflorescences of few-flowered corymbs. Female flowers with sepals similar to those of the male ; petals 3 ; staminodes 3, hairy ; carpels 6 ; ovary 1·2–1·6 mm. long, densely tomentellous. Drupelets c. 1·5–1·8 × 1 cm., subglobular, puberulous to pubescent.

S. Rhodesia. W : Gwaai, fr., *Eyles* 6448 (K ; SRGH). S : Nuanetsi Distr., Nyamasikana R., fr. viii.1956, *Mowbray* 122 (K ; SRGH). **Mozambique.** MS : Chiramba, fr. iii.1859, *Kirk* (K). Known only from S. Rhodesia and Mozambique. A species of riverine forest.

3. TILIACORA Colebr.

Tiliacora Colebr. in Trans. Linn. Soc. **13** : 53 (1822).

Robust lianes. Leaves simple, lamina entire, penninerved. Male inflorescences of false racemes of condensed cymules, few-flowered, axillary or from the old stems, or of axillary, solitary, sometimes 1-flowered cymules. Male flowers with 6–12 sepals, glabrous or rarely partly puberulous, the outer bract-like, the inner much larger, obovate or elliptic, somewhat fleshy or coriaceous ; petals 3–6 ; stamens

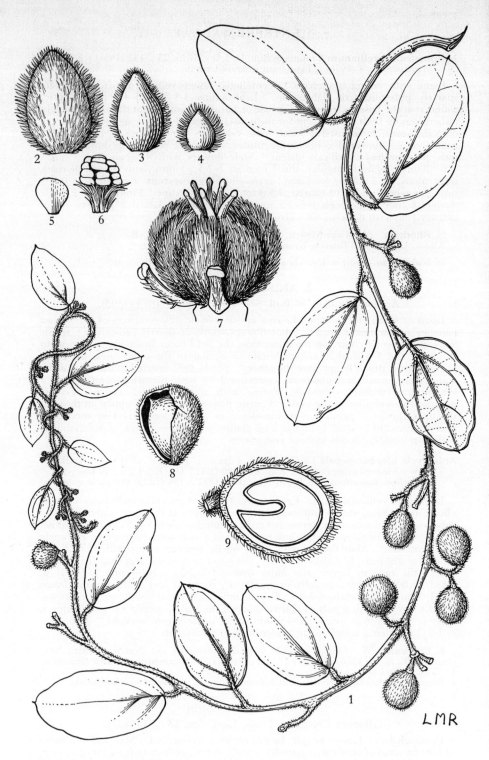

Tab. 18. ANISOCYCLA BLEPHAROSEPALA. 1, fruiting shoot (× ⅔) *Kirk* s.n. ; 2–6, male flower
(×14) *Swynnerton* 1346 (2, outer sepal, 3, middle sepal ,4, inner sepal, 5, petal, 6,
synandrium) ; 7, female flower with perianth removed (×14) *Menyhart* 780 ; 8,
drupelet (×2) and 9, seed (×2), both from *Mowbray* 122.

3–9, free or somewhat united ; anthers introrse ; thecae dehiscing longitudinally. Female inflorescences similar to the male, or more simple, sometimes as spikes of solitary flowers. Female flowers with sepals and petals similar to those of the male, staminodes absent ; carpels 6–30, borne on an apparent gynophore. Drupelets ovoid, stipitate ; the remains of the stigma visible near the stipe ; exocarp glabrescent, smooth or verrucose, endocarp compressed, woody, furrowed. Endosperm ruminate, scanty or absent.

Tiliacora funifera (Miers) Oliv., F.T.A. **1** : 44 (1868).—Gibbs in Journ. Linn. Soc., Bot. **37** : 429 (1906).—Diels in Engl., Pflanzenr. IV, 94 : 64 (1910).—Eyles in Trans. Roy. Soc. S. Afr. **5** : 353 (1916).—Troupin, F.T.E.A. Menisperm. : 9 (1956). Type : Nyasaland, Manganja Hills, *Meller* (K, holotype).
   Hypserpa funifera Miers, Contr. Bot. **3** : 104 (1871). Type as above.
   Tiliacora warneckei Engl. ex Diels, loc. cit. Type from Togoland.
   Tiliacora pynaertii De Wild. in Bull. Inst. Roy. Col. Belg. **2** : 573 (1931). Type from the Belgian Congo.
   Tiliacora glycosmantha Diels in Notizbl. Bot. Gart. Berl. **11** : 662 (1932). Type from Tanganyika (Iringa).
   Tiliacora johannis Exell in Journ. of Bot. **73**, Suppl. Polypet. Addend. : 7 (1935).— Exell & Mendonça, C.F.A. **1**, 1 : 34 (1937). Type from Angola (Cabinda).

Liane. Leaf-lamina 5–20 × 3–10 cm., ovate-lanceolate, ovate-oblong or broadly ovate, subcordate to rounded or somewhat obtuse at the base, obtuse to acute or acuminate at the apex, papery or coriaceous, glabrous, nerves in 3–5 pairs, sometimes sparsely puberulous beneath, petiole 1·5–5 cm. long, slender, puberulous or glabrescent. Male inflorescences either of axillary solitary cymules on peduncles 1–1·2 cm. long, or of 3–9-flowered cymules arranged in false racemes which are axillary or springing from the old stems, solitary or clustered, up to 15 cm. long ; axes and peduncles puberulous. Male flowers with 6–9 sepals, the 3–6 outer ones 0·8–1·5 × 0·8–1·5 mm., triangular to orbicular, thickened and ciliolate, the 3 inner 3·5–4 × 1·8–2·3 mm., obovate-elliptic ; petals 6, 1·5–2·5 mm. long, clawed, thickened on the margins ; stamens 3·5 mm. long, exserted, free or slightly united at the base. Female inflorescences similar to the male. Female flowers with 6–9 sepals, the outer 1·2 mm. long, lanceolate to ovate, the inner 2·5 mm. in diam., suborbicular ; petals (5) 6, 1–1·5 mm. long ; carpels 8–12, c. 1 mm. long. Drupelets 5–7 mm. long ; obovoid to nearly globose ; stipe 1·5–3 mm. long.

N. Rhodesia. N : Isoka Distr., Mafinga Mts., st. 20.xi.1952, *White* 3735 (K). C : Chingombe, fr. 27.ix.1957, *Fanshawe* 3742 (K). S : Mumbwa Distr., SW. of Chanobi Concession, fl. 16.ix.1947, *Brenan* 7861 (BR ; EA ; FHO ; K). **S. Rhodesia.** W : Victoria Falls, vii.1911, *Fries* 36 (UPS). E : Umtali, Commonage, st. 25.x.1949, *Chase* 1799 (BM ; SRGH). S : Sabi River, Chiribira Falls, st. 10.vi.1950, *Chase* 2420 (BM ; SRGH). **Nyasaland.** N : Vipya, Chamambo Forest, fr. 22.i.1956, *Chapman* 271 (FHO ; K). S : Manganja Hills, 17.ix.1861, *Meller* (K). **Mozambique.** N : Cabo Delgado, fl. 20.viii.1948, *Barbosa* 1829 (LISC). Z : Massingire, 1000 m., st. 8.viii.1942, *Torre* 4535 (BM ; LISC). MS : Mt. Maruma, ix.1906, *Swynnerton* 1347 (BM ; K). SS : Massinga, Inhachengo, fl. 30.viii.1944, *Mendonça* 1893 (BM ; LISC).
   Also in Togoland, Ghana, Belgian Congo and Angola. Lowland and upland rain-forest, riverine forest and moist shady places in woodland, 220–1,250 m.

The variation within this species is shown particularly in the shape and consistency of the leaves, which has led to the recognition of various alleged taxa.

4. TRICLISIA Benth.

Triclisia Benth. in Benth. & Hook., Gen. Pl. **1** : 39 (1862).

Robust, twining lianes. Leaves simple, lamina entire, rarely serrate. Male inflorescences of corymb-like cymules or of small, many-flowered panicles. Male flowers with 9–18 sepals, all densely pubescent outside, the outer ones bract-like, the inner ones suborbicular and somewhat united ; petals 6, very small or absent, glabrous and somewhat fleshy ; stamens 3–6, free ; anther-thecae with obliquely longitudinal dehiscence ; gynoecium replaced by a tuft of rusty hairs. Female inflorescences similar to the male. Female flowers with sepals similar to those of the male, petals 3–6 or absent ; staminodes reduced ; carpels 6–40, pubescent, narrowed into a cylindrical style. Drupelets obovoid, flattened, stipitate ; exocarp usually velvety, sparsely puberulous ; endocarp rugose and fibrous-hairy. Seeds without endosperm.

Triclisia sacleuxii (Pierre) Diels in Engl., Pflanzenr. IV, 94 : 72 (1910).—Troupin, F.T.E.A. Menisperm. : 5 (1956). Syntypes from Zanzibar.

    Pycnostylis sacleuxii Pierre in Bull. Soc. Linn. Par., N.S. **10** : 81 (1898). Syntypes as above.

Liane with young branchlets rusty-tomentellous. Leaf-lamina 7–18 × 5–13 cm., elliptic, ovate-elliptic or ovate-lanceolate, rounded to subcordate at the base, obtuse to shortly cuspidate or long-acuminate at the apex, rigid to subcoriaceous, glabrescent above, rusty-tomentose to puberulous on the nerves beneath, nerves in 3–6 pairs, petiole 2–10 cm. long. Male inflorescences axillary or springing from the old stems, of pedunculate cymules arranged in corymb-like clusters ; peduncles 0·5–1 cm. long, greyish-tomentellous. Male flowers with greyish-tomentellous pedicels 4–5 mm. long and furnished with 3 bracteoles ; sepals 12–18, the 9–12 outer 2–4 × 1·5–3·5 mm., the 3–6 inner 5–7 × 3–4 mm. ; petals 0·5–1 mm. long ; stamens 6, 3–5 mm. long. Female inflorescences similar to the male ; peduncles of the cymules up to 2 cm. long and becoming robust in fruit. Female flowers with 20–25 carpels c. 2 mm. long. Drupelets up to 3 × 1·7 cm., truncate at the base. Seeds 2–2·5 cm. long.

Var. **ovalifolia** Troupin in Bull. Jard. Bot. Brux. **25** : 134 (1955) ; F.T.E.A. Menisperm. : 7 (1956). Type from Tanganyika (Rungwe Distr.).

    Triclisia sp.—Brenan, T.T.C.L. : 328 (1949).

Lamina of leaf ovate-lanceolate, markedly cordate at the base, long-acuminate at the apex.

Mozambique. MS : Chimoio, Gondala, R. Nhamissenguere, i.1948, *Mendonça* 3739 (BM ; COI ; LISC).

Also in Tanganyika. Lowland rain-forest and riverine forest, 0–1700 m.

Var. *sacleuxii* has the laminae ovate-elliptic to broadly elliptic, rounded to subcordate at the base, obtuse to emarginate or shortly cuspidate at the apex. It occurs in French Moyen Congo, Angola, Kenya, Tanganyika and Zanzibar.

5. JATEORHIZA Miers.

Jateorhiza Miers in Hook., Niger Fl. : 212 (1849).

Somewhat woody lianes with dense indumentum. Leaves with long petioles, 3–5 (7)-lobed. Male inflorescences of elongate axillary panicles, the lateral axes bearing 3–7-flowered clusters. Male flowers with 6 sepals, the 3 outer elongate to elliptic, the 3 inner obovate ; petals 6, somewhat concave, mostly abruptly bent inwards at the apex and with their margins incurved and enveloping the androcium ; stamens 6, free or connate ; anthers introrse, globular ; thecae with transverse dehiscence. Female inflorescences of axillary racemes. Female flowers with sepals and petals more or less similar to those of the male ; staminodes 6, tongue-shaped. Carpels 3, subovoid ; styles small, recurved, the broad stigma produced into 2–3-cleft lamellae. Drupelets ovoid or subovoid, exocarp strigose-hispid or setulose ; endocarp ovoid, flattened, the ventral side more or less smooth, the dorsal side clothed with numerous fibrillose hfirs. Seeds with fleshy, ruminate endosperm.

Jateorhiza palmata (Lam.) Miers in Hook., Niger Fl. : 214 (1849).—Diels in Engl., Pflanzenr. IV, 94 : 166 (1910).—Brenan, T.T.C.L. : 327 (1949).—Troupin, F.T.E.A. Menisperm. : 15 (1950). TAB. 19 fig. B. Type from Mauritius (from a cultivated plant).

    Menispermum palmatum Lam., Encycl. Méth. **4** : 99 (1797). Type as above.

    Menispermum columba Roxb., Fl. Ind. **3** : 807 (1832). Type from a cultivated plant in a garden at Madras probably originating from Mozambique.

    Jateorhiza columba (Roxb.) Oliv., F.T.A. **1** : 42 (1868). Type as above.

    Jateorhiza miersii Oliv., loc. cit. *nom. illegit.* Type as for *J. palmata.*

Liane with branchlets densely pubescent at first, later strigose. Rootstock tuberous. Leaf-lamina 15–35 × 16–40 cm., broadly rounded, deeply cordate at the base, generally with 5 broadly ovate lobes, acuminate at the apex, sometimes angular, membranous, with strigose hairs on both sides, rarely glabrescent ; basal nerves 5–7, palmate, petiole 18–25 cm. long, strigose. Male inflorescences 40 cm. long ; lateral branches 2–10 cm. long ; main axes strigose ; secondary axes sometimes glabrous, with a linear-lanceolate ciliate bract at the base ; pedicels absent.

Male flowers with greenish sepals 2·7–3·2 × 1·3–1·6 mm. ; petals 1·8–2·2 mm. long ; stamens 1–1·8 mm. long, free, slightly adnate to the base of the petals. Female inflorescences 8–10 cm. long. Female flowers with carpels 1–1·5 mm. long, rusty-pubescent. Drupelets 2–2·5 × 1·5–2 cm.

S. Rhodesia. N : Urungwe, Msukwe R., fl. xi.1953, *Wild* 4169 (K ; SRGH). **Nyasaland.** S : Shire Highlands, fl. 1891, *Buchanan* 242 (K). **Mozambique.** MS : Sena, fr. i.1860, *Kirk* 168 (K).
Also in Kenya, Tanganyika and Mauritius ; probably introduced in Ghana. Lowland rain-forests and riverine forests, 0–1500 m.

6. DIOSCOREOPHYLLUM Engl.

Dioscoreophyllum Engl., Pflanzenw. Ost-Afr. **C** : 181 (1895).

Herbaceous twiners : Leaves simple, with long petioles, entire or lobed. Male inflorescences of axillary, long-pedunculate racemes. Male flowers with 6–8 sepals in 2 whorls ; petals absent ; stamens 3–6, fused into a cylindric or subglobular, sometimes flattened, sessile or stipitate synandrium ; anther-thecae oblong. parallel, dehiscing longitudinally. Female inflorescences similar to the male. Female flowers with 6 sepals ; petals absent ; carpels 3–6, with thickened, recurved stigmas. Drupelets subovoid, topped by the remains of the style and stigma ; exocarp smooth and sometimes shining ; endocarp crustaceous, verrucose or covered on the upper side with short prickles much enlarged at the base. Seeds with thick fleshy endosperm.

Dioscoreophyllum cumminsii (Stapf) Diels in Engl., Pflanzenr. IV, 94 : 181, fig. 64 a–f. (1910). TAB. **19** fig. A. Type from Ghana.
Rhopalandria cumminsii Stapf in Kew Bull. **1898** : 71 (1898). Type as above.

Twiner with more or less slender, sparsely pubescent or hirsute branchlets. Leaf-lamina 9–20 × 9–20 cm., ovate-triangular, sagittate-cordate at the base, sharply acuminate or apiculate at the apex, herbaceous or submembranous, sparsely hairy to glabrescent, basal nerves 7–9, palmate, petiole 6–15 cm. long. Male inflorescences up to 30 cm. long on a peduncle 6–15 cm. long ; axis hairy or pubescent ; bracts 2·5–3 mm. long, ovate ; pedicels 2–4 mm. long. Male flowers with oblong-elliptic sepals, 3–4 × 1·5–2 mm., generally glabrescent ; synandrium wider than high, sessile or shortly stipitate, with a thickened stipe up to 0·5 mm. long. Female inflorescences 8–10 cm. long. Female flowers with carpels 1·5–2 mm. long. Drupelets up to 3·5 cm. long ; peduncle 1 cm. long, glabrescent. Seeds 1·5–3 cm. long.

Var. **leptotrichos** Troupin in F.W.T.A., ed. 2, **1**, 2 : 758 (1958). Type : S. Rhodesia, Chirinda Forest, *Swynnerton* 100 (BM).
Dioscoreophyllum volkensii sensu Diels in Engl., Pflanzenr. IV, 94 : 183 (1910) pro parte quoad *Swynnerton* 100 & 100a, non Engl.
Dioscoreophyllum chirindense Swynnerton in Journ. Linn. Soc., Bot. **40** : 19 (1911). —Eyles in Trans. Roy. Soc. S. Afr. **5** : 353 (1916). Type as above.

Plant sparsely clothed with slender pale hairs.

S. Rhodesia. E : Chirinda, i.1939, *Hopkins* in GHS 7107 (PRE ; SRGH). **Mozambique.** Z : Massingire, Metalola Mts., fl. 23.v.1943, *Torre* 5371 (BM ; LISC). MS : Manica, Dombe, fl. 24.i.1948, *Barbosa* 897 (BM).
Also in Sierra Leone and Ghana. Evergreen forests.

7. CHASMANTHERA Hochst.

Chasmanthera Hochst. in Flora, **27** : 21 (1844).

Liane with verrucose bark. Leaves simple, with long petioles, suborbicular-subangular, densely hairy ; nerves palmate. Male inflorescences of false racemes of 3–5-flowered cymules ; bracts filiform, persistent. Male flowers with 6–9 sepals, the 3 outer bract-like and hairy, the 3–6 inner larger, membranous to papyraceous, concave, pubescent outside ; petals 6, fleshy, ribbed inside ; stamens 6, erect, with long connate filaments ; anther-thecae with longitudinal dehiscence. Female inflorescences of pendulous racemes. Female flowers with sepals and petals similar to those of the male, sometimes larger ; staminodes 6, small, elongate ; carpels 3, subovoid, narrowed into a short style ; stigma membranous,

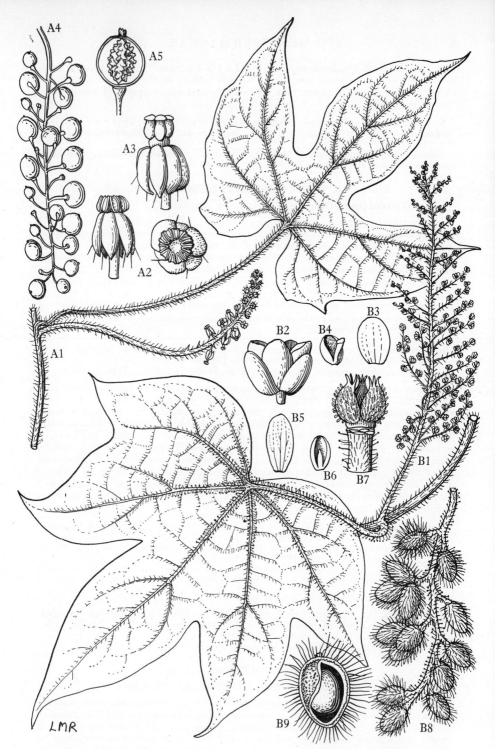

Tab. 19. A.—DIOSCOREOPHYLLUM CUMMINSII. A1, young leaf and male infl. (× ⅔) *Swyn-*
nerton 100A (leaf) and *Chase* 431 (infl.); A2, male fl. 2 views (×5) *Chase* 431;
A3, female fl. (×5) *Swynnerton* 6521 ; A4, infruct. (possibly immature) (× ⅔) *Torre*
5371A and *Swynnerton* 6521; A5, section of drupelet showing endocarp (×2) *Torre*
5371A. B.—JATEORHIZA PALMATA. B1, leaf and male infl. (× ⅔) *Wild* 4169 ; B2–B4,
male fl., all (×5) *Wild* 4169 (B2, flower, B3, sepal, B4, petal and stamen); B5–B7,
female fl., all (×5) *Bojer* s.n. (B5, sepal, B6, petal and staminode, B7, carpels); B8, infruct. (× ⅔) ; and B9, drupelet, showing endocarp (×1) both from *Kirk* s.n.

recurved, longitudinally cleft. Drupelets 3, ellipsoid and unequal-sided, apiculate ; exocarp coriaceous ; endocarp with a dorsal, median, slightly tuberculate ridge and 3 apical teeth, and 2 ventral narrow marginal wings. Seeds with ruminate endosperm.

Chasmanthera dependens Hochst. in Flora, **27**: 21 (1844).—Oliv., F.T.A. **1**: 41 (1868).—Diels in Engl., Pflanzenr. IV, 94 : 152, fig. 51 A–F & J–M (1910).—Troupin in Keay, F.W.T.A. ed. 2, **1**, 1 : 74 (1954).—Cufod. in Bull. Jard. Bot. Brux. **24**, suppl. : 115 (1954).—Troupin, F.T.E.A. Menisperm. : 17 (1956). TAB. **20**. Type from Ethiopia.

Liane ; mature branches with flaking bark ; young branchlets densely pubescent. Leaf-lamina 7–20 cm. long and wide, clearly cordate at base, acuminate or subobtuse at apex, membranous to subpapyraceous, silky-tomentellous when young, later pubescent, basal nerves 5–7, palmate, petiole 7–14 cm. long. Male inflorescences 10–30 cm. long, 1·5–2 cm. wide ; pedicels 3–6 mm. long ; bracts linear-filiform, pubescent or subtomentellous. Male flowers with lanceolate outer sepals 1·5–2 × 0·5–1 mm. ; inner sepals 2·5–3·5 × 1·5–2 mm., obovate with a tuft of hairs at the apex and sometimes also down the median line outside ; petals subequal, 2–2·5 × 1·5–2 mm., obovate, glabrous ; stamens 2·5–3 mm. long. Female inflorescences 10–18 cm. long. Female flowers with staminodes about 1 mm. long ; carpels 1·8–2 mm. long, ± united at the apex by the stigma. Drupelets up to 2 × 1·2 cm. Seeds 1–1·8 cm. long.

N. Rhodesia. N : Abercorn Distr., Kambole Escarpment, fl. 2.i.1955, *Richards* 3868 (K).
Widely spread from Sierra Leone east to Somaliland and south to Tanganyika, the eastern Belgian Congo and N. Rhodesia. Lowland rain-forest, riverine forest, and in drier country on termite hills, near rock outcrops and in dried-up water-courses, 800–1500 m.

C. welwitschii Troupin of the rain-forests and riverine forests of the Congo Basin is closely related to this species. It differs in the longer inflorescences and the stiffer and fewer hairs. The two species appear to be ecologically distinct.

8. TINOSPORA Miers

Tinospora Miers in Ann. Mag. Nat. Hist., Ser. 2, **7** : 35, 38 (1851).
Desmonema Miers, op. cit., Ser. 3, **20** : 260 (1867) non Rafin. (1833).
Hyalosepalum Troupin in Bull. Jard. Bot. Brux. **19** : 430 (1949).

Herbaceous twiners or woody lianes or sometimes small, rambling shrubs. Leaves simple, entire ; nerves palmate, rarely pinnate. Male inflorescences of false racemes, which are either simple or compound (i.e. panicles) of 2–4-flowered cymules. Male flowers with 6 sepals, the 3 inner much larger than the outer, hyaline or membranous ; petals 6, rarely 3, fleshy, with inrolled margins, the inner smaller than the outer ; stamens 3–6, quite free, or with their filaments connate either at the base, half-way up, or for their entire length ; anther-thecae with longitudinal dehiscence. Female inflorescences of false racemes of solitary flowers. Female flowers with sepals and petals similar to those of the male ; carpels 3, obliquely ovoid. Drupelets 3 or fewer ; exocarp pulpy ; endocarp more or less verrucose outside, with a condyle on its inner side which makes a large subglobular cavity. Seeds with fleshy ruminate endosperm.

Leaves subcordate at the base, the central part of the base attenuate-cuneate ; stamens 6, with filaments connate at the base or to half-way up - - 1. *mossambicensis*
Leaves not as above, ovate-cordate or suborbicular :
    Stamens 3, filaments connate to the apex ; leaves ovate or ovate-cordate - 2. *caffra*
    Stamens 6, filaments connate to half-way up ; leaves broadly ovate to suborbicular, normally pale green - - - - - - - - - - 3. *tenera*

1. **Tinospora mossambicensis** Engl., Bot. Jahrb. **26** : 404 (1899).—Troupin, F.T.E.A. Menisperm. : 18 (1956). Type : Mozambique, without locality, *Stuhlmann* 731 (B, holotype).
    Desmonema mossambicense (Engl.) Diels in Engl., Pflanzenr. IV, 94 : 153 (1910). Type as above.

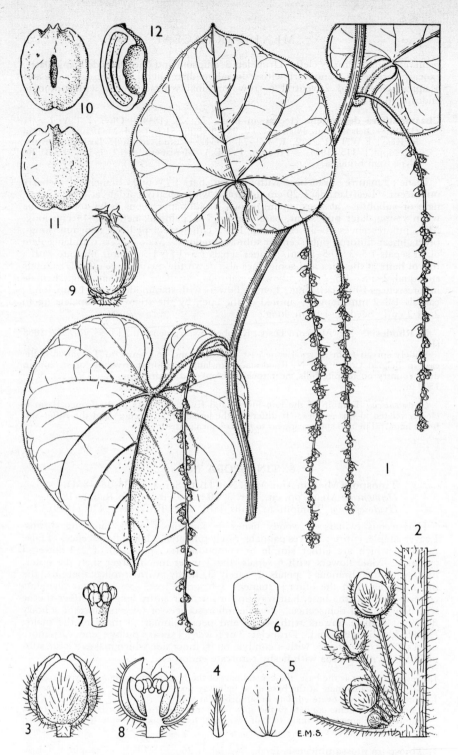

Tab. 20. CHASMANTHERA DEPENDENS. 1, habit (× ⅔) ; 2, cymule of ♂ inflorescence (× 5) ;
3, ♂ flower (× 10) ; 4, outer sepal (× 10) ; 5, inner sepal (× 10) ; 6, petal (× 10) ;
7, stamens (× 10) ; 8, section of ♂ flower (× 10) ; 9, carpels (× 10) ; 10, ventral side
of drupelet (× 2) ; 11, dorsal side of drupelet (× 2) ; 12, section of drupelet (× 2).
From F.T.E.A.

Slender liane with glabrous striate branchlets. Leaf-lamina broadly ovate-triangular, slightly cordate at the base ; the lateral lobes united in the middle by a projection of the lamina towards the petiole giving the central part of the lamina a cuneate appearance, glabrous, nerves 5–7, palmate, petiole 4–6 cm. long. Male inflorescences 20–35 cm. long ; pedicels 1·5–2 mm. long. Male flowers with outer sepals 0·8–1·2 × 1·5–2 mm. ; inner sepals 2–2·3 × 1·5–2 mm. ; petals 1·5–2 mm. long, the inner narrower ; stamens 6, their filaments connate at the base or to half-way up, 0·8–1·2 mm. long. Female inflorescences 20–30 cm. long ; pedicels 4–5 mm. long. Female flowers with carpels 1·5–2 mm. long. Drupelets unknown.

Mozambique. Without precise locality, fl. ♀, i.1889, *Stuhlmann* 731 (B, holotype). Also in Tanganyika. Lowland rain-forests.

A species easily recognizable by its leaves. It seems to be relatively scarce. Specimens have also been collected by Schlieben in Tanganyika.

2. **Tinospora caffra** (Miers) Troupin in Bull. Jard. Bot. Brux. **25** : 137 (1955) ; F.T.E.A. Menisperm. : 19 (1956). TAB. **21** fig. C. Type from Natal.
Desmonema caffrum Miers in Ann. Mag. Nat. Hist., Ser. 3, **20** : 261 (1867).—Diels in Engl., Pflanzenr. IV, 94 : 156 (1910). Type as above.
Hyalosepalum caffrum (Miers) Troupin in Bull. Jard. Bot. Brux. **19** : 431 (1949).—Troupin, F.C.B. **2** : 230, t. 20 (1951). Type as above.
Desmonema pallide-aurantiacum sensu Suesseng. in Proc. & Trans. Rhod. Sci. Ass. **43** : 86 (1951).

Climber or somewhat woody liane ; stem verrucose, bark sometimes scaling. Leaf-lamina 2·5–8 × 1·5–7 cm., ovate, ovate-cordate or almost orbicular, cordate or rounded at the base, abruptly apiculate and mucronate or subacuminate at the apex, glabrous, petiole 2·5–10 cm. long. Male inflorescences usually of false racemes of cymules, sometimes panicled, 7–30 cm. long ; pedicels 0·3–1 cm. long. Male flowers with triangular to subovate outer sepals 0·8–1·5 × 2–4 mm., the inner oblong to oblanceolate, and narrower ; stamens 3, filaments connate to the apex, 1·5–3 mm. long. Female inflorescences 7–15 cm. long. Petals of female flowers with a small membranous appendage, perhaps a staminode, at the base ; carpels 2–3 mm. long. Drupelets 0·8–1·2 × 0·4–0·7 cm.

N. Rhodesia. W : Ndola, fr. 13.iii.1954, *Fanshawe* 956 (BR ; K). S : Mazabuka, xii.1930, *C.R.S.* 151 (PRE). **S. Rhodesia.** N : Darwin Distr., Umvukwes, fl. 22.xii.1952, *Wild* 3956 (BR ; K ; SRGH). W : Bulawayo, xi.1943, *Brain* 5805 (SRGH). C : Salisbury, fl. 22.xi.1926, *Eyles* 4553 (BM ; K ; SRGH). E : Inyanga, st. 27.ii.1930, *F.N. & W.* 3193 (BM ; BR ; UPS). **Nyasaland.** N : Lupemba, 24 km. S. of Karonga, fl. 2.ii.1953, *Williamson* 145 (BM). **Mozambique.** Z : between Nhamacurra and Vila de Maganja da Costa, fr. iii.1943, *Torre* 4985 (BM ; LISC). MS : Chimoio, fl. 4.ii.1948, *Andrada* 1041 (LISC). SS : Gaza, near Vila João Belo, fl. 10.iii.1942, *Torre* 3996 (BM ; LISC).
Also in Ubangi-Chari, Sudan, Belgian Congo, Urundi, Uganda, Kenya, Tanganyika, Angola, Transvaal and Natal. Lowland and upland rain-forest and deciduous bushland, often near rock outcrops, 0–2,000 m.

The species *T. caffra* is here treated in a wide sense—in other words the differences in leaf shape and development of inflorescences seem not to be of taxonomic importance but rather due to variations in shade and moisture. These morphological variations of the plant may well be no more than ecological adaptations of neither specific nor varietal significance.

3. **Tinospora tenera** Miers, Contr. Bot. **3** : 37 (1871).—Troupin, F.T.E.A. Menisperm. : 20 (1956). TAB. **21** fig. D. Type : Mozambique, lower Shire Valley, *Kirk* (K).
Desmonema tenerum (Miers) Diels in Engl., Pflanzenr. IV, 94 : 154 (1910).—Brenan, T.T.C.L. : 327 (1949). Type as above.
Hyalosepalum tenerum (Miers) Troupin in Bull. Jard. Bot. Brux. **19** : 431 (1949). Type as above.

Liane with yellow-brown glabrous branchlets. Leaf-lamina 4–7 × 2·5–6 cm., broadly ovate to suborbicular, cordate or obtuse at the base, acuminate, apiculate and sometimes mucronulate at the apex, glabrous on both sides, papery, pale green, nerves 5, palmate, petiole 2–3·5 cm. long, glabrous. Male inflorescences 10–35 cm. long ; pedicels 2–3 mm. long. Male flowers with oblong-obovate outer sepals 0·5–1 × 0·4–3·7 mm. ; inner sepals 1· 1·5 mm. long, obovate-spathulate ; petals

F

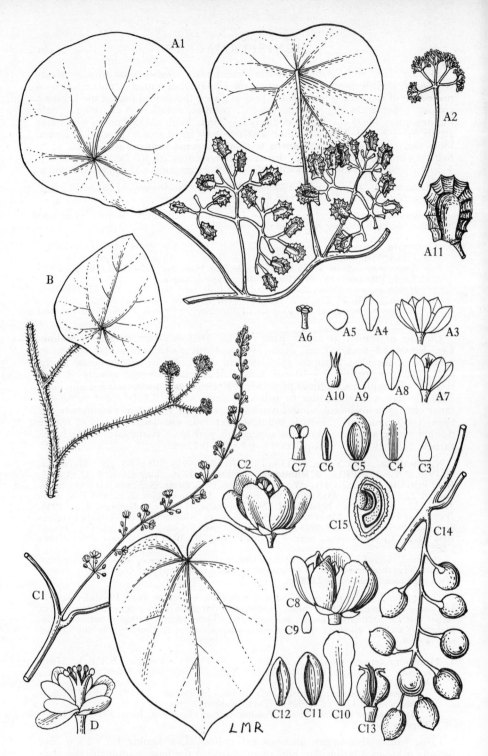

Tab. 21. A.—STEPHANIA ABYSSINICA VAR. ABYSSINICA. A1, leaves and infruct. (× ⅔) *Milne-Redhead* 3327
A (leaves) and *Geesteranus* 5631 (infruct.) ; A2, male infl. (× ⅔) ; A3–A6, male fl. (× 6) *Milne-Redhead*
3224 (A3, fl., A4, sepal, A5, petal, A6, synandrium) ; A7–A10, female fl. (× 6) *Richards* 243 (A7, fl., A8,
sepal, A9, petal, A10, gynoecium) ; A11, fruit (× 2) *Geesteranus* 5631. B.—STEPHANIA ABYSSINICA VAR.
TOMENTELLA, part of shoot (× ⅔) *Buchanan* 1459. C.—TINOSPORA CAFFRA. C1, leaf and male infl. (× ⅔)
Faulkner 407 ; C2–C7, male fl. (× 6) *Faulkner* 407 (C2, fl., C3, outer sepal, C4, inner sepal, C5, outer
sepal, C6, inner petal, C7, synandrium) ; C8–C13, female fl. (× 6) *Greatrex* in GHS 26450 (C8, fl.,
C9, outer sepal C10, inner sepal, C11, outer petal, C12, inner petal, C13, gynoecium) ; C14, infruct.
(× ⅔) and C15, drupelet showing condyle (× ⅔), both from *Fanshawe* 956. D.—TINOSPORA TENERA, male fl.
(× 6) *Robinson* 390.

1–1·5 mm. long, keeled ; stamens 6, their filaments 1–1·5 mm. long, connate to half-way up. Female inflorescences, female flowers and fruits unknown.

N. Rhodesia. S : Mapanza E. fl. 6.xii.1953, *Robinson* 390 (K). **Mozambique.** Z : Quelimane, *Stuhlmann* 742 (B). T : Lower Shire Valley, fl. 3.i.1862, *Kirk* (K). MS : Chizombero, fr. 14.iii.1948, *Garcia* in *Mendonça* 608 (BM ; LISC). LM : Lourenço Marques, fl. 9.xii.1897, *Schlechter* 11675 (BM ; K).
Also in Tanganyika and the Transvaal. Lowland rain-forests.

Chase 1381, (S. Rhodesia (E), Hot Springs, Melsetter, fr. 29.xii.1948) may also belong to this species. The leaves of this specimen are truncate or only very slightly cordate, and so resemble *T. tenera* rather than *T. caffra*. However, male flowers will have to be obtained before *T. tenera* can be recorded with certainty from S. Rhodesia.

9. COCCULUS DC.

Cocculus DC., Syst. **1** : 515 (1817) *nom. conserv.*

Prostrate or erect climbers or suffrutices ; branchlets rambling, or reduced to cladodes.* Leaves simple, various in form, entire or lobed. Male inflorescences of cymules which are either axillary and clustered 1–3-together, or solitary and arising from leafless branches, or grouped in ± condensed clusters on the cladodes* ; pedicels present or absent. Male flowers with 6 sepals, the 3 outer ones reduced, the 3 inner concave ; petals 6, furnished at base with more or less fleshy inflexed auricles surrounding the stamen-filaments, often bifid or deeply emarginate at the apex ; stamens 6–9, free ; anther-thecae with transverse dehiscence. Female inflorescences similar to the male or more simple or reduced to solitary or clustered flowers. Female flowers with sepals similar to those of the male ; base of petals much less inflexed ; staminodes 6, linear-filiform or absent ; carpels 3–6, subovoid and compressed, with cylindric style and recurved-spathulate stigma. Drupes obovoid or flattened and rounded ; the remains of the style visible near the base. Seeds with scanty endosperm.

Cocculus hirsutus (L.) Diels in Engl., Pflanzenr. IV, 94 : 236 (1910).— Brenan, T.T.C.L. : 326 (1949).—Milne-Redh. in Mem. N.Y.Bot. Gard. **8**, 3 : 215 (1953).— Williamson, Useful Pl. Nyasal. : 37 (1955).—Troupin, F.T.E.A. Menisperm. : 12 (1956). TAB. **22**. Type from India.
Menispermum hirsutum L., Sp. Pl. **1** : 341 (1753). Type as above.
Menispermum villosum Lam., Encycl. Méth. Bot. **4** : 97 (1797) *nom. illegit.* Type as above.
Cocculus villosus DC., Syst. Nat. **1** : 525 (1817) *nom. illegit.*—Oliv., F.T.A. **1** : 45 (1868).—Eyles in Trans. Roy. Soc. S. Afr. **5** : 353 (1916). Type as above.

Climber reaching several metres in length ; young branchlets ± densely pubescent, tomentose. Leaves yellowish-tomentose ; lamina 4–8 (9) × 2·5–6 (7) cm., variable in shape, that of the leaves in the lower part of the main branches clearly 3–5-lobed, that of the other leaves narrowly to broadly ovate, ovate-oblong or obovate, base broadened, cuneate, more or less rounded or rarely cordate, apex obtuse to rounded and mucronulate, bract-like and then 0·5 cm. long on the lateral and flowering branchlets, densely tomentellous when young, later sparsely tomentellous to glabrescent, basal nerves 5, petiole 0·5–2·5 cm. long. Male inflorescences of many-flowered cymules clustered 2–3 together or rarely solitary, 1–2·5 cm. long ; peduncle up to 1·5 cm. long ; pedicels 0·5–1 cm. long. Male flowers with pilose sepals, the 3 inner 1·5–2·5 × 1·7–2 mm., broadly ovate or obovate, the 3 outer 1·4–2 × 0·4–0·8 mm., oblong to lanceolate ; petals 0·5–1·5 × 0·3–0·6 mm., ovate-oblong, sparsely pubescent to glabrescent ; stamens 0·7–1 mm. long. Female inflorescences 0·5–2·5 cm. long. Female flowers with staminodes 0·5 cm. long ; carpels 0·7–1 mm. long. Drupelets 4–8 × 4–5 mm. ; endocarp clearly ribbed on the lateral faces and with a prominent dorsal crest ; septum of the condyle perforated.

Caprivi Strip. Ngamiland, E. of Kwando R., x.1945, *Curson* 1175 (PRE). **Bechuanaland Prot.** N : Thamalakane R., fl. 27.vi.1937, *Erens* 313 (K ; PRE ; SRGH). **N. Rhodesia.** B : near New Sesheke Boma, fl. 9.viii.1947, *Brenan & Keay* 1660 (EA ; FHO ; K). N : Luangwa Valley, Munyamadzi R., fl. 7.x.1933, *Michelmore* 654 (K). C : Kafue Gorge, viii.1929, *Burtt Davy* 20806 (FHO). **S. Rhodesia.** N : Sebungwe

* *C. balfourii* Schweinf., confined to the island of Socotra.

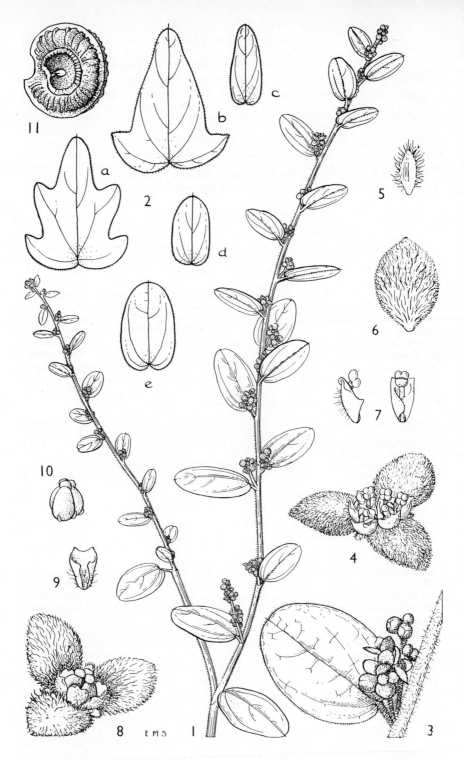

Tab. 22. COCCULUS HIRSUTUS. 1, habit (×1) ; 2a—e, leaf variation (×1) ; 3, ♂ inflorescence (×3) ; 4, ♂ flower (×10) ; 5, outer sepal (×10) ; 6, inner sepal (×10) ; 7, petals and stamens (×10) ; 8, ♀ flower (×8) ; 9, petal (×10) ; 10, carpels and staminodes (×10) ; 11, drupelet (×5). From F.T.E.A.

Distr., fl. ix.1955, *Davies* 1409 (BR ; K ; SRGH). E : Chirinda, fl. ix.1907, *Johnson* 273 (BR ; K). S : Chibi Distr., Beitbridge-Lundi, fl. 18.x.1932, *F.N. & W.* 2052 (BM ; BR ; S ; SRGH). **Nyasaland**. C : Kota Kota Distr., fl. 7.ix.1946, *Brass* 17562 (BM ; K). S : Chikwawa, fl. 2.x.1946, *Brass* 17901 (K). **Mozambique.** N : Nacala, fr. 16.x.1948, *Barbosa* 2446 (LISC) ; Porto Amelia, fl. 7.ix.1948, *Barbosa* 2017 (LISC). T : Changara, fl. 18.viii.1943, *Torre* 5770 (BM ; LISC). MS : Sone, x.1928, *Surcouf* B 149 (P). SS : Inhambane, Massinga, fl. 30.viii.1942, *Mendonça* 1896 (BM) ; Limpopo, Uanetze River, fl. vii.1932, *Smuts* P 329 (K ; PRE). LM : Marracuene, fl. 6.ix.1945, *Pedro* 49 (K ; PRE).

From the Sudan and Eritrea south to Natal, Angola and SW. Africa ; also in Asia from Central Arabia to Southern China. Bushland and semi-desert shrub, up to 1140 m.

According to Mrs. Williamson (loc. cit.) the fruits are used with pods of *Acacia subalata* Vatke (= *A. nilotica* subsp. *subalata* (Vatke) Brenan) to give a blue dye and the stems are used for making basket chairs.

10. CISSAMPELOS L.

Cissampelos L., Sp. Pl. **2** : 1031 (1753) ; Gen. Pl. ed. 5 : 455 (1754)

Twining lianes or sometimes scandent shrubs. Leaves simple, peltate or sub-peltate, entire or angular. Male inflorescences of corymbose cymules which are either solitary or clustered or arranged in more or less developed false racemes. Male flowers with 4 (5) obovate often spreading sepals ; petals usually connate into a patelliform or cup-shaped corolla, sometimes incompletely connate ; stamens connate into a 4–10-locular synandrium ; anther-thecae with longi-tudinal dehiscence. Female inflorescences less developed than the male cymules, 3–9-flowered, axillary or in false racemes, arising from the axils of leaves or of accrescent bracts. Female flowers with 1 sepal, rarely more ; petal 1, rarely 2–4, smaller than the sepal, sometimes broader than long ; carpel 1. Drupe with hairy or glabrous exocarp ; mesocarp fleshy and thin ; endocarp woody, with 1 dorsal ridge, the sides with small transverse often verrucose ribs. Seeds with scanty endosperm.

Leaves subpeltate :
  Branchlets sparsely pilose (hairs ± spreading) or glabrescent :
    Petiole 1–2·5 cm. long, ± stout, lamina ovate-triangular to cordate, acute at apex ;
      synandrium 4 (–6)-locular - - - - - - - *1. hirta*
    Petiole 2–7 cm. long, slender ; leaf-lamina sometimes cordate, more often sub-reniform, auriculate and attenuate at the base, sometimes subcordate, clearly rounded at apex ; synandrium 4-locular - - - - *2. torulosa*
  Branchlets pubescent to puberulous (hairs crisped and subappressed) :
    Leaves normally suborbicular, rounded or sometimes subcordate at the base, rounded at the apex ; male inflorescences of fascicled and axillary corymbose cymules, rarely of cymules arranged along an axis not exceeding 10 cm. in length ; synandrium 4-locular - - - - - *3. pareira* var. *orbiculata*
    Leaves heart-shaped to subreniform, obtuse to somewhat rounded at apex, ± deeply cordate at base ; male inflorescences usually of cymules arranged along an axis up to 15 cm. long ; synandrium 6–8-locular - - - - *4. mucronata*
Leaves clearly peltate :
  Petiole of adult leaves 4–16 cm. long ; lamina generally membranous to somewhat papery, generally pentagonal or angular, sometimes triangular to broadly ovate, obtuse or acute or emarginate and mucronulate at apex ; male inflorescences up to 40 cm. long - - - - - - - *5. owariensis*
  Petiole of adult leaves 1–3·5 (6) cm. long ; lamina papery or subcoriaceous :
    Leaves not orbicular, heart-shaped to pentagonal, densely puberulous or tomen-tellous beneath ; male inflorescences up to 30 cm. long - - *6. rigidifolia*
    Leaves orbicular, puberulous beneath ; male inflorescences not exceeding 5 cm. in length - - - - - - - *3. pareira* var. *orbiculata*

1. **Cissampelos hirta** Klotzsch in Peters, Reise Mossamb. Bot. **1** : 174 (1861).—Diels in Engl., Pflanzenr. IV, 94 : 291 (1910) non Miers (1866). Type : Mozambique, Inhambane, *Peters* (B, holotype).
    Cissampelos tamnifolia Miers, Contr. Bot. **3** : 185 (1871). Type : Mozambique, Delagoa Bay, *Forbes* 11 (K, holotype).
    Cissampelos pareira var. *klotzschii* Dur. & Schinz, Consp. Fl. Afr. **1**, 2 : 51 (1898). Type as for *C. hirta*.
    Cissampelos pareira var. *mucronata* subvar. *hirta* (Klotzsch) Engl., Bot. Jahrb. **26** : 395 (1899). Type as for *C. hirta*.

Slender liane with branchlets striate, glabrescent. Leaf-lamina 2–4·5 × 3–5 cm., ovate-subtriangular, broadly cordate, sometimes deeply cordate, subobtuse or subtruncate at the base, acute and slightly mucronate at the apex, puberulous or glabrescent on the lower side, except the nerves, sparsely puberulous on the upper side, nerves 5–7-palmate, petiole 1–1·5 cm. long. Male inflorescences of corymbose cymes, solitary or 2–4-fasciculate, sometimes arranged in false axillary racemes 3–6 cm. long ; pedicels slender, c. 1 mm. long. Male flowers with 4 sepals, obovate-unguiculate, c. 1 × 0·7 mm., glabrous, with black spots when dry ; synandrium 4(6)-locular. Female inflorescences often solitary, of 5–10-flowered cymules arranged on an axis 4–10 cm. long ; bracts triangular, ciliate. Female flowers with sepals 1·2–1·5 × 1 mm., ovate-elliptic, glabrous ; carpels 0·7 mm. long. Drupe 5–6 × 4–5 mm., compressed, sparsely hairy or glabrescent.

Mozambique. MS : Manica, x.1945, *Pedro* 172 (PRE). SS : Massinga, fl. ix.1946, *Gomes e Sousa* 1851 (BR ; K). LM : Delagoa Bay, 1893, *Junod* 478 (BR ; P) ; Delagoa Bay, fl. 30.xi.1897, *Schlechter* 11548 (BM ; K ; PRE).
Not known elsewhere.

2. **Cissampelos torulosa** E. Mey. ex Harv. in Harv. & Sond., F.C. **1** : 11 (1860).— Oliv., F.T.A. **1** : 46 (1868).—Diels in Engl., Pflanzenr. IV, 94 : 297 (1910).—Eyles in Trans. Roy. Soc. S. Afr. **5** : 353 (1916).—Steedman, Trees, etc. S. Rhod. : 7 (1933). Type from S. Africa.
  Menispermum capense Thunb., Prodr. Pl. Cap. **2** : 85 (1800) non *Cissampelos capensis* L. f. (1781). Type from S. Africa.
  Cissampelos wildemaniana Van de Bossche ex De Wild., Pl. Nov. Herb. Hort. Then. **1** : 5, t. 2 (1904). Type : Mozambique, Morrumbala, *Luja* 473 (BR).

Liane, sparsely hairy to glabrescent. Leaf-lamina 2–7 × 2–9 cm., subreniform, sometimes ovate-orbicular, generally auriculate and attenuate near the petiole or ± broadly cordate at the base, obtuse or rounded at the apex, discolorous, slightly pubescent or glabrescent on both sides but with a few yellowish hairs at the base of the nerves, nerves 3–5-palmate, petiole 2–7 cm. long, generally inserted at the base of the lamina or very near (1 mm.). Male inflorescences of axillary cymes, solitary or two together ; peduncle and pedicels sparsely pubescent to glabrescent ; bract linear, 0·5 mm. long. Male flowers with 4 sepals 1–1·2 × 0·5–0·7 mm., ovate-elliptic or obovate, slightly connate at the base, glabrous, 1-nerved ; petals 4, very often connate into a cup-shaped corolla, sometimes imperfectly connate and one petal free or the cup-shaped corolla once or twice deeply indented ; synandrium 4-locular, ± 1 mm. long. Female inflorescences of cymes of 1–4-flowered cymules 2–10 cm. long ; pedicels 0·3–1 cm. long. Female flowers generally with 2 sepals, sometimes 3 or 4, 1·5 × 0·5 mm., glabrous, 1-nerved ; petals very often 2, sometimes 3 or 4, 0·5–0·7 × 0·5–1·5 mm., more or less fleshy, glabrous ; carpel 1–1·2 mm. long ; stigma 3-fid. Drupe obovate-compressed, 4–7 × 3–5 mm. Seeds 6–8 mm. long.

S. Rhodesia. E : Chipete forest patch, fl. i.1906, *Swynnerton* 219 (BM ; K). **Nyasaland.** S : Shire Highlands, fl. xii.1893, *Scott Elliot* 8584 (K). **Mozambique.** Z : Morrumbala, fl. 30.xii.1858, *Kirk* (K). MS ; Mafusi, Mossurize, fr. 23.ii.1907, *Johnson* 154 (K). SS : between Morrumbene and Massinga, fl. 26.ii.1955, *E.M. & W.* 656 (BM ; LISC ; SRGH). LM : Maputo, fr. 18.xi.1944, *Mendonça* 2925 (LISC).
Also in the Transvaal, Swaziland, Natal and Cape Province.

3. **Cissampelos pareira** L., Sp. Pl. **2** : 1031 (1753). Type from Brazil.

Liane with stem somewhat woody at the base. Leaves sometimes peltate, lamina 2–12 × 2–12 cm., broadly cordate to ovate, apex obtusely or emarginately mucronate or rarely acuminate, base truncate, to deeply reniform-cordate, membranous to papery, ± pilose or glabrescent above, puberulous to tomentose below, with 3–7 basal nerves, palmate, petiole 1–7 cm. long, puberulous to subtomentose. Male inflorescences of axillary, generally solitary or paired, corymbose cymules not exceeding 4 cm. in length, sometimes arranged in the axils of bracts along an axis up to 10 cm. long ; axes, peduncles and pedicels whitish-pubescent. Male flowers with 4–5 ovate or obovate keeled sepals 1·2–1·5 × 0·7 mm., tubercled and pubescent ; synandrium 4-locular. Female inflorescences of 5–9-flowered cymules arranged in axillary, false racemes 5–10 cm. long, solitary or clustered 2–3 together ; bracts up to 1·5 cm. in diam., suborbicular-reniform, pubescent or tomentose. Female flowers with sepals similar to those of the male ; petals 1·5–1·7 × 2 mm., obtriangular to subreniform, very sparsely pubescent. Drupe 4–6 × 3–4 mm. wide, hairy-pubescent.

Widely distributed throughout the tropics.

Var. **orbiculata** (DC.) Miq. in Ann. Lugd. Bat. **4** : 85 (1868).—Troupin, F.T.E.A. Menisperm. : 26 (1956). TAB. **23**. Type from the East Indies.

Cissamp·los orbiculata DC., Syst. Nat. **1** : 537 (1817). Type as above.

Cissampelos pareira var. *typica* Diels in Engl., Pflanzenr. IV, 94 : 288 (1910) non var. *pareira*.

Petiole inserted 1–4 mm. from the base of the lamina ; lamina suborbicular or broadly ovate, rounded or subcordate or subtruncate at the base, rounded or emarginate and mucronulate at the apex, densely puberulous to tomentose beneath.

S. Rhodesia. E : Umtali, Maranka Reserve, fl. & fr. 10.ii.1953, *Chase* 4762 (BR ; K ; SRGH). S : Nuanetsi distr., Mwambe R., fl. xii.1955, *Davies* 1778 (K ; SRGH). **Mozambique.** MS : Meringua, fr. 28.vi.1950, *Chase* 2455 (BM ; SRGH). SS : Massinga, fl. 30.viii.1942, *Mendonça* 34 (LISC).

Also in Uganda, Kenya, Tanganyika, Zanzibar, Ethiopia, and throughout tropical Asia to Indonesia. Upland and lowland rain-forest, coastal evergreen bush and deciduous bushland, often persisting on cleared ground and in plantations ; also in secondary vegetation and near rock-outcrops, 0–2300 m.

4. **Cissampelos mucronata** A. Rich. in Guill., Perr. & Rich., Fl. Senegamb. Tent. **1** : 11 (1831).—Diels in Engl., Pflanzenr. IV, 94 : 300 (1910).—Suesseng. in Proc. & Trans. Rhod. Sci. Ass. **43** : 86 (1951).—Troupin, F.C.B. **2** : 250, t. 22 (1951) ; F.T.E.A. Menisperm.: 27 (1956).—Milne-Redh. in Mem. N.Y. Bot. Gard. **8**, 3 : 215 (1953). Type from Senegal.

Cissampelos pareira var. *mucronata* (A. Rich.) Engl., Bot. Jahrb. **26** : 394 (1899) pro parte.—Gibbs in Journ. Linn. Soc., Bot. **37** : 429 (1906).—Eyles in Trans. Roy. Soc. S. Afr. **5** : 353 (1916). Type as above.

Cissampelos macrostachya Klotzsch in Peters, Reise Mossamb. Bot. **1** : 172 (1861). Type : Mozambique, Sena, *Peters* (B, holotype †).

Cissampelos senensis Klotzsch, tom. cit.: 173 (1861). Type : Mozambique, Sena, *Peters* (B, holotype †).

Liane with woody rootstock ; branchlets pubescent to subtomentose. Leaves subpeltate, lamina 4–12 × 4–13 cm., ovate-cordate, broadly or narrowly cordate at the base, normally obtuse of somewhat rounded and mucronulate at the apex, tomentellous or puberulous or glabrescent on both sides with a bright grey or yellowish indumentum, basal nerves 5–7-palmate, petioles 2–4·5 cm. long and inserted 0·5–3 mm. above the base of the lamina. Male inflorescences of corymbose cymules clustered 3–6-together and either axillary or arranged in false racemes 5–15 cm. long ; pedicels 1–2 mm. long. Male flowers with 4–5 sepals 1–1·5 × 0·7–1 mm., pubescent outside ; corolla cup-shaped, 1–1·5 mm. long and in diam., spreading after flowering ; synandrium 6–10-locular, 1–1·5 mm. long. Female inflorescences 5–16 cm. long ; bracts accrescent, often elongate-mucronulate. Female flowers with sepal 1·5 × 1 mm., pubescent ; petal 0·7 × 2 mm., glabrous ; carpel 1–1·2 mm. long, glabrescent. Drupe 0·4–0·7 × 0·3–0·5 mm., pubescent.

Caprivi Strip. Lisikili, 24 km. E. of Katima Mulilo, fl. 17.vii.1952, *Codd* 7105 (BM ; K). **Bechuanaland Prot.** N : Ngamiland, 1930–1, *Curson* 435 (BR ; PRE). **N. Rhodesia.** B : Shangombo, Mashi R., fl. 8.viii.1952, *Codd* 7449 (BM ; K). N : Abercorn Distr., Ulungu, xi.1942, *Glover* in Hb. Bredo 6189 (BR). W : Mwinilunga Distr., Kalenda Plain, fl. 18.x.1937, *Milne-Redhead* 2826 (BM ; BR ; K ; PRE). C : Chilanga, fl. 1.xi.1909, *Rogers* 8441 (K). S : Namwala Distr., N. of Namwala, fr. 24.vi.1952, *White* 2980 (FHO ; K). **S. Rhodesia.** N : Sebungwe, fl. ix.1945, *Davies* 1525 (BR ; K ; SRGH). W : Bulawayo, Matopos Nat. Park, x.1938, *Wall* 8 (S). C : Makoni Distr., Maidstone, fr. 29.xi.1930, *F.N. & W.* 3279 (BM ; BR ; S). E : Umtali, xi.1947, *Chase* 2574 (PRE). S : Sabi R., fl. & fr. xi.1906, *Swynnerton* 2108 (BM ; K). **Nyasaland.** C : Kota Kota, Kaombe R., fl. 22.x.1943, *Benson* 602 (K ; PRE). Lilongwe, fl. 28.xi.1952, *Jackson* 998 (BM ; K). S : Chikwawa Distr., Mwanza River, fr. 6.x.1946, *Brass* 17997 (BR ; K ; NY ; SRGH). **Mozambique.** N : Mandimba, bud. 10.iv.1952, *Gomes e Sousa* 4531 (K). T : opposite Sena, fl. i.1859, *Kirk* (K). Z : Mocuba, R. Licungo, fr. 19.x.1942, *Torre* 4620 (BM). MS : Chimoio, Gondola, R. Nhamissanguere, fr. 17.ii.1948, *Garcia* 265 (BM ; LISC). LM : Sabié, fl. 7.vi.1948, *Torre* 7956 (LISC).

From Senegal eastwards to Ethiopia and southwards to SW. Africa, Transvaal and

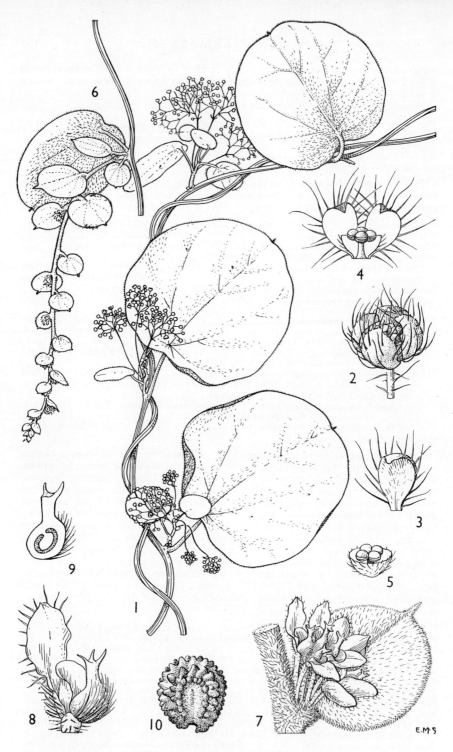

Tab. 23. CISSAMPELOS PAREIRA VAR. ORBICULATA. 1, habit of ♂ plant (×1); 2, ♂ flower (×20); 3, sepal (×10); 4, section of flower (×20); 5, corolla and stamens (×20); 6, habit of ♀ plant (×1); 7, cymule of ♀ inflorescence and bracts (×10); 8, ♀ flower (×20); 9, ovary (×20); 10, seed (×5). From F.T.E.A.

Natal. Deciduous bushland often on termite hills and near rock-outcrops, riverine forest, in swamps and often persisting on cultivated land, 0-1800 m.

5. **Cissampelos owariensis** Beauv. ex DC., Prodr. **1**: 100 (1824).—Diels in Engl. Pflanzenr. IV, 94: 302 (1910).—Hutch. & Dalz., F.W.T.A., **1**: 75 (1927).— Troupin, F.C.B. **2**: 249 (1951) pro parte; in F.T.E.A. Menisperm.: 28 (1950). Type from Nigeria.

Liane with stem and branchlets more or less densely pubescent, rarely puberulous, sometimes ultimately glabrescent; indumentum normally composed of spreading hairs. Leaf-lamina broadly ovate, generally suborbicular and angular (with the angles more or less prominent) or broadly triangular, truncate, rounded or subcordate at the base, obtuse and mucronulate at the apex, with variable indumentum, very rarely glabrous beneath, basal nerves 5–7, palmate, petiole 4–16 cm. long, inserted 0·8–2 cm. from the base of the lamina. Male inflorescences up to 40 cm. long; cymules with pubescent peduncles 0·5–3 cm. long. Male inflorescences with sepals 1–1·5 × 0·7–1·2 mm., hairy outside; corolla 1–1·2 mm. long, cup-shaped. Female inflorescences up to 35 cm. long; bracts up to 4 cm. in diam., suborbicular or reniform, mucronulate, hairy, long-ciliolate; pedicels 1–1·2 mm. long. Female flowers with sepal 1·2–2 × 0·7–1 mm.; petals 1–1·2 × 1–1·2 mm., truncate or subreniform; carpel 1–1·3 mm. long, hairy. Drupe 4–6 × 4–5 mm., hairy.

N. Rhodesia. N: Abercorn Distr., Tasker's Deviation, fl. 9.i.1952, *Richards* 424 (K). W: Mwinilunga Distr., Matonchi R., fl. 11.xi.1937, *Milne-Redhead* 3190 (BR; K). S: Mumbwa, fr. *Macaulay* 1126 (K).
Also from Sierra Leone to the Belgian Congo, Tanganyika and Angola. Lowland rain-forest and riverine forest, up to 900 m.

6. **Cissampelos rigidifolia** (Engl.) Diels in Engl., Pflanzenr. IV, 94: 303 (1910).— Troupin, F.T.E.A. Menisperm.: 30 (1956). Type from Sudan (Equatoria).
   Cissampelos pareira var. *transitoria* subvar. *rigidifolia* Engl., Bot. Jahrb. **26**: 395 (1899). Type as above.

Liane with sparingly to shortly pubescent branchlets. Leaf-lamina 2·5–10 × 2·5–9 cm., broadly ovate to somewhat rounded, subcordate to emarginate at the base, obtuse to subacute and mucronulate at the apex, densely woolly-tomentellous or tomentellous beneath, sparingly pubescent to glabrescent above, clearly discolorous, papery to subcoriaceous, basal nerves 5–10, palmate, petiole 1·8–5 cm. long, densely pubescent, inserted 0·4–1 cm. from the base of the lamina. Male inflorescences 8–30 cm. long; bracts suborbicular, apiculate, densely pubescent, tomentose; pedicels 0·5–1·5 cm. long. Male flowers with sparsely pilose sepals 1·2–1·5 (2) × 0·5–0·8 mm.; corolla cup-shaped, often incompletely connate, 0·5–1 mm. long, sparingly pubescent; synandrium 4 (6) locular, 0·5–1 mm. long. Female inflorescences 10–20 mm. long; pedicels 1–1·5 mm. long. Female flowers with sepal 1–1·2 × 0·5–0·6 mm.; petal 0·5–0·7 × 1–1·2 mm.; carpel 0·5–0·7 mm. long, densely pubescent. Drupe 4–5 × 2·5–3 mm., pubescent.

Var. **rigidifolia**

Leaves tomentellous beneath, glabrous above; stems sparsely puberulous to glabrescent.

N. Rhodesia. N: Isoka Distr., xi.1952, *White* 3728 (BR; BRLU; FHO).
Also in Ubangi-Chari, the Sudan, Belgian Congo, Uganda and Tanganyika. Woodland or wooded grassland; 700–1800 m.

Var. *lanuginosa* Troupin has the leaves densely woolly-tomentellous and greyish-white beneath, sparsely pubescent above. It is as yet known only from Tanganyika.

11. STEPHANIA Lour.

Stephania Lour., Fl. Cochinch. **2**: 608 (1790).

Herbaceous or woody lianes, sometimes with succulent leaves. Leaves simple, peltate; lamina triangular, ovate or suborbicular. Male inflorescences of panicles, false umbels or hemispherical umbel-like cymes. Male flowers with 6–8 sepals; petals 3–4, free, rarely absent; stamens 2–6, arranged in a disk-shaped stalked synandrium, round the margin of which the anthers form a horizontal ring; thecae with transverse dehiscence. Female inflorescences similar to the male. Female flowers with 3–6 free sepals ; petals 2–4, free ; staminodes usually absent ; carpel 1 ; style small ; stigma slightly lobed, or laciniate with divaricate divisions. Drupe with smooth, glabrous or hairy exocarp, endocarp subovate and compressed or reniform, truncate at the base, with 2–4 rows of tubercles or more or less projecting prickles, or merely with transverse ribs ; the condyle somewhat concave on either side with its septum either perforated or not. Seeds with endosperm.

Stem succulent ; male inflorescences of false umbels of hemispherical cymes arising from leafless stems, very shortly pedunculate ; drupes subreniform ; endocarp transversely ribbed   -   -   -   -   -   -   -   -   -   -   -   **1.** *cyanantha*
Stem woody or herbaceous, not succulent ; male inflorescences ± long-pedunculate, in false umbels of cymules, axillary, solitary or clustered 2–4 together ; drupes obovate ; endocarp with little prickles or scattered tubercles -   -   -   -   **2.** *abyssinica*

1. **Stephania cyanantha** Welw. ex Hiern, Cat. Afr. Pl. Welw. **1** : 20 (1896).—Diels in Engl., Pflanzenr. IV, 94 : 276 (1910).—Exell & Mendonça, C.F.A. **1,** 1 : 42 (1937).— Troupin, F.C.B. **2** : 244 (1951) ; F.T.E.A. Menisperm. : 21 (1956). Type from Angola (Cuanza Norte).

Somewhat woody, twining, completely glabrous liane ; stem and branches succulent. Leaf-lamina 4·5–6·5 × 4·5 cm., triangular-orbicular to orbicular, rarely reniform, acuminate to obtuse at the apex, entire or with the margins slightly undulate, membranous ; basal nerves 10–12, palmate, petiole slender, 3–5 cm. long. Male inflorescences of false umbels or subglobular cymes, usually arising from leafless stems, 1–2·5 cm. in diam. ; peduncles and pedicels very short. Male flowers with 6 obovate, incurved, single-nerved sepals 1·5–2 × 1–1·5 mm. ; petals 3–5, 0·6–1 × 1–1·5 mm., broadly subreniform, fleshy ; synandrium 6–9-locular, 1·2–1·6 mm. in diam. on a stipe 0·5–1 mm. long. Female inflorescences similar to the male. Female flowers unknown. Drupe subreniform, 5–7 mm. long ; exocarp like a thin skin, red when fresh ; endocarp transversely ribbed and furrowed. Seeds 5–7 mm. long.

N. Rhodesia. W : Solwezi River Gorge, fl. 13.ix.1952, *Angus* 430 (FHO ; K).
Also in Fernando Po, French Cameroons, Belgian Congo, Ruanda, Angola, Kenya and Tanganyika.

This rather rare species is usually found at the higher altitudes up to 2200 m., especially in swamps and on volcanic lava. Its precise ecological requirements are still unknown. According to Welwitsch, it is monoecious. The androecium is quite typical and character-istic of *Stephania*, but the inflorescences and the fruits are very different from those of the other species at present placed in this genus.

2. **Stephania abyssinica** (Dill. & Rich.) Walp., Repert. **1** : 96 (1842).—Oliv., F.T.A. **1** : 47 (1868).—Diels in Engl., Pflanzenr. IV, 94 : 268, t. 89 (1910).—Exell & Mendonça, C.F.A. **1,** 1 : 43 (1937).—Troupin, F.C.B. **2** : 245 (1951) ; F.T.E.A. Menisperm. 22 (1956).—Milne-Redh. in Mem. N.Y.Bot. Gard. **8,** 3 : 215 (1953). Type from Ethiopia (Adowa).
     Clypea abyssinica Dill. & Rich. in Ann. Sci. Nat., Sér. 2, Bot. **14** : 263 (1840). Type as above.

Twining liane, woody at the base ; stem covered with a thin bark ; branchlets glabrous or more or less densely pubescent to tomentose when young. Leaf-lamina 5–20 × 4–13 cm., ovate to broadly ovate, rarely suborbicular, rounded at the base, obtuse or subacute at the apex, membranous or papery, slightly discolorous, glabrous or tomentellous, basal nerves 8–10, palmate, petiole 4–12 cm. long. Male inflorescences of false compound umbels, axillary, solitary or clustered 2–4 to-gether ; axes glabrous or tomentellous ; peduncle 4–10 cm. long with 3–6 rays ending in umbel-like cymes ; involucre of 3–5 caducous bracts. Male flowers with 6–8 obovate or subobovate sepals 1·2–2·5 × 0·6–1·2 mm., purplish, their bases often violet ; petals 3–4, 0·8–1·2 mm. long, broadly ovate or suborbicular ; synandrium 6–8-locular. Female inflorescences similar to the male. Female flowers with 3–4 sepals ; carpel glabrous. Drupe subspherical-flattened, 0·5–0·8 cm. in diam., glabrous ; endocarp with small prickles or thick tubercles arranged in three lines ; condyle not perforated. Seeds up to 0·8 cm. long.

A species with varied ecological requirements, not however penetrating into the rain-forest; in grassland or wooded grassland up to 3500 m., preferably in moist shady places, especially on edges of rivers and swamps.

Var. **abyssinica.** TAB. **21** fig. A.

Young branchlets, petioles, lower sides of leaves, inflorescences and (partly) the sepals glabrous.

N. Rhodesia. N : Abercorn Distr., Chila Lake, fl. 4.ii.1952, *Richards* 243 (K). W : Mwinilunga Distr., Matonchi Farm, fl. 20.xi.1937, *Milne-Redhead* 3327 (BM ; BR ; K). S : Mumbwa, fl. *Macaulay* 1115 (K). **S. Rhodesia.** E : Inyanga, fr. 29.i.1931, *F.N. & W.* 4712 (BM ; BR ; PRE ; S ; SRGH). **Nyasaland.** N : Matipa Forest, fl. viii.1954, *Chapman* 323 (BM). S : Mt. Mlanje, Nayawani Forest, fl. 22.viii.1956, *Newman & Whitmore* 517 (BM). **Mozambique.** N : Tungue, Nangade, fl. 19.x.1942, *Mendonça* 957 (LISC). Z : Quelimane Distr., Lugela-Mocuba, Namagoa, fl. 15.ii.1949, *Faulkner* 197 (BR ; K). MS : Chimoio, Gondola, Nhamissenguere, R. Nhamouare, fr. 21.i.1948, *Mendonça* 3656 (BM).

Widely spread, from French Guinea east to Ethiopia and south through the Belgian Congo to Angola, Basutoland and Natal.

Var. **tomentella** (Oliv.) Diels in Engl., Pflanzenr. IV, 94 : 270 (1910).—Brenan, T.T.C.L. : 328 (1949).—Troupin, F.C.B. **2** : 246 (1951) ; F.T.E.A. Menisperm. 22 (1956). TAB. **21** fig. B. Type from Tanganyika (Kilimanjaro).

Stephania hernandifolia var. *tomentella* Oliv. in Trans. Linn. Soc., Bot., Ser. 2 : 328 (1887). Type as above.

Young branchlets, petioles, lower surfaces of leaves, inflorescences and the sepals more or less densely pubescent to tomentose.

S. Rhodesia. E : Umtali, ix.1947, *Chase* 582 (SRGH). **Nyasaland.** Without precise locality, fl. 1891, *Buchanan* 1459 (BM ; K). **Mozambique.** Z : Milange, fl. & fr. 1200 m., 26.ii.1943, *Torre* 4862 (BM).

Widespread from the British Cameroons north-east to Ethiopia and south to the Cape Province.

8. BERBERIDACEAE

By H. Wild

Shrubs or herbs. Leaves alternate or radical, simple or compound. Stipules usually absent. Flowers bisexual, in panicles, racemes, fascicles or solitary. Sepals and petals similar or dissimilar, in 2 to several series, free, hypogynous, imbricate or the outer valvate, caducous, rarely absent. Stamens 4–9, opposite the petals, hypogynous, free ; anthers 2-thecous, opening lengthwise or by valves. Ovary 1-locular ; ovules few, ascending, or more rarely numerous ; style short or absent. Fruit a berry, achene or capsule. Seeds with copious endosperm and small or long embryo ; cotyledons short.

BERBERIS L.

Berberis L., Sp. Pl. **1** : 330 (1753) ; Gen. Pl. ed. 5 : 153 (1754).

Evergreen or deciduous shrubs ; branches with usually 3-partite spines and short axillary shoots bearing the leaves. Stipules absent. Leaves apparently simple but (in our species at least) derived from 3-foliolate leaves, the side leaflets represented by vestigial leaflets and the terminal leaflet articulated at the base. Flowers in panicles, racemes, fascicles or solitary, yellowish or orange. Sepals c. 9, petaloid, free, imbricate, the outer smaller than the inner. Petals 6, free, imbricate, in two series, similar to the sepals but biglandular near the base. Stamens 6, free ; anthers opening by two valves. Ovules few, basal, erect. Fruit a berry.

Berberis holstii Engl., Pflanzenw. Ost-Afr. **C** : 181 (1895).—Sprague in Hook., Ic. Pl.: t. 3021 (1922).—Brenan, T.T.C.L.: 70 (1949).—Milne-Redh. in Mem. N.Y.Bot. Gard. **8, 3** : 215 (1953). TAB. **24.** Type from Tanganyika.

Berberis aristata var. *subintegra* Engl., Bot. Jahrb. **28** : 389 (1900). Type from Tanganyika.

Berberis petitiana C. K. Schneider in Bull. Herb. Boiss., Sér. 2, **5** : 455 (1905).— R.E.Fr. in Notizbl. Bot. Gart. Berl. **9** : 319 (1925). Type from Ethiopia.

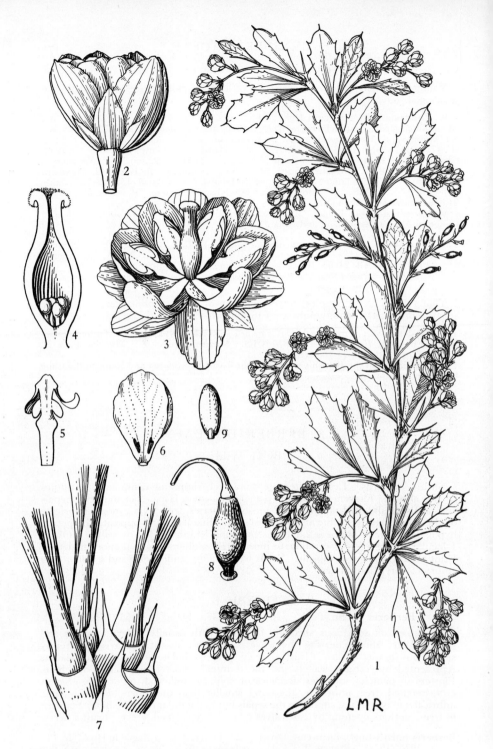

Tab. 24. BERBERIS HOLSTII. 1, branch with flowers ($\times \frac{2}{3}$) *Rogers* 415 ; 2, flower ($\times 4$) *Rogers* 415 ; 3, flower expanded ($\times 4$) *Rogers* 415 ; 4, longitudinal section of gynoecium ($\times 8$) *Rogers* 415 ; 5, stamen ($\times 4$) *Rogers* 415 ; 6, petal ($\times 4$) *Rogers* 415 ; 7, leaf-bases shewing vestigial lateral leaflets ($\times 4$) *Brass* 17325 ; 8, fruit ($\times 2$) *Drummond & Hemsley* 1553 ; 9, seed ($\times 2$) *Drummond & Hemsley* 1553.

Glabrous shrub up to 3 m. tall ; branches purplish when young, sulcate, with 3-partite spines up to 4 cm. long. Leaves usually clustered on short lateral shoots, almost sessile, apparently simple but in reality 3-foliolate with a normal, terminal leaflet articulated at its base and a petiole of c. 1 mm. long bearing 2 subulate lateral leaflets 1–3 mm. long at its apex. The petioles and subulate leaflets are characteristically persistent on the short shoots ; lamina of terminal leaflet up to 6 × 2·7 cm., coriaceous, oblong or oblanceolate to obovate, midrib produced as a short prickle, margin prickly-dentate or more rarely entire, purplish when young and somewhat glaucous below. Flowers yellow, in axillary racemes or cymes up to 5 cm. long ; bracts c. 3 mm. long, lanceolate-acuminate. Sepals increasing in size towards the interior, smallest c. 2 × 2 mm., largest c. 6·5 × 4 mm., ovate to broadly ovate, rounded at the apex. Petals somewhat smaller than the inner sepals, obovate, with two linear glands near the base. Stamens on stout filaments c. 2 mm. long ; anthers c. 2·5 mm. long, oblong, opening by a pair of spreading wing-like valves 1·5 mm. long, hinged at the apex of the anther-thecae. Ovary narrowly ellipsoid, c. 4-ovulate, with a broad capitate subsessile stigma. Berry c. 1·3 × 0·7 cm., ellipsoid, dark blue, pruinose, with a persistent stigma. Seed c. 6 × 3 mm., usually solitary, ellipsoid, brown, rugulose.

Nyasaland. N : Nyika Plateau, fl. xi.1903, *Henderson* (BM).

From the mountains of Ethiopia to the Nyika Plateau. A species of ericoid scrub, grassland or forest edges at 2–3000 m. The Nyika Plateau is its most southerly locality recorded so far.

B. holstii is very near the Himalayan *B. aristata* DC. and has been considered a variety of this latter species by Engler (loc. cit.). However, in *B. aristata* the vestigial leaflets are even less well developed and not so characteristically persistent on the short shoots, in addition the ovules are usually 2 rather than 4 and the branches tend to have whitish rather than purplish bark. Although nearly related therefore, they are probably distinct species.

9. CABOMBACEAE

By H. Wild

Aquatic herbs with perennial rhizomes ; stems coated with mucilage. Leaves alternate, floating leaves peltate, sometimes with finely dissected submerged leaves in addition. Flowers axillary, solitary, actinomorphic. Sepals 3. Petals 3, hypogynous. Stamens 3–18 ; anthers extrorse, opening lengthwise. Carpels 2–18, free ; style very short or absent but with an attenuated, entire stigma ; ovules 1–3, pendulous, parietal. Fruiting carpels indehiscent. Seeds 1–3 with a fleshy endosperm.

BRASENIA Schreb.

Brasenia Schreb. in L., Gen. Pl. ed. 8, **1** : 372 (1789).

Leaves floating and peltate, submerged dissected leaves lacking. Stamens 12–18. Carpels 6–18.

Brasenia schreberi J. F. Gmel. in L., Syst. Nat. ed. 13, **2** : 853 (1791).—Exell & Mendonça, C.F.A. **1**, 1 : 45 (1937). TAB. 25. Type from United States of America (N. Jersey).

Hydropeltis purpurea Michx., Fl. Bor. Am. **1** : 324, t. 29 (1803). Syntypes from United States of America (Tennessee and Carolina).

Brasenia peltata Pursh, Fl. Am. Sept. **2** : 389 (1814) *nom. illegit.*—Oliv., F.T.A. **1** : 52 (1868). Types as for *Hydropeltis purpurea*.

Brasenia purpurea (Michx.) Casp. in Jorn. Sci. Acad. Lisb. **4** : 312 (1873).— R.E.Fr., Wiss. Ergebn. Schwed. Rhod.-Kongo-Exped. **1** : 39 (1914). Types as above.

Glabrous aquatic plant with all its submerged parts covered with a profuse, clear, slimy mucilage. Leaves floating, rotund, peltate, 7·5 (10) × 5–7 cm., margin entire, green above, reddish-brown or purplish below, nerves 12, radiating from the apex of the petiole, rather faint, petiole reddish-brown, attached at the centre of the leaf-blade, up to 1 m. or more long according to the depth of the

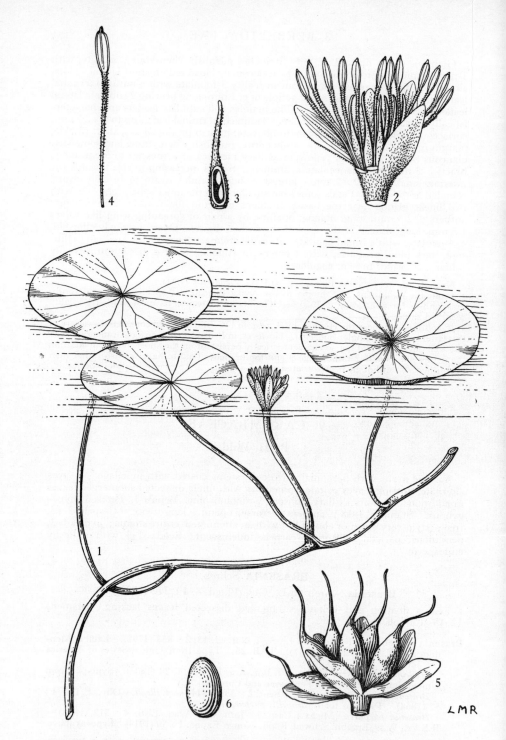

Tab. 25. BRASENIA SCHREBERI. 1, portion of plant with flower (× ⅔) *Story* 4786 ; 2, flower with one sepal and one petal removed (× 2) *Story* 4786 ; 3, carpel opened to show ovules (× 3) *Story* 4786 ; 4, stamen (× 3) *Story* 4786 ; 5, fruit (× 2) *Gilges* 7 ; 6, seed (× 6) *Gilges* 7.

water, c. 1·5 mm. in diam. Flowers solitary, axillary on reddish-brown peduncles up to c. 8 cm. long, c. 2 mm. in diam. Sepals purplish-brownish-red, petaloid, c. 1·4 × 0·3 cm., narrowly oblong, apex rounded. Petals similar to the sepals but slightly larger. Stamens with slender papillose filaments up to 1 cm. long ; anthers reddish, 3·5 mm. long, narrowly oblong, apiculate at the apex. Carpels 5 × 2 mm., narrowly lanceolate-ovoid, sparsely papillose, stigma simple, 3 mm. long, papillose. Ripe carpels 7 × 2 mm., ellipsoid with a persistent stigma. Seeds 3·5 × 2·3 mm., ellipsoid, testa pale brown.

Bechuanaland Prot. N : Okovango Swamps, near 22° 30′ E, 19° 5′ S., fl. 28.ix.1954, *Story* 4786 (K ; PRE). **N. Rhodesia.** N : Lake Bangweulu, Kamindas, fl. 13.x.1911, *Fries* 1001 (UPS). B : Ngambi (Ngambwe) Swamps, fr., *Gilges* 7 (PRE).

Also in Angola, Kenya, Tanganyika, Uganda, N. America, India, Japan, and Australia. An aquatic herb of moderately deep water, often found with *Nymphaea* spp.

The outside of the sepals and petals as well as the submerged undersides of the leaves, pedicels, petioles etc. are covered with minute purplish glandular hairs. These may be the hairs which secrete the mucilage.

The original description of this species gives neither collector nor country of origin but the type is in the Schreber collection which is in the Botanische Staatssammlung, Munich, and Dr. Merxmüller, the Director of that institution, tells me that it bears the following information in Schreber's own hand : Floyd No. 1 (presumably the collector), a Latin description and the type locality " Hope in N. Jersey ".

10. NYMPHAEACEAE

By F. A. Mendonça

Aquatic rhizomatous herbs rooted in the ground. Leaves petiolate, exstipulate, floating, emergent or rarely submerged, usually ± peltate, vernation involute. Flowers bisexual, actinomorphic, solitary, large and handsome, pedunculate, floating, emergent or rarely submerged. Sepals 4–6. Petals (6–10) numerous, some occasionally ± sepaloid. Stamens numerous, hypo- or perigynous ; anthers introrse, dehiscing by longitudinal slits. Carpels (6–10) numerous, immersed in the torus ; ovules 1 or numerous in each carpel, pendulous from the walls or apex of the carpel ; styles free. Fruit fleshy or spongy. Seeds with fleshy arils.

NYMPHAEA L.

Nymphaea L., Sp. Pl. **1** : 510 (1753) ; Gen. Pl. ed. 5 : 227 (1754) emend.— Sm., Fl. Graec. Prodr. **1** : 360 (1808) *nom. conserv.*

Robust or delicate aquatic herbs, rhizome often tuberous. Leaves large or medium, floating, emergent or submerged, lamina suborbicular to elliptic, deeply cordate, peltate, palmatinerved. Sepals 4 (sometimes appearing more numerous owing to the presence of ± sepaloid petals). Petals 5–numerous. Stamens perigynous ; filaments petaloid ; anthers obtuse or appendiculate owing to a prolongation of the connective. Carpels numerous, pluriovulate. Fruit fleshy, ripening under water. Seeds small, surrounded by a pulpy, sack-like aril open at the top.

I fully agree with the observations made by Hauman (F.C.B. **2** : 154 (1951)) about the polymorphism of the species of this genus but I differ somewhat from his nomenclatural treatment and from his conclusions about the distribution of the species which seem to have been taken largely from Gilg (in Engl., Bot. Jahrb. **41** (1908)).

Flowers white or cream ; stamens not prolonged beyond the anthers by an appendage of the connective ; leaves repand, dentate-mucronate by the protrusion at the margin of convergent prominent nerves - - - - - - - - 1. *lotus*
Flowers blue, pink or yellow (rarely white) ; stamens with a prolongation of the connective ; leaves not dentate-mucronate :
    Flowers blue or pink (rarely white) :
    Leaves floating, deeply incised-cordate, orbicular to elliptic :

Leaves with a sinuate-dentate or sinuate-lobulate margin, sometimes entire towards
the apex :
  Leaves with a sinuate-dentate margin, teeth acute, primary lateral nerves 7–9 on
  each side of the midrib, prominent reticulation forming closed areas less than
  half way to the margin ; stamens c. 100   -   -   -   2. *petersiana*
  Leaves with a sinuate-lobulate margin, lobules rounded or obtuse, sometimes
  entire towards the apex, primary nerves 8–11 on each side of the midrib,
  raised or flat, reticulation forming closed areas more than $\frac{2}{3}$ of the way towards
  the margin ; stamens considerably more than 100   -   -   - 3. *capensis*
Leaves with entire margin, rarely undulate towards the base :
  Leaves large or medium in size, 6–30 cm. in diam., orbicular to elliptic, green,
  red or purple beneath, primary nerves 5–8 on each side of the midrib ;
  flowers 6–20 cm. in diam. ; stamens very variable in number, sometimes more
  than 100   -   -   -   -   -   -   -   - 4. *caerulea*
  Leaves smaller, not more than 5 cm. in diam., primary nerves 4–5 on each side
  of the midrib, impressed or inconspicuous ; stamens 8–20   -   5. *maculata*
Leaves submerged, rarely some of them floating, very broadly bilobed-divaricate
  with the lobes subelliptic with a rather narrow isthmus along the midrib
                                                                6. *divaricata*
Flowers yellow ; leaves floating, chartaceous, incised-cordate, orbicular or suborbicular
                                                                7. *sulphurea*

1. **Nymphaea lotus** L., Sp. Pl. **1** : 511 (1753) emend.—Oliv., F.T.A. **1** : 52 (1868).—
Eyles in Trans. Roy. Soc. S. Afr. **5** : 352 (1916).—Exell & Mendonça, C.F.A. **1, 1** :
46 (1937).—Hauman, F.C.B. **2** : 156 (1951).—Milne-Redh. in Mem. N.Y.Bot.
Gard. **8,** 3 : 216 (1953). Syntypes from Ceylon, India, Egypt and Jamaica.

Robust aquatic herb with tuberous rhizome. Leaves 10–32 × 11–28 cm., coriaceous, orbicular or suborbicular, incised-cordate, somewhat peltate, lobes nearly
closed or slightly overlapping, margin ± repand, dentate-mucronate, teeth
formed by the convergence at the margin of (2) 3 nerves, upper surface smooth,
under surface prominently nerved to the edge, primary lateral nerves 7–9 on each
side of the midrib, forking dichotomously 3–4 times and not themselves forming a
closed reticulation, secondary nerves 7–9 pairs arising from the midrib. Flowers
white or cream, 10–18 cm. in diam., peduncle stout, glabrous (in specimens from
our area). Sepals 4, 4·5–9 × 2–3·5 cm., ovate-oblong or oblong-lanceolate, obtuse.
Petals c. 20, the outermost as long as the sepals, oblong or oblong-lanceolate,
rounded or acute at the apex. Stamens 40–60 ; anthers obtuse, without prolongation of the connective. Carpels 20–30 ; style-appendage 7–10 mm. long.
Fruit 4–6 cm. in diam., depressed-globose. Seeds 1·2 mm. long, ellipsoid, with
longitudinal lines of hairs.

Bechuanaland Prot. N : Maun, fl. 2.vi.1930, *van Son* in Herb. Transv. Mus. 28954
(BM ; K ; SRGH). **N. Rhodesia.** S : Mazabuka, Lochinvar, fl. 7.iv.1955, *E.M. & W.*
1438 (BM ; LISC ; SRGH). **S. Rhodesia.** N : Lomagundi, Umfuli R., fl. 5.ix.1952,
Whellan 675 (K ; SRGH). W : Bulawayo, *Engler* (*fide* Eyles). **Nyasaland.** N :
Lagoons near Dwambasi R. (" Roangwa "), fl. ix.1861, *Kirk* (K). C : Kota Kota,
Kaombe R., fl. 2.x.1943, *Benson* 413 (BM ; K ; SRGH). S : Chikwawa, lower Mwanza
R., fl. & fr. 4.x.1946, *Brass* 17960 (K ; SRGH). **Mozambique.** N : Nametil, fl.
12.vii.1948, *Pedro & Pedrógão* 4439 (BM). Z : Mopeia, fl. 12.xi.1942, *Mendonça* (LISC).
MS : R. Punguè, between Vila Pery and Gouveia, fl. 23.v.1949, *Pedro & Pedrógão* 5860
(SRGH). LM : Marracuene, fl. 10.v.1946, *Gomes e Sousa* 4329 (COI ; K).
Widespread in the Old World mainly in tropical and subtropical regions. Aquatic plant
of rivers, lakes and pools.

2. **Nymphaea petersiana** Klotzsch in Peters, Reise Mossamb. Bot. **1** : 152 (1861).—
Gilg in Engl., Bot. Jahrb. **41** : 364 (1908).—Burtt Davy, F.P.F.T. **1** : 112 (1926).
TAB. 26. Type : Mozambique, Tete, *Peters* (B†).
  Nymphaea stellata sensu Oliv., F.T.A. **1** : 52 (1868) pro parte quoad syn. *N.
  petersiana.*
  Nymphaea capensis sensu Conard, Waterlilies : 153 (1905) pro parte quoad syn.
  N. petersiana.

Robust aquatic herb with tuberous rhizome. Leaves 17–40 × 16–32 cm., chartaceous, orbicular or subelliptic, incised-cordate, peltate, lobes slightly divergent,
acute or caudate, margin sinuate-dentate, teeth acute or blunt, upper surface
smooth, under surface with prominent reticulate nervation, primary lateral nerves
7–9 on each side of the midrib, forming closed areas less than half way to the
margin, secondary nerves 6–8 pairs arising from the midrib. Flowers blue or

pinkish, 6–12 cm. in diam. Sepals 4, lanceolate, obtuse. Petals 14–20, as long as the sepals, oblong-lanceolate, obtuse or subacute. Stamens c. 100. Carpels 16–20. Fruit 2 × 3 cm., depressed-globose. Seeds 1 mm. long, subellipsoid, with longitudinal lines of hairs.

S. Rhodesia. N : Urungwe Distr., Zambezi Valley, Naodza R., Sungulu vlei, fl. 20.iv.1953, *Lovemore* 361 (K ; SRGH). **Nyasaland.** S : Shire R., Elephant Marsh, fl. 1863, *Kirk* (K). **Mozambique.** Z : R. Chire, Chicomo, fl. 10.iii.1938, *Lawrence* 630 (K). T : Lupata, fl. vii.1859, *Kirk* (K). MS : Chigogo, fl. 16.iv.1860, *Kirk* (K).

Also in Angola, Tanganyika and possibly the Transvaal. Aquatic plant of rivers, lakes and pools.

3. **Nymphaea capensis** Thunb., Prodr. Pl. Cap. **2** : 92 (1800).—Casp. in Bot. Zeit. **35** : 203 (1877).—Conard, Waterlilies : 153 (1905) pro parte excl. syn. *N. petersiana.*— Gilg in Engl., Bot. Jahrb. **41** : 364 (1908).—Burtt Davy, F.P.F.T. **1** : 112 (1926).— Hauman, F.C.B. **2** : 163 (1951). Type from the Cape.

Nymphaea stellata sensu Harv. in Harv. & Sond., F.C. **1** : 14 (1860) pro parte excl. syn. *N. caerulea.*

Nymphaea zanzibariensis Casp., loc. cit. Type from Zanzibar.

Aquatic herb with tuberous rhizome. Leaves 8–35 × 7·5–42 cm., chartaceous, with long petioles, glabrous or densely pubescent, orbicular or subelliptic, incised-cordate, peltate, lobes slightly divergent or somewhat overlapping, acute or caudate, margin ± irregularly sinuate-lobulate or entire towards the apex, upper surface smooth, under surface with conspicuously raised nerves, green or rarely reddish or reddish-purple, primary lateral nerves 8–11 on each side of the midrib, forming elongated closed areas stretching to more than ⅔ of the way to the margin, 6–8 pairs of secondary nerves arising from the midrib. Flowers blue or pinkish, 8–12 cm. in diam. Sepals 4, 4–10 × 1·5–3·5 cm., green or purple at the margins. Petals 14–20, outermost as long as the sepals, oblong, obtuse or subacute. Stamens very numerous, 100–200 or more, densely congested, the outermost with long appendages. Carpels 14–24 ; style very short. Fruit 2·2 × 3·2 cm. (in the specimens seen), depressed-globose. Seeds 1·2 mm. long, ellipsoid.

Bechuanaland Prot. N : Kabulabula, fl. 11.vii.1937, *Erens* 395 (K), 396 (SRGH). **N. Rhodesia.** N : Lake Bangweulu, Samfya, fl. 7.x.1944, *Greenway & Brenan* 8177 (K). **Mozambique.** N : Mocojo, Angoane, fl. 12.ix.1948, *Barbosa* 2060 (LISC). Z : Maganja da Costa, Maanha marsh, fl. 14.ix.1944, *Mendonça* 2053 (BM ; LISC ; SRGH). SS : Vilanculos, Mapinhane, fl. 1.ix.1942, *Mendonça* 79 (BM ; LISC). LM : Maputo, Ponta do Ouro, fl. 18.xi.1944, *Mendonça* 2934 (BM ; LISC).

Also in S. Africa, Tanganyika, Zanzibar, Madagascar and Comoro Is. Aquatic plant of rivers, lakes and pools.

4. **Nymphaea caerulea** Savigny in Déc. Egypt. **1** : 74 (1798).—Casp. in Bot. Zeit. **35** : 203 (1877).—Conard, Waterlilies : 141, t. 8 (1905).—Exell & Mendonça, C.F.A. **1**, 1 : 46 (1937).—Milne-Redh. in Mem. N.Y.Bot. Gard. **8**, 3 : 216 (1953). Type from Egypt.

Nymphaea stellata sensu Harv. in Harv. & Sond., F.C. **1** : 14 (1860) pro parte quoad specim. Burke.—Oliv., F.T.A. **1** : 52 (1868) pro parte quoad syn. *N. caerulea.* —Eyles in Trans. Roy. Soc. S. Afr. **5** : 352 (1916).

Nymphaea calliantha Conard, in Ann. Conserv. Jard. Bot. Genève, **7-8** : 19 (1904).—Hauman, F.C.B. **2** : 161 (1951). Type from Angola (Bié).

Nymphaea muschlerana Gilg in Engl., Bot. Jahrb. **41** : 357 (1908).—Hauman, tom. cit. : 158 (1951). Type from Angola (Bié).

Nymphaea magnifica Gilg, tom. cit. : 359 (1908).—R.E.Fr., Wiss. Ergebn. Schwed. Rhod.-Kongo-Exped. **1** : 40 (1914). Non *Castalia magnifica* Salisb. Type from Ruanda.

Nymphaea engleri Gilg, tom. cit. : 360 (1908).—R.E.Fr., tom. cit. : 41 (1914). Type from Angola (Huila).

Nymphaea mildbraedii Gilg, tom. cit. : 361.—R.E.Fr., loc. cit. Type from Ruanda.

Nymphaea sp.—J. R. Drumm. in Journ. Linn. Soc., Bot. **40** : 20 (1911).

Nymphaea cyclophylla R.E.Fr., loc. cit. Type : N. Rhodesia, Mukanshi, *Fries* 1122 (UPS, holotype).

Nymphaea vernayi Bremek. & Oberm. in Ann. Transv. Mus. **16** : 412 (1935). Type : Bechuanaland Prot., Maun, *van Son* in Herb. Transv. Mus. 28955 (BM ; PRE, holotype ; SRGH).

Nymphaea maculata sensu Hauman, tom. cit. : 162 (1951).

Stout or weak aquatic herb with tuberous rhizome. Leaves 8–30 × 6–28 cm., orbicular, suborbicular or elliptic, coriaceous, incised-cordate, peltate, lobes obtuse

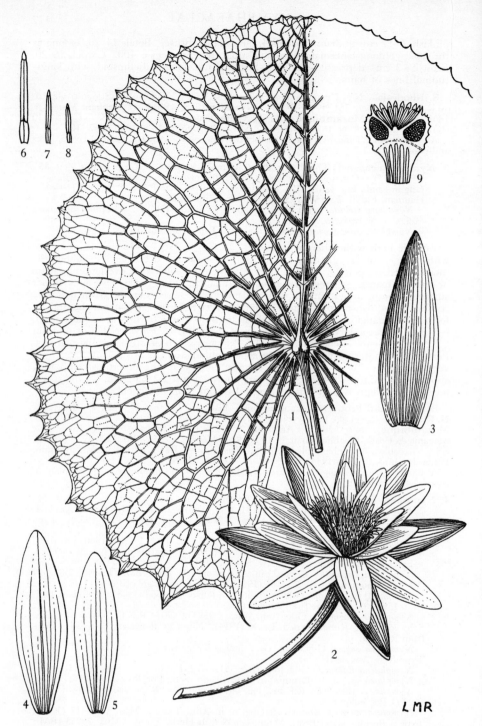

Tab. 26. NYMPHAEA PETERSIANA. 1, under-surface of leaf (×1) *Lovemore* 226 ; 2, flower
(×1) ; 3, sepal (×1½) ; 4, outer petal (×1½) ; 5, inner petal (×1½) ; 6, outer
stamen (×1½) ; 7, middle stamen (×1½) ; 8, inner stamen (×1½) ; 9, longitudinal
sect. of gynoecium (×1½). 2–9 from *Lovemore* 361.

or acute, divergent, nearly closed or overlapping, margin entire or slightly undulate towards the base, upper surface green, smooth, under surface green, red or purple, nervation raised or flat, primary lateral nerves 5–8 on each side of the midrib, 4–5 pairs of secondary nerves arising from the midrib. Flowers blue or pink, 6–20 cm. in diam. Sepals normally 4, 3–8 × 1·1–2·5 cm., oblong-ovate or oblong-lanceolate, green, sometimes marked with dark purple lines or dots, sometimes with reddish-purple margins. Petals 12–24, as long the sepals, oblong-lanceolate, obtuse or acute, some outer ones occasionally sepaloid. Stamens 30–100 or more. Carpels 14–24 ; style short. Fruit 2–4 cm. in diam., depressed-globose. Seeds 1·1 mm. long, ellipsoid, with longitudinal lines of hairs.

Bechuanaland Prot. N : Maun, *van Son* in Herb. Mus. Transv. 28955 (BM ; PRE ; SRGH). SW : lat. 23° S., without precise locality, fl. 8.iv.1864, *Chapman & Baines* (K). **N. Rhodesia.** B : Nangweshi, fl. 23.vii.1952, *Codd* 7162 (K ; SRGH). N : Abercorn, Nimkola, fl. 23.ii.1955, *Richards* 4665 (K). W : Mwinilunga, fl. 22.xi.1937, *Milne-Redhead* 3337 (K). C : Broken Hill, fl. xi.1909, *Rogers* 8596 (K ; SRGH). S : Choma, Siamambo Forest Reserve, fl. 23.vii.1953, *Angus* 14 (K). **S. Rhodesia.** N : Sebungwe, fl. 14.xi.1935, *Lovemore* 495 (SRGH). W : Matopo Hills, fl. iv.1904, *Eyles* 35 (BM ; SRGH). C : Salisbury, Hunyani R., fl. 2.ii.1950, *Wild* 3226 (K ; SRGH). S : Zimbabwe, fl. 1.vii.1930, *Hutchinson & Gillett* 3383 (K). **Nyasaland.** N : Lake Nyasa, fl. 13.vii.1936, *Burtt* 6087 (BM ; K). C : Kota Kota, Chia, fl. 5.ix.1946, *Brass* 17544 (K). S : Fort Johnston, fl. 9.xi.1929, *Burtt Davy* 1764 (K). **Mozambique.** N : Vila Cabral, fl. vi.1934, *Torre* 181 (BM ; COI). Z : Maganja da Costa, fl. 14.ix.1944, *Mendonça* 2072 (BM ; LISC). T : Furancungo, fl. 19.x.1943, *Torre* 6067 (LISC). MS : Mavita, fl. & fr. 20.iv.1948, *Barbosa* 1521 (BM ; LISC).

Egypt, widespread in tropical Africa and in the Transvaal. Aquatic plant of rivers, lakes and pools.

N. vernayi was described as having 8 sepals but the number must necessarily be a matter of some uncertainty because of the frequent presence of transitions between petals and sepals.

5. **Nymphaea maculata** Schumach. apud Schumach. & Thonn. in Kongel. Dansk. Vid. Selsk. Naturvid. Math. Afh. **4** : 21 (1829).—Keay, F.W.T.A. ed. 2, **1** : 66 (1954). Type from Ghana.
 Nymphaea heudelotii Planch. in Ann. Sci. Nat., Sér. 3, Bot. **19** : 41 (1853).—Conard, Waterlilies : 147, fig. 56 (1905).—Gilg in Engl., Bot. Jahrb. **41** : 356 (1908).—R.E.Fr., Wiss. Ergebn. Schwed. Rhod.-Kongo-Exped. **1** : 39 (1914).—Exell & Mendonça, C.F.A. **1**, 1 : 43 (1937).—Hauman, F.C.B. **2** : 157, t. 13 (1951). Type from French Guinea.
 Nymphaea heudelotii var. *nana* Conard, tom. cit. : 149, fig. 57 (1905). Type from Angola (Bié).
 Nymphaea erici-rosenii R.E.Fr. tom. cit. : 40, t. 5 fig. 3 (1914).—Hauman, tom. cit. : 158 (1951). Type : N. Rhodesia, Maumba, *Fries* 629 (UPS, holotype).

Delicate aquatic herb with elongate rhizome 1 cm. thick. Leaves 2–5 × 1·8–4·5 cm., with long slender petiole, chartaceous, orbicular or suborbicular, incised-cordate, slightly peltate, lobes rounded, obtuse or acute, divergent, nearly closed or sometimes overlapping, smooth on both surfaces, upper surface dull green, under surface green, reddish or brownish-purple, primary lateral nerves 4–5 on each side of the midrib, impressed, slightly raised or inconspicuous. Flowers blue or pink (rarely white), 3–8 cm. in diam. Sepals 4, 2–4 × 0·6–1 cm., oblong-lanceolate, green, sometimes with dark purple dots. Petals 7–14, as long as the sepals, lanceolate, acute. Stamens 8–20, connective appendage acute. Carpels 6–10 ; stigma subsessile. Fruit 1–1·5 cm. in diam., globose. Seeds 1 mm. in diam., subglobose, smooth.

Bechuanaland Prot. SE : eastern Bamangwato Territory, fl. 3.iii.1876, *Holub* (K). **N. Rhodesia.** N : Abercorn, Kawimbe, fl. 4.ii.1955, *Richards* 1124 (K). W : Mwinilunga, Sinkabolo Dambo, fl. 12.xi.1937, *Milne-Redhead* 3209 (K).

Also in west tropical Africa, Belgian Congo, Uganda and Angola. Aquatic plant of rivers, lakes and pools.

6. **Nymphaea divaricata** Hutch. in Kew Bull. **1931** : 246, cum fig. (1931) ; Botanist in S. Afr. : 525 cum fig., 528 (1946). Type : N. Rhodesia, Kasama, *Hutchinson & Gillett* 4045 (K, holotype).

Aquatic herb with elongate rhizome 1–2 cm. thick. Leaves 8–24 cm. broad, submerged or rarely some of them floating (fide A. A. Bullock), thinly papyraceous, bilobed-divaricate, cordate at the base, retuse at the apex, with a ± narrow isthmus along the midrib, slightly or not peltate, margin entire, smooth on both surfaces, primary lateral nerves 5–6 on each side of the midrib, 1–2 pairs of secondary nerves arising from the midrib (or absent). Flowers blue or pink (yellow in specimens from Angola), 4–10 cm. in diam., floating or slightly emergent. Sepals 4 (8 reported in the original description), 2·6–4·5 × 0·8–1·2 cm., ovate-oblong or lanceolate, acute. Petals 12–15, as long as the sepals, lanceolate, obtuse or acute. Stamens 20–30, connective-appendage acute. Carpels 12–18 ; style short. Fruit not seen.

N. Rhodesia. N : Abercorn, Kambole, fl. 30.v.1936, *Burtt* 6228 (BM ; K) ; Pansa R., fl. 2.x.1949, *Bullock* 1155 (K ; SRGH).
Also in Angola and the Belgian Congo. Aquatic herb in rivers, lakes and pools.

7. **Nymphaea sulphurea** Gilg in Warb., Kunene-Samb.-Exped. Baum : 235 (1903).—Conard, Waterlilies : 161, t. 12 (1905).—Exell & Mendonça, C.F.A. **1**, 1 : 48 (1937).
Type from Angola (Bié).
Nymphaea primulina Hutch. in Kew Bull. **1931** : 245, cum fig. (1931). Type : N. Rhodesia, Kasama, *Hutchinson & Gillett* 4048 (K, holotype).

Moderately stout aquatic herb with elongate rhizome 1–2 cm. thick. Leaves, 3–13 × 3–12 cm., chartaceous, orbicular or suborbicular, incised-cordate, peltate, lobes obtuse or rounded, nearly closed or divergent, smooth on both surfaces, green or reddish beneath, primary lateral nerves 6–7 on each side of the midrib, impressed. Flowers yellow, 4–8 cm. in diam. Sepals 4, 2·5–5 × 0·8–1·6 cm., ovate-lanceolate. Petals 8–14, as long as the sepals, oblong-lanceolate, obtuse or acute. Stamens 30–50. Carpels 12–16. Fruit 2 × 2·6 cm., subglobose. Seeds 1·5 mm. long, ellipsoid, smooth.

N. Rhodesia. N : Kasama, fl. 23.viii.1930, *Hutchinson & Gillett* 4048 (K). W : Mwinilunga, Sinkabolo Dambo, fl. 30.x.1937, *Milne-Redhead* 2871 (BM ; K).
Also in the Belgian Congo and the south of Angola. Aquatic plant of rivers, lakes and pools.

11. PAPAVERACEAE

By A. W. Exell

Annual, biennial or perennial herbs (rarely shrubby), usually with white or yellowish latex, with alternate, exstipulate leaves. Flowers actinomorphic, bisexual, usually hypogynous. Sepals 2–3, imbricate, free or calyptrate, caducous. Petals 4–6 (12) free, imbricate, fugacious. Stamens usually numerous. Ovary syncarpous, 1-locular with parietal placentas (rarely multilocular or spuriously 2-locular) and numerous ovules. Fruit usually a capsule dehiscing by valves or pores. Seeds small, numerous ; endosperm oily.
There are no indigenous species of this family in our area but *Argemone mexicana* is an established alien.

ARGEMONE L.

Argemone L., Sp. Pl. **1** : 508 (1753) ; Gen. Pl. ed. 5 : 225 (1754).

Annual or biennial herbs (rarely perennial and somewhat shrubby) with yellow latex. Leaves pinnate-lobed or incised, usually spiny. Sepals 3, horned at the apex. Petals 6. Stamens numerous. Carpels 3 (4–6) ; stigmas the same number as the carpels and alternating with the placentas. Capsule usually spiny.

Argemone mexicana L., Sp. Pl. **1** : 508 (1753).—Klotzsch in Peters, Reise Mossamb. Bot. **1** : 169 (1861).—Oliv., F.T.A. **1** : 54 (1868).—Fedde in Engl., Pflanzenr. IV, 104 : 273, fig. 36 B–D (1909).—Eyles in Trans. Roy. Soc. S. Afr. **5**, 4 : 355 (1916).—Burtt Davy, F.P.F.T. **1** : 116 (1926).—Exell & Mendonça, C.F.A. **1**, 1 : 48 (1937).—Martineau, Rhod. Wild Fl. : 28 (1954).—Wild, Common Rhod. Weeds : fig. 11 (1955). Type from tropical America.

Erect, glaucous herb up to 1 m. tall. Leaves sessile, ± amplexicaul, sinuate-pinnatifid with prickly teeth, often variegated. Flowers 3·5–4·5 cm. in diam., bright yellow, cream or white. Capsule 3 × 1·5 cm., ellipsoid.

Bechuanaland Prot. SE : 120 km. WNW. of Francistown on Maun Rd., fl. & fr. 2.v.1957, *Drummond* 5293 (SRGH). **S. Rhodesia.** N : Mazoe, Maryvale, fl. viii.1922, *Eyles* in GHS 3245 (SRGH). W : Bulawayo Commonage, fl. & fr. iv.1918, *Eyles* 961 (BM ; SRGH). C : Salisbury, fl. viii.1921, *Eyles* 6117 (SRGH). **Nyasaland.** Without locality, fl. 23.v.1919, *Johnson* (K). **Mozambique.** N : Mecaloja, fl. & fr. 4.ix.1934, *Torre* 418 (BM ; COI). Z : without precise locality, *Stewart* (BM). T : Tete, fl. 24.vii.1931, *Pomba Guerra* 32 (COI). SS : Inhambane, fl. xi.1955, *Gomes e Sousa* 1689 (COI). LM : Maputo, Catuane, fl. 20.xi.1944, *Mendonça* 2977 (BM ; LISC).

Native of Central and tropical S. America. Widespread as an introduction in all warm regions. A common weed of arable and waste land at the lower altitudes and in Matabeleland.

Mexican Poppy.

12. FUMARIACEAE

By A. W. Exell

Herbs, sometimes climbing, with alternate or radical, exstipulate, usually finely divided leaves. Flowers in racemes or spikes, rarely solitary, usually zygomorphic, bisexual, hypogynous. Sepals 2 (rarely 0), caducous. Petals 4 (6 or more), one or both of the outer ones spurred or saccate, inner ones often cohering at the apex. Stamens 6, perhaps to be regarded as 2 tripartite elements, the central branch of each bearing a 2-thecous anther and each lateral branch a 1-thecous anther. Ovary 1-locular, usually with 2 parietal placentas, each with 1–∞ anatropous ovules. Fruit a capsule or nutlet. Seeds with copious endosperm.

FUMARIA L.

Fumaria L., Sp. Pl. **2** : 699 (1753) ; Gen. Pl. ed. 5 : 314 (1754).

Annual herbs. Leaves all cauline. Flowers zygomorphic in leaf-opposed racemes. Only the upper petal spurred. Ovules 1 or 1 on each placenta. Fruit a nutlet.

Fumaria muralis Sond. ex Koch, Syn. Fl. Germ. ed. 2, **3** : 1017 (1845).—Burtt Davy, F.P.F.T. **1** : 117 (1926).—Exell & Mendonça, C.F.A. **1**, 1 : 49 (1937). Type from Germany.

Fumaria officinalis var. *capensis* Harv. in Harv. & Sond., F.C. **1** : 18 (1860). Type from S. Africa.

Straggling annual herb with flexuous stems up to 50 cm. long. Leaves petiolate, lamina bipinnatisect with cuneate lobes, glabrous. Flowers 7–12 mm. long, pinkish or purplish, tipped with blackish-red or dark purple, pedicellate, pedicel 2–3 mm. long, in few-flowered bracteolate racemes, bracteoles usually shorter than the pedicels. Sepals 1·5–2 mm. long, ovate, dentate towards the base. Lower petal with narrow, erect margins, not spathulate. Fruit 1·5–2 mm. in diam., subglobose to ovoid, smooth or very finely rugulose.

S. Rhodesia. C : Salisbury, fl. & fr. ix.1956, *Whellan* 1155 (SRGH).

Western Europe, N. Africa, Canary Is. and Madeira. An introduced weed of cultivated ground.

13. CRUCIFERAE

By A. W. Exell

Annual, biennial or perennial herbs (rarely somewhat shrubby) with alternate (rarely opposite or verticillate) exstipulate, simple or compound leaves sometimes forming a basal rosette. Inflorescence usually racemose. Flowers actinomorphic (except for the stamens) usually bisexual, hypogynous. Sepals 4, free, in two series, often somewhat saccate. Petals 4 (rarely fewer or absent). Stamens usually 6, tetradynamous (rarely fewer or numerous) ; anthers 2- (rarely 1-) thecous, opening lengthwise. Ovary sessile (rarely stipitate), syncarpous of 2 carpels, 1-locular with 1–2 parietal placentas or divided into 2 chambers by a false septum. Fruit usually a dehiscent silique or silicule (more rarely indehiscent or transversely or longitudinally jointed). Seeds 1 to numerous, with no or very little endosperm.

Fruit dehiscent (except sometimes for the tip) :
  Fruit a silique (at least 3 times as long as broad) :
    Fruit terete (though possibly appearing flattened in dried, immature specimens) with convex valves or 4-angled :
      Seeds 2-seriate (in species from our area) :
        Fruit not-beaked but with a persistent style -   -   -   -   **1. Rorippa**
        Fruit with a broad, flat beak   -   -   -   -   -   -   **2. Eruca**
      Seeds 1-seriate :
        Fruit terete, ± circular in cross-section :
          Valves of fruit each with a conspicuous dorsal vein, appearing 1-veined ; seeds subspherical   -   -   -   -   -   -   **3. Brassica**
          Valves of fruit each with 3 or more veins :
            Fruit beaked ; seeds subspherical -   -   -   -   **4. Sinapis**
            Fruit not or scarcely beaked ; seeds ovoid or ellipsoid   - **5. Sisymbrium**
      Fruit 4-angled in cross-section, valves keeled ; seeds ovoid or ellipsoid
                                  6. Erucastrum
    Fruit laterally compressed with flattened valves ; seed uniseriate, cylindric (in our species)   -   -   -   -   -   -   -   -   **7. Cardamine**
  Fruit a silicule (up to 1½ times as long as broad) :
    Each loculus of the fruit 1-seeded   -   -   -   -   -   **8. Lepidium**
    Each loculus of the fruit several–many-seeded ; silicule triangular-obcordate
                                  9. Capsella
Fruit indehiscent, usually eventually separating into 2 or more 1-seeded joints :
  Fruit short, compressed, 2-lobed, wrinkled, dividing longitudinally into 2 joints
                                10. Coronopus
  Fruit elongated, terete or moniliform, usually eventually dividing transversely into several joints   -   -   -   -   -   -   **11. Raphanus**

1. RORIPPA Scop.

Rorippa Scop., Fl. Carniol. : 520 (1760).

Annual, biennial or perennial herbs sometimes almost acaulescent, often aquatic or semi-aquatic, glabrous or with simple hairs. Leaves alternate, entire or variously pinnatifid or pinnatisect or pinnate. Flowers yellow, white or pinkish in terminal or axillary racemes or panicles or sometimes solitary and axillary. Petals usually 4, occasionally reduced in number or absent. Stamens usually 6, tetradynamous, occasionally fewer. Fruit cylindric to ellipsoid. Seeds 1–2-seriate (2-seriate in our area) lenticular to subspherical ; cotyledons accumbent.

Fruit 3–4 times as long as broad ; flowers solitary, axillary, pale yellow   1. *cryptantha*
Fruit 5–10 times as long as broad ; flowers in terminal or axillary racemes or panicles :
  Leaves toothed to pinnatisect ; flowers yellow :
    Sepals 2·5–3 mm. long ; petals 3–6 mm. long ; fruit tapering gradually into the persistent style :
      Silique up to 18 mm. long ; petals 4–6 mm. long ; persistent style up to 2 mm. long
                                2. *fluviatilis*
      Silique up to 38 mm. long ; petals 3–4 mm. long ; persistent style up to 1 mm. long
                                3. *nudiuscula*
    Sepals 1·5–2 mm. long ; petals 2–2·5 mm. long ; fruit with an apiculate persistent style -   -   -   -   -   -   -   -   -   4. *madagascariensis*
  Leaves pinnate (when well developed) ; flowers white :
    Petals 4–5 mm. long, longer than the sepals   -   -   5. *nasturtium-aquaticum*
    Petals about 1·5 mm. long, shorter than the sepals   -   -   6. *humifusa*

1. **Rorippa cryptantha** (A. Rich.) Robyns & Boutique, F.C.B. **2** : 535 (1951). Type from Ethiopia.
    Nasturtium cryptanthum A. Rich., Tent. Fl. Abyss. **1** : 15 (1847). Type as above.

An annual semi-aquatic herb with radiating, prostrate-ascending, glabrous stems up to about 40 cm. long. Leaves up to about 10 × 5 cm., pinnatipartite, glabrous, lobes often irregularly incised or coarsely toothed. Flowers pale yellow, solitary in the axils of the leaves or reduced leaves. Sepals 2 × 0·7 mm., oblong-elliptic, glabrous. Petals 1·5 × 0·6 mm., obovate. Stamens 6, 1·5–1·8 mm. long. Silique up to 12 × 3 mm., shortly pedicellate, pedicel 1·5–2 mm. long, cylindric-ellipsoid, slightly broader at the base, glabrous, persistent style less than 1 mm. long. Seeds 2-seriate, 0·8 mm. in diam., subspherical, slightly flattened, glabrous.

N. Rhodesia. W : Lunga R., Mwinilunga, fr. 1.xii.1938, *Milne-Redhead* 3483 (BM ; K).
Also in Ethiopia, Angola (*Mendes* 1085) and the Belgian Congo. Damp ground.

2. **Rorippa fluviatilis** (E. Mey. ex Sond.) R. A. Dyer in Bot. Surv. S. Afr., Mem. **17** : 138 (1937). Type from S. Africa.
  Nasturtium fluviatile E. Mey. ex Sond. in Linnaea **23** : 2 (1850).—Sond. in Harv. & Sond., F.C. **1** : 21 (1860) pro parte.—Eyles in Trans. Roy. Soc.S. Afr., **5** : 355 (1916).—Burtt Davy, F.P.F.T. **1** : 125 (1926). Type as above.
  Nasturtium palustre sensu Eckl. & Zeyh., Enum. Pl. Austr.-Afr. Extratrop. **1**, 5 (1835) non DC.

Annual, semi-aquatic herb with erect, glabrous stems up to 40 cm. tall. Lower leaves petiolate, upper subsessile, up to 10 × 2·5 cm., narrowly elliptic, narrowly obovate-elliptic or narrowly oblong-elliptic in outline, pinnatifid or pinnatisect, lobes sometimes bluntly toothed, glabrous. Flowers yellow, pedicellate, pedicels up to 15 mm. long, in terminal panicles or in racemes in the axils of the upper leaves. Sepals 3 × 1–1·5 mm., petaloid, elliptic. Petals 4–6 mm. long. Stamens 6, filaments 2·5–3 mm. long, anthers 1·3 mm. long. Ovary 1·5 × 0·3 mm., ellipsoid to cylindric, ovules numerous. Silique up to 18 × 2 mm., cylindric, glabrous, persistent style up to 2 mm. long. Seeds 2-seriate, 0·5 mm. in diam., lenticular, glabrous.

Bechuanaland Prot. SE : Mochudi, fr. i.1914, *Rogers* 6417 (K). **S. Rhodesia.** W : Bulawayo, fl. & fr. i.1898, *Rand* 21 (BM).
Also in S. Africa. Growing on mud in pans etc.

3. **Rorippa nudiuscula** (E. Mey. ex Sond.) Thell. in Viert. Naturf. Ges. Zür. **56** : 259 (1911). Type from S. Africa.
  Arabis? nudiuscula E. Mey. [in Drège, Zwei Pflanz.-Docum. : 54 (1843) *nom. nud.*] ex Sond. in Harv. & Sond., F.C. **1** : 22 (1860) (sphalm. *nudicaulis*). Type as above.
  Nasturtium nudiusculum (E. Mey. ex Sond.) O. E. Schulz in Fedde, Repert. **33** : 274 (1934). Type as above.

Subsp. **serrata** (Burtt Davy) Exell, comb. nov. TAB. **27** fig. A. Type from Transvaal, *Wilms* 18 (K).
  Nasturtium elongatum var. *serratum* Burtt Davy, F.P.F.T. **1** : 47 (1926). Type as above.

Perennial herb with an upright, glabrous or very sparsely pubescent stem up to about 50 cm. tall. Basal leaves in a rosette, stalked, lamina up to 7 × 2·5 cm., obovate to obovate-lanceolate, glabrous, apex rounded, margins irregularly serrulate or denticulate, base cuneate, decurrent into the stalk ; cauline leaves sessile, narrowly elliptic with a few teeth near the apex, slightly auriculate at the base. Flowers yellow, pedicellate, pedicels up to 4 mm. long, glabrous, in a terminal ebracteate raceme up to 14 cm. long. Sepals 2–3 mm. long, petaloid, glabrous. Petals 4, 3–3·5 mm. long, obovate. Stamens 6, filaments 2 mm. long, anthers 1 mm. long. Ovary cylindric, ovules numerous. Silique up to 38 × 1·8 mm., cylindric, glabrous with persistent style up to 1 mm. long. Seeds 2-seriate (immature in our specimens).

S. Rhodesia. C : Selukwe Distr., fl. & fr. 8.xii.1953, *Wild* 4309 (SRGH). S : Zimbabwe Ruins, fl. & fr. 14.ii.1955, *E.M. & W.* 362 (BM ; LISC ; SRGH).
Species : S. Rhodesia and S. Africa ; subspecies : S. Rhodesia and the Transvaal.

The type-subspecies has pinnate leaves and seems to be restricted to the Cape Province.

4. **Rorippa madagascariensis** (DC.) Hara in Journ. Jap. Bot. **30** : 198 (1955). Type from Madagascar.
  Nasturtium madagascariensis DC., Syst. Nat. **2** : 192 (1821). Type as above.
  Nasturtium indicum sensu Oliv., F.T.A. **1** : 58 (1868).—Burtt Davy, F.P.F.T. **1** : 125 (1926).

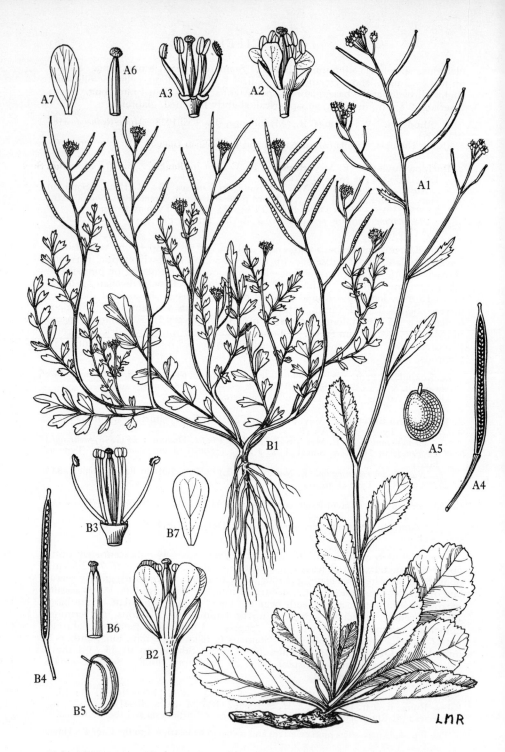

Tab. 27. A.—RORIPPA NUDIUSCULA SUBSP. SERRATA. A1, plant (×1) *Wild* 4309; A2,
flower (×6) *Wild* 4309; A3, flower with sepals and petals removed (×9) *Greatrex*
s.n.; A4, fruit opened to show seeds (×2) *Wilms* 19; A5, seed (×15) *Wilms* 19;
A6, style (×9) *Wild* 4309; A7, petal (×9) *Wild* 4309. B.—CARDAMINE FLEXUOSA.
B1, plant (×1); B2, flower (×15); B3, flower with sepals and petals removed
(×15); B4, fruit opened to show seeds (×3); B5, seed (×24); B6, style (×15);
B7, petal (×15). All from *Arnold* in GHS 12601.

Rorippa indica sensu Exell & Mendonça, C.F.A. **1**, 1 : 50 (1937).—Robyns & Boutique, F.C.B. **2** : 532 (1951).

Annual herb with branched, erect, glabrous stems up to 40–50 cm. tall. Leaves sessile, up to about 8 × 5 cm., somewhat amplexicaul at the base, obovate, oblong or ovate in outline, pinnatipartite or lyrate-pinnatipartite, lobes often toothed or crenate, glabrous or very sparsely pubescent. Flowers yellow, pedicellate, pedicels 1·5–3 mm. long, in terminal racemes up to 15 cm. long. Sepals 1·5–2 × 1 mm., oblong. Petals 2–2·5 × 0·7–0·8 mm., narrowly oblong. Stamens 6, filaments 1·6–2 mm. long, anthers 0·5 mm. long. Ovary 1·5 × 0·8 mm., ellipsoid, with a short, stout style and capitate stigma. Silique up to 18 × 2 mm., cylindric, glabrous, style persistent, 0·5–1 mm. Seeds 2-seriate, about 0·7 mm. in diam., lenticular.

N. Rhodesia. E : Petauke Distr., Luangwa River Bridge, fl. & fr. 5.ix.1947, *Greenway & Brenan* 8049 (K ; PRE). S : Mapanza N., fl. & fr. 20.ix.1953, *Robinson* 323 (K). **S. Rhodesia.** N : Mtoko Distr., Mkota Reserve, fl. & fr. 11.vi.1950, *Whellan* 464 (SRGH). S : Bubye R., fl. & fr. 9.v.1958, *Drummond* 5727 (SRGH). **Nyasaland.** S : banks of Shire, fl. & fr. 8.x.1887, *Scott* (K). **Mozambique.** N : Rio Lúrio, Cuamba, fl. & fr. 2.viii.1934, *Torre* 546 (BM ; COI ; LISC). T : Boroma, fl. & fr. 22.ix.1944, *Mendonça* 355 (BM ; LISC). Z : Quelimane, fl. & fr. 1908, *Sim* 20742 (PRE). MS : Gorongosa, fl. & fr. 13.ix.1944, *Mendonça* 2459 (BM ; LISC).
Widespread in tropical Africa and Madagascar. Banks of rivers.

5. **Rorippa nasturtium-aquaticum** (L.) Hayek, Sched. Fl. Stir. Exs. **3–4** : 22 (1905). Type from Europe.
    Sisymbrium nasturtium-aquaticum L., Pl. **2** : 657 (1953). Type as above.
    Nasturtium officinale R. Br. apud Ait. f. in Ait., Hort. Kew. ed. 2, **4** : 110 (1812).—Sond., in Harv. & Sond., F.C. **1** : 21 (1860).—Oliv., F.T.A. **1** : 58 (1868).—Burtt Davy, F.P.F.T. **1** : 125 (1926).—Exell & Mendonça, C.F.A. **1**, 1 : 50 (1937). Type as above.

Aquatic or semi-aquatic herb with glabrous stems, often rooting at the nodes, varying greatly in length according to conditions. Lower leaves up to about 13 × 5·5 cm. (reduced when not growing in water), lyrate-pinnate, the upper sessile and auricled, glabrous or with a few sparse hairs, leaflets oblong-elliptic to sub-orbicular, obtuse, margins usually sinuate. Flowers white, pedicellate, pedicel up to 7–10 mm. long, in short, terminal, somewhat corymbose racemes. Sepals 2–2·5 × 0·8 mm., oblong. Petals 4–5 mm. long. Stamens 6. Silique 10–18 × 1·5–3 mm., terete, cylindric, somewhat curved, with persistent style 0·5–1 mm. long. Seeds 2-seriate, 0·7–1 × 0·5–0·6 mm., lenticular.

S. Rhodesia. N : Mazoe, fl. & fr. xi.1949, *Whellan* in GHS 26781 (SRGH). C : Makabusi R., fl. & fr. 21.ix.1947, *Wild* 2013 (SRGH). E : Honde Valley, fl. & fr. 1934, *Gilliland* 1163 (BM). **Mozambique.** MS : R. Zambusi, fl. 17.ix.1950, *Chase* 2904 (BM ; SRGH).
Cosmopolitan. Streams and pools.

Common Watercress.

6. **Rorippa humifusa** (Guill. & Perr.) Hiern, Cat. Afr. Pl. Welw. **1** : xxvi (1896).—Robyns & Boutique, F.C.B. **2** : 534 (1951). Type from Senegal.
    Nasturtium humifusum Guill. & Perr. in Guill., Perr. & Rich., Fl. Senegamb. Tent. **1** : 19 (1831).—Oliv., F.T.A. **1** : 58 (1868).—Exell & Mendonça, C.F.A. **1**, 1 : 50 (1937). Type as above.

Almost acaulescent herb with leaves forming a rosette. Leaves numerous, up to 16 cm. long, pinnate or pinnatipartite, glabrous, rhachis winged, leaflets irregularly suborbicular to broadly ovate, crenate or sometimes deeply divided. Flowers white, pedicellate, with pedicels 1–1·5 mm. long, in decumbent racemes up to 10 cm. long in the axils of the basal leaves. Sepals 1–1·3 × 0·5 mm., oblong. Petals usually 2, less than 1 mm. long. Stamens 6. Ovary cylindric-ellipsoid, style short and thick, stigma entire. Silique 6–10 × 1–1·5 mm., glabrous, style persistent, 0·5–1 mm. long. Seeds 0·5–0·6 × 0·3–0·5 mm., 2-seriate, numerous, bright orange, flattened-ellipsoid to lenticular.

N. Rhodesia. S : Mapanza W., fl. & fr. 2.ix.1953, *Robinson* 299 (K).
Also from Senegal to Angola and in Uganda, Kenya and Madagascar. On damp sand on river banks.

2. ERUCA Mill.

Eruca Mill., Gard. Dict. Abridg. ed. 4 (1754).

Annual, biennial or perennial herbs. Flowers yellowish, cream or purplish in terminal, ebracteate racemes. Sepals erect, inner pair saccate at the base. Petals exceeding the sepals, unguiculate. Stamens 6. Stigma bilobed. Silique comparatively broad with prominently 1-veined valves and a broad, flat, seedless beak Seeds 2-seriate, spherical or ovoid. Cotyledons conduplicate.

Eruca sativa Mill., Gard. Dict., ed. 8 : n. 1 (1768).—Oliv., F.T.A. **1** : (1868).—Burtt Davy, F.P.F.T. **1** : 128 (1926). TAB. **28** fig. A. Type a cultivated plant.

Annual or perennial glaucous herb with slender tap-root and stiff, erect, pubescent stem up to 80 cm. tall. Lower leaves stalked, lyrate-pinnatifid or pinnate, terminal leaflet or lobe oblong or obovate, lateral ones narrower, all coarsely toothed or lobed or sometimes entire. Flowers yellow or whitish veined with purple, pedicellate, pedicels 2–4 mm. long. Sepals 8–10 mm. long, oblong. Petals up to 25 mm. long, obovate-cuneate. Silique 12–25 × 3–5 mm., with an elongate-triangular beak up to 7–8 mm. long and 4 mm. broad at the base. Seeds brownish, 2 × 1 mm., flattened-ellipsoid.

S. Rhodesia. E : Umtali Distr., fl. & fr. 7.viii.1956, *Chase* 6164 (SRGH). Mediterranean Region and Asia. Probably an escape from cultivation.

Rocket Salad.

3. BRASSICA L.

Brassica L., Sp. Pl. **2** : 666 (1753) ; Gen. Pl. ed. 5 : 299 (1754).

Annual or biennial (rarely perennial) herbs, glabrous or with simple hairs. Flowers usually yellow, pedicellate, in ebracteate racemes. Sepals erect, inner pair ± saccate. Petals 4, unguiculate. Stamens 6. Fruit a beaked silique with convex valves, each valve with 1 prominent vein, beak with 0–3 seeds. Seeds 1–seriate, subspherical. Cotyledons conduplicate.

In addition to the species dealt with below, forms of *B. oleracea* L. (Cabbage etc.), glabrous and glaucous with sessile but not amplexicaul cauline leaves, may be found as escapes from cultivation and *B. tournefortii* Gouan has been found as an impurity in seed from Australia.

Upper stem-leaves narrowed into a stalk-like base (not amplexicaul) ; flowers pale yellow
　　　　　　　　　　　　　　　　　　　　　　　　　　　　　　　1. *juncea*
Upper stem-leaves amplexicaul ; flowers bright yellow　-　-　-　-　2. *rapa*

1. **Brassica juncea** (L.) Czern., Consp. Pl. Chark. : 8 (1859).—Oliv., F.T.A. **1** : 65 (1868).—O. E. Schulz in Engl., Pflanzenr. IV, 105, 1 : 55 (1919).—Burtt Davy, F.P.F.T. **1** : 128 (1926).—Exell & Mendonça, C.F.A. **1**, 1 : 51 (1937).—Robyns & Boutique, F.C.B. **2** : 530 (1951). TAB. **28** fig. B. Type from Asia.
　　Sinapis juncea L., Sp. Pl. **2** : 668 (1753). Type as above.

Annual herb 60–70 cm. tall with purplish nearly glabrous stems. Leaves stalked, lyrate-pinnatifid to nearly entire, somewhat glaucous. Flowers pale yellow with pedicels up to 8 mm. long in terminal racemes. Sepals somewhat spreading. Petals 4, 5–9 mm. long. Stamens 6. Silique 2·5–5 cm. long (usually about 3 cm. in our material), narrowly cylindric, valves with reticulate nervation, with a tapering, seedless beak 5–10 mm. long, at its tip narrower than the stigma. Seeds 1–1·3 mm. in diam., subspherical, yellowish- or reddish-brown.

S. Rhodesia. C : Marandellas, fl. & fr. 30.i.1942, *Dehn* 585 (SRGH). **Nyasaland.** N : Masuku Plateau, fl. & fr. vii.1896, *Whyte* (K). C : Angoniland, fl. & fr. 1901, *Sharpe* 77 (K). S : Ncheu Distr., Lower Kirk Range, Chipusiri, fl. & fr. 17.iii.1955, *E.M. & W.* 942 (BM ; LISC ; SRGH). **Mozambique.** Z : Zimbo, R. Chinde, fl. 11.vii.1889, *Sarmento* 103 (COI).

Native of Asia. Probably escapes from cultivation. Indian or Chinese Mustard.

Our specimens would probably come under subsp. *integrifolia* (West) Thell. (see under *Brassica integrifolia* (West) O. E. Schulz, tom. cit. : 58 (1918), where a specimen from Zomba is cited, and see also Suesseng. in Proc. & Trans. Rhod. Sci. Ass. **43** : 87 (1951)) but the infra-specific taxa are not always very distinct in this species.

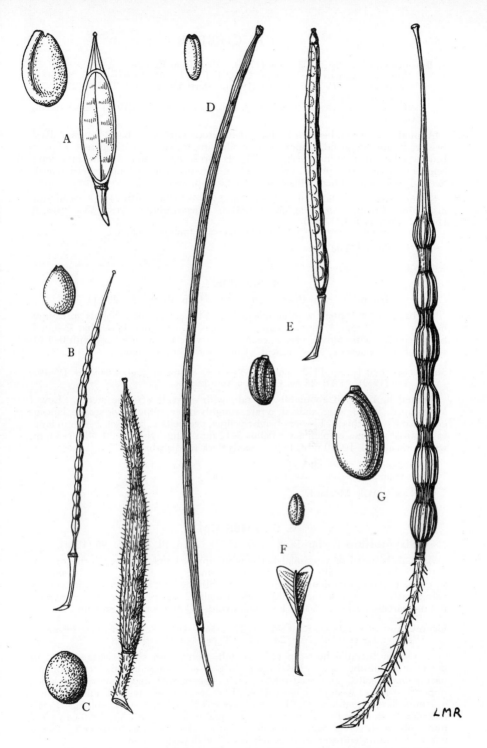

Tab. 28. A.—ERUCA SATIVA, fruit (×3), seed (×12). B.—BRASSICA JUNCEA, fruit (×3),
seed (×12). C.—SINAPIS ARVENSIS, fruit (×3), seed (×12). D.—SISYMBRIUM
ORIENTALE, fruit (×3), seed (×12). E.—ERUCASTRUM ARABICUM, fruit (×3), seed
(×12). F.—CAPSELLA BURSA-PASTORIS, fruit (×3), seed (×12). G.—RAPHANUS
RAPHANISTRUM, fruit (×3), seed (×12).

2. **Brassica rapa** L., Sp. Pl. **2** : 666 (1753). Type from Europe.

> *Brassica campestris* L., loc. cit.—Oliv., F.T.A. **1** : 66 (1868).—O. E. Schulz in Engl., Pflanzenr. IV, 105, 1 : 45 (1919).—Burtt Davy, F.P.F.T. **1** : 127 (1926). Type from Europe.
> *Brassica* sp.—Eyles in Trans. Roy. Soc. S. Afr. **5** : 355 (1916) quoad specim. *Eyles* 295.

Annual or biennial herb with stout or tuberous tap-root. Basal leaves stalked, bristly, lyrate-pinnatifid, upper ones sessile, amplexicaul, ± glaucous. Flowers bright yellow, in terminal racemes. Sepals spreading. Petals 4, 6–10 mm. long. Stamens 6. Silique curved, erect, slightly flattened with a long, tapering beak. Seeds 1·5–2 mm. in diam., blackish or reddish-brown.

S. Rhodesia. N : Mazoe, fl. & fr. iii.1906, *Eyles* 295 (BM ; SRGH). E : Chimanimani Mts., Musapa Gap, *Wild* 3520 (SRGH). **Mozambique.** MS : Vila Paiva, fl. viii.1945, *Pimenta* 241 (LISC).
North temperate regions. Occasional as an escape from cultivation.

The Turnip and Wild Rape.

4. SINAPIS L.

Sinapis L., Sp. Pl. **2** : 668 (1753) ; Gen. Pl. ed. 5 : 299 (1754).

Herbs, usually annual, with simple hairs. Flowers yellow. Sepals spreading, not saccate. Petals 4, unguiculate. Stamens 6. Stigma somewhat 2-lobed. Silique terete, with convex valves, each 3–7-nerved, and a long beak with 0–4 (9) seeds. Seeds 1-seriate, subspherical. Cotyledons conduplicate.

Sinapis arvensis L., Sp. Pl. **2** : 668 (1753).—O. E. Schulz in Engl., Pflanzenr. IV, 105, 1 : 119 (1919). TAB. **28** fig. C. Type from Europe.

Annual herb up to about 80 cm. tall, with sparsely strigose stems. Lower leaves up to 20 cm. long, stalked, lyrate, roughly hairy, upper ones sessile, lanceolate, coarsely toothed. Flowers bright yellow, pedicellate, pedicels 2–4 mm. long, in terminal racemes or panicles. Petals 5–12 mm. long. Silique 20–40 × 1–3 mm. with conical beak 10–20 mm. long. Seeds dark reddish-brown.

S. Rhodesia. E : Odzani Valley, fl. & fr. 1915, *Teague* 320 (K).
North temperate regions. An introduced weed.

Charlock or Wild Mustard.

5. SISYMBRIUM L.

Sisymbrium L., Sp. Pl. **2** : 657 (1753) ; Gen. Pl. ed. 5 : 296 (1754).

Annual, biennial or perennial herbs with simple hairs. Leaves entire or pinnatifid. Flowers usually yellow in lax racemes. Sepals not or only slightly saccate. Petals 4, unguiculate. Stamens 6. Silique long and slender with convex valves each with a prominent midrib and usually 2 less prominent lateral nerves, not or scarcely beaked. Seeds 1-seriate, ovoid. Cotyledons incumbent.

Sisymbrium orientale L., Cent. Pl. **2** : 24 (1756).—O. E. Schulz in Engl., Pflanzenr. IV, 105 : 122 (1924). TAB. **28** fig. D. Type from the Orient.

Annual or biennial herb up to 1 m. tall, stems with short, weak, patent or downward-pointing hairs. Basal leaves in a rosette, long-stalked, lyrate-pinnatipartite or pinnate, appressed-pilose ; stem leaves with hastate terminal lobe ; uppermost leaves hastate or lanceolate. Flowers pale yellow, pedicellate, pedicel 4–7 mm. long, in ebracteate racemes. Sepals 4·5 × 1 mm., erect, oblong. Petals 5–8 mm. long, obovate. Stamens very unequal in length. Silique 4–10 cm. × 1–2 mm., obliquely erect with pedicels 3–7 mm. long, valves 3-nerved. Seeds 0·6–1 × 0·5 mm., ellipsoid or narrowly ovoid, yellowish-brown.

N. Rhodesia. S : Muckle Neuk, 20 km. N. of Choma, fl. & fr. 11.x.1954, *Robinson* 927 (K ; SRGH). **S. Rhodesia.** W : Bulawayo Sewage Farm, fl. & fr. 20.ix.1957, *Cronin* in GHS 79962 (K ; SRGH).

From the Canary Is. to the Caucasus. Weed of cultivation.

6. ERUCASTRUM C. Presl

Erucastrum C. Presl, Fl. Sic. **1** : 92 (1826).

Annual, biennial or perennial herbs usually ± hairy with simple hairs. Lower leaves lyrate-pinnatifid, upper ones usually linear. Flowers yellow in bracteate or ebracteate racemes. Sepals erect or spreading, inner pair somewhat saccate. Petals 4, unguiculate. Stamens 6. Silique quadrangular with keeled valves, each with 1 prominent dorsal nerve and a conical beak with 0–3 seeds. Seeds 1-seriate, ovoid or ellipsoid. Cotyledons longitudinally conduplicate.

Erucastrum arabicum Fisch. & Mey., Animadv. Bot. in Index Quint. Sem. Petrop. : 35 (1839) ; in Linnaea **13**, Lit.-Ber. : 109 (1839).—O. E. Schulz in Engl., Pflanzenr. IV, 105, 1 : 96 (1919). Type from Arabia.

Var. **arabicum.** TAB. **28** fig. E.
Brassica sp.—Eyles in Trans. Roy. Soc. S. Afr. **5** : 355 (1916) quoad specim. *Rogers* 5756.

Annual herb up to 1 m. tall with sparsely strigose stems. Lower leaves stalked, lyrate-pinnatipartite, upper irregularly toothed, sessile, up to 15 cm. long, sparsely strigulose. Flowers pale purplish, white or yellow, pedicellate, pedicels up to 5 mm. long, in terminal racemes or panicles. Sepals 3–3·5 × 1–1·5 mm., oblong to broadly elliptic, erect, very sparsely hispidulous. Petals 4–6 mm. long, obovate-oblong. Silique 20–45 × 1·5–2 mm., linear-oblong, with short, conical beak, 1·5–2 mm. long. Seeds 1–1·5 × 0·7–0·8 mm., ellipsoid, dark brown, surface minutely reticulate.

Bechuanaland Prot. N : Ngamiland, fl. & fr. ii.1897, *Lugard* 174 (K). **S. Rhodesia.** W : Bulawayo, fl. & fr. 18.ii.1912, *Rogers* 5756 (SRGH).
Arabia and eastern and central tropical Africa. Weed of cultivation.

7. CARDAMINE L.

Cardamine L., Sp. Pl. **2** : 654 (1753) ; Gen. Pl. ed. 5 : 295 (1754).

Annual or perennial herbs. Leaves simple, trifoliolate or pinnate. Flowers white or purple in terminal, subumbellate racemes. Inner sepals not or scarcely saccate. Petals 4, occasionally absent. Stamens 6 (4). Silique laterally compressed, often opening explosively, the valves coiling spirally from the base. Seeds 1-seriate, smooth or finely tuberculate. Cotyledons accumbent.

Leaves trifoliolate - - - - - - - - - - - 1. *africana*
Leaves pinnate - - - - - - - - - - - 2. *flexuosa*

1. **Cardamine africana** L., Sp. Pl. **2** : 655 (1753).—Sond. in Harv. & Sond., F.C. **1** : 23 (1860).—Oliv., F.T.A. **1** : 60 (1868).—Burtt Davy, F.P.F.T. **1** : 126 (1926).— Robyns & Boutique, F.C.B. **2** : 535, t. 52 (1951). Type from S. Africa.

Slender perennial herb up to 40–60 cm. tall, stems often rooting at the nodes, sparsely pubescent. Leaves trifoliolate, petiolate, petioles slender, up to 5–10 cm. long, glabrous or sparsely pubescent, leaflets up to 4 × 3 cm., ovate to ovate-lanceolate, apex acute to rounded, margins crenate-dentate, base truncate, rounded or slightly cordate, sparsely pubescent or strigulose. Flowers white, greenish-white or greenish-yellow, pedicellate, pedicels up to 4 mm. long, in corymbose racemes. Sepals 2–3 × 1–1·5 mm., suberect, obovate-oblong. Petals 4, 5 × 1·5 mm., narrowly oblanceolate or cuneate. Stamens 6. Silique 25–45 × 1·5–2 mm., straight, linear, flattened, tapering gradually, with pedicel 3–12 mm. long. Seeds 1·8–2 × 1·2–1·5 mm., flattened-ellipsoid, brownish.

S. Rhodesia. E : Umtali Distr., Engwa, fl. & fr. 2.ii.1955, *E.M. & W.* 139 (BM ; LISC ; SRGH).
Widespread on the mountains of tropical and S. Africa, India and Brazil. Forest-floor species of submontane evergreen forest.

2. **Cardamine flexuosa** With., Bot. Arr. Brit. Pl. ed. 3, **3** : 578, t. 3 (1796). TAB. **27** fig. B. Type from Great Britain.

Annual or perennial herb, 10–50 cm. tall, with slender, branched stems, somewhat pubescent towards the base. Leaves petiolate, pinnate with rounded reniform or obovate-cuneate leaflets up to about 10 × 8 mm. (in material from our area) usually lobed or toothed ; cauline leaves sessile or shortly petiolate. Flowers white in terminal, subcorymbose racemes. Sepals 1·5 × 0·5 mm., elliptic-oblong, greenish with a white margin. Petals 1·5–2·5 × 0·7 mm., oblong. Stamens 6, filaments 0·5 mm. long. Ovary 1 × 0·3 mm., cylindric. Silique 10–25 × 0·7–1 mm., linear-oblong, glabrous. Seeds 2·5–1·5 mm., flattened-ellipsoid.

S. Rhodesia. C : Salisbury Experimental Station, fl. & fr. 31.vii.1944, *Arnold* in GHS 12601 (SRGH).

North temperate regions. An introduced weed.

Bitter-cress.

8. LEPIDIUM L.

Lepidium L., Sp. Pl. **2** : 643 (1753) ; Gen. Pl. ed. 5 : 291 (1754).

Annual, biennial or perennial herbs, occasionally somewhat shrubby, glabrous or with simple hairs, usually with a slender tap-root. Leaves entire to pinnate. Flowers small, whitish, in terminal and axillary, ebracteate racemes. Sepals not saccate. Petals 4, sometimes vestigial or 0. Stamens 2, 4 or 6. Ovary 2-ovulate. Fruit a silicule with keeled or winged valves. Seeds usually 1 in each loculus, pendent from the apex.

The Pepper-cresses.

Silicule up to 2 (2·8) mm. broad, clearly emarginate at the apex -   -   1. *bonariense*
Silicule not more than 1·7 mm. broad, clearly or very slightly emarginate at the apex :
 Style protruding slightly beyond the emargination of the silicule ; plant densely
  crisped-pubescent   -   -   -   -   -   -   -   -   -   2. *africanum*
 Style included in the emargination of the silicule ; plant minutely pubescent or almost
  glabrous   -   -   -   -   -   -   -   -   -   -   3. *divaricatum*

1. **Lepidium bonariense** L., Sp. Pl. **2** : 645 (1753).—Thell. in Mitt. Bot. Mus. Zür. **28** : 256 (1906). TAB. **29** fig. B1. Type from Argentina.

Annual or perennial herb up to about 50 cm. tall with a single, erect rather sparsely and minutely pubescent stem. Lower cauline leaves up to about 7 cm. long, pinnatipartite to bipinnatipartite, petiolate, pubescent ; upper leaves pinnatipartite to subentire with 2–4 acute teeth. Flowers greenish-white at first, pedicellate, pedicels 1–2 mm. long, in dense corymbose racemes elongating later until up to 17 cm. long. Sepals c. 1 mm. long. Petals minute, less than 0·5 mm. long. Stamens 2. Silicule up to 3·5 × 2·8 mm., broadly elliptic to suborbicular, slightly winged at the apex and clearly emarginate with the remains of the style included in the emargination. Seeds 1·5 × 0·9 mm., orange-brown, flattened-ellipsoid, with a narrow hyaline margin.

S. Rhodesia. C : Selukwe Distr., fl. & fr. 8.xii.1953, *Wild* 4282 (SRGH). E : Umtali Distr., Christmas Pass, fl. & fr. 29.xi.1955, *Drummond* 5052 (SRGH).
Cosmopolitan weed of S. American origin. Introduced.

2. **Lepidium africanum** (Burm. f.) DC., Syst. Nat. **2** : 552 (1821).—Sond. in Harv. & Sond., F.C. **1** : 29 (1860). TAB. **29** fig. B2. Type from S. Africa.
  Thlaspi africanum Burm. f., Fl. Cap. Prodr.: 17 (1768). Type as above.

Var. **aethiopicum** (Hiern) Thell. apud Schinz in Viert. Naturf. Ges. Zür. **51**, 1 : 187 (1906) ; in Mitt. Bot. Mus. Zür. **28** : 182 (1906).—Exell & Mendonça, C.F.A. **1**, 1 : 52 (1937). Type from Angola (Huila).
  Lepidium ruderale var. *aethiopicum* Hiern, Cat. Afr. Pl. Welw. **1** : 25 (1896). Type as above.

Perennial herb with arcuate-ascending, pubescent stems up to 20 cm. tall. Basal leaves petiolate, up to 25 × 3 mm., narrowly oblanceolate, shallowly toothed, pubescent ; cauline leaves subsessile, otherwise similar. Flowers very small, greenish-white, in dense terminal racemes elongating to up to 15 cm. long. Sepals 1·4 × 0·5 mm., ovate-elliptic. Petals minute. Stamens 2. Silicule 2·5–3 × 1·5–1·7 mm., flattened-ellipsoid, apex very slightly emarginate, style protruding slightly beyond the emargination. Seeds 1·5 × 0·8 mm., brown, ellipsoid.

S. Rhodesia. E : Inyanga, 9–10 km. N. of Troutbeck, fl. & fr. 20.xi.1957, *Robinson* 1969 (K ; SRGH). **Mozambique.** SS : Inhambane Velho, fr. ix.1937, *Gomes e Sousa* 2031 (K).

Tab. 29. A.—LEPIDIUM DIVARICATUM SUBSP. LINOIDES. A1, plant (×1) ; A2, flower with young fruit (×24) ; A3, fruit (×9). All from *Mendonça* 2797. B1.—LEPIDIUM BONARIENSE, fruit (×9) *Wild* 4282. B2.—LEPIDIUM AFRICANUM, fruit (×9) *Robinson* 1969. C.—CORONOPUS INTEGRIFOLIUS. C1, plant (×1) ; C2, flower (×24) ; C3, fruit (×21). All from *Drummond & Seagrief* 5168.

Species : Angola, S. Rhodesia, Mozambique and S. Africa ; variety : Angola, S. Rhodesia and Mozambique.

3. **Lepidium divaricatum** Ait., Hort. Kew. **2** : 375 (1789).—Burtt Davy, F.P.F.T. **1** : 129 (1926). Type from S. Africa.

Annual or perennial herb with erect, branched, glabrous or pubescent stems up to 50 cm. tall. Basal leaves ± pinnatifid ; cauline leaves subentire to pinnati-partite or pinnatisect, glabrous or pubescent. Flowers greenish-white, pedicellate, pedicels up to 3 mm. long, in elongated terminal and lateral racemes up to 25 cm. long. Sepals c. 0·6 mm. long. Petals minute or absent, 0·4–0·5 mm. long. Stamens 2, filaments 1 mm. long. Silicule 2·8–3 × 1·5–1·6 mm., flattened-ellipsoid, slightly winged near the apex, clearly emarginate at the apex, style included in the emargination. Seeds 1·4 × 0·6 mm., ellipsoid, brown.

Cauline leaves pinnatifid to pinnatisect   -   -   -   -   -   subsp. *divaricatum*
Cauline leaves subentire to shallowly toothed   -   -   -   -   -   subsp. *linoides*

Subsp. **divaricatum**
> *Lepidium pinnatum* sensu Sond. in Harv. & Sond., F.C. **1** : 30 (1860) pro parte.
> *Lepidium divaricatum* subsp. *eudivaricatum* Thell. in Viert. Naturf. Ges. Zür. **51** : 167 (1906).

Mozambique. LM : Maputo, Quinta da Pedra, fl. & fr. 9.i.1948, *Gomes e Sousa* 3654 (PRE).
Mozambique and S. Africa. A weed of cultivated ground.

Subsp. **linoides** (Thunb.) Thell. in Viert. Naturf. Ges. Zür. **51** : 167 (1906). TAB. **29** fig. A. Type from S. Africa.
> *Lepidium linoides* Thunb., Prodr. Pl. Cap. **2** : 107 (1800).—Sond. in Harv. & Sond., F.C. **1** : 28 (1860).—Burtt Davy, F.P.F.T. **1** : 129 (1926). Type as above.

S. Rhodesia. C : Salisbury, Agricultural Exp. Station, fr. x.1955, *Drummond* 4933 (SRGH). E : Umtali Golf Course, fl. & fr. 9.vii.1952, *Chase* 4572 (BM ; SRGH). **Mozambique.** SS : Vila João Belo, fl. & fr. 7.iii.1948, *Torre* 7510 (BM ; LISC). LM : Magude, Xinavane, fl. & fr. 7.xi.1944, *Mendonça* 2797a (BM ; LISC).
From Ethiopia to S. Africa and in Madagascar. A weed of cultivated ground.

Those who wish to delve into the classification below the subspecific level should consult Thellung (loc. cit.). We appear to have subsp. *linoides* var. *linoides* and var. *subdentatum* (Burch.) Sond. in our area.

9. CAPSELLA Medic.

Capsella Medic., Pflanz.-Gatt. : 85 (1792) *nom. conserv.*

Annual or biennial herbs with slender tap-root. Basal leaves entire or pinnatifid, cauline amplexicaul. Flowers usually white. Sepals not saccate. Petals 4. Stamens 6. Silicule flattened-obcordate (rarely ovoid) with keeled, net-veined valves. Seeds several in each loculus.

Capsella bursa-pastoris (L.) Medic., Pflanz.-Gatt.: 85 (1792).—Sond. in Harv. & Sond., F.C. **1** : 31 (1860).—Burtt Davy, F.P.F.T. **1** : 128 (1926).—Robyns & Boutique, F.C.B. **2** : 524 (1951). TAB. **28** fig. F. Type from Europe.
> *Thlaspi bursa-pastoris* L., Sp. Pl. **2** : 647 (1753). Type as above.

Annual or biennial herb up to 30 cm. tall, glabrous or with simple and stellate hairs. Basal leaves in a rosette, up to 20 × 4·5 cm., ± oblanceolate in outline, narrowed into the stalk and varying from deeply pinnatifid to entire ; cauline leaves very variable in shape but always amplexicaul with acute auricles. Flowers white, pedicellate, pedicels up to 6 mm. long, in elongated racemes up to 20 cm. long. Sepals 1·5 × 0·6 mm., oblong-elliptic. Petals up to about twice as long as the sepals. Silicule up to 9 × 6 mm., triangular-obcordate, emarginate at the apex, with a pedicel up to 2 cm. long. Seeds c. 0·8 × 0·4 mm., ellipsoid, up to 12 in each loculus.

S. Rhodesia. W : Shangani, Gwampa Forest Reserve, fl. & fr. ii.1957, *Goldsmith* 24/58 (K ; SRGH). C : Salisbury, fl. & fr. 3.vi.1942, *Hopkins* in GHS 9174 (SRGH). E : Inyanga, fl. & fr. 22.i.1942, *Hopkins* in GHS 8637 (SRGH).
Cosmopolitan. Weed of cultivation.

Shepherd's Purse.

10. CORONOPUS Zinn

Coronopus Zinn, Cat. Pl. Hort. Gott. : 325 (1757) *nom. conserv.*
Senebiera DC. in Mém. Soc. Hist. Nat. Par. : 140 (1799).

Annual, biennial or perennial herbs usually ± pilose with simple hairs. Leaves pinnatisect or entire. Flowers small, whitish, in terminal racemes which appear to be leaf-opposed owing to sympodial growth. Sepals ± spreading, not saccate. Petals 4 or 0. Stamens 6, 5, 4 or 2. Ovary with 2 ovules. Fruit 2-lobed, dividing vertically into 1-seeded halves.
The Wart-cresses.

Leaves pinnatipartite ; stems pilose - - - - - - - - 1. *didymus*
Leaves entire ; stems pubescent - - - - - - - 2. *integrifolius*

1. **Coronopus didymus** (L.) Sm., Fl. Brit. **2** : 691 (1800).—Muschl. in Engl., Bot. Jahrb. **41** : 134 (1908).—Burtt Davy, F.P.F.T. **1** : 129 (1926). Locality of type not stated.
    Lepidium didymum L., Syst. Nat. ed. 12, **2** : 433 et Mant. Pl. : 92 (1767). Type as above.
    Senebiera didyma (L.) Pers., Syn. Pl. **2** : 185 (1806).—Sond. in Harv. & Sond., F.C. **1** : 27 (1860).—Type as above.

Annual or biennial, foetid herb with slender tap-root and prostrate or ascending, much branched, shortly pilose stems up to 40–50 cm. tall. Lower leaves petiolate, pinnatisect, lobes pinnatifid ; cauline leaves sessile, with narrow almost entire segments, all sparsely pubescent or nearly glabrous. Flowers small, whitish, pedicellate, pedicels 1–2 mm. long, in rather dense racemes elongating to about 6 cm., appearing extra-axillary or leaf-opposed but really terminal. Sepals c. 0·5 mm. long. Petals 4, shorter than the sepals. Stamens 2 (rarely 4 or 5). Fruit 1·5–1·8 × 2·5 mm. constricted at the septum into 2 subspherical, reticulate-pitted 1-seeded mericarps. Seeds 1 × 0·5 mm., orange-brown, flattened-ellipsoid.

S. Rhodesia. C : Salisbury, fl. & fr. 7.iv.1953, *Wild* 4110 (SRGH). E : Umtali, Commonage, fl. & fr. 24.vii.1953, *Chase* 5013 (BM ; SRGH).
Cosmopolitan (perhaps of American origin). Weed of cultivation.

Our plants belong to subsp. *didymus*. For a detailed infra-specific classification see Muschler (loc. cit.).

2. **Coronopus integrifolius** (DC.) Spreng. in L., Syst. Veg. ed. 16, **2** : 853 (1825).—Muschl. in Engl., Bot. Jahrb. **41** : 138 (1908).—Burtt Davy, F.P.F.T. **1** : 129 (1926). TAB. **29** fig. C. Type from Madagascar.
    Senebiera integrifolia DC., in Mém. Soc. Hist. Nat. Par. : 144, t. 8 (*pinnatifida* sphalm.) (1799).—Klotzsch in Peters, Reise Mossamb. Bot. **1** : 169 (1861).—Oliv., F.T.A. **1** : 70 (1868). Type as above.
    Coronopus engleranus Muschl., tom. cit. : 139 (1908). Type : Mozambique, Zambezi Estuary, *Peters* (B†).

Perennial herb with slender, prostrate-ascending, branched, pubescent stems up to 20 cm. tall. Basal leaves greyish, entire or subentire, petiolate, petiole 1–2·5 cm. long, linear-lanceolate, acute, minutely pubescent or nearly glabrous ; cauline leaves sessile or subsessile. Flowers small, pedicellate, pedicels 1–2 mm. long, in racemes up to 5 cm. long, which appear extra-axillary or leaf-opposed owing to sympodial growth but are really terminal. Sepals c. 0·5 mm. long, elliptic, with white margins. Petals usually a little longer than the sepals. Stamens 4–6. Fruit pedicellate with pedicel 2–3 mm. long, 1 × 2·5 mm., reticulate-pitted, greyish, dividing vertically into two 1-seeded nutlets. Seeds 1–1·2 × 0·5–0·7 mm., flattened-ellipsoid.

Bechuanaland Prot. N : Mumpswe Pan, 40 km. NNW. of Nata R., fl. & fr. 21.iv.1957, *Drummond & Seagrief* 5168 (K ; SRGH). SW : Chukudu Pan, fl. & fr. 22.vi.1955, *Story* 4956 (K). **Mozambique.** MS : Expedition I., fl. & fr. viii.1858, *Kirk* (K).
Widespread but sporadic in the warmer regions of the Old World. Edges of pans and coastal lagoons and swamps usually at the lower altitudes.

11. RAPHANUS L.

Raphanus L., Sp. Pl. **2** : 669 (1753) ; Gen. Pl. ed. 5 : 300 (1754).

Annual, biennial or perennial herbs with lyrate-pinnatifid, somewhat glaucous leaves. Flowers in terminal, ebracteate racemes. Sepals erect. Petals 4, un-guiculate. Stamens 6. Fruit jointed, the lowest joint (corresponding to the valves of a typical silique) short, slender, seedless, the upper joints containing the seeds, cylindric or moniliform and usually breaking transversely when ripe into 1-seeded joints, apex narrowed into a seedless beak. Seeds subspherical or ovoid, reddish-brown, pitted.

Fruit not or little constricted between the seeds ; tap-root tuberous -   - 1. *sativus*
Fruit constricted between the seeds ; tap-root not tuberous   -   - 2. *raphanistrum*

1. **Raphanus sativus** L., Sp. Pl. **2** : 669 (1753).—Oliv., F.T.A. **1** : 73 (1868).—Burtt Davy, F.P.F.T. **1** : 130 (1926).—Exell & Mendonça, C.F.A. **1**, 1 : 52 (1937). Syntypes from plants cultivated in Europe.

Annual or biennial herb with a tuberous white, pink or red tap-root and erect, bristly stem up to 100 cm. tall. Flowers white or purplish. Fruit inflated, up to 15 mm. in diam., not or little constricted between the seeds and not breaking trans-versely into joints, beak long and conical.

S. Rhodesia. E : Umtali Distr., Engwa, 1980 m., fl. 10.ii.1955, *E.M. & W.* 357 (BM ; LISC ; SRGH). C : Salisbury, fl. iii.1918, *Walters* 2483 (SRGH).
North temperate regions. An escape from cultivation.

The Radish.

2. **Raphanus raphanistrum** L., Sp. Pl. **2** : 669 (1753).—Burtt Davy, F.P.F.T. **1** : 130 (1926). TAB. **28** fig G. Type from Europe.

Annual herb with a slender tap-root and erect, bristly stem up to 20–60 cm. tall. Lower leaves lyrate-pinnatifid with a large, rounded terminal lobe and 1–4 pairs of smaller, distant, lateral lobes. Sepals 6–10 mm. long. Petals yellow, lilac or white, usually dark-veined. Fruit cylindric, 5–8 mm. in diam. constricted between the seeds and breaking transversely into 1-seeded joints, with slender beak. Seeds 1·5–3 mm. in diam., ovoid-spherical.

S. Rhodesia. E : Umtali Distr., Engwa, 8.ii.1955, *E.M. & W.* 288 (BM ; LISC ; SRGH).
Cosmopolitan weed of cultivation.

Wild Radish or White Charlock.

14. CAPPARIDACEAE
By H. Wild

Herbs, shrubs, trees or lianes. Leaves alternate, simple or digitately 3–9-foliolate ; stipules absent or rudimentary, rarely spiny. Inflorescences terminal or axillary, of racemes, corymbs or panicles, or flowers solitary and axillary. Flowers actinomorphic or zygomorphic, bisexual or unisexual by abortion, usually 4-merous, hypogynous. Receptacle cupular, funnel-shaped or cylindric with an entire, undulate, dentate or fimbriate margin, sometimes very short. Sepals 3–4 (5). Petals 4 (0, 5, 6 or more). Stamens 5–∞ usually borne on a short or elongated androphore, sometimes accompanied by staminodes. Ovary usually borne on a more or less elongated gynophore, usually 1-locular with 2 parietal placentas but sometimes 2-locular by the intrusion of the placentas or multi-locular ; ovules 4–∞. Style short or absent. Fruit a capsule or a berry. Seeds reniform or subglobose, without endosperm ; embryo usually curved.

Fruit an elongated capsule with 2 valves separating from a persistent replum ; annual or
  perennial herbs :
  Androphore absent or very short, i.e. stamens not inserted on an elongated stalk
                                                                    1. Cleome
  Androphore longer than the corolla, i.e. stamens inserted on an elongated stalk
                                                              2. Gynandropsis
Fruit various, often baccate, rarely dehiscent but then without a replum ; woody plants,
  or, if apparently herbaceous, then with a woody rootstock :

Androphore elongated, longer than the sepals :
  Upper and lower sepals enclosing the lateral ones ; androphore with a conspicuous
    nectary at its base ; fruit narrowly cylindric ; leaves simple -   - **3. Cadaba**
  Upper and lower sepals not enclosing the lateral ones ; androphore with a bundle
    of staminodes at its base; fruit globose, large; leaves 3-foliolate **4. Cladostemon**
Androphore short, not as long as the sepals :
  Petals absent :
    Calyx rupturing transversely on expansion ; fruit with 8–10 longitudinal ribs
                                      5. Thylacium
    Calyx not rupturing transversely ; fruit without ribs :
      Sepals 3 ; flowers solitary in the axils -   -   -   -   **6. Courbonia**
      Sepals 4 ; flowers usually racemose or corymbose :
        Receptacle cylindric or funnel-shaped -   -   -   **7. Maerua**
        Receptacle apparently absent or shallowly concave :
          Leaves 3-foliolate (male flowers large and showy, with petals)
                                        10. Crateva
          Leaves simple - - - - - - - - **8. Boscia**
  Petals present :
    Branches spiny - - - - - - - - **9. Capparis**
    Branches not spiny :
      Petals with a long claw and a broadly elliptic or ovate lamina ; sepals open in
        aestivation (not touching in bud) - - - - **10. Crateva**
      Petals not clawed, or if clawed with a narrowly oblanceolate lamina; sepals
        valvate :
        Fruit tardily dehiscent with 4 coriaceous valves ; receptacle cupular or
        broadly funnel-shaped ; petals ribbon-like, as long as or longer than the
        sepals - - - - - - - **11. Ritchiea**
        Fruit not dehiscent ; receptacle cylindric or narrowly funnel-shaped ; petals
        shorter than the sepals or, if longer (*M. bussei*), oblanceolate **7. Maerua**

1. CLEOME L.

Cleome L., Sp. Pl. **2** : 671 (1753) ; Gen. Pl. ed. 5 : 302 (1754).

Annual or perennial herbs or occasionally small shrubs. Leaves usually petiolate, simple or digitately 3–9-foliolate ; leaflets usually narrow, linear to lanceolate, entire or serrulate. Flowers ± zygomorphic, in terminal racemes. Sepals 4, usually free or almost so, usually narrow and often glandular or setulose-pubescent. Petals 4, sessile or clawed, equal or unequal. Stamens 2–many, all fertile or some sterile, usually borne on a small torus or receptacle ; filaments equal or unequal and declinate, occasionally thickened at the apex ; anthers 2-thecous, usually oblong or narrowly oblong. Ovary sessile or with a short gynophore, with many ovules on two parietal placentas ; style short or absent ; stigma capitate or truncate. Fruit an oblong or linear capsule often borne on an elongated gynophore, with two valves separating from the seed-bearing placentas, often tipped with the persistent style ; valves glabrous or pubescent, smooth or strongly longitudinally nerved. Seeds reniform, smooth or ridged transversely, radially or longitudinally, or reticulate-tuberculate, glabrous or pubescent.

Leaves simple :
  Racemes lax ; leaves widely spaced ; stems, ovary and fruit rather sparsely and shortly
    pubescent - - - - - - - - - **1. *monophylla***
  Racemes dense ; leaves crowded ; stems, ovary and fruit with long, patent, purplish
    hairs and shorter glandular hairs - - - - - **2. *densifolia***
Leaves compound :
  Stamens both fertile and sterile* :
    Flowers golden-yellow ; fertile stamens 2, staminodes 6–12 ; plant glaucous ;
      leaves 3–9-foliolate ; seeds usually pubescent - - - **11. *diandra***
    Flowers pink to violet ; fertile stamens 2–4, staminodes 2–4 ; leaves 3–5-foliolate,
      seeds glabrous - - -   - **12. *maculata***
  Stamens all fertile :
    Bracts of inflorescence minute, caducous :
      Leaflets linear, up to 2 (3) mm. wide ; mature capsule 4–7 cm. long
                  13. *macrophylla* var. *maculatiflora*
      Leaflets more than 2 mm. wide ; mature capsule 7–10 cm. long
                  13. *macrophylla* var. *macrophylla*

* Flowers should be examined in bud as the anthers are often caducous.

Bracts conspicuous, passing gradually into the leaves and often compound :
  Leaflets linear (occasionally broader in *C. iberidella* but then the 2 upper petals
    with a purple-margined yellow spot) :
  Flowers sulphur-yellow ; seeds tuberculate-reticulate, c. 1 mm. in diam.
                                                                              10. *sulfurea*
  Flowers pinkish or purple ; seeds transversely rugose or ridged, 1·3 mm. or
    more in diam. :
    Stamens 6–9 ; petals 5–9 mm. long ; annual herbs up to 30 cm. tall :
      Leaves 5–7 (9)-foliolate - - - - - - - 7. *rubella*
      Leaves 3–5-foliolate - - - - - - - - 8. *iberidella*
    Stamens 10–12 ; petals 10–20 mm. long ; a coarse herb up to 1·6 m. tall
                                                                              6. *hirta*
Leaflets obovate, oblanceolate, oblong-lanceolate, rhomboid-elliptic, elliptic or
    lanceolate ; petals sometimes bicoloured but not with a margined spot :
  Seeds smooth, minutely longitudinally tessellate ; leaflets sparsely pubescent
    with some scattered sessile glands - - - - 3. *bororensis*
  Seeds transversely ridged or striate ; leaflets sparsely setulose-, harshly strigillose-
    or densely glandular-pubescent :
    Stamens 6 ; leaflets sparsely setulose-pubescent or glabrous - 4. *ciliata*
    Stamens 8–13 ; leaflets densely strigillose- or glandular-pubescent :
      Gynophore and capsule glandular; valves with 8–9 longitudinal nerves; stems
        and leaves densely glandular-pubescent ; leaflets 3–7 - 9. *oxyphylla*
      Gynophore glabrous ; capsule minutely puberulous with about 3 longi-
        tudinal nerves ; stems and leaves with a coarse strigillose pubescence ;
        leaflets 3–5 - - - - - - - - - 5. *stricta*

1. **Cleome monophylla** L., Sp. Pl. 2 : 672 (1753).—Sond. in Harv. & Sond., F.C. 1 :
  56 (1860).—Oliv., F.T.A. 1 : 76 (1868).—Bak. f. in Journ. Linn. Soc., Bot. 40 :
  20 (1911).—R.E.Fr., Wiss. Ergebn. Schwed. Rhod.-Kongo-Exped. 1 : 49 (1914).—
  Gilg & Bened. in Engl., Bot. Jahrb. 53 : 153 (1915).—Eyles in Trans. Roy. Soc. S.
  Afr. 5 : 355 (1916).—Burtt Davy, F.P.F.T. 1 : 120 (1926).—Arwidss. in Bot.
  Notis. 1935 : 357 (1935).—Exell & Mendonça, C.F.A. 1, 1 : 54 (1937).—Hauman &
  Wilczek, F.C.B. 2 : 511 (1951). Type from India.
    Cleome epilobioides Bak. in Kew Bull. 1897 : 243 (1897). Type : Nyasaland,
  Mt. Mlanje, *Whyte* (K, holotype).

Erect annual herb to 0·6 m. tall, usually branched. Stems pubescent with some
hairs gland-tipped. Leaves petiolate ; lamina 2–7·5 × 0·3–2·5 cm., linear-lanceolate
to oblong, acute or subacute at the apex, rounded or slightly cordate at the base,
entire, pubescent on both sides with hairs often glandular; petiole up to 3 cm. long
(often less), pubescent. Inflorescence a raceme, elongating in fruit ; bracts sessile,
similar to the leaves but slightly more cordate at the base and smaller ; pedicels
slender, glandular-pubescent, up to 1 cm. long, elongating in fruit up to 2 cm.
Sepals up to 5 mm. long, linear-lanceolate, caudate or long-acuminate at the apex,
pubescent outside with some hairs glandular. Petals pale rose or mauve, with a
yellow band bordered with purple on the upper two, rarely white, lamina up to
9 mm. long, oblong, rounded at the apex, narrowing into a basal claw somewhat
shorter than the lamina. Stamens 6, unequal, all fertile ; filaments slender, up to
1 cm. long, glabrous ; anthers oblong, c. 1·5 mm. long. Ovary almost sessile,
c. 1·5 mm. long, linear-oblong, puberulous ; style very short ; stigma capitate.
Capsule up to c. 10 cm. long, narrowly linear, puberulous, straight or slightly
curved, narrowed to both ends with up to 6 longitudinal, anastomosing nerves on
each valve ; gynophore very short, up to 2 mm. long ; style persistent. Seeds
dark brown, c. 1·8 mm. in diam., radially ridged with the ridges bearing minute
puberulous incrustations visible only at a ×40 magnification.

Bechuanaland Prot. N : Ngamiland, fl. & fr. xii.1896, *Lugard* 95 (K). SE : Mochudi, fl. & fr. i.1915, *Harbor* (PRE ; SRGH). **N. Rhodesia.** N : Miloko, Luapula R.,
fl. & fr. 6.ix.1953, *Fanshawe* 276 (K ; SRGH). W : Mwinilunga, fl. & fr. 26.xii.1937,
Milne-Redhead 3829 (BM ; K). C : 10 km. E. of Lusaka, fl. 30.i.1956, *King* 294 (K).
E : Ft. Jameson, Msoro Mission, fl. & fr. xii.1930, *Bush* 12 (K). S : Dundwa, 10 km.
S. of Mapanza, fl. & fr. 6.iv.1953, *Robinson* 160 (K). **S. Rhodesia.** N : Shamva, fl. & fr.
xii.1932, *Leviseur* (PRE ; SRGH). W : Bulawayo, fr. 18.ii.1912, *Rogers* 5757 (K ;
SRGH). C : Rusape, fl. & fr. 1952, *Dehn* 140/52 (K ; M ; PRE ; SRGH). E : Inyanga,
Cheshire, fl. & fr. 4.ii.1931, *Norlindh & Weimarck* 4854 (BM ; LD ; PRE ; SRGH).
S : Chibi Distr., fl. 5.xii.1954, *Robinson* 454 (SRGH). **Nyasaland.** N : Mzimba,
Mkawa Exp. Sta., fl. & fr. 5.iv.1955, *Jackson* 1583 (K ; SRGH). C : Dowa, Mwera Hill,
fl. & fr. 26.i.1954, *Jackson* 1197 (BM). S : Blantyre, Lirangwe, fl. 13.i.1956, *Jackson* 1810

(K ; SRGH). **Mozambique.** N : Massangulo, fl. & fr. i.1933, *Gomes e Sousa* 1231
(COI). Z : Namagoa, Mocuba, fl. 17.iii.1949, *Faulkner* 413 (COI ; K ; SRGH). T :
Tete, fl. & fr. ii.1859, *Kirk* (K). MS : Chimoio, Garuso, fl. & fr. 3.iii.1948, *Barbosa* 1079
(BM ; LISC). SS : Lower Buzi, fl. & fr. xii.1906, *Swynnerton* 2111 (BM ; K). LM :
Vila Luiza, fl. & fr. 7.iv.1947, *Barbosa* 130 (COI ; LM ; SRGH).

Widely distributed through Africa, India and Ceylon. Often occurs as an annual weed
but is also a common constituent of many vegetation types at all altitudes.

It has been recorded as a host of Tobacco Aphis in Nyasaland and is widely used by
natives as a relish.

2. **Cleome densifolia** C. H. Wright in Kew Bull. **1907** : 360 (1907).—Gilg & Bened.
in Engl., Bot. Jahrb. **53** : 153 (1915).—Milne-Redh. in Mem. N.Y.Bot. Gard. **8,** 3 :
218 (1953). Type : Nyasaland, Mt. Mlanje, Tuchila Plateau, *Purves* 94 (K,
holotype)

Shrubby, branching, viscid perennial 1–2 m. tall ; branches striate, straw-
coloured with dark, purplish, long hairs interspersed with short, paler, glandular
hairs. Leaves petiolate with a pubescence similar to that of the branches ; lamina
1·5–5 × 0·7–2·0 cm., oblong-elliptic to narrowly ovate, acute at the apex, rounded or
slightly cordate at the base, entire, with both long hairs and short glandular hairs
on both surfaces with the longer hairs tending to be confined to the nerves of the
underside ; petiole up to 1 cm. long ; the upper leaves progressively smaller, more
nearly sessile and grading into bracts each subtending a single flower of the terminal
racemes. Raceme dense at first but internodes elongating in fruit ; pedicels
slender, purplish, glandular-pubescent. Sepals c. 7 mm. long, linear-lanceolate,
acuminate at the apex with a gland at the tip, glandular-pubescent outside. Petals
0·7–1·2 × 0·3 cm., pink with the two upper having a cordate, purple-margined
spot, narrowly obovate, rounded at the apex, narrowed into a basal claw c. ⅓ the
length of the lamina. Stamens 6, all fertile ; filaments slightly wider at the base,
attenuated towards the apex, slightly shorter than the petals ; anther-thecae
c. 2·5 mm. long, oblong. Ovary linear, subsessile, narrowing into a short purplish
style glandular-pubescent at its base ; gynophore increasing rapidly after anthesis
to about 2·5 mm. in length and the style to about 7 mm. ; stigma hardly wider than
the style. Capsule up to 3·5 × 0·3 cm., spindle-shaped, with about 8 longitudinal
anastomosing nerves on each valve, with longer purple hairs and shorter glandular
hairs ; style persistent. Seeds brown, not ridged, about 2·2 mm. in diam., the cells
of the testa arranged longitudinally.

Nyasaland. S : Mt. Mlanje, fl. & fr. 9.vii.1946, *Brass* 16754 (BM ; K ; SRGH).
Endemic in Nyasaland and known so far only from Mt. Mlanje. High-altitude grasslands and forest edges.

3. **Cleome bororensis** (Klotzsch) Oliv. F.T.A. **1** : 81 (1868).—Gilg & Bened. in Engl.,
Bot. Jahrb. **53** : 155 (1915). Type : Mozambique, Boror, *Peters* (B, holotype).
Anomalostemon bororensis Klotzsch in Peters, Reise Mossamb. Bot. **1** : 162 (1861).
Type as above.
Polanisia bororensis (Klotzsch) Pax ex Gilg in Engl., Pflanzenw. Ost-Afr. **C** : 184
(1895). Type as above.

Annual herb up to about 1·3 m. tall ; stems greenish, striate, pubescent. Leaves
3-foliolate, petiolate ; leaflets 1·5–5 × 0·6–2·5 cm., membranous, subsessile, elliptic
to broadly elliptic, acute and mucronate at the apex, cuneate at the base, sparsely
pubescent on both sides with some scattered, sessile glands beneath ; petiole 4 cm.
long, pubescent ; upper leaves becoming reduced in size up the stem and grading
into sessile, entire bracts, c. 8 × 3·5 mm., subtending the flowers. Inflorescence a
raceme, dense at first but soon elongating ; pedicels slender, pubescent, up to
1·3 cm. long. Sepals up to 7 mm. long, linear-lanceolate with a subulate apex,
with both scattered subsessile glands and longer glandular hairs, hairs sometimes
purplish. Petals yellow, up to 1·2 cm. long, with an oblong lamina rounded
and apiculate at the apex, and a claw about half the length of the lamina, the
two outer ones slightly narrower. Stamens 6–8, slightly shorter than the petals,
two somewhat shorter than the rest and sometimes sterile ; anther-thecae c. 2 mm.
long, oblong, bluntly mucronate at the apex. Ovary linear, flattened, pubescent,
subsessile ; style somewhat flattened, up to 1 mm. long ; stigma subcapitate.
Capsule c. 8 × 0·5 cm., on an elongated, striate, pubescent gynophore up to 1·5 cm.
long ; valves flattened, tapering to both ends, pubescent, with 8–9 longitudinal

anastomosing nerves ; persistent style up to 5 mm. long. Seeds shining brown, c. 1·7 mm. in diam., almost smooth but minutely longitudinally tessellated.

Mozambique. Z : Boror, fl. *Peters* (B). MS : Beira, fl. & fr. 29.v.1948, *Mendonça* 4432 (BM ; LISC). LM : Macocololo, fl. & fr. 19.i.1898, *Schlechter* 12060 (BM ; COI ; K ; PRE).
Known only from Mozambique. Ecology unknown.

4. **Cleome ciliata** Schumach. in Kongel. Dansk. Vid. Selsk. Naturvid. Math. Afh. **4** : 68 (1829).—Oliv., F.T.A. **1** : 78 (1868).—Gilg & Bened. in Engl., Bot. Jahrb. **53** : 159 (1915).—Exell & Mendonça, C.F.A. **1**, 1 : 54 (1937).—Hauman & Wilczek, F.C.B. **2** : 514 (1951). Type from Ghana.

Annual herb up to about 1 m. tall, usually with spreading branches or sometimes procumbent ; stems greenish, striate and sparsely setulose-pubescent. Leaves 3-foliolate or rarely 5-foliolate, petiolate ; leaflets 0·75–4·0 × 0·4–1·75 cm., rhomboid-elliptic to lanceolate, acute or acuminate at the apex, cuneate at the base, minutely and sparsely setulose-pubescent on both sides or glabrescent, decreasing in size up the stem and becoming gradually more nearly sessile. Inflorescence a rather lax terminal raceme ; bracts usually 3-foliolate, sessile at the top of the stem, passing imperceptibly into the leaves ; pedicels up to 2·5 cm. long or up to 3 cm. in fruit, very slender, puberulous. Sepals c. 5 mm. long, linear-lanceolate or linear-subulate, caudate at the apex, somewhat keeled, sparsely puberulous on the back and margins. Petals up to 1·2 × 0·2 cm., pinkish- or bluish-purple, with a lanceolate or oblong-lanceolate, mucronate lamina and a basal claw about ⅓ the length of the lamina. Stamens 6, filaments slender, slightly shorter than the petals, the two lower slightly longer than the rest, anther-thecae up to 2·5 mm. long, narrowly linear, connective minutely apiculate. Ovary linear, subsessile, glandular-puberulous ; style very short ; stigma capitate. Capsule up to 7 × 0·4 cm., linear, tapering to both ends, on a glabrous gynophore c. 6 mm. long ; valves glabrous or with a few scattered subsessile glands, with about 8 longitudinal anastomosing nerves ; style persistent and elongating to 2–3 mm. Seeds c. 1·5 mm. in diam., brown, transversely ridged.

N. Rhodesia. N : Abercorn Distr., Lake Tanganyika, Kavala I., fl. & fr. vii.1890, *Carson* 30 (K). W : Ndola, fl. & fr. 11.iv.1954, *Fanshawe* 1076 (K).
Common through W. Africa, Cameroons, Belgian Congo, Angola, Uganda and the Sudan but reaches the edge of its range in the northern part of N. Rhodesia. Favours damp situations and soon becomes a weed. It has also been recorded as an introduced weed in the West Indies and Singapore.

5. **Cleome stricta** (Klotzsch) R. A. Graham in Kew Bull. **1958** : 31 (1958). Type : Mozambique, *Peters* (B, holotype).
　　Symphyostemon strictus Klotzsch in Peters, Reise Mossamb. Bot. **1** : 159 (1861). Type as above.
　　Cleome strigosa sensu Oliv., F.T.A. **1** : 80 (1868) pro parte.—Gilg & Bened. in Engl., Bot. Jahrb. **53** : 159 (1915).
　　Polanisia viscosa sensu Bak. f. in Journ. Linn. Soc., Bot. **40** : 20 (1911).

Erect or spreading herb c. 0·6 m. tall : branches striate with a coarse, whitish pubescence of longish hairs, some shorter hairs and scattered, subsessile, glandular hairs. Leaves 3–5-foliolate, petiolate ; leaflets 0·75–2·5 × 0·3–1·5 cm., reducing in size upwards and passing gradually into the sessile and finally simple bracts, narrowly obovate to obovate, rounded or truncate at the apex, cuneate at the base, harshly strigillose-pubescent on both sides, nerves somewhat impressed above and prominent below, petiole up to 2·5 cm. long, pubescent like the branches. Inflorescence racemose and terminal on the branches, moderately dense at first but the internodes elongating in fruit ; pedicels up to 1 cm. long, slender, with short glandular hairs. Sepals up to 8 mm. long, linear-lanceolate, acuminate or caudate-acuminate at the apex, with short, glandular hairs on both sides. Petals up to 2·5 × 0·5 cm., rose-coloured or purplish, becoming yellow towards the base, with an oblong-lanceolate lamina gradually narrowing into a basal claw of about the same length. Stamens 10–13, about the same length as the petals ; filaments slender, sometimes purplish towards the base and with the shortest one, or occasionally the shortest two, swollen just below the anther ; anther-thecae up to 2·5 mm. long, linear-oblong. Ovary linear, with a very minute pubescence visible only above a magnification of ×25 ; style very short ; stigma hardly wider than the style. Capsule up to 7 × 0·3 cm., on a glabrous gynophore up to 1·8 cm. long, linear, tapering at both ends, with a persistent style up to 2·5 mm. long ; valves minutely strigillose or pubescent, with about 3 longitudinal nerves. Seeds c. 1·5 mm. in diam., dark brown, almost black, transversely ridged.

Mozambique. N : Cabo Delgado, Mecufi, fl. & fr. 29.x.1942, *Mendonça* 1105 (BM ; LISC). MS : Beira, fl. & fr. vi.1919, *Nobbs* in Herb. Eyles 1768 (BM ; SRGH). SS : Bazaruto Is., fl. & fr. viii.1936, *Gomes e Sousa* 1830 (COI ; K). LM : João Belo, fl. 10.xii.1940, *Torre* 2299 (BM ; LISC).
Also in the coastal areas of Somaliland, Kenya, Zanzibar and Tanganyika. A plant of sandy coastal soils which sometimes becomes a weed of cultivation in coastal areas.

6. **Cleome hirta** (Klotzsch) Oliv., F.T.A. **1** : 81 (1868).—Briq. in Ann. Conserv. Jard. Bot. Genève, **17** : 363 (1914).—R.E.Fr., Wiss. Ergebn. Schwed. Rhod.-Kongo-Exped. **1** : 49 (1914).—Gilg & Bened. in Engl., Bot. Jahrb. **53** : 162 (1915).—Eyles in Trans. Roy. Soc. S. Afr. **5** : 355 (1916).—Burtt Davy, F.P.F.T. **1** : 121 (1926).— Bremek. & Oberm. in Ann. Transv. Mus. **16** : 414 (1935).—Exell & Mendonça, C.F.A. **1**, 1 : 56 (1937).—Hauman & Wilczek, F.C.B. **2** : 515 (1951). Type : Mozambique, Sena, *Peters* (B, holotype).
Decastemon hirtus Klotzsch in Peters, Reise Mossamb. Bot. **1** : 157 (1861). Type as above.
Polanisia hirta (Klotzsch) Pax in Engl., Bot. Jahrb. **10** : 14 (1888). Type as above.
Cleome pulcherrima Busc. & Muschl. in Engl., Bot. Jahrb. **49** : 467 (1913).—Gilg & Bened., loc. cit. Type from Tanganyika.
Cleome bechuanensis Bremek. & Oberm. in Ann. Transv. Mus. **16** : 413 (1935). Type : Bechuanaland Prot., Chobe R., Kabulabula, *van Son* (PRE, holotype ; SRGH).
Cleome pentaphylla sensu Hopkins, Bacon & Gyde, Comm. Veld Fl. : 59 (1940).

Coarse, aromatic, viscid, erect herb up to 1·6 m. tall ; branches striate, pubescent with gland-tipped hairs. Leaves 5–7–9-foliolate, on glandular, hairy petioles up to 7 cm. long but often less : leaflets 0·5–5·5 × 0·05–0·5 cm., linear, linear-lanceolate or narrowly oblong-lanceolate, subacute or rounded at the apex, cuneate at the base, glandular-hairy on both sides or sometimes glabrescent, decreasing in size upwards and passing into the sessile bracts which are commonly 3-foliolate but occasionally simple or 5-foliolate. Inflorescence a terminal raceme which is moderately dense at first but eventually becomes greatly elongated in fruit ; pedicels slender, glandular-pubescent. Sepals up to 1·2 cm. long, linear, subulate towards the apex, glandular-hairy outside. Petals up to 1·9 × 0·4 cm., purplish or pink, paler towards the base, with a lanceolate-oblong lamina rounded at the apex and with a distinct basal claw from ⅓ to almost as long as the lamina. Stamens 10–12, filaments up to c. 2·5 cm. long, purplish, very slender, often glandular-pubescent towards the base, exserted ; anther-thecae c. 2 mm. long, linear-oblong. Ovary linear, glandular-pubescent, with a short, glandular gynophore c. 2 mm. long even when young ; style c. 2 mm. long, glabrous ; stigma subcapitate. Capsule up to 16 × 0·4 cm., linear, on a glandular gynophore 0·5–2·0 cm. long ; valves glandular-pubescent or glabrescent, with 7–9 anastomosing, longitudinal nerves ; persistent style 3–5 mm. long. Seeds 2–2·5 mm. in diam., transversely tuberculate-rugose with a very minute, mealy-glandular pubescence visible at a × 40 magnification.

Bechuanaland Prot. N : Chobe R., 60 km. N. of Kachikau on road to Kazane, fl. & fr. 11.vii.1937, *Erens* 393 (K ; PRE ; SRGH). SE : Kwena, 20 km. W. of Gaberones, fl. & fr. 5.v.1955, *Reineke* 328 (K ; PRE). **N. Rhodesia.** B : near Senanga, fl. & fr. 30.vii.1952, *Codd* 7241 (K ; PRE ; SRGH). N : Luapula R., Miloki, fl. & fr. 6.ix.1953, *Fanshawe* 272 (K). E : Katete, fl. 11.ii.1957, *Wright* 148 (K). S : Livingstone, fl. i.1910, *Rogers* 7257 (K ; SRGH). **S. Rhodesia.** N : Darwin, fl. & fr. 14.vi.1921, *Eyles* 3103 (SRGH). W : Bulawayo, fl. & fr. 15.ii.1944, *Hopkins* in GHS 11741 (K ; SRGH). C : Que Que Reserve, fl. & fr. 8.i.1955, *Ward* 7 (SRGH). E : Sabi Valley, Mtema, fl. 28.i.1948, *Wild* 2394 (K ; SRGH). S : Gwanda, Bubye R., fl. & fr. v. 1955, *Davies* 1277 (SRGH). **Nyasaland.** N : Mzimba, Mbawa Exp. Station, fl. & fr. 5.iv.1955, *Jackson* 1590 (K ; SRGH). C : Kasungu, fl. & fr. 26.viii.1946, *Brass* 17431 (K). S : Blantyre, fl. & fr. 1876, *Simons* (BM). **Mozambique.** N : east coast of Lake Nyasa, fl. & fr. *Johnson* 28 (K). T : Boroma, Msusa, fl. & fr. *Chase* 2687 (BM ; K ; PRE ; SRGH). MS : Maringa, fl. & fr. 26.vi.1950, *Chase* 2498 (BM ; K ; SRGH). SS : Guijá Distr., R. Limpopo, fl. & fr. vii.1915, *Gazaland Exped.* (PRE ; SRGH). LM : Manhiça, Mugaba, fl. 15.iv.1945, *Gomes e Sousa* 175 (BM ; LISC ; PRE).
Also in Somaliland, Kenya, Uganda, Tanganyika, Belgian Congo, Angola, SW. Africa and the Transvaal. A plant of open woodlands in rainfall areas of usually less than 75 cm.

p.a., i.e. in *Acacia* or *Colophospermum mopane* woodland rather than in *Brachystegia* woodland.

The plant described as *C. bechuanensis* represents the dry-season form which has narrower leaves and smaller flowers than typical *C. hirta*. *C. pulcherrima* was described from a specimen collected allegedly by the Duchess of Aosta near Broken Hill in N. Rhodesia but Gilg and Benedict have shown (loc. cit.) that in reality it was based on a specimen collected in Tanganyika by von Prittwitz, so it does not really concern us here.

7. **Cleome rubella** Burch., Trav. Int. S. Afr. **1** : 543 (1822).—Sond. in Harv. & Sond., F.C. **1** : 56 (1860).—Gilg & Bened. in Engl., Bot. Jahrb. **53** : 157 (1915).—Burtt Davy, F.P.F.T. **1** : 121 (1926).—Exell & Mendonça, C.F.A. **1, 1** : 56 (1937). Type from Cape Province (Griqualand West, Asbestos Mts.).

Annual herb c. 0·3 m. tall, much branched as a rule, with densely glandular-pubescent stems. Leaves 5–7 (9)-foliolate, with glandular-pubescent petioles about 2 cm. long but becoming shorter towards the inflorescences ; leaflets 1–3 × 0·15–0·2 cm., linear, often channelled above, subacute or obtuse at the apex, glaucous, sparsely glandular-pubescent or glabrous, very shortly petiolulate. Inflorescence of terminal, rather lax racemes ; bracts similar to the leaves but sessile or subsessile and with progressively fewer leaflets up the stem ; pedicels 0·5–1·0 cm. long, slender, glandular-pubescent. Sepals 3–4 mm. long, linear-subulate with hyaline margins, sparsely glandular-puberulous. Petals 5–8 × 2–2·5 mm., from violet to rose-pink, oblong-lanceolate, blunt at the apex, with a basal claw up to ⅓ the length of the lamina. Stamens 6–9 ; filaments about as long as the petals, slender, glabrous, purplish ; anther-thecae c. 1·5 mm. long, linear. Ovary linear, minutely glandular-pubescent, on a glandular-pubescent gynophore c. 1 mm. long ; style c. 1 mm. long, glabrous ; stigma subcapitate. Capsule c. 2–3·5 × 0·2–0·3 cm., on a glandular-pubescent gynophore 2–3 mm. long ; valves glandular-pubescent, with about 9 longitudinal, anastomosing nerves ; persistent style about 2 mm. long. Seeds c. 2 mm. in diam., dark brown, transversely rugose.

Bechuanaland Prot. N : Ngamiland, Botletle valley, fl., & fr., *Lugard* 189 (K). SE : Bakwena country, fr. 1876, *Holub* (K). **N. Rhodesia.** B : Sesheke, fl. & fr. 1911, *Macaulay* 15 (K).

Also in Angola, SW. Africa, Cape Province, Orange Free State, Natal and the Transvaal. Recorded only from our driest areas to the west.

This species is related to *C. hirta* but is consistently smaller in all parts.

8. **Cleome iberidella** Welw. ex Oliv., F.T.A. **1** : 79 (1868).—Gilg & Bened. in Engl., Bot. Jahrb. **53** : 160 (1915).—Exell & Mendonça, C.F.A. **1, 1** : 54 (1937). Type from Angola (Cuanza Norte).

Erect, sparsely glandular-setose herb up to 0·5 m. tall ; branches slender, striate. Leaves 3–5-foliolate ; leaflets 0·5–2·0 × 0·1–0·8 cm., linear to lanceolate or oblanceolate, apex rounded, mucronulate, cuneate at the base ; petiole up to 3 cm. long. Inflorescence a terminal elongated raceme to 15 cm. long, with many flowers ; bracts 3-foliolate or simple, similar to the upper leaves but sessile ; pedicels up to 1·3 cm. long, glandular-puberulous, capillary. Sepals c. 2·3 × 0·5 mm., lanceolate to linear-lanceolate, acuminate at the apex, glandular-puberulous, with a keeled midrib. Petals c. 6 × 1·5 mm., subequal, mauve, the two upper petals with a sulphur-yellow blotch with deep mauve edges, oblanceolate or narrowly elliptic, acute or obtuse at the apex, with a claw c. ½ of the length of the lamina. Stamens 6, all fertile ; filaments c. 7 mm. long ; anthers c. 2 × 0·4 mm., narrowly linear, arcuate. Ovary linear, glandular-puberulous, multiovulate, with a small, subsessile, capitate stigma. Capsule c. 2·5 × 0·2 cm., spindle-shaped, reflexed or decurved ; valves with c. 12 longitudinal striations, glandular-puberulous, with a persistent style c. 2 mm. long. Seeds c. 1·3 mm. in diam., dark brown, transversely ridged.

Bechuanaland Prot. N : Ngamiland, fl. xii.1930, *Curson* 711 (PRE).
Also in Angola. Recorded from roadsides and as a weed of cultivation ; its ecology is not well known.

9. **Cleome oxyphylla** Burch., Trav. Int. S. Afr. **2** : 226 (1824).—Gilg & Bened. in Engl., Bot. Jahrb. **53** : 158 (1915). Type from Cape Province (Griquatown).
　　Polanisia oxyphylla (Burch.) DC., Prodr. **1** : 242 (1824).—Sond. in Harv. & Sond., F.C. **1** : 57 (1860). Type as above.

Erect, branching herb 0·3–0·6 m. tall ; stems striate, glandular-pubescent, some-times purplish. Leaves 3–7-foliolate ; leaflets 0·5–2·5 × 0·25–0·8 cm., oblong-lanceolate, narrowly oblanceolate or oblanceolate, acute or rather blunt and often mucronate at the apex, cuneate at the base, somewhat glaucous and glandular-pubescent on both sides, merging upwards into the subsessile or sessile, usually 3-foliolate bracts ; petiole up to 6 cm. long. Inflorescence of rather lax terminal racemes ; pedicels up to 1·8 cm. long, glandular-pubescent. Sepals 4–6·5 mm. long, narrowly lanceolate, glandular-pubescent outside, acuminate at the apex. Petals c. 1·5 × 0·5–0·7 cm., rose-pink or purplish, yellow towards the base, narrowly obovate, rounded at the apex, with a short claw less than ¼ the length of the lamina. Stamens 8–12 ; filaments slender, the lower four longer than the remainder and with longer anthers, the longer about the same length as the petals, glabrous ; anther-thecae c. 2 mm. long, linear-oblong. Ovary linear, glandular-pubescent, on a very short, glandular-pubescent gynophore c. 1 mm. long ; style very short ; stigma subcapitate. Capsule c. 8 × 0·4 cm., on a glandular-pubescent gynophore c. 0·8 cm. long ; valves glandular-pubescent, with 8–9 longitudinal anastomosing nerves ; persistent style c. 2·5 mm. long, glabrous. Seeds c. 2·0 mm. in diam., dark brown, transversely striate.

S. Rhodesia. W : Matobo, Besna Kobila, fl. & fr. i.1954, *Miller* 2064 (K ; PRE ; SRGH).
Also in the northern parts of the Cape Province and the Transvaal.

In our area this plant has not been collected outside the Matobo District, and is appar-ently very rare. It is closely related to *C. hirta* but can be distinguished by the very short claws of the petals and the oblanceolate rather than linear leaves. The original description gives the flower colour as yellow but pinkish pigment as well as yellow can still be seen in the flowers of the type-specimen.

10. **Cleome sulfurea** Bremek. & Oberm. in Ann. Transv. Mus. **16** : 414 (1935). Type : Bechuanaland Prot., Kaotwe, *van Son* (BM ; K ; PRE, holotype ; SRGH).

Annual herb 20–40 cm. tall, little branched ; stems striate, glandular-pubescent. Leaves 3-foliolate, petiolate ; leaflets up to 2·0 × 0·2 cm., gradually diminishing upwards and passing into the sessile, 3-foliolate bracts ; petiole glandular-pubescent, c. 0·5 cm. long. Inflorescence a terminal lax raceme ; pedicels up to 1·5 cm. long or 2·0 cm. in fruit, glandular-pubescent. Sepals 5–6 mm. long, membranous, linear or linear-oblanceolate, acuminate at the apex, glandular-pubescent outside. Petals sulphur-yellow, 9–10 × 4–6 mm., narrowly obovate to oblanceolate, tapering gradually to the base but lacking a distinct claw. Stamens all fertile ; filaments slender and purplish towards the base, the 4–5 lower c. 1·5 cm. long and 13–14 upper c. 9 mm., long, glabrous ; anther-thecae up to 3 mm. long, linear-oblong, minutely apiculate at the apex. Ovary linear with minute, subsessile glands visible only at magnifications of × 50 upwards ; gynophore glabrous, c. 1 mm. long ; style c. 1 mm. long, purplish ; stigma subcapitate. Capsule 4–5 × 0·2 cm., on a glabrous gynophore up to 1 cm. long ; valves glandular-pubescent, with 11–13 longitudinal, anastomosing nerves ; persistent style up to 3 mm. long. Seeds c. 1 mm. in diam., almost black, tuberculate-reticulate.

Bechuanaland Prot. SW : Kaotwe, fl. & fr. 10.iv.1930, *van Son* in Herb. Transv. Mus. 28809 (BM ; K ; PRE, holotype : SRGH).
Endemic in Bechuanaland Prot. and known so far from the type-collection only.

This is a very distinct species related to the SW. African species *C. suffruticosa* Schinz and *C. luederitziana* Schinz and to the Angolan *C. foliosa* Hook. f. All have yellow flowers and small blackish seeds barely more than 1 mm. in diam. as in *C. sulfurea*, but the latter is unique in this group in having linear leaves.

11. **Cleome diandra** Burch., Trav. Int. S. Afr. **1** : 548 (1822).—Oliv., F.T.A. **1** : 79 (1868).—Gilg & Bened. in Engl., Bot. Jahrb. **53** : 166 (1915).—Burtt Davy, F.P.F.T. **1** : 121 (1926).—Bremek. & Oberm. in Ann. Transv. Mus. **16** : 414 (1935).—Exell & Mendonça, C.F.A. **1**, 1 : 58 (1937). FRONTISP. & TAB. **30** fig. C. Type from Cape Province (near Asbestos Mts.).
Polanisia dianthera DC., Prodr. **1** : 242 (1824) *nom. illegit.* Type as above.
Dianthera petersiana Klotzsch ex Sond. in Harv. & Sond., F.C. **1** : 57 (1860) ; in Peters, Reise Mossamb. Bot. **1** : 160, t. 27 (1861). Syntypes from S. Africa and Mozambique : Tete, *Peters* (B) ; Sena, *Peters* (B).

Dianthera burchelliana Klotzsch ex Sond. in Harv. & Sond., F.C. **1** : 58 (1860) ;
in Peters, Reise Mossamb. Bot. **1** : 161 (1861) *nom. illegit.* Type from S. Africa.
  Polanisia petersiana (Klotzsch) Pax in Engl., Bot. Jahrb. **19** : 134 (1894). Syn-
types as for *Dianthera petersiana*.
  Polanisia diandra (Burch.) Dur. & Schinz, Consp. Fl. Afr. **1**, 2 : 162 (1898).
Type as for *Cleome diandra*.
  Cleome petersiana (Klotzsch) Briq., in Ann. Conserv. Jard. Bot. Genève, **17** :
364 (1914). Syntypes as for *Dianthera petersiana*.

Slender, glaucous, erect herb up to 75 cm. tall with striate, smooth or aculeolate-
glandular stems, sometimes with a spindle-shaped swelling in the main stem,
branches few. Leaves 3–5–7–9-foliolate ; leaflets 1–4 × 0·05–0·2 cm., very
narrowly linear, minutely petiolulate, blunt or subacute at the apex, glabrous,
diminishing upwards and passing into the sessile or subsessile, usually 3-foliolate
bracts ; petioles slender, up to c. 6 cm. long, glabrous. Inflorescence of lax terminal
racemes : pedicels up to c. 3·5 cm. long in fruit, slender, glabrous. Sepals
0·7–1·0 × 0·2–0·4 cm., narrowly ovate, lanceolate, or linear-lanceolate, acuminate
at the apex, glabrous, pale green with darker green nerves. Petals bright golden-
yellow with purple spots at the base, the two larger c. 2·4 × 1·4 cm., with a short
claw c. 2·5 mm. long, broadly obovate or obovate, rounded at the apex, the two
shorter petals c. 1 × 0·1–0·25 cm., with a very short claw up to 2 mm. long, linear-
oblong to narrowly oblong, the lamina rounded or subacute at the apex and rounded
or subcordate at the base. Fertile stamens 2 with the filaments incurved, up to
4 cm. long ; anther-thecae c. 4 mm. long, linear-oblong ; staminodes 6–12, 3–6
mm. long, with minute, sterile, caducous anther-thecae less than 0·5 mm. long and
the filaments often swollen at the apices. Ovary linear, glabrous, with a distinct
gynophore c. 4 mm. long ; style c. 1 mm. long, glabrous ; stigma capitate. Capsule 5–10 × 0·3–0·4 cm. on a glabrous gynophore up to 1·7 cm. long ; valves
glabrous, tapering to both ends, with c. 9 longitudinal, anastomosing nerves ;
persistent style up to 4 mm. long, glabrous. Seeds c. 1·5 mm. in diam., brown,
pubescent or some seeds glabrescent ; testa tessellated.

Bechuanaland Prot. N : Ngamiland, Kwebe, fl. & fr. i.1897, *Lugard* 130 (K).
SW : Matapa Pan, fl. & fr. iv.1930, *van Son* in Herb. Transv. Mus. 28810 (PRE). SE :
Kanye, fl. & fr. iii.1950, *Miller* B/996 (PRE). **S. Rhodesia.** N : Sebungwe, fl. & fr.
vi.1956, *Davies* 2016 (SRGH). S : 3 km. W. of Beitbridge, fl. & fr. 15.ii.1955, *E.M. & W.*
389 (BM ; LISC ; SRGH). **Mozambique.** MS : Inhaminga, fl. 20.x.1949, *Pedro &
Pedrógão* 8808 (LMJ ; SRGH). SS : Caniçado, Regulo Chirunzo, fl. & fr. 18.vi.1947,
Pedrógão 322 (K ; LMJ ; PRE). LM : Lourenço Marques, fl. & fr. 18.i.1948, *Faulkner*
190 (COI ; K ; SRGH).
  Also in Kenya, Tanganyika, Angola, SW. Africa, Cape Province, Natal and the Transvaal. In low-altitude, low-rainfall areas with about 20–45 cm. per annum; often found in
Colophospermum mopane woodland and with a tendency to become a weed of cultivation.

A handsome and very distinct species.

12. **Cleome maculata** (Sond.) Szyszyl., Polypet. Thalam. Rehm. : 109 (1887).—
    Briq. in Ann. Conserv. Jard. Bot. Genève, **17** : 363 (1914).—Eyles in Trans. Roy.
    Soc. S. Afr. **5** : 355 (1916). TAB. **30** fig. B. Type from Natal (Mooi R.).
      Polanisia maculata Sond. in Linnaea **23** : 6 (1850). Type as above.
      Tetratelia maculata (Sond.) Sond. in Harv. & Sond., F.C. **1** : 58 (1860). Type
    as above.
      Chilocalyx maculatus (Sond.) Gilg & Bened. in Engl., Bot. Jahrb. **53** : 168 (1915).
    Type as above.
      Cleome maculata (Sond.) Burtt Davy in Kew Bull. **1924** : 224 (1924) ; F.P.F.T.
    1 : 121 (1926). Type as above.

Erect annual herb c. 0·3 m. tall, usually branched ; stems striate, glabrous or
minutely and sparsely asperous. Leaves 3–5-foliolate, leaflets 1–3·5 × 0·05–0·15
cm., linear, subacute, minutely and sparsely asperous-puberulous or glabrous ;
petioles up to 2·5 cm. long. Inflorescence of lax terminal racemes ; bracts up to
1 mm. long, linear, caducous ; pedicels up to 7 mm. long, slender, sparsely
setulose-puberulous. Sepals 1·5–2 mm. long, narrowly lanceolate, acute, setulose-
puberulous on the keel and margins. Petals pink to violet, the two slightly shorter
upper ones having a median, bright yellow, dark violet or purple-margined band,
1–1·5 × 0·15–0·25 cm. ; lamina oblong-elliptic, blunt at the apex or somewhat
apiculate, with a basal claw as long as or almost as long as the lamina. Stamens 6–8,

slightly connate at the base ; 2–4 longer and fertile on slender filaments, as long as the petals ; anther-thecae c. 2 mm. long, linear-oblong ; 2–4 sterile stamens slightly shorter with a clavate swelling at the top of the filaments and reduced, abortive anther-thecae up to 0·75 mm. long which are occasionally partially fertile. Ovary linear, sparsely puberulous, on a gynophore c. 0·5 mm. long ; style c. 1 mm. long ; stigma subcapitate. Capsule 4–10 × 0·25 cm., on a glabrous gynophore up to 1 cm. long ; valves linear, tapering to both ends, very sparsely puberulous or glabrous, with 3–4 longitudinal nerves ; persistent style up to 7 mm. long. Seeds brown, c. 1·5 mm. in diam., transversely ridged.

Bechuanaland Prot. SE : Mochudi, fl. & fr., *Harbor* (PRE ; SRGH). **S. Rhodesia.** S : Shashi R., fl. & fr. i.1898, *Rand* 22 (BM).
Also in SW. Africa, Natal and the Transvaal. Reaching only the dry western part of our area.

This species has very much the appearance of small specimens of *C. macrophylla*, but the clavate tips of the filaments of the sterile stamens distinguish it. Because of this character it was placed by Sonder in a distinct genus *Tetratelia* but he is not followed here as *C. diandra* sometimes shows the same character and so also does the tropical American genus *Physostemon* Mast. In the case of this latter genus the sterile stamens differ from those of *Tetratelia* in being longer than the fertile ones. *Physostemon* is itself often united with *Cleome* and if *Tetratelia* were to be retained as a distinct genus it would be logically necessary to revive a number of other genera such as those named *Chilocalyx* and *Dianthera* by Klotzsch. *Cleome*, however, forms such a compact and readily recognizable genus, that it would be quite pointless to do this.

13. **Cleome macrophylla** (Klotzsch) Briq. in Ann. Conserv. Jard. Bot. Genève, **17** : 365 (1914). Syntypes : Mozambique, Sena, *Peters* (B) ; Tete, *Peters* (B).
    Chilocalyx macrophyllus Klotzsch in Peters, Reise Mossamb. Bot. **1** : 155 (1861). —Gilg & Bened. in Engl., Bot. Jahrb. **53** : 168 (1915). Types as above.
    Chilocalyx tenuifolius Klotzsch, loc. cit.—Gilg & Bened., loc. cit. Type : Mozambique, Sena, *Peters* (B).
    Cleome chilocalyx Oliv., F.T.A. **1** : 81 (1868) *nom. illegit.* Type as for *Cleome macrophylla*.
    Cleome chilocalyx var. *tenuifolia* (Klotzsch) Oliv., loc. cit. Type as for *Chilocalyx tenuifolius*.
    Cleome inconcinna Briq., tom. cit. : 364 (1914). Type : Mozambique, Lourenço Marques, *Schlechter* 11516 (BM ; G, holotype ; PRE).
    Tetratelia tenuifolia (Klotzsch) Arwidss. in Bot. Notis. **1935** : 357 (1935). Type as for *Chilocalyx tenuifolius*.

Erect, branching, annual herb to 0·6 m. tall ; stems striate, glabrous or very sparsely glandular-puberulous when very young. Leaves 3-foliolate ; leaflets 1–6·5 × 0·05–2·5 cm., linear, narrowly lanceolate or narrowly ovate, acuminate at the apex, cuneate at the base, sparsely and minutely asperulous at least on the margins ; petiole up to 6·5 cm. long, glabrous. Inflorescence of moderately dense terminal racemes, rapidly elongating in fruit ; bracts c. 1 mm. long, caducous, subulate ; pedicels up to 4 mm. long, glandular-puberulous, slender. Sepals purplish at the tips, up to 5 mm. long, linear-subulate, unequal, glabrous or sparsely setulose on the keel and margins. Petals mauve, the two upper ones with a bright yellow spot surrounded by a short, dark purple line, 8–14 × 1·5–2·5 mm. ; lamina narrowly elliptic-oblong, acute or rather blunt, basal claw almost as long as the lamina. Stamens 6–12, all fertile though some have smaller anthers ; filaments slender, rather unequal, the longer at length slightly exceeding the petals ; anther-thecae up to 2 mm. long, linear-oblong. Ovary linear, glabrous, on a gynophore c. 1·5 mm. long ; style 0·5 mm. long ; stigma capitate. Capsule 4–9 × 0·15–0·2 cm., on a glabrous gynophore up to 1·3 cm. long ; valves linear, tapering to both ends, glabrous, with about 3 longitudinal nerves ; persistent style 2–4 mm. long. Seeds c. 1·5 mm. in diam., brown, transversely ridged.

var. **macrophylla.** TAB. 30 fig. A.
    Cleome nationae Burtt Davy in Kew Bull. **1924** : 224 (1924) pro parte quoad specim. Schlechter.
    Leaflets more than 2 mm. wide ; mature capsule 7–10 mm. long.

N. Rhodesia. S : Mumbwa, fl. & fr. 1911, *Macaulay* 360 (K). **S. Rhodesia.** N. Urungwe, Kariba Gorge, fl. & fr. 25.ii.1953, *Wild* 4033 (K ; SRGH). W : Wankie, fl. *Levy* 1117 (PRE). E : Sabi R., E. bank, Giriwayo, 19.i.1957, *Phipps* 47 (K ; SRGH).

Tab. 30. A.—CLEOME MACROPHYLLA VAR. MACROPHYLLA. A1, flowering and fruiting
branches (× ⅔) *Kirk* s.n. ; A2, flower (× 2) *Wild* 2430. B.—CLEOME MACULATA,
flower (× 2) *Harbor* s.n. C.—CLEOME DIANDRA, flower (× 2) *Lugard* 130.

S : Mtilikwe R., Bangara Falls, fl. 13.xii.1953, *Wild* 4375 (K ; SRGH). **Nyasaland.** N : Nyika Plateau, fl. & fr. ii-iii.1903, *McClounie* (K). S : Shire Highlands, fl. *Buchanan* (K). **Mozambique.** Z : Lower Shire, fl. & fr. v.1861, *Meller* (K). T : Mutarara, fl. & fr. 6.v.1943, *Torre* 5301 (BM ; LISC). MS : Chimoio, fl. & fr. 26.ii.1948, *Garcia* 374 (BM ; LISC). SS : Homoine, fl. & fr. 25.ii.1955, *E.M. & W.* 574 (BM ; LISC ; SRGH). LM : Marracuene, fl. & fr. 17.iii.1945, *Sousa* 76 (BM ; LISC ; PRE).

Also in Tanganyika. Often found by rivers and on hillsides growing in rock clefts or in shallow soil overlying rock masses.

var. **maculatiflora** (Merxm.) Wild in Bol. Soc. Brot., Sér. 2, **32** : 38 (1958). Type : S. Rhodesia, Marandellas, *Dehn* 140a (M, holotype ; SRGH).
  Cleome nationae Burtt Davy in Kew Bull. **1924** : 224 (1924) pro pa.te ; F.P.F.T. **1** : 121 (1926). Type from the Transvaal (Rustenberg).
  Tetratelia nationae (Burtt Davy) Pax & K. Hoffm. in Engl. & Prantl., Nat. Pflanzenfam. ed. 2, **17b** : 219 (1936).—Hauman & Wilczek, F.C.B. **2** : 518 (1951). Type as for *Cleome nationae.*
  Tetratelia tenuifolia var. *maculatiflora* Merxm. in Proc. & Trans. Rhod. Sci. Ass. **43** : 86 (1951). Type as for *Cleome macrophylla* var. *maculatiflora.*

Leaflets not more than 2 mm. wide, always linear, the petals c. 1 cm. long and all other parts proportionately smaller than in var. *macrophylla.* Stamens also often fewer than in var. *macrophylla* and occasionally as few as 6. Mature capsule 4–7 cm. long.

S. Rhodesia. N : Mtoko, fl. & fr. 28.i.1949, *Hopkins* in GHS 7891 (K ; SRGH). W. Matopos, fl. & fr. 3.xi.1930, *Hill* (K ; SRGH). C : Salisbury, fl. & fr. 10.i.1948, *Wild* 2279 (K ; SRGH). E : Umtali, Mt. Sheni, fl. & fr. 5.iv.1950, *Chase* 1985 (K ; SRGH). S : Sabi R., west bank, fl. & fr. 29.i.1948, *Wild* 2430 (K ; SRGH). **Mozambique.** LM : Magude, fl. & fr. 30.xi.1944, *Mendonça* 3187 (BM ; LISC).
Also in the Transvaal. Very commonly found in the shallow soils of granite outcrops and at rather higher altitudes (up to 2000 m.) than var. *macrophylla.*

Arwidsson (loc. cit.) included these two varieties under his new combination *Tetratelia tenuifolia.* *C. macrophylla,* however, is not really related to *C. maculata,* the only true *Tetratelia* in its restricted sense, for, although some anthers in the former species may be smaller than the others, they are never sterile and there is no clavate swelling of the filament. It is probably more nearly related to *C. ciliata.*

2. GYNANDROPSIS DC.

Gynandropsis DC., Prodr., **1** : 237 (1824) *nom. conserv.*

Annual herb or rarely rather shrubby. Leaves petiolate, 3–7-foliolate. Inflorescences of terminal many-flowered racemes with foliaceous bracts. Flowers zygomorphic, usually bisexual ; sepals 4, free ; petals 4, clawed, spathulate or oblong. Stamens 6, free, inserted on a filiform androgynophore at the base of the gynophore ; anthers basifixed, somewhat curved ; gynophore short ; ovary 1-locular, ovules many on two parietal placentas ; style persistent, short ; stigma capitate. Capsule dehiscent with two valves and a persistent replum. Seeds numerous, reniform ; testa reticulate or rugose.

Gynandropsis speciosa DC., a native of Central and S. America, has been introduced in our area as a garden plant (e.g. *Brander* in GHS 7988 (SRGH)). It differs from *G. gynandra* in being almost glabrous and having larger leaves and flowers.

Gynandropsis gynandra (L). Briq. in Ann. Conserv. Jard. Bot. Genève, **17** : 382 (1914).—Arwidss. in Bot. Notis. **1935** : 357 (1935).—Exell & Mendonça, C.F.A. **1, 1** : 58 (1937).—Hauman & Wilczek, F.C.B. **2** : 519 (1951).—Martineau, Rhod. Wild Fl. : 30, t. 7 (1954). TAB. **31.** Type a cultivated specimen.
  Cleome gynandra L., Sp. Pl. **2** : 671 (1753). Type as above.
  Cleome pentaphylla L., Sp. Pl. ed. 2, **2** : 938 (1763) *nom. illegit.* Type as above.
  Pedicellaria pentaphylla Schrank in Roem. & Ust., Mag. Bot. **3**, 8 : 11 (1790) *nom. illegit.*—Gilg & Bened. in Engl., Bot. Jahrb. **53** : 167 (1915). Type as above.
  Gynandropsis pentaphylla DC., Prodr. **1** : 238 (1824) *nom. illegit.*—Oliv., F.T.A. **1** : 82 (1868).—Ficalho, Pl. Ut. Afr. Port. : 91 (1884).—Eyles in Trans. Roy. Soc. S. Afr. **5** : 356 (1916).—Burtt Davy, F.P.F.T. **1** : 121 (1926). Type as above.

Glandular-puberulous or glabrescent annual herb c. 0·6 m. tall, sometimes becoming rather shrubby below. Leaves 5-foliolate or 3-foliolate above ; leaflets 3–10 × 1–4·5 cm., narrowly obovate or oblanceolate, rounded or acute to acuminate

Tab 31. GYNANDROPSIS GYNANDRA. 1, flowering stem (× ⅔) *Lugard* 75 ; 2, flower (× ⅘)
Lugard 108 ; 3, fruit (× ⅔) *Macaulay* 46 ; 4, seed (×4) *Macaulay* 46.

LMR

at the apex, cuneate at the base, repand-denticulate or entire, glabrescent or glabrous ; petioles up to 12 cm. long ; petiolules c. 2 mm. long. Inflorescence of terminal racemes with sessile or subsessile foliaceous bracts 3-foliolate below simple above ; pedicels 1–2 cm. long. Sepals 2–4 mm. long, lanceolate to narrowly ovate, acute at the apex, glandular. Petals white or rose-pink, 1–2 × 0·3–0·5 cm., with an oblong blade, rounded at the apex and narrowing abruptly into a claw rather longer than the lamina. Androgynophore c. 2 cm. long, purplish or pinkish ; filaments very slender, purplish or pinkish, 2·5–4 cm. long ; anthers c. 2·5 mm. long, linear-oblong ; gynophore 0·5–2 cm. long ; ovary linear-oblong, glandular ; style 1–2 mm. long, glabrous ; stigma capitate. Capsule 3–15 × 0·4–1·2 cm., linear or linear-oblong, compressed, with many longitudinally anastomosing nerves, glandular or glabrescent, terminated by the persistent style. Seeds brown, numerous, c. 1·5 mm. in diam., rugose-reticulate.

Bechuanaland Prot. N : Ngami, Mothlatlogo, fl. & fr. 11.v.1930, *van Son* (BM ; K ; PRE ; SRGH). SE : Mochudi, fl. i.–iv.1914, *Harbor* in Herb. Rogers 6479 (K). **N. Rhodesia.** B : Sitoti Ferry, fl. & fr. 6.vii.1952, *Codd* 7412 (BM ; K ; PRE). N : Abercorn Distr., Mwandwisi valley, fl. & fr. iv.1937, *Trapnell* 1768 (K). S : Namwala, fl. 1933–35, *Read* 33 (BM ; K ; PRE). **S. Rhodesia.** N : Mtoko, Mkota Reserve, fl. & fr. 11.viii.1950, *Whellan* 466 (SRGH). W : Matopos, fl. 21.iii.1943, *Hopkins* in GHS 9901 (K ; SRGH). C : Salisbury, fl. & fr. i.1919, *Eyles* 1466 (BM ; PRE ; SRGH). E : Umtali, fl. & fr. 26.iii.1955, *Chase* 5529 (K ; SRGH). S : Gwanda, Shashi Plain, fl. & fr. xii.1954, *Davies* 919 (SRGH). **Nyasaland.** N : North Mzimba, fl. & fr. 15.iii.1954, *Jackson* 1239 (BM). S : Port Herald, fl. 1.iii.1933, *Lawrence* 23 (K). **Mozambique.** N : Nampula, fl. & fr. 21.i.1937, *Torre* 1198 (COI). Z : Namagoa, Mocuba, fl. & fr. vii–viii. *Faulkner* 11 (COI ; K). T : Tete, Ulandi, fl. i.1932, *Guerra* 79 (COI). MS : Vila Machado, fl. & fr. 1.iii.1948, *Mendonça* 3837 (BM ; LISC). SS : Vila de João Belo, fl. & fr. 8.x.1945, *Pedro* 224 (K ; PRE). LM : Lourenço Marques, fl. & fr. 25.ii.1945, *Sousa* 45 (BM ; LISC).

Widely distributed in Africa, Madagascar, Mediterranean Region, Asia, Polynesia and tropical America. A weed of cultivation or ruderal.

Used as a relish by the Bantu.

3. CADABA Forsk.

Cadaba Forsk., Fl. Aegypt.-Arab. : CVI, 67 (1775).

Shrubs, unarmed or the branches becoming spine-tipped. Leaves usually entire, occasionally rudimentary or absent. Flowers somewhat zygomorphic, solitary or fascicled in the leaf-axils or in terminal, leafless racemes. Sepals 4, subequal or unequal in 2 series, the upper and lower valvate and enclosing the lateral sepals. Petals 2–4 or 0, clawed. Disk large and conspicuous forming a nectary adherent to the base of the androgynophore, urceolate or infundibuliform or tubular, sometimes with a recurved neck, sometimes dentate at the mouth. Androgynophore elongated and often declinate. Stamens 4–8. Ovary on an elongated gynophore, 1-locular with 2–(4) placentas, multiovulate ; stigma sessile, capitate or forming an apiculus on the ovary. Fruit cylindric, indehiscent or tardily dehiscent by 2 valves, glabrous or glandular, many-seeded. Seeds subglobose or subreniform ; testa ridged or smooth, often embedded in a bright orange or scarlet, floury matrix.

Leafless shrub ; branches with subspinous apices   -   -   -   -   - 1. *aphylla*
Leafy shrubs :
  Flowers solitary in the leaf axils, often on abbreviated side shoots :
    Sepals, petioles, young leaves and all young parts papillose-farinose ; buds ovate,
      acute in outline ; nectary with a straight neck   -   -   - 2. *termitaria*
    Sepals, petioles, young leaves and all young parts glabrous or pubescent, never
      farinose ; buds globose ; nectary with a recurved neck   -   -   3. *natalensis*
  Flowers in terminal, leafless racemes :
    Inflorescence and all young parts densely viscid-glandular -   -   - 4. *kirkii*
    Inflorescence and all young parts, except for the ovary, glabrous -   5. *glaberrima*

1. **Cadaba aphylla** (Thunb.) Wild, comb. nov. TAB. 32 fig. B. Type from Cape Province, *Sparrman* (UPS).
     Cleome juncea Sparrm. in Nov. Act. Soc. Sci. Ups. 3 : 192 (1780) *nom. illegit.*, non Berg. (1767). Type as above.

Cleome aphylla Thunb., Prodr. Pl. Cap. **2** : 109 (1800). Type as above.
Schepperia juncea DC., Prodr. **1** : 245 (1824) *nom. illegit.*—Sond. in Harv. &
Sond., F.C. **1** : 59 (1860). Type as above.
Cadaba juncea Harv. ex Hook. f., Gen. S. Afr. Pl. ed. 2 : 13 (1868) *nom. illegit.*—
Gilg & Bened. in Engl., Bot. Jahrb. **53** : 224 (1915).—Burtt Davy, F.P.F.T. **1** :
123 (1926). Type as above.

Leafless, twiggy shrub to 2 m. tall ; branches rather virgate, stiff, smooth, green
or glaucous with subspinous apices, glabrous, young branches with subulate leaf
scales up to 2 mm. long. Flowers in short, corymbose, axillary racemes ; rhachis
0·3–2·3 cm. long, glabrous or glandular-pubescent ; bracts subulate, c. 1 mm. long ;
pedicels glabrous or glandular-pubescent, up to 1·3 cm. long. Sepals 4, yellow or
reddish-purple, 1–1·7 × 0·7–1 cm., the lowest rather larger than the remainder,
connate at the base into a very shallow receptacle c. 1 mm. in depth, broadly
elliptic, obtuse at the apex, with capitate, glandular hairs densely or sparsely scattered on both sides. Petals 0. Androgynophore 2·5–3 mm. long, glabrous,
shallowly declinate with a hooded nectary at its base c. 4 mm. broad. Stamens 8 ;
filaments 1 cm. long ; anthers 2·3 × 0·75 mm., oblong. Gynophore c. 1 cm. long,
glabrous or glandular-pubescent. Ovary narrowly cylindric, glabrous or glandular-
pubescent ; ovules numerous, on 2 placentas ; stigma capitate, sessile. Fruit up
to 8 × 0·4 cm., cylindric, subtorulose, glandular or minutely verrucose, many-
seeded. Seeds brown, c. 0·3 cm. in diam., subglobose.

Bechuanaland Prot. SE : Mochudi, fl. viii.1914, *Harbor* (PRE ; SRGH).
Also in Cape Province, Transvaal and SW Africa. Dry bushland or semi-desert
conditions.

2. **Cadaba termitaria** N.E.Br. in Hook., Ic. Pl. **26** : t. 2527 (1897).—Gilg & Bened. in
Engl., Bot. Jahrb. **53** : 228 (1915). TAB. **32** fig. A. Type : S. Rhodesia, Hartley,
Marshall (K, holotype).
Cadaba macropoda Gilg in Engl., Bot. Jahrb. **33** : 222 (1903). Type : Bechuana-
land Prot., Mahalapye, *Passarge* 48 (B, holotype).
Cadaba juncea sensu Eyles in Trans. Roy. Soc. S. Afr. **5** : 356 (1916).
Cadaba natalensis sensu Eyles, loc. cit.

Much-branched bush to c. 3 m. tall ; branches straggling or occasionally
scandent, farinose-papillose when very young, soon becoming glabrous and blackish. Leaves often crowded on short side-shoots ; lamina 0·1–3 × 0·3–1·5 cm.,
oblong, oblong-elliptic or narrowly obovate, apex rounded or emarginate, often
apiculate, rounded or broadly cuneate at the base, papillose-farinose on both sides
at first, glabrous later; petiole c. 1·5 mm. long, papillose-farinose. Flowers solitary
in the axils of the leaves towards the apices of the short side-shoots ; pedicels
c. 0·5 cm. long, papillose-farinose. Sepals greenish, c. 1·2 × 0·75 cm., free or
almost free, subequal, ovate-oblong, the upper and lower sepals longitudinally
folded, the laterals with a keeled midrib, acute at the apex, papillose-puberulous on
both sides, margins of lateral sepals pubescent. Petals 0. Androgynophore reddish, c. 3 cm. long, glabrous, shallowly declinate, with flask-shaped nectary up to
1 × 0·4 cm. adnate to its base for the greater part of the nectary's length. Stamens
5 ; filaments 0·7–0·8 cm. long, glabrous ; anthers 2·5 × 0·75 mm., oblong. Gynophore c. 1 cm. long, glandular-puberulous. Ovary narrowly cylindric, glandular-
puberulous ; ovules numerous, on two placentas ; stigma sessile, capitate. Fruit
dark green, up to 9 × 0·4 cm., cylindric, subtorulose, papillose-farinose when
young. Seeds subglobose with a rugose dark brown testa, embedded in an orange
or scarlet powdery matrix.

Bechuanaland Prot. N : Kwebe, fl. xii.1896, *Lugard* 71 (K). SE : Gaberones
Reserve, fl. viii.1945, *Miller* B 376 (PRE ; SRGH). **N. Rhodesia.** S : Mazabuka,
fl. 1.viii.1952, *Angus* 127 (FHO ; K). **S. Rhodesia.** N : Lomagundi, fl. x.1910, *Bell* in
GHS 993 (SRGH). W : Shangani, fl. 6.viii.1952, *Wild* 3844 (K ; SRGH). C : Salisbury, Hunyani R., fl. 28.viii.1927, *Eyles* 5063 (BM ; K ; SRGH). E : Hot Springs,
fl. & fr. 21.x.1948, *Chase* 1197 (BM ; K ; SRGH). S : Fort Victoria, Umshandige Dam,
fl. & fr. 9.x.1949, *Wild* 3008 (K ; SRGH). **Mozambique.** T : Boroma, Msusa, fl.
27.vii.1950, *Chase* 2827 (BM ; K ; SRGH). MS : Chibabava, Machase Rd., fl.
30.vii.1949, *Pedro & Pedrógão* 7899 (LMJ ; SRGH). SS : Inhambane, Saúte, fl.
19.v.1941, *Torre* 2688 (BM ; LISC).
Also in the Transvaal. Dry woodland or bushland in the hotter and drier areas or on
termite mounds in *Brachystegia* woodland.

Tab. 32. A.—CADABA TERMITARIA. A1, flowering branch (× ⅔) *Eyles* 5063 ; A2, flower with one sepal and 2 half sepals removed to show nectary (× 2) *Eyles* 5063. B.—CADABA APHYLLA, flower with one sepal and 2 half sepals removed to show nectary (× 2) *Reyneke* 412. C.—CADABA NATALENSIS, flower with one sepal and 2 half sepals removed to show nectary (× 2) *Gomes e Sousa* 3615. D.—CADABA KIRKII, flower with one sepal and 2 half sepals and one petal removed to show nectary (× 2) *Wild* 4184.

3. **Cadaba natalensis** Sond. in Linnaea **23**: 8 (1850); in Harv. & Sond., F.C. **1**: 59 (1860).—Gilg & Bened. in Engl., Bot. Jahrb. **53**: 228 (1915).—Burtt Davy, F.P.F.T. **1**: 123 (1926). TAB. **32** fig. C. Type from Natal.

Small shrub 1–2 m. tall with stiff, greyish-brown branches, glabrous or occasionally with a short, coarse pubescence. Leaves alternate or crowded on short lateral spurs; lamina 1·2–3·5 × 0·4–1·5 cm., elliptic or oblanceolate, apex rounded or emarginate, apiculate, cuneate at the base, glabrous on both sides or sparsely pubescent; petiole up to 5 mm. long, pubescent, at least on the upper side. Flowers solitary in the axils of the leaves but becoming racemose in appearance on the short side shoots; pedicels up to 2 cm. long, glabrous or pubescent. Sepals c. 0·9 × 0·8 cm., very broadly ovate, subacute at the apex, subequal, the upper and lower very concave, pubescent on the margins, sometimes pubescent outside; lateral sepals pubescent on the margins. Petals 0. Androgynophore c. 2 cm. long, slightly declinate, with a basal flask-shaped nectary c. 7 × 3·5 mm. having a recurved neck and dentate mouth. Stamens 5–6, filaments c. 1 cm. long, glabrous; anthers 3 × 1·5 mm., oblong. Gynophore c. 1 cm. long, glabrous or minutely puberulous. Ovary narrowly cylindric, glabrous or minutely puberulous; ovules numerous, on 2 placentas; stigma sessile, capitate. Fruit 3–4 × 0·4 cm., narrowly cylindric, subtorulose, minutely verrucose, many-seeded. Seeds c. 2·5 mm. in diam., subreniform, dark brown, with concentric rugose ridges, embedded in a bright orange or scarlet, powdery matrix.

Mozambique. SS: Macia, Muianga, fl. 11.vii.1947, *Pedro & Pedrógão* 1445 (K; LMJ; PRE; SRGH). LM: between Maputo and Porto Henrique, fl. 22.xi.1947, *Mendonça* 3502 (BM; LISC; SRGH).

Also in the Transvaal, Natal and Swaziland. Dry woodland or bush, often with *Acacia* spp.

Gomes e Sousa 3763 (COI; K; PRE) from the Maputo district of Lourenço Marques Province is noticeably pubescent compared with other Mozambique material. It can be matched with material from Natal, however, and intermediate, slightly pubescent forms exist in the Union of S. Africa, so this Mozambique pubescent form is not worthy of separation even at varietal level.

4. **Cadaba kirkii** Oliv., F.T.A. **1**: 90 (1868).—Gilg & Bened. in Engl., Bot. Jahrb. **53**: 231 (1915).—Brenan, T.T.C.L.: 111 (1949).—Milne-Redh. in Mem. N.Y.Bot. Gard. **8**, 3: 216 (1953). TAB. **32** fig. D. Syntypes: Nyasaland, Upper Shire valley, *Kirk* (K); Mozambique, Niassa, near Roangwa, *Kirk* (K).

Shrub up to 5 m. tall with the young branches covered with golden, viscid glands, older branches glabrescent, brown. Leaf-lamina 2·5–9 × 1–5 cm., ovate or broadly elliptic, obtuse at the apex, mucronate, rounded or subcordate at the base, glandular-pubescent on both sides or soon glabrescent, midrib pale and prominent below. Inflorescence of many-flowered, terminal, leafless racemes; rhachis up to 12 cm. long; pedicels up to 1·7 cm. long, with viscid, glandular hairs; bracts up to 1 cm. long subulate or linear-spathulate, glandular. Sepals yellowish or yellowish-green, c. 1·2 × 0·6 cm., free, subequal, elliptic, the upper and lower folded longitudinally, the lateral sepals with a keeled midrib, acute at the apex, densely glandular-pubescent outside, puberulous inside. Petals yellowish or cream, c. 1·7 × 0·15 cm., linear-spathulate with a claw as long as the lamina, sparsely puberulous, tapering to an acuminate apex. Androgynophore 6 mm. long, glabrous, with a basal narrowly funnel-shaped nectary 4 × 2 mm. with a dentate margin at the mouth. Stamens 5; filaments unequal, 0·5–1·2 cm. long, glabrous; anthers dark purple, 3 × 1·5 mm., oblong. Gynophore c. 6 mm. long, glandular-pubescent. Ovary cylindric, glandular-pubescent; ovules numerous, on 2 placentas; stigma subsessile, subapiculate, narrower than the ovary. Fruit c. 6 × 0·5 mm., many-seeded, narrowly cylindric, subtorulose, densely viscid. Seeds c. 3·5 mm. in diam., subglobose or irregularly compressed, with a smooth brown testa.

N. Rhodesia. N: Abercorn, fl. & fr. 2.xi.1952, *Robertson* 204 (K; PRE; SRGH). C: Feira, fl. 30.v.1952, *White* 2909 (FHO; K). S: Mazabuka Rd., 32 km. W. of Zambezi R., *Obermeyer* in Herb. Transv. Mus. 36543 (K; PRE; SRGH). **S. Rhodesia.** N: Urungwe, Msukwe R., fl. & fr. 18.xii.1953, *Wild* 4184 (K; SRGH). W: Wankie, fl. & fr. 18.vi.1934, *Eyles* 8014 (BM; K; SRGH). **Nyasaland.** N: Nyika, fl. vii.1896, *Whyte* 120 (K). C: Domira Bay, fl. 9.vii.1936, *Burtt* 6077 (BM; K).

S : Chikwawa, fl. 2.x.1946, *Brass* 17889 (K ; PRE ; SRGH). **Mozambique.** N : between Macomia and Mipande, fl. & fr. 30.ix.1948, *Barbosa* 2306 (BM ; LISC ; SRGH). Z : Morrumbala, between Aguas Quentes and Metalola, fl. & fr. 12.viii.1942, *Torre* 4554 (BM ; LISC ; SRGH). T : between Changara and R. Mazoe, fl. & fr. 4.ix.1949, *Pedro & Pedrógão* 8206 (LMJ ; SRGH).

Also in Tanganyika. In lower-altitude, drier types of woodland or in thickets in river valleys or on termite mounds in *Brachystegia* woodland.

SPECIES NOT SUFFICIENTLY KNOWN

5. **Cadaba glaberrima** Gilg & Bened. in Engl., Bot. Jahrb. **53** : 230 (1915). Type : Nyasaland, Tanganyika Plateau, fl. vii.1896, *Whyte* (K, holotype, †).

From the original description this species seems to be distinguished from *C. kirkii* by being glabrous in all parts except for the ovary which is glandular-pilose. In all other respects the description would fit that of *C. kirkii*. The fruit was not known when the species was described.

Nyasaland. N : Tanganyika Plateau, fl. vii.1896, *Whyte* (†).
Known only from the type gathering which should be in the Kew Herbarium but cannot be traced. This may be a glabrescent form of *C. kirkii* but no material of the latter species has been seen which is entirely devoid of glandular hairs on the inflorescence-rhachis and pedicels.

4. CLADOSTEMON A. Braun & Vatke

Cladostemon A. Braun & Vatke in Monatsber. Königl. Preuss. Akad. Wiss. Berl. **1876** : 866 (1877).

Shrubs or small trees with 3-foliolate, petiolate leaves with articulated petioles and petiolules ; stipules minute and inconspicuous. Flowers in rather loose, corymbose racemes. Sepals 4, free or almost free. Petals 4 with the two upper larger, unguiculate. Gynophore with an appendix near its base of 5–6 filiform segments. Stamens 5–9 adhering to the gynophore for the greater part of their length, unequal ; anthers linear, curved. Ovary 1-locular with 2 multiovulate parietal placentas ; stigma peltate, subsessile. Fruit large, globose with a longish stalk ; pericarp coriaceous. Seeds numerous, subglobose ; cotyledons large, fleshy, curved.

Cladostemon kirkii (Oliv.) Pax & Gilg in Engl., Pflanzenw. Ost-Afr. **C** : 185 (1895).— Gilg & Bened. in Engl., Bot. Jahrb. **53** : 183 (1915).—Engl., Pflanzenw. Afr. **3**, 1 : 233, t. 147 (1915).—Milne-Redh. in Kew Bull. **1948** : 449 (1949). TAB. **33** fig. A. Syntypes : Mozambique, Tete, Lupata, *Kirk* (K, †). Lectotype : Kirk's drawing of a specimen from Lupata, (K).

Euadenia? kirkii Oliv., F.T.A. **1** : 91 (1868). Type as above.

Cladostemon paxianus Gilg in Engl., Pflanzenw. Ost-Afr. **C** : 185 (1895). Type : Mozambique, Niassa Province, Mossuril, Cabaceira, *Carvalho* (COI, holotype).

Glabrous bush or small tree not more than 6 m. tall in our area ; young branches rather soft and green becoming pale brown. Leaves 3-foliolate ; leaflets 3·5–13 × 0·5–7 cm., membranous, elliptic or narrowly ovate, acute or acuminate at the apex, cuneate at the base, the lateral leaflets often asymmetric, undulate and crisped at the margin at least when young ; petiole slender, up to 12 cm. long ; petiolules up to 4 mm. long, articulate at the base. Flowers in lax, terminal or axillary, corymbose racemes ; pedicels up to 7·5 cm. long ; bracts similar to the leaves but smaller. Sepals c. 1·5 × 0·4 cm., free or almost free, lanceolate, acuminate, with a strong mid-nerve. Petals pale yellowish, yellowish-green or almost white, the two upper larger, up to 6 × 3 cm., elliptic, attenuate-acuminate at the apex, cuneate with a short claw at the base, undulate and crisped at the margins, with a distinct midrib, lower petals markedly smaller and narrower, 2–3 cm. long. Appendage diverging from the base of the gynophore, equalling or exceeding the sepals and divided into 5–6 filiform segments each terminating in an oblong appendage. Gynophore up to 12 cm. long with the 5–9 stamens adherent for $\frac{2}{3}$–$\frac{8}{9}$ of its length ; free portion of filaments up to 2·5 cm. long, slender, arising irregularly from the gynophore ; anthers c. 0·9 mm. long, linear. Ovary c. 5 × 3·5 mm., ovoid. Fruit up to 9 cm. in diam., with a tough coriaceous pericarp, pale brown, globose above, narrowing abruptly below into a stalk 10–20 cm. long, which is thicker for the first 3–4 cm. below the globose portion. Seeds numerous, 1–1·3 cm. in diam., subglobose, brown, irregularly compressed.

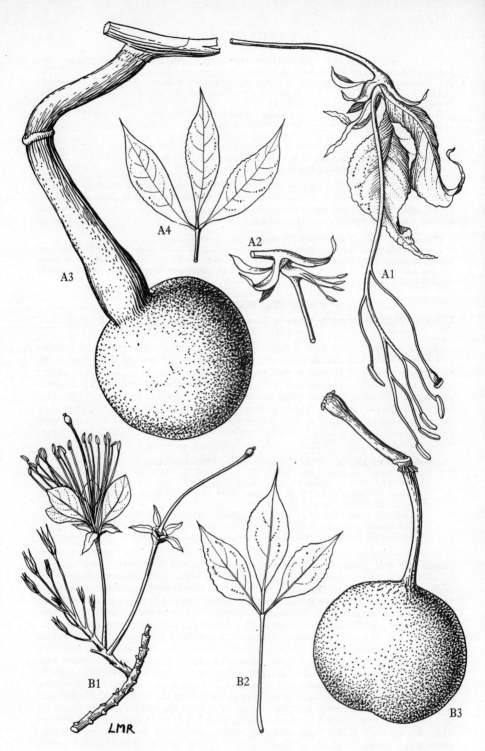

Tab. 33. A.—CLADOSTEMON KIRKII. A1, flower (× ⅔) *Kirk* s.n. ; A2, base of gynophore with appendage (× ⅔) *Kirk* s.n. ; A3, fruit (× 5) *Chase* 2212 ; A4, young leaf (× ½) *Kirk* s.n. B.—CRATEVA RELIGIOSA. B1, inflorescence (× ⅔) *Richards* 6281 ; B2, leaf (× ½) *Speke & Grant* 748 ; B3, fruit (× ⅔) *Barter* 252.

N. Rhodesia. N : Mkupa-Chiengi, fl. 10–12.x.1949, *Bullock* 1228 (K ; PRE ; SRGH). **S. Rhodesia.** N : Mtoko, Mazoe-Nyaderi R. Junction, fl. & fr. 26.ix.1955, *Lovemore* 435 (K ; SRGH). E : Chipinga, Sabi Valley, fl. xi.1956, *Mowbray* 130 (K ; SRGH). S : Sabi-Lundi Junction, Chitsa's Kraal, fr. 14.vi.1950, *Chase* 2212 (K ; SRGH). **Nyasaland.** S : Zomba, fl. 1901, *Sharpe* 180 (K). **Mozambique.** N : between Mossuril and Cabaceira, fl. 1884–5, *Carvalho* (COI). Z : between Marral and Mopeia, fl. 15.x.1941, *Torre* 3673 A (BM ; LISC). T : Lupata, fl. & fr. x.1858, *Kirk* (K). MS : Dombe, between Guza and Côa, fl. 21.x.1953, *Pedro* 4374 (K ; LMJ ; PRE). SS : R. Save, 50 km. S. of Massangena, fl. 12.x.1946, *Hornby* 2492 (K ; SRGH). LM : Maputo, Maziminhuana, fl. 22.x.1947, *Gomes e Sousa* 3637 (COI ; K ; LM ; PRE ; SRGH).

Also in Kenya, Tanganyika and Natal (Zululand). Low-altitude, dry, open woodland or bush.

Apart from the one record from N. Rhodesia, near the shore of Lake Mweru, it is very much a species of the East African coastal plain. A very handsome species with sweet-scented flowers.

5. THYLACIUM Lour.

Thylacium Lour., Fl. Cochinch. : 342 (1790) (" Thilachium ").

Shrubs or small trees with alternate, simple or 3-foliolate leaves. Flowers in corymbose racemes, terminal or on short axillary shoots. Calyx entire in bud and closed at the apex, splitting transversely on expansion, the upper portion falling off. Petals 0. Stamens many on a columnar androphore. Ovary on a long gynophore with 4–10 multiovulate placentas, 1-locular or partially 4–10-locular by the intrusion of spurious dissepiments from the placentas ; stigma sessile. Fruit baccate. Seeds numerous, rather large.

In the original description the genus is printed as *Thilachium* and in the description of the species as *Thilakium* but as the author states that he is basing the name on the shape of the fruit (Θυλακος—a bag or pouch) the name is correctly latinized as *Thylacium*.

Thylacium africanum Lour., Fl. Cochinch.: 342 (1790) ("Thilakium").—Oliv., F.T.A., **1** : 82 (1868).—Sim, For. Fl. Port. E. Afr.: 9 (1909).—Gilg & Bened. in Engl., Bot. Jahrb. **53** : 263 (1915).—Brenan, T.T.C.L.: 120 (1949).—Milne-Redh. in Mem. N.Y.Bot. Gard. **8**, 3 : 218 (1953). TAB. **34** fig. B. Type : Mozambique, *Loureiro* (LISU † ; P).

Thylacium ovalifolium Juss. in Ann. Mus. Nation. Hist. Nat. Par. **12** : 71 (1808) *nom. illegit.* Type as above.

Thylacium querimbense Klotzsch in Peters, Reise Mossamb. Bot. **1** : 163 (1861). Type : Mozambique, Niassa, Querimba I., *Peters* (B, holotype).

Thylacium verrucosum Klotzsch, tom. cit.: 164 (1861).—Bak. f. in Journ. Linn. Soc., Bot. **40** : 21 (1911).—Eyles in Trans. Roy. Soc. S. Afr. **5** : 356 (1916). Type : Mozambique, Sena, Tambare, *Peters* (B, holotype).

Bush or small tree up to 5 m. tall with the young branches pale brown or grey and often punctate-scabrous. Leaves 3-foliolate or some simple ; leaflets 3–10 × 1–6·5 cm., elliptic, narrowly lanceolate, oblanceolate or broadly oblong, obtuse and mucronate at the apex, rounded or subcuneate at the base, margin somewhat revolute, glabrous on both sides ; petiole up to 8 cm. long, scabrous ; petiolules up to 6 mm. long, verruculose or smooth, articulated at their bases like the petiole. Inflorescence of few-flowered, terminal or axillary, corymbose racemes ; pedicels scabrid-puberulous, up to 2·5 cm. long ; bracts usually simple, similar to the leaves but smaller. Calyx c. 2 × 1·7 cm., greenish, obovoid or turbiniform before dehiscence, with an apiculus at the apex, glabrous with 12–16 longitudinal nerves, circumscissile at the broadest part, persistent base campanulate. Androgynophore stout, glabrous, c. 4 mm. long. Stamens up to 4·5 cm. long, numerous, free or slightly connate and broadened at the base ; filaments white ; anther-thecae 2·5–3 mm. long, linear-oblong, slightly divergent at the base. Ovary on a glabrous gynophore up to 3·5 cm. long, cylindric, longitudinally 5–10-ribbed, stigma sessile, capitate. Fruit 5–6 × 2·5–3 cm., cylindric-oblong, blunt at the apex, longitudinally 8–10-ribbed, indehiscent with a tough pericarp. Seeds numerous, up to 8 mm. in diam., subglobose or somewhat compressed, pale brown, verrucose.

S. Rhodesia. E : Sabi valley, Hot Springs, fl. 18.viii.1947, *McGregor* 39/47 (SRGH). S : Sabi valley, near Birchenough Bridge, fl. 22.x.1948, *Chase* 940 (K ; SRGH). **Nyasaland.** S : Chikwawa, fl. 2.x.1946, *Brass* 17891 (K ; PRE ; SRGH). **Mozambique.**

N : Praia do Lumbo, fr. 3.x.1936, *Torre* 885 (COI ; LISC). Z : Mopeia, fl. 12.ix.1944, *Mendonça* 2044 (BM ; LISC). T : between Mandié and Changara, fl. 31.viii.1949, *Pedro & Pedrógão* 8169 (LMJ ; SRGH). MS : between Maringué and Chimoio, fl. 1.x.1949, *Pedro & Pedrógão* 8454 (LMJ ; SRGH). SS : Inhambane, Mavume, fl. x.1938, *Gomes e Sousa* 2172 (K). LM : between Maputo R. and Porto Henrique, fr. 22.xi.1947, *Mendonça* 3506 (BM ; LISC).

Also in Kenya, Tanganyika and Madagascar. In *Acacia* or *Colophospermum mopane* woodland at low altitudes and in coastal thickets or on termite mounds. It is confined to the eastern part of our area and enters S. Rhodesia only along the lower reaches of the Sabi, Limpopo and probably the Zambezi valleys.

6. COURBONIA Brongn.

Courbonia Brongn. in Bull. Soc. Bot. Fr. **7** : 901 (1860).

Small shrubs or occasionally perennial herbs. Leaves simple, entire, glaucous. Flowers borne singly in the leaf axils. Sepals 3, rarely 2 or 4, valvate. Petals 0. Disk with a short, toothed, free margin, adnate to the mouth of the receptacle. Stamens numerous, borne on a columnar androgynophore equalling or exceeding the receptacle ; filaments free ; anthers ovoid or oblong, apiculate. Ovary fusiform, on a long gynophore, 2-locular (or rarely 3-locular) by fusion of the placentas ; ovules 2 on each placenta, parietal or inserted on the spurious dissepiment. Fruit globose, coriaceous, indehiscent, slightly fleshy inside, 1–4-seeded. Seeds large, without endosperm, cotyledons very thick ; radicle deeply included.

Courbonia glauca (Klotzsch) Gilg & Bened. in Engl., Bot. Jahrb. **53** : 221 (1915).—Milne-Redh. in Mem. N.Y.Bot. Gard. **8**, 3 : 217 (1953). TAB. **34** fig. A. Type : Mozambique, Sena, *Peters* (BM, holotype).
Physanthemum glaucum Klotzsch in Peters, Reise Mossamb. Bot. **1** : 167, t. 29 (1861). Type as above.
Courbonia decumbens sensu Oliv., F.T.A. **1** : 88 (1868).—Sim, For. Fl. Port. E. Afr.: 10 (1909).—Bak. f. in Journ. Linn. Soc., Bot. **40** : 21 (1911).—Eyles in Trans. Roy. Soc. S. Afr. **5** : 356 (1916).

Small, glabrous, glaucous shrub up to 2 m. tall but usually c. 1 m. tall, with spreading branches from ground level or branching somewhat above the base ; branches smooth and green. Leaf-lamina 1·5–5·5 ×0·75–5·5 cm., broadly elliptic, ovate or rotund, apex obtuse or subacute, mucronate, rounded or slightly cordate at the base, 3–5-nerved from near the base ; petiole 3–6 mm. long, slender. Flowers axillary, solitary, towards the ends of the branches on pedicels 1–2·5 cm. long. Sepals yellowish-green to whitish-green, c. 2 × 1·5 cm., narrowly ovate to ovate, acute at the apex, with three principal nerves ; receptacle c. 2 mm. long ; disk sinuate-dentate at the margin, 1–2 mm. long. Stamens numerous on a short columnar androgynophore 2–3 mm. long ; filaments c. 2 cm. long, slender ; anthers oblong, apiculate. Ovary 2- or rarely 3-locular, fusiform, glabrous, on a long slender gynophore 1·7–3 cm. long ; style 0 ; stigma capitate. Fruit c. 2 cm. in diam., globose, glaucous-green, bluntly apiculate at the apex, with a coriaceous pericarp and somewhat fleshy inside, borne on a usually recurved gynophore which thickens in fruit. Seeds solitary or occasionally 2–3-together, up to 1·8 ×0·8 cm., thick and discoid, fleshy with a papery testa, with two large cortyledons and a deeply embedded radicle.

N. Rhodesia. N : Lake Tanganyika, Sumbu valley, fl. 24.vi.1933, *Michelmore* 445 (K). S : Gwembe valley, fr. 1932, *Macrae* 9 (K). **S. Rhodesia.** N : Urungwe, fr. 24.ix.1954, *Lovemore* 404 (K ; PRE ; SRGH). W : Wankie, Sebungwe R., fl. 11.v.1955, *Plowes* 1830 (K ; PRE ; SRGH). E : Chipinga Distr., Sabi valley, fr. 27.vi.1955, *Mowbray* 40 (SRGH). S : Nuanetsi, fl. v.1953. *Davies* 592 (K ; SRGH). **Nyasaland.** N : Nyika plateau, fr. vi.1896, *Whyte* (K). S : Chikwawa, fl. 2.x.1946, *Brass* 17890 (K ; PRE ; SRGH). **Mozambique.** N : Porto Amelia, fl. & fr. 9.x.1948, *Barbosa* 2028 (BM ; LISC). T : between Panhame and Maluvira, fr. 18.ix.1949, *Pedro & Pedrógão* 8313 (LMJ ; SRGH). MS : between Mungari and Mandiè, fl. *Pedro & Pedrógão* 8155 (LMJ ; SRGH). SS : Limpopo-Nuanetsi Junction, fr. vii.1932, *Smuts* 316 (K ; PRE). LM : Sabié and Magude, fl. 3.v.1944, *Torre* 6553 (BM ; LISC).

Also in the Transvaal, Zululand, Kenya, Uganda, Tanganyika and the Belgian Congo. In low-altitude, low-rainfall woodlands, often associated with *Colophospermum mopane*.

The fruit is said to be edible although some tribes in the Zambezi valley consider it necessary to cook it and throw away the water before eating and even then only eat it

Tab. 34. A.—COURBONIA GLAUCA. A1, flowering and fruiting branch (× ⅔) *Rodin* 4487 ;
A2, flower (× 3) *Rodin* 4487. B.—THYLACIUM AFRICANUM. B1, leaf (× ½) *Chase*
940 ; B2, flower (× ⅔) *Chase* 940 ; B3, fruit (× ⅔) *Davies* 1664.

when food is short. This species is near *C. decumbens* Brongn. of Ethiopia and Somaliland but the latter has a longer receptacle (3–4 mm.) and yellowish rather than greenish stems ; the leaves are also cuneate at the base.

7. MAERUA Forsk.

Maerua Forsk., Fl. Aegypt.-Arab. : CXIII, 104 (1775).

Shrublets, shrubs, small trees or climbers without spines. Leaves alternate or crowded on abbreviated side shoots, simple or 3 (5)-foliolate, sessile or petiolate. Flowers solitary or fasciculate in the leaf axils or in terminal or axillary racemes or panicles. Flowers bisexual. Sepals (3) 4 ; receptacle cylindric or funnel-shaped, its inner margin often produced into a disc with an entire, undulate, dentate or fimbriate margin. Petals 0 or (3) 4 (8), inserted at the mouth of the receptacle. Androgynophore equalling or exceeding the receptacle. Stamens usually indefinite and exserted ; anthers basifixed and with longitudinal dehiscence. Ovary borne on an elongated gynophore ; ovary globose, ovoid, ellipsoid or cylindric with 1–2 loculi, ovules 4–∞ ; stigma sessile or nearly so. Fruit globose, ovoid, ellipsoid, oblong-ellipsoid or cylindric, sometimes torulose, smooth or verrucose. Seeds subglobose or oblong-ellipsoid, smooth or rugose.

Leaves 3–5-foliolate or 3-foliolate and simple on the same branch, if simple then petiole
    always jointed at the base :
  Leaflets linear or lanceolate-linear :
    Sepals c. 5 × 3·5 mm. ; leaflets with revolute margins -   -   - 1. *rosmarinoides*
    Sepals c. 10 × 5 mm. ; leaflets not or hardly revolute -   -   - 2. *grantii*
  Leaflets elliptic, oblong, lanceolate, ovate or obovate :
    Branches persistently smooth, striate and usually persistently green :
      Ovary and fruit globose, ovules c. 10 - - - - - 3. *cerasicarpa*
      Ovary and fruit oblong-cylindric, ovules c. 20 or more :
        Inner margin of receptacle slightly thickened but not undulate or dentate
                                           6. *scandens*
      Inner margin of receptacle undulate-dentate :
        Leaflets oblong-elliptic, elliptic or lanceolate, apex rounded or subacute, often
          rather glaucous :
          Fruit smooth -   -   -   -   -   - 4. *juncea* subsp. *juncea*
          Fruit rough -   -   -   -   -   - 4. *juncea* subsp. *crustata*
        Leaflets narrowly ovate to ovate, apex acuminate, not glaucous 5. *acuminata*
    Branches not persistently smooth and striate, soon corticate and often densely
      lenticelled :
      Ovary and fruit ovoid or globose :
        Leaves glabrous - - - - - - - - - 19. *bussei*
        Leaves lanate or densely yellowish-hairy at least when young :
          Sepals c. 2 cm. long ; indumentum of stems and leaves etc., coarse and
          yellowish -   -   -   -   -   -   -   - 7. *schliebenii*
          Sepals c. 1 cm. long ; indumentum of stems and leaves lanate and greyish
                                           8. *prittwitzii*
      Ovary and fruit oblong-cylindric, torulose or oblong-ellipsoid :
        Petals present :
          Fruit oblong-ellipsoid, smooth -   -   -   -   - 9. *nervosa*
          Fruit oblong-cylindric with 4 longitudinal ridges or torulose :
            Leaves usually simple, shining above, glandular-puberulous below or
              glabrous ; petals elliptic - - - - - - 11. *friesii*
            Leaves usually 3-foliolate, dull, pubescent or glabrous ; petals orbicular,
              glabrous - - - - - - - - 12. *pubescens*
        Petals absent :
          Shrublet flowering before the leaves -   -   -   - 13. *pygmaea*
          Shrubs, small trees, or climbers flowering with the leaves :
            Climber with membranous 3–5-foliolate leaves ; flowers in leafless panicles
                                        14. *paniculata*
            Shrubs or small trees, leaves subcoriaceous, 3-foliolate, flowers in axillary,
              subcorymbose racemes -   -   -   -   - 10. *cafra*
Leaves always simple ; petiole rarely jointed at the base :
  Leaves fasciculate, small, usually less than 2 cm. long :
    Petioles glabrous -   -   -   -   -   -   -   - 16. *buxifolia*
    Petioles variously hairy :
      Leaves glabrous and smooth on both sides ; sepals 1·2–1·4 cm. long 17. *endlichii*
      Leaves pubescent, puberulous or papillose on both sides ; sepals 7 mm. long
                                          15. *parvifolia*

Leaves not fasciculate ; adult leaves moderate-sized or large :
  Ovary and fruit globose, smooth :
    Inflorescence terminal, corymbose ; bracts all filamentous or subulate :
      Leaves scabrous-pubescent - - - - - - - - 18. *kirkii*
      Leaves glabrous - - - - - - - - - - 19. *bussei*
    Inflorescences racemose or flowers in the axils of the upper leaves ; bracts similar
      to the leaves - - - - - - - - - 20. *andradae*
  Ovary and fruit cylindric, ovoid or oblong-ovoid, torulose, verrucose or smooth :
    Leaves narrowly elliptic ; fruit verrucose - - - - 22. *salicifolia*
    Leaves narrowly ovate, ovate, oblong, elliptic or lanceolate :
      Fruit torulose - - - - - - - - - 23. *angolensis*
      Fruit not torulose :
        Branches glabrous, brown ; racemes up to 20 cm. long - 21. *brunnescens*
        Branches puberulous, blackish ; racemes up to 4 cm. long - - 11. *friesii*

1. **Maerua rosmarinoides** (Sond.) Gilg & Bened. in Engl., Bot. Jahrb. **53** : 240 (1915).
—Burtt Davy, F.P.F.T. **1** : 122 (1926). Type from Natal (Durban).
  Niebuhria rosmarinoides Sond. in Linnaea, **23** : 7 (1850) ; in Harv. & Sond., F.C.
1 : 60 (1860). Type as above.

Much-branched shrub up to 2·6 m. tall, quite glabrous except for the sepals and petioles ; branchlets slender, finely striate, greenish becoming grey-brown. Leaves (1) 3 (5)-foliolate, petiolate ; leaflets 1·5–6·5 × 0·2–0·25 cm., laterals somewhat shorter, linear, apex mucronate, base narrowly rounded, margin somewhat revolute, paler beneath ; petiole up to 2·5 cm. long ; petiolules 1–2 mm. long, channelled and papillose above. Inflorescences of few-flowered, terminal, corymbose racemes on short side branches ; pedicels up to 1·5 cm. long, slender ; bracts c. 1 mm. long, trifid and subulate or the lower foliose but usually simple. Receptacle c. 2·5 × 1·75 mm., cylindric, glabrous ; sepals 5–6 × 3–3·5 mm., oblong, blunt at the apex, margin papillose. Petals 2·5–3 × 1·5 mm., elliptic, blunt at the apex, very shortly clawed at the base. Androgynophore slightly longer than the receptacle. Stamens 10–16, c. 1·8 cm. long, sometimes some sterile with shorter filaments ; anthers 1·2 × 0·9 mm., ovate. Ovary on a gynophore up to 2·2 cm. long, oblong ; ovules c. 20 on 2 placentas ; stigma sessile, capitate. Fruit up to 2 × 0·9 cm., oblong-cylindric, sometimes slightly torulose, minutely verrucose. Seeds several, c. 7 mm. in diam., compressed-globose, dark brown, testa rugose-verrucose.

Mozambique. LM : Namaacha, fl. 22.xii.1944, *Torre* 6933 (BM ; LISC).
Also in Natal and the Transvaal. In low-altitude and coastal woodland and bushland.

2. **Maerua grantii** Oliv., F.T.A. **1** : 84 (1868).—Gilg & Bened. in Engl., Bot. Jahrb. **53** :
240 (1915).—Brenan, T.T.C.L. : 116 (1949). Type from Tanganyika.

Much-branched, glabrous shrub up to 2·6 m. tall with slender, virgate, striate, greenish branchlets. Leaves 3-foliolate or simple above ; leaflets 2·5–7 × 0·1–1 cm., the lateral leaflets smaller, linear, narrowed to the acuminate apex and cuneate at the base, margin plane or slightly revolute ; petiole up to 2·5 cm. long ; petiolules up to 5 mm. long, papillose on the upper side. Inflorescence of few-flowered, lax racemes at the ends of the branches or on short lateral shoots or solitary and axillary ; pedicels up to 5 cm. long, slender. Receptacle c. 5 × 1·75 mm., glabrous, cylindric but somewhat four-sided, with 2–3 very short, irregular or subulate lobes at its mouth and between the petals ; sepals c. 1 × 0·5 cm., oblong-elliptic, blunt at the apex, minutely woolly at the margins. Petals whitish, 5 × 2 mm., narrowly obovate, acute at the apex, very shortly clawed at the base. Androgynophore slightly longer than the receptacle. Stamens c. 35 ; filaments c. 2 cm. long ; anthers 1·5 × 0·75 mm., oblong-elliptic. Ovary on a gynophore up to 2–3 cm. long, ovoid ; ovules 10–12 on 2 placentas ; stigma subsessile, capitate. Fruit up to 1·75 cm. in diam., globose, minutely verrucose. Seeds 2–3.

Mozambique. N : Macondes Distr. between Muêda and Nairoto, fl. & fr. 17.ix.1948.
Pedro & Pedrógão 5266 (EA ; LMJ ; SRGH).
Also in Kenya and Tanganyika. In dry woodland or bushland or in transitional areas in *Brachystegia* woodland or in grassland.
The root is reported to be used as an abortifacient.

3. **Maerua cerasicarpa** Gilg in Engl., Bot. Jahrb. **33** : 227 (1903).—Gilg & Bened. in
Engl., Bot. Jahrb. **53** : 244 (1915). TAB. **35** fig. B. Type from Tanganyika.

Maerua sphaerogyna Gilg & Bened. in Engl., Bot. Jahrb. **53**: 244 (1915).—
Brenan, T.T.C.L.: 117 (1949).—Hauman & Wilczek, F.C.B. **2**: 492 (1951).
Type from Belgian Congo (Lake Kivu).

Shrub up to 7 m. tall, glabrous except for the sepal margins, with green, striate branches. Leaves 3-foliolate; leaflets 1·5–4 × 1–2·4 cm., broadly elliptic, ovate or oblong-lanceolate, apex subacute, mucronulate, base broadly cuneate or rounded, nerves slightly prominent below, veins laxly reticulate; petiole up to 3 cm. long, striate; petiolules up to 6 mm. long, channelled and papillose-verrucose above. Inflorescence of lax, few-flowered racemes, terminal or axillary; pedicels up to 1·7 cm. long; bracts up to 1·5 mm. long, subulate, trifid. Receptacle 3–4 × 1·5 mm., cylindric, scarcely ribbed, with subulate teeth at the mouth within; sepals greenish, 6–8 × 3–5 mm., oblong, apex rounded and acuminate, margins densely white-puberulous. Petals 4·5 × 2 mm., broadly elliptic, acute at the apex, shortly clawed at the base. Androgynophore distinctly longer than the receptacle. Stamens c. 25, filaments up to 2 cm. long; anthers c. 2·3 × 1 mm. Ovary on a gynophore up to 2·2 cm. long, ovoid; ovules c. 10 on 2 placentas; stigma subsessile, capitate. Fruit up to 1·75 cm. in diam., globose, shining, almost smooth. Seeds several, c. 6 mm. in diam., subglobose.

N. Rhodesia. N: Kalambo Falls, fl. & fr. 2.xi.1952, *Robertson* 203 (K; PRE; SRGH). **Mozambique.** N: Mecufi, fl. & fr. 11.x.1948, *Barbosa* 2350 (BM; LISC). Also in Tanganyika and the Belgian Congo. In thickets or on termite mounds.

4. **Maerua juncea** Pax in Engl., Bot. Jahrb. **14**: 302 (1891).—Gilg & Bened. in Engl., Bot. Jahrb. **53**: 240 (1915). Type from Tanganyika.
    Maerua nervosa var. *flagellaris* Oliv., F.T.A. **1**: 84 (1868).—Gibbs in Journ. Linn. Soc., Bot. **37**: 429 (1906).—Eyles in Trans. Roy. Soc., S. Afr. **5**: 356 (1916). Syntypes: Nyasaland, R. Shire, near Longane, *Kirk* (K); Papinbeji, shore of Lake Nyasa, *Kirk* (K).
    Maerua flagellaris (Oliv.) Gilg & Bened. in Engl., Bot. Jahrb. **53**: 244 (1915).—
Brenan, T.T.C.L.: 116 (1949).—Milne-Redh. in Mem. N.Y.Bot. Gard. **8,** 3: 217 (1953). Syntypes as above.

Small shrub or climber up to 15 m. tall, glabrous except for the sepal margins and petiolules; young branches green and striate. Leaves simple or 3-foliolate; leaflets 1–4·5 × 0·4–2 cm., oblong-elliptic, elliptic or lanceolate, apex rounded or subacute, mucronate, broadly cuneate or rounded at the base, often rather glaucous, nerves in 5–7 pairs, not conspicuous; petiole up to 2 cm. long; petiolules up to 5 mm. long, minutely puberulous on the upper side. Inflorescence of 2–6-flowered lax racemes at the ends of the branches or solitary in the axils of the upper leaves; pedicels up to 4·5 cm. long; bracts trifid, setiform, caducous or like the leaves but usually simple and smaller. Receptacle c. 5 mm. long and 4 mm. wide at the mouth with an undulate dentate disc at the mouth within, funnel-shaped; sepals 0·9–1·6 × 0·4–0·8 cm., narrowly obovate-oblong to oblong-elliptic, rounded at the apex and shortly mucronate, margin densely puberulous. Petals white, 5–8 × 3·2–6·5 mm., broadly elliptic to orbicular or rotund, rounded at the apex or acute, shortly clawed at the base. Androgynophore the same length as the receptacle. Stamens 20–30, on filaments up to 2·5 cm. long; anthers 2·5 × 1 mm., oblong. Ovary on a gynophore up to 2·3 cm. long, oblong-cylindric; ovules c. 36 on 2 placentas; stigma sessile, capitate. Fruit up to 3 × 2·5 cm., ellipsoid, smooth or roughened. Seeds 10–20, up to 8 mm. in diam., subglobose, testa rugose.

Subsp. **juncea**
    Maerua maschonica Gilg in Engl., Bot. Jahrb. **53**: 240 (1915) pro parte excl. specim. *Galpin* 1063 et 7455.—Eyles in Trans. Roy. Soc. S. Afr. **5**: 356 (1916).—
O. B. Mill. in Journ. S. Afr. Bot. **18**: 16 (1952).
    Maerua caffra sensu Eyles, loc. cit.
    Capparis sp. sensu Steedman, Trees etc. S. Rhod.: 9 (1933).

Petals rotund, ratio of length to breadth 1 : 0·73–0·96. Fruit always smooth.

Bechuanaland Prot. N: Chobe R., Serondela, fr. xi.1947, *Miller* B 526 (PRE). **N. Rhodesia.** C: Chilanga, fl. 10.x.1909, *Rogers* 8550 (K; SRGH). S: Kalomo, fl. vi.1920, *Rogers* 26010 (K; PRE). **S. Rhodesia.** N: Urungwe, fl. 4.viii.1956, *Goodier* 88 (SRGH). W: Matopos, fl. & fr. 5.x.1952, *Plowes* 1488 (SRGH). C: Gwelo, fl. & fr. 9.x.1924, *Eyles* in GHS 1300 (K; SRGH). E: Umtali, fl. 19.ix.1954, *Chase* 5286

(BM ; K ; PRE ; SRGH). **Nyasaland.** N : Mwanembe Mt., fr. ix.1902, *McClounie* 166 (K). C : Kasungu, fl. 25.viii.1946, *Brass* 17420 (BM ; K ; SRGH). S : Chikwawa, fl. 3.x.1946, *Brass* 17911 (K ; PRE ; SRGH). **Mozambique.** N : Mecufi, fl. 11.x.1948, *Barbosa* 2350 (BM ; LISC ; SRGH). T : between Tete and Changara, fr. 30.x.1944, *Torre* 3714 (BM ; LISC). MS : between Sena and Sangadze, fr. 9.x.1949, *Pedro & Pedrógão* 8558 (LMJ ; SRGH). SS : Inharrime, fr. 16.ix.1948, *Myre & Carvalho* 221 (LM ; SRGH).

Also in Tanganyika and the Belgian Congo. Found in the drier types of low-altitude open woodland and often associated with *Acacia* spp. but also occurring on termitaria at higher altitudes in *Brachystegia* country.

Subsp. **crustata** (Wild) Wild, comb. nov. Type : Transvaal, Pretoria, *Codd* 6227 (K, holotype ; PRE ; SRGH).
    Maerua maschonica Gilg in Engl., Bot. Jahrb. **53** : 240 (1915) pro parte quoad specim. Galpin.—Burtt Davy, F.P.F.T. **1** : 122 (1926) pro parte excl. specim. Engler.
    Maerua flagellaris subsp. *crustata* Wild in Bol. Soc. Brot., Sér. 2, **32** : 48 (1958). Type as above.
    Maerua nervosa sensu Bak. f. in Journ. Linn. Soc., Bot. **40** : 20 (1911).
    Maerua flagellaris sensu O. B. Mill. in Journ. S. Afr. Bot. **18** : 16 (1952) pro parte quoad specim. Lugard. " flagellaria ".

Petals broadly elliptic, ratio of length to breadth 1 : 0·43–0·81. Fruit with a very rough pericarp.

Bechuanaland Prot. N : Kwebe Hills, fl. & fr. 14.xii.1898, *Lugard* 135a (K). SE : Mochudi, fl. & fr. i.1915, *Harbor* in Herb. Transv. Mus. 17021 (PRE). **S. Rhodesia.** E : Hot Springs, fr. 22.x.1948, *Chase* 1489 (BM ; K ; SRGH). S : Ndanga, Mtilikwe R., fr. 11.x.1951, *Seymour-Hall* 41/51 (K ; SRGH). **Mozambique.** SS : between Caniçado and Xirunzú, fr. 12.xii.1940, *Torre* 2366 (BM ; LISC ; SRGH). LM : Maputo, Porto Henrique, fl. & fr. *Barbosa* 640 (BM ; LISC).

Also in the Transvaal and Zululand.

The ecology and habit of this subspecies are similar to those of subsp. *juncea*. It is readily distinguished in fruit, but in flower only the petals offer a diagnostic feature of any value and even here there is an overlap. The areas of distribution of the two can be divided by a line running through the more northerly part of SE. Bechuanaland, the Southern Division of S. Rhodesia and the Sul do Save Province of Mozambique with subsp. *juncea* to the north and subsp. *crustata* to the south.

5. **Maerua acuminata** Oliv., F.T.A. **1** : 85 (1868).—Gilg & Bened. in Engl., Bot. Jahrb. **53** : 246 (1915).—Brenan, T.T.C.L. : 117 (1949). Type : Tanganyika (or Mozambique?), R. Rovuma, *Kirk* (K, holotype).

Small shrub, glabrous except for the sepal margins ; branches slender, finely striate. Leaves 3-foliolate ; leaflets 4–6·4 × 1·5–3 cm., lateral leaflets somewhat smaller, narrowly ovate to ovate, apex acuminate, mucronate, base rounded or obtuse, nerves in 6–7 pairs, not prominent ; petiole up to 3·3 cm. long ; petiolules up to 5·5 mm. long. Inflorescence of axillary, corymbose racemes collected towards the ends of the branches ; peduncles 1–2 cm. long, slender ; pedicels up to 2 cm. long, slender ; bracts trifid, c. 1 mm. long, subulate. Receptacle 3·5 mm. long, 1·5 mm. wide at the mouth, cylindric, with a denticulate-undulate rim 0·5 mm. wide at its lip within ; sepals c. 8 × 3·5 mm., oblong-elliptic, rounded at the apex, mucronulate, densely papillose at the margin. Petals 0. Androgynophore the same length as the receptacle. Stamens 25–30 ; anthers 1·5 × 0·75 mm., oblong. Ovary on a gynophore c. 1·75 cm. long, oblong-cylindric ; ovules c. 20 on 2 placentas ; stigma sessile, capitate. Fruit not so far known.

Mozambique. N : R. Rovuma, fl. iii.1861, *Kirk* (K).

Known so far only from the type collection which may have been collected on either the Tanganyika or Mozambique side of the R. Rovuma. Ecology unknown.

6. **Maerua scandens** (Klotzsch) Gilg in Engl., Bot. Jahrb. **33** : 223 (1902).—Gilg & Bened., op. cit. **53** : 246 (1915). Type : Mozambique, Boror, *Peters* (B, holotype).
    Streblocarpus scandens Klotzsch in Peters, Reise Mossamb. Bot. **1** : 165 (1861). Type as above.
    Maerua nervosa (Hochst.) Oliv., F.T.A. **1** : 84 (1868) pro parte quoad specim. Peters ex Boror.

Small shrub or climber, glabrous except for the sepal margins ; branches green, smooth and longitudinally striate. Leaves 3-foliolate ; leaflets 2–5 × 0·7–2 cm., lanceolate to ovate, attenuate to a mucronate and sometimes recurved apex, base broadly cuneate or narrowly rounded ; lateral nerves in 5–7 pairs, prominent below, looped within the margin, veins laxly reticulate ; petiole c. 2·5 cm. long ; petiolules c. 2 mm. long. Inflorescence of 2–5-flowered, lax, terminal racemes, often borne on short side branches ; pedicels up to 2 cm. long ; bracts trifid, up to 2 mm. long, subulate. Receptacle 3 mm. long, 2·5–3 mm. in diam. at the mouth, funnel-shaped ; sepals 1·2–1·7 × 0·35–0·7 cm., oblong-elliptic, rounded at the minutely mucronulate apex, margins densely puberulous. Petals whitish, 7·5 × 2·5–4 mm., elliptic to obovate, blunt or acute at the apex, shortly clawed at the base. Androgynophore as long as the receptacle. Stamens 40–50 ; filaments up to 2·5 cm. long. Ovary on a gynophore up to 2·5 cm. long, oblong-cylindric ; ovules 36–40 on 2 placentas ; stigma sessile, capitate. Fruit not seen.

Mozambique. Z : Boror, fl. *Peters* (B). SS : Chibuto, fl. 16.vii.1944, *Torre* 6781 (BM ; LISC ; SRGH).

Known only from Mozambique. In dense coastal *Brachystegia* woodland, apparently rather rare.

7. **Maerua schliebenii** Gilg-Bened. in Notizbl. Bot. Gart. Berl. **13** : 273 (1938).— Brenan, T.T.C.L. : 117 (1949). Type from Tanganyika (Lindi Distr.).

Small shrub 1–2 m. tall or with sarmentose branches to 8 m. long ; young branches with long, dense, yellowish hairs, older branches glabrescent and pale brown. Leaves 3-foliolate or rarely simple above ; leaflets 1–9·5 × 0·5–5 cm., lateral leaflets smaller, elliptic or broadly elliptic, blunt or subacute at the apex, mucronulate, broadly cuneate at the base, glabrescent above, densely yellowish-hairy below or later glabrescent ; petiole up to 5 cm. long, hairy ; petiolules 2–4 mm. long, densely hairy. Flowers greenish, solitary in the upper axils on densely pilose pedicels up to 4 cm. long. Receptacle 6–10 × 2 mm. densely pilose outside, with c. 10 longitudinal ribs ; sepals up to 2·2 × 0·75 cm., lanceolate, acute or acuminate at the apex, pubescent on both sides, more densely so outside, margin very shortly and densely lanate. Petals up to 7·5 × 5 mm., obovate, ovate or broadly elliptic, apex blunt or acuminate, shortly clawed at the base, margin somewhat undulate, Androgynophore slightly longer than the receptacle. Stamens c. 40 ; filaments up to 4 cm. long, slender ; anthers 2 × 1·2 mm., narrowly ovate. Ovary on a gynophore up to 4 cm. long, ovoid, densely grey-tomentose ; ovules c. 30 on 2 placentas. Fruit so far unknown.

Mozambique. N : between Nampula and Mecuburi, fl. 21.xi.1936, *Torre* 1085 (COI ; LISC).

Also in Tanganyika (Lindi Distr.). In coastal woodland, apparently rare.

8. **Maerua prittwitzii** Gilg & Bened. in Engl., Bot. Jahrb. **53** : 254 (1915). TAB. **35** fig. D. Type from Tanganyika.
    Maerua rhodesiana Wild in Bol. Soc. Brot., Sér. 2, **32** : 44 (1958). Type : S. Rhodesia, Urungwe, *Wild* 4188 (K, holotype ; PRE ; SRGH).

Shrub or small tree up to 6 m. tall with erect or occasionally sarmentose branches which are grey-lanate when young. Leaves 3-foliolate or the upper ones simple ; leaflets 2–12 × 1–6 cm., lateral leaflets about ½ the median leaflet in size, lanceolate, ovate or broadly elliptic, apex obtuse, retuse or subacute, mucronulate, base cuneate or obtuse, rugose and glabrescent above with impressed nerves and veins, greyish and shortly lanate below with prominent nerves and veins ; petiole up to 4 cm. long. Inflorescence of short, dense, terminal racemes ; pedicels pilose, up to 2 cm. long ; bracts 3 mm. long, subulate, pilose, or the lower ones like the leaves but always simple and smaller. Receptacle cylindric, pilose outside, slightly widened at the mouth, 2·75 mm. long, 2·5 mm. in diam. at the mouth ; sepals 10 × 4 mm., oblong-ovate, apex acute, pilose outside, margin lanate. Petals white, 7 × 3·5 mm., obovate, rounded at the apex, shortly clawed at the base. Androgynophore c. 3·5 mm. long, shortly exserted from the receptacle. Stamens c. 30 ; filaments c. 2·2 cm. long ; anthers 1·5 × 0·75 mm., oblong. Ovary on a gynophore c. 2·5 cm. long, ovoid, glabrous at first, soon becoming densely tomentose ; ovules 14–16 on 2 placentas ; stigma sessile, capitate. Fruit c. 2·2 × 2 cm., ovoid, strongly verrucose, densely and shortly tomentose ; exocarp fleshy but becoming fibrous, c. 4 mm. thick. Seeds 1–2, c. 7 mm. in diam., subglobose ; testa pale brown, slightly rugose.

N. Rhodesia. C : 72 km. N. of Mazabuka on Great North Road, fr. 2.v.1953, *Bainbridge* 16/53 (FHO ; K). S : Gwembe valley, fl. 30.iii.1952, *White* 2373 (FHO). **S. Rhodesia.** N : Urungwe, fr. 9.v.1951, *Lovemore* 32 (K ; SRGH). W : Gwaai, Nata Reserve, fl. xi.1952, *Davies* 398 (SRGH).

Also in Tanganyika. In low-altitude woodlands of the drier type or in *Brachystegia* woodland on termite mounds.

9. **Maerua nervosa** (Hochst.) Oliv., F.T.A. **1** : 84 (1868) excl. descr. et specim. cit.— Gilg & Bened. in Engl., Bot. Jahrb. **53** : 244 (1915).—Sim, For. Flor. Port. E. Afr. : 9 (1909). Type from Natal (Durban).

    Niebuhria nervosa Hochst. in Flora **27** : 289 (1844).—Sond. in Harv. & Sond., F.C. **1** : 60 (1860). Type as above.

    Maerua floribunda Sim, For. Flor. Port. E. Afr. : 10 (1909).—Gilg & Bened. in Engl., Bot. Jahrb. **53** : 261 (1915). Type : Mozambique, Lourenço Marques, *Sim* 5044 (PRE, holotype).

Shrub or small tree up to 3 m. tall, quite glabrous except for the bracts and sepal margins ; branchlets striate with some cuticular cells often golden and shining. Leaves 3-foliolate ; leaflets 2–5 × 1–3 cm., ovate-elliptic or obovate, apex rounded or emarginate, mucronulate, broadly cuneate at the base, both sides rather shining, nerves slightly raised below, veins laxly reticulate below ; petiole up to 2·5 cm. long ; petiolules up to 3 mm. long. Inflorescence of axillary, corymbose racemes, 4–6-flowered and collected towards the ends of the branches ; peduncles c. 1·5 cm. long ; pedicels c. 1 cm. long ; bracts trifid, 1 mm. long, subulate, minutely papillate-pubescent. Receptacle 5–6 mm. long, 2·5 mm. wide at the apex, cylindric, longitudinally c.12-ribbed, with 4 quadrate dentate lobes 1·5 mm. long at the mouth within ; sepals c. 1 × 0·4–0·5 cm., oblong-lanceolate to obovate-concave, apex rounded, apiculate, margin densely puberulous. Petals white, c. 5 × 3 mm., narrowly obovate to obovate, apex apiculate, base shortly clawed. Androgynophore up to 9 mm. long, exserted from the mouth of the receptacle. Stamens 10–20 ; filaments c. 2 cm. long ; anthers 2 × 1·25 mm., oblong. Ovary on a gynophore up to 2·5 cm. long, oblong-ellipsoid ; ovules c. 20 on 2 placentas ; stigma sessile, capitate, narrower than the ovary. Fruit up to 3·3 × 2 cm., oblong-ellipsoid, rather shining, verrucose. Seeds 10–15, c. 6 mm. in diam., subglobose, verrucose.

Mozambique. LM : Polana, fl. 16.viii.1942, *Mendonça* 1 (BM ; LISC).

Also in Natal. A distinctly coastal species.

10. **Maerua cafra** (DC.) Pax in Engl. & Prantl, Nat. Pflanzenfam. **3**, 2 : 234 (1891).— Sim, For. & For. Fl. Col. Cap. Good Hope : 122, t. 9 fig. 1 (1907) ; For. Fl. Port. E. Afr. : 9 (1909). TAB. 35 fig. C. Type from Cape Province.

    Capparis triphylla Thunb., Prodr. Pl. Cap. : 92 (1800). Type from Cape Province.

    Niebuhria cafra DC., Prodr. **1** : 243 (1824). Type as for *Maerua cafra*.

    Niebuhria triphylla (Thunb.) Wendl. in Bartl. & Wendl., Beytr. **2** : 29 (1825).— Sond. in Harv. & Sond., F.C. **1** : 60 (1860). Type as for *Capparis triphylla*.

    Maerua triphylla (Thunb.) Dur. & Schinz, Consp. Fl. Afr. **1**, 2 : 168 (1898) non *M. triphylla* A. Rich. (1847).—Gilg & Bened. in Engl., Bot. Jahrb. **53** : 244 (1915).—Burtt Davy, F.P.F.T. **1** : 122 (1926). Type as for *Capparis triphylla*.

Glabrous shrub or small tree up to 4 m. tall very like *M. nervosa* superficially but differs in the following ways : the leaflets are often acute or acuminate rather than rounded at the apices, the venation is not reticulate or prominent below ; the receptacle is c. 4 mm. long and widens gradually to a diameter of 4 mm. at the mouth and lacks quadrate lobes ; the sepals are narrowly ovate-acuminate and up to 2 × 1 cm. ; the petals are absent ; the androgynophore is not or barely exserted ; the stamens are 35–40 ; the fruit reaches 4·5 × 2·7 cm.

S. Rhodesia. E : Umtali, fl. & fr. 11.x.1951, *Chase* 4140 (BM ; K ; SRGH).

Also in Natal, Cape Province and the Transvaal. In forest patches and open woodland.

This species is known only from one area around Umtali in S. Rhodesia as far as our flora is concerned. Sim, however, reports it as common around Lourenço Marques and Umbeluzi but unfortunately he quotes no number and did not apparently collect it. No one else has done so since and the record needs to be confirmed but Sim probably knew the species well and should not have been mistaken.

11. **Maerua friesii** Gilg & Bened. in Wiss. Ergebn. Schwed. Rhod.-Kongo-Exped. **1** : 53, t. 6 fig. 3–4 (1914) ; in Engl., Bot. Jahrb. **53** : 246 (1915).—Hauman & Wilczek,

F.C.B. **2** : 493 (1951). TAB. **35** fig. A. Type : Northern Rhodesia, Lake Bangweulu, *Fries* 706 (B, holotype ; K ; UPS).

   Maerua monticola Gilg & Bened. in Engl., Bot. Jahrb. 53 : 260 (1915). Type : Nyasaland, Mwanembe Mt., *McClounie* 174 (K, holotype).

Evergreen climber or scrambling shrub with blackish puberulous stems attaining 10 cm. in diam. Leaves simple or 3-foliolate on the sterile branches ; leaflets 2·5–10·5 × 1·2–4·5 cm. narrowly ovate or ovate, apex blunt or subacute, mucronulate, base rounded or subcordate, shining and glabrous above except for the puberulous midrib, glandular-puberulous below or glabrous ; nerves in 5–6 pairs, prominent below ; petiole 1–4 cm. long in 3-foliolate leaves, c. 5 mm. long in simple leaves, puberulous. Flowers solitary in the upper axils or in short axillary and terminal racemes ; rhachis 0·2–4 cm. long, puberulous ; pedicels up to 2·5 cm. long, glabrous ; bracts 1 mm. long, trifid, subulate, or leaf-like. Receptacle c. 5 mm. long, up to 2·5 mm. in diam. at the mouth, cylindric, glabrous, with a thickened, undulate, dentate disk, 0·5 mm. long at the mouth within ; sepals 7 × 3–4 mm., broadly oblong-elliptic, acute at the apex, puberulous at the margins and within. Petals whitish, 3–5 × 2 mm., elliptic, acute at the apex, clawed at the base, margin slightly undulate and puberulous. Androgynophore c. 6·5 mm. long, exserted from the receptacle. Stamens 50–60 ; filaments 1·5–1·9 cm. long ; anthers 1·25 × 0·5 mm., oblong. Ovary on a gynophore c. 1·6 cm. long, oblong-ellipsoid ; ovules c. 16 on 2 placentas ; stigma sessile, capitate. Fruit 1·5–3 × 0·6–1·1 cm., oblong-cylindric or ovoid, narrowed to a blunt apex, with 4 shallow, longitudinal ridges, finely verrucose. Seeds several, c. 5 mm. in diam., subglobose, pale brown.

N. Rhodesia. N : Mweru Wantipa, fl. & fr. 22.ix.1937, *Trapnell* 1787 (K). W : Mufulira, fl. 11.viii.1954, *Fanshawe* 1450 (K ; SRGH). C : Broken Hill, fr. 10.ix.1947, *Brenan & Greenway* 7852 (FHO ; K). S : Mazabuka, fl. 3.viii.1952, *Angus* 161 (FHO ; K). **Nyasaland.** N : Mwanembe Mt., fl. ix.1902, *McClounie* 174 (K).
   Also in Tanganyika and the Belgian Congo. In forest patches, *Baikiaea plurijuga* woodland, semi-evergreen thickets and on termite mounds.

12. **Maerua pubescens** (Klotzsch) Gilg in Engl., Bot. Jahrb. **33** : 223 (1903).—Gilg & Bened., op. cit. **53** : 243 (1915).—Brenan, T.T.C.L. : 116 (1949). TAB. **35** fig. E. Type : Mozambique, Sena, *Peters* (B, holotype).
   Streblocarpus pubescens Klotzsch in Peters, Reise Mossamb. Bot. **1** : 165 (1861). Type as above.
   Maerua nervosa Oliv., F.T.A. **1** : 84 (1868) pro parte quoad specim. Peters ex Sena.—Eyles in Trans. Roy. Soc. S. Afr. **5** : 356 (1916). Non *Niebuhria nervosa* Höchst. Type as for *Maerua pubescens*.
   Maerua racemosa Sim, For. Flor. Port. E. Afr. : 9 (1909). Type : Mozambique, Lourenço Marques, *Sim* 6389 (K, holotype).
   Maerua cylindricarpa Gilg & Bened. in Engl., Bot. Jahrb. **53** : 241 (1915) *nom. illegit.*—Brenan, T.T.C.L. : 116 (1949). Type as for *M. pubescens*.
   Maerua stenogyna Gilg & Bened. in Engl., Bot. Jahrb. **53** : 243 (1915).—Arwidss. in Bot. Notis. **1935** : 359 (1935).—Brenan, T.T.C.L. : 116 (1949). Type from Tanganyika.

Much-branched shrub, scrambler or small tree up to 5 m. tall, branchlets pubescent or glabrescent. Leaves 3-foliolate ; leaflets 2–7 × 1–3 cm., lateral leaflets from about half the size of the middle leaflet, elliptic or broadly elliptic, rounded or emarginate at the apex, mucronulate, broadly cuneate at the base, pubescent or glabrous on both sides, nerves fairly prominent below, in 3–4 pairs, looping within the margin, veins rather widely reticulate ; petiole up to 2·5 cm. long, pubescent or glabrous ; petiolules up to 5 mm. long, pubescent or papillose on the upper surface. Inflorescence of short, terminal or axillary corymbose racemes ; pedicels c. 2 cm. long, pubescent or puberulous ; bracts c. 1 mm. long, subulate, trifid, puberulous or like the leaves but smaller and usually simple. Receptacle c. 3 mm. long, 2·5 mm. wide at the mouth, cylindric with c. 12 longitudinal ribs, puberulous or glabrescent outside with minutely puberulous denticulations just within the mouth. Sepals green, up to 1·2 × 0·35 cm., elliptic-oblong, acute at the apex, densely puberulous at the margins, remainder glabrous or puberulous outside. Petals white, c. 6 × 5·5 mm., orbicular with a short claw, margin somewhat undulate. Androgynophore slightly longer than the receptacle. Stamens 12–16 ; filaments up to 2 cm. long ; anthers c. 1·7 × 1 mm., oblong. Ovary on a gynophore up to 2·5 cm. long, cylindric, glabrous or puberulous ; ovules

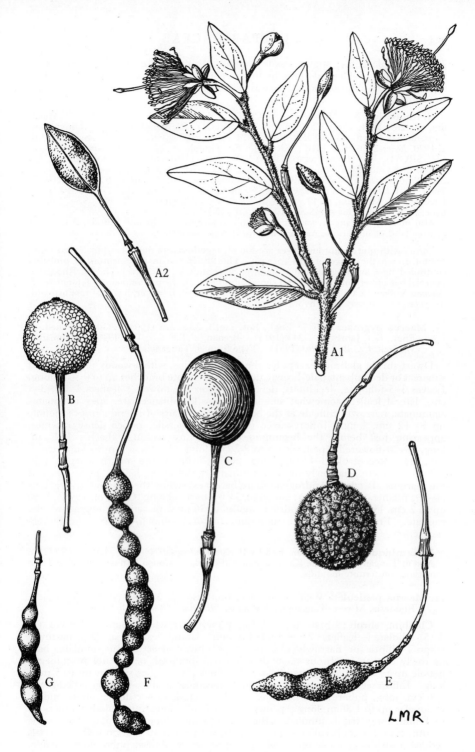

Tab. 35. A.—MAERUA FRIESII. A1, flowering branch (×⅔) *Richards* 6217 ; A2, fruit (×1) *Fanshawe* 1621. B.—MAERUA CERASICARPA, fruit (×1) *Robertson* 203. C.—MAERUA CAFRA, fruit (×1) *Chase* 944. D.—MAERUA PRITTWITZII, fruit (×1) *Lovemore* 32. E.—MAERUA PUBESCENS, fruit (×1) *Eyles* 4382. F.—MAERUA ANGOLENSIS, fruit (×1) *Richards* 6837. G.—MAERUA PARVIFOLIA, fruit (×1) *Obermeyer* 2428.

20–30 on 2 placentas; stigma sessile, capitate. Fruit up to c. 10 × 1 cm., cylindric, torulose, with 15–20 seeds or sometimes reduced and ellipsoid with one or two seeds, pericarp almost smooth. Seeds pale brown, 5–6 mm. in diam., subglobose; testa rugose.

N. Rhodesia. N : Abercorn, Lake Chila, fl. 24.x.1955, *Kafuli* 42 (BM). W : Mufulira, fr. 28.v.1934, *Eyles* 8203 (BM ; K ; SRGH). **S. Rhodesia.** N : Umvukwes, Umsengedzi R., fl. 23.xii.1952, *Wild* 3972 (K ; PRE ; SRGH). C : Salisbury, fl. xi.1920, *Eyles* 2756 (K ; PRE ; SRGH). E : Umtali, fr. 22.ii.1949, *Chase* 1379 (BM ; K ; SRGH). S : Fort Victoria, fl. xi.1952, *Davies* 361 (SRGH). **Nyasaland.** C : Dowa, fl. 6.viii.1951, *Chase* 3879 (BM ; K ; SRGH). S : Mlanje, Likubula Gorge, fr. 20.vi.1946, *Brass* 16378 (K ; SRGH). **Mozambique.** N : Nampula, fl. & fr. 4.iv.1937, *Torre* 1301 (BM ; COI ; LISC). Z : Chinde, fl. & fr. 13.x.1941, *Torre* 3640 (BM ; LISC). T : Mutarara, fr. 17.v.1948, *Torre* 4277 (BM ; LISC ; SRGH). SS : Inhambane, fr. viii.1936, *Gomes e Sousa* 1906 (COI ; K).

Also in Kenya, Tanganyika, Zanzibar and Madagascar. In low-altitude, drier types of woodland or on termite mounds in *Brachystegia* woodland.

Forms with quite glabrous leaves are found, together with the pubescent-leaved form, right through the range of this species. This difference in indumentum is probably controlled by a single gene and is of little taxonomic significance. The N. Rhodesian material has puberulous ovaries and shows affinities with the predominantly 1-foliolate *M. johannis* Volkens & Gilg. In view of its 3-foliolate leaves, however, it is best retained with *M. pubescens* : it may be of hybrid origin.

13. **Maerua pygmaea** Gilg in Engl., Bot. Jahrb. **33** : 228 (1903).—Gilg & Bened. in Engl., Bot. Jahrb. **53** : 243 (1915).—Brenan, T.T.C.L. : 117 (1949).—Hauman & Wilczek, F.C.B. **2** : 490 (1951). Type from S. Tanganyika.

Dwarf shrub, glabrous except for the sepal margins, with a woody rootstock and suberect or decumbent, pale-brown, densely lenticellate branches up to 0·5 m. long. Leaves 3–5-foliolate, deciduous, developing after the flowers ; leaflets 4–11 × 3–5 cm., lateral leaflets somewhat smaller, subsessile, oblanceolate, apex attenuate-acuminate, narrowly cuneate at the base, with 4–5 pairs of lateral nerves ; petiole up to 12 cm. long. Inflorescence of short, subsessile, rather dense racemes, appearing just before the beginning of the rainy season ; rhachis 1–4 cm. long, 15–20-flowered ; pedicels up to 2·2 cm. long ; bracts c. 1 mm. long, trifid, subulate. Receptacle 3–4 mm. long, 2 mm. wide at the mouth ; sepals 1·2–2·0 × 0·5–0·7 cm., narrowly ovate, acuminate at the apex, densely puberulous at the margins. Petals 0. Androgynophore barely exceeding the receptacle. Stamens 50–60; filaments c. 2 cm. long; anthers 1·3 × 0·8 mm., oblong. Ovary on a gynophore up to 2 cm. long, oblong-cylindric ; ovules c. 36 on 2 placentas ; stigma sessile, capitate. Fruit 2·5 × 0·75 cm. (not quite mature), oblong-cylindric, finely verrucose.

Mozambique. N : Tungue, fl. 19.x.1942, *Mendonça* 971 (BM ; LISC ; SRGH). Also in Tanganyika and the Katanga area of the Belgian Congo. In open woodland and sometimes on termite mounds.

14. **Maerua paniculata** Wild in Bol. Soc. Brot., Sér. 2, **32** : 43 (1958). Type : N. Rhodesia, Mweru-Tanganyika lowlands, *Michelmore* 446 (K, holotype).

Climbing shrub ; branches reddish, puberulous when young. Leaves (1) 3 (5)-foliolate ; leaflets 2·3–4·5 × 1–1·8 cm., lateral leaflets smaller, narrowly elliptic, elliptic or narrowly obovate, apex obtuse or acute, mucronulate, base narrowly cuneate, glabrous on both sides, membranous, nerves not prominent ; petiole up to 2·3 cm. long, papillose-puberulous ; petiolules similar, up to 5 mm. long. Inflorescence of terminal or axillary, racemose, leafless panicles ; peduncles 2–6 cm. long, puberulous ; racemes c. 10-flowered, rather dense, on secondary peduncles up to 1·5 cm. long ; pedicels up to 7 mm. long, puberulous ; bracts up to 1 mm. long, trifid, subulate. Receptacle 3 mm. long × 1·5 mm. wide at the mouth, cylindric, puberulous-pilose outside, with a dentate margin within ; sepals reddish outside, c. 7 × 4 mm., oblong-elliptic, apex obtuse, puberulous outside, particularly towards the margin. Petals 0. Androgynophore about the same length as the receptacle. Stamens c. 40, on reddish filaments c. 1·3 cm. long ; anthers 1·3 × 1·0 mm., oblong-ovate. Ovary on a glabrous gynophore up to 1·3 cm. long, oblong-ellipsoid, glabrous ; ovules c. 16 on 2 placentas ; stigma sessile, capitate, narrower than the ovary. Fruit c. 1×0.8 cm., ellipsoid-globose, longitudinally 4–6-ribbed.

N. Rhodesia. N: Mweru-Tanganyika lowlands, Chisyela-Chikuku watershed, fl. 24.vi.1933, *Michelmore* 446 (K).
Known only from the type collection. In dense deciduous scrub.

15. **Maerua parvifolia** Pax in Engl., Bot. Jahrb. **19**: 135 (1894).—Gilg & Bened. in Engl., Bot. Jahrb. **53**: 250 (1915).—Arwidss. in Bot. Notis. **1935**: 359 (1935). TAB. **35** fig. G. Type from SW. Africa.
   Maerua hirticaulis Gilg & Bened. in Engl., Bot. Jahrb. **53**: 251 (1915). Type: Mozambique, Sena, *Kirk* (K, holotype).
   Maerua legatii Burtt Davy in Kew Bull. **1924**: 225 (1924); F.P.F.T. **1**: 122 (1926).—O. B. Mill. in Journ. S. Afr. Bot. **18**: 16 (1952). Type from Transvaal.
   ? *Maerua crassifolia* sensu O. B. Mill., loc. cit.

Small shrub up to c. 2 m. tall ; branches stiff and rigid, closely pubescent, glabrescent or with patent, rather long hairs on the young parts. Leaves alternate or fasciculate on abbreviated side shoots, simple ; lamina $0.2–1.5 \times 0.15–0.65$ cm., narrowly oblong-obovate to elliptic, apex rounded or emarginate, mucronulate, base rounded or broadly cuneate, both sides hairy with a dense crisped pubescence, or with longer patent hairs particularly on the midrib, or almost glabrous with a dense or sparse scattering of minute papillae on both sides most readily visible in the more glabrescent forms, nerves in 4–5 pairs, not prominent ; petiole c. 1 mm. long, invariably pubescent or with longer patent hairs. Flowers solitary or in pairs on the abbreviated lateral shoots; pedicels up to 1.6 cm. long, pubescent or glabrescent; bracts minute, setiform. Receptacle c. 3.5 mm. long, 1.5–2 mm. wide at the mouth. cylindric, pubescent outside, slightly thickened and with a few dentate processes at the margin within ; sepals c. 7×4 mm., oblong-elliptic, apex rounded and apiculate, more or less pubescent outside. Petals white, 2.5×1.5 mm., elliptic, shortly clawed at the base, margin puberulous. Androgynophore c. 4 mm. long, shortly exserted from the receptacle. Stamens 15–25 ; filaments c. 1.5 cm. long ; anthers 1.75×0.75 mm., oblong. Ovary on a gynophore c. 3 cm. long, narrowly oblong-cylindric; ovules c. 10 on 2 placentas ; stigma sessile, subcapitate. Fruit c. 4.5×0.7 cm., cylindric, torulose, narrowed abruptly to a pointed apex, minutely but densely puberulous. Seeds 2–several, c. 6 mm. in diam., subglobose ; testa pale brown, roughened.

Bechuanaland Prot. N: Ngamiland, Tamasetse, fl. 18.ix.1949, *Pole-Evans* 4623 (PRE). **N. Rhodesia.** S: Bombwe, fl. 10.ix.1932, *Martin* 44/33 (FHO). **S. Rhodesia.** N. Sebungwe, st. 27.v.1947, *West* 2333 (K ; SRGH). W: Bulawayo, fl. 15.ix.1943, *Feiertag* in GHS 45567 (K ; SRGH). C: Hartley, fl. 12.x.1947, *Hornby* 3132 (SRGH). E: Hot Springs, fl. 21.x.1948, *Chase* 1196 (BM ; SRGH). S: Nuanetsi, fr. 1.xi.1955, *Wild* 4686 (K ; SRGH). **Nyasaland.** S: Utale, near Balaka, fl. 8.x.1952, *Carey* 6 (BM). **Mozambique.** T: Chicôa, Chioco, fl. 25.ix.1942, *Mendonça* 429 (BM ; LISC). MS: between Tambara and Chemba, fl. 25.ix.1949, *Pedro & Pedrógão* 8372 (LMJ ; SRGH). SS: Alto Limpopo, Mapai, fl. & fr. 3.xi.1944, *Mendonça* 2728 (BM ; LISC). LM: Goba, fr. 23.x.1944, *Mendonça* 3063 (BM ; LISC).
Also in SW. Africa and the Transvaal. In lower-altitude, drier types of woodland or on termite mounds up to c. 1200 m.

This species has two forms, one with patent rather long hairs on the young stem and leaves represented by the synonym *M. hirticaulis*, and the other with shorter crisped hairs only represented by the type of *M. parvifolia* and the other synonym *M. legatii*. The former is usually found in the north of the area in Nyasaland and the Tete and Manica and Sofala Provinces of Mozambique and the latter in S. Rhodesia, the Sul do Save and Lourenço Marques Provinces of Mozambique, the Transvaal and SW. Africa. However, in the Transvaal and S. Rhodesia are a proportion of intermediates and even a few specimens of the northern form ; so *M. parvifolia* is here looked upon as a species variable as regards pubescence, which cannot be divided satisfactorily into subspecific taxa.

16. **Maerua buxifolia** (Welw. ex Oliv.) Gilg & Bened. in Engl., Bot. Jahrb. **53**: 249 (1915).—Exell & Mendonça, C.F.A. **1**, 1: 60 (1937).—Type from Angola (Mossamedes).
   Maerua rigida var. *buxifolia* Welw. ex Oliv., F.T.A. **1**: 86 (1868). Type as above.

Small shrub very like *M. parvifolia* but the leaves are entirely glabrous, free from papillae and more glaucous ; but the most important diagnostic character is that the petioles are always entirely glabrous and smooth. This is never the case with *M. parvifolia* even in the more glabrescent forms.

S. Rhodesia. N : Sebungwe, near Sinansanga, fl. x. 1955, *Davies* 1441 (K ; SRGH). C : Gatooma, fr. 20.xii.1927, *Eyles* 5078 (K ; SRGH).
Also in southern Angola. A species of *Colophospermum mopane* woodland or mixed woodland in the hotter and drier areas.

17. **Maerua endlichii** Gilg & Bened. in Engl., Bot. Jahrb. **53** : 249 (1915).—Brenan, T.T.C.L. : 118 (1949). Syntypes from Tanganyika.

A straggling branched shrub up to 4 m. tall; branches glabrous, becoming densely lenticelled; bark pale grey. Leaves alternate or in fascicles on abbreviated side shoots, simple ; lamina 0·7–3·7 × 0·25–0·7 cm., very narrowly elliptic or narrowly oblanceolate, apex rounded or subacute, mucronulate, base very narrowly rounded or cuneate, glabrous on both sides, upper surface smooth and rather polished, dull below ; nerves in 5–6 pairs, inconspicuous above, slightly raised below ; petioles up to 2 mm. long, channelled above and puberulous along the sides of the furrow. Flowers solitary or occasionally in pairs on the abbreviated side shoots ; pedicels up to 3 cm. long, glabrous, slender ; bracts up to 1 mm. long, trifid, subulate. Receptacle 3·5–5·0 mm. long, 2·5–3 mm. wide at the mouth, narrowly funnel-shaped, glabrous, with a slightly thickened very slightly dentate disk at the inner margin ; sepals 1·2–1·4 × 0·5 cm., oblong-lanceolate, apex apiculate or acuminate, glabrous on both sides, margins puberulous. Petals 2 × 1 mm., oblong, blunt at the apex, or absent. Androgynophore slightly longer than the receptacle. Stamens 15–20 ; filaments c. 2 cm. long ; anthers 3 × 1 mm., narrowly oblong. Ovary on a gynophore up to 1·6 cm. long, narrowly oblong ; ovules c. 20 on 2 placentas which carry a single row of ovules on each placenta not two rows as is usual in this genus ; style up to 1 mm. long ; stigma subcapitate. Fruit c. 4 cm. long, cylindric, somewhat torulose, smooth, abruptly narrowed at the apex.

N. Rhodesia. N : Lake Tanganyika, Sumbu, fl. 27.vi.1933, *Michelmore* 451 (K).
Also in Tanganyika. Found in dense deciduous scrub near the shore of Lake Tanganyika.

Michelmore's specimen, unlike Endlich's specimen, is apetalous and in this respect matches *Volkens* 1740 (B† ; BM) from Kilimanjaro quoted as a syntype by Gilg and Benedict. I agree with them that although the lack of petals would in some *Maerua* spp. indicate a specific difference it does not do so here as the plants are otherwise so much alike and the petals, when they do occur, so minute.

18. **Maerua kirkii** (Oliv.) F. White in Bol. Soc. Brot., Sér. 2, **32** : 33 (1958). Syntypes : Nyasaland, Lake Nyasa, Cape McClear, *Kirk* (K) ; Upper Shire R., Mitonda, *Kirk* (K).
    Capparis kirkii Oliv., F.T.A. **1** : 98 (1868).—Gilg & Bened. in Engl., Bot. Jahrb. **53** : 199 (1915).—Brenan, T.T.C.L. 112 (1949).—Hauman & Wilczek, F.C.B. **2** : 458 (1951). Syntypes as above.

Evergreen bush or small tree up to 5 m. tall ; young branches scabrous-puberulous, sulcate. Leaves simple ; lamina 4–12 × 2–5 cm., coriaceous, oblong-obovate, narrowly obovate-oblong or oblanceolate, apex rounded, obtuse or acute, mucronate, narrowed to an obtuse or subcordate base, scabrous-pubescent on both sides, rather shining above, dull below and with the nerves strongly raised, venation strongly reticulate, on puberulous petioles up to 5 mm. long. Flowers numerous in terminal, dense, corymbose racemes ; rhachis c. 1 cm. long, pubescent ; pedicels up to 3 cm. long, pubescent ; bracts 1–2·5 mm. long, subulate, trifid, pilose, persistent. Receptacle 2·0 mm. long, 2·25 mm. wide at the mouth, funnel-shaped, pubescent outside, with a disk at the mouth inside which is rather thick, undulately lobed and 2 mm. long ; sepals c. 9 × 5 mm., oblong, concave, rounded at the apex, pubescent on both sides. Petals whitish, c. 0·6 × 3·5 mm., obovate-oblong, rounded at the apex, rounded at the base. Androgynophore exserted from the receptacle, 2·5–3 mm. long. Stamens c. 60 on filaments up to 3 cm. long; anthers 1·3 × 1 mm., broadly oblong. Ovary on a gynophore up to 2·5 cm. long, ovoid ; ovules c. 4 on 2 placentas ; stigma subsessile, capitate. Fruit up to 1·7 cm. in diam., globose, bluntly apiculate, shining, minutely and smoothly verrucose. Seeds 2–4, c. 1 × 0·5 cm., dark brown, oblong, ellipsoid ; testa smooth.

N. Rhodesia. N : Mweru marsh, fl. 24.ix.1937, *Trapnell* 1790 (K). C : Luangwa-Mpamadzi Junction, fl. 3.x.1933, *Michelmore* 629 (K). S : Gwembe valley, S. of Masuku

Mission, fl. *White* 2348 (FHO). **S. Rhodesia.** N : Urungwe, Msukwe R., fr. 18.xi.1953, *Wild* 4185 (K ; SRGH). C : Gatooma, fr. 2.xi.1929, *Jack* in GHS 4036 (SRGH). S : Ndanga, fl. 25.vii.1955, *Mowbray* 52 (K ; SRGH). **Nyasaland.** S : Shire R., Matope Bridge, fl. 30.ix.1937, *Lawrence* 491 (K). **Mozambique.** N : Macondes, between Muêda and Nairoto, fr. 20.ix.1948, *Mendonça* 1376 (BM ; LISC). MS : between Maringuè and Macossa, fl. 30.ix.1949, *Pedro & Pedrógão* 8432 (LMJ ; SRGH). SS : Madanda Forest, fl. ix.1911, *Dawe* 411 (K).

Also in Tanganyika. In the drier type of low-altitude woodland, occasionally penetrating *Brachystegia* woodland up to about 1300 m. but then only on termitaria.

19. **Maerua bussei** (Gilg & Bened.) Wilczek in Hauman & Wilczek, F.C.B. **2** : 494, t. 47 (1951). Syntypes from Tanganyika.

 Capparis bussei Gilg & Bened. in Engl., Bot. Jahrb. **53** : 200 (1915).—Gilg & Bened. in Notizbl. Bot. Gart. Berl. **13** : 30 (1938).—Brenan, T.T.C.L. : 112 (1949). Syntypes as above.

Dense shrub 2–4 m. high with glabrous, sulcate branchlets. Leaves simple or 3-foliolate ; leaf-lamina (or leaflet) 7–12 × 2·5–6 cm., coriaceous, oblanceolate oblanceolate-elliptic or elliptic, apex acute, retuse or rounded, mucronate, base cuneate or very narrowly rounded, glabrous and finely reticulately veined on both sides; simple leaves on wrinkled puberulous or glabrescent petioles up to 6 mm. long, 3-foliolate leaves on petioles up to 10 cm. long ; petiolules 2–3 mm. long, wrinkled. Inflorescence of terminal corymbose racemes ; rhachis 1–3 cm. long, glabrous ; pedicels up to 6 cm. long, glabrous ; bracts c. 4 mm. long, trifid, middle lobe caducous, filamentous lateral lobes 1 mm. long, persistent as scales at the base of old inflorescences. Receptacle 1·5 mm. tall, 1·5 mm. wide at the mouth, funnel-shaped, inner margin thickened and slightly undulate ; sepals c. 1 × 0·4 cm., lanceolate, apex acuminate. Petals c. 1·6 × 0·5 cm., oblanceolate, apex acuminate. Androgynophore slightly exserted from the receptacle. Stamens c. 60 ; filaments c. 3 cm., long ; anthers 1·5 × 0·75 mm., broadly oblong. Ovary on a gynophore c. 3·5 cm. long, ovoid-globose ; ovules 4 on 2 placentas ; style 0·5 mm. long ; stigma capitate. Fruit c. 2 cm. in diam., globose, smooth. Seeds up to 4, c. 10 mm. long, oblong-ellipsoid.

 N. Rhodesia. N : Lake Tanganyika, Kamba, fl. 27.vi.1933, *Michelmore* 450 (K).

Also in Tanganyika and the Belgian Congo. In low-altitude dry woodland, thickets, or sometimes at the edge of forest patches.

The occurrence of simple and 3-foliolate leaves in this species is peculiar as it does not follow the normal pattern in *Maerua* of replacement of 3-foliolate leaves by simple leaves towards the inflorescence but instead the two kinds are mixed on the branches. This is presumably due to the fact that the leaves do not fall till the end of their second year at least, so that the simple leaves of last year's inflorescence are still on the second year wood when the new inflorescences develop.

20. **Maerua andradae** Wild in Bol. Soc. Brot., Sér. 2, **32** : 41 (1958). Type : Mozambique, Porto Amelia, *Andrada* 1293 (BM, holotype : COI ; LISC ; SRGH).

Small shrub or perennial herb up to 30 cm. tall, glabrous except for the sepal-margins ; branches greenish, longitudinally striate. Leaves simple ; lamina 5–10 × 1–2·5 cm., narrowly oblong-elliptic or elliptic, apex acuminate, base narrowly cuneate, nerves in 4–5 pairs, fairly prominent on both sides ; petioles up to 5 cm. long. Inflorescence racemose, terminal or axillary, or flowers single in the axils of the upper leaves ; pedicels up to 2·5 cm. long ; bracts like the leaves but smaller. Receptacle c. 4 mm. long and 2 mm. wide, cylindric, slightly wider at the mouth, longitudinally 10–12-ribbed ; sepals 0·8–1·0 × 0·4–0·5 cm., oblong-spathulate, apex obtuse, mucronulate, margin shortly and densely lanate. Petals white, c. 7 × 3 mm., narrowly obovate, apex acute, margin undulate. Androgynophore very slightly exceeding the receptacle. Stamens c. 30 ; filaments c. 1·8 cm. long ; anthers 1·75 × 0·75 mm., oblong. Ovary on a slightly sulcate gynophore 2–2·7 cm. long, ovoid ; ovules 10–12 on 2 placentas ; stigma subsessile, capitate. Fruit c. 1 cm. in diam., globose, with a persistent stigma, smooth. Seeds 5–8, c. 5 mm. in diam., irregularly ellipsoid, testa smooth.

 Mozambique. N : between Macomia and Mipande, fr. 30.ix.1948, *Barbosa* 2305 (BM ; LISC ; SRGH).

Known only from the Niassa Province of Mozambique. In low-altitude *Acacia* woodland.

21. **Maerua brunnescens** Wild in Bol. Soc. Brot., Sér. 2, **32**: 42 (1958). Type: Mozambique, Manica e Sofala, Lacerdonia, *Pedro & Pedrógão* 8577 (LMJ; SRGH, holotype).

Small shrub up to 2 m. tall, glabrous except for the sepal-margins; branches brownish, smooth, erect or sometimes sarmentose. Leaves simple; lamina 2·5–7 × 1·8–3·5 cm., ovate, broadly oblong or broadly elliptic, apex usually obtuse or retuse, mucronulate, base rounded, margin cartilaginous, becoming brown when dry, nerves in 5–6 pairs, slightly prominent on both sides; petioles up to 9 mm. long. Inflorescence racemose, terminal or axillary, up to 20 cm. long; pedicels 5–6 mm. long; bracts up to 6 mm. long, subulate or the lower ones like the leaves but smaller. Receptacle 4·5 mm. long, 2·5 mm. wide at the mouth, cylindric, widening slightly at the mouth, with c. 12 longitudinal ribs; sepals c. 8·5 × 3·5 mm., narrowly oblong-ovate, apex acute, mucronate, margin shortly lanate. Petals 0. Androgynophore exserted from the receptacle, 6 mm. long. Stamens c. 70; filaments 1 cm. long; anthers 1·5 × 0·7 mm., elliptic. Ovary on a gynophore 1–1·2 cm. long and slightly sulcate below, oblong-ovoid; ovules 26–30 on 2 placentas; stigma sessile, capitate. Fruit c. 2·8 × 1·8 cm., oblong-ovoid, slightly verrucose. Seeds 7–10, c. 6·5 mm. in diam., subglobose, pale brown, slightly rugose.

Mozambique. Z: between Mopeia and Marral, fl. & fr. 15.x.1941, *Torre* 3661 (BM; LISC; SRGH). MS: Gorongosa, fl. 12.x.1946, *Simão* 1064 (LISC; LM; SRGH). SS: Vilanculos, fl. 2.xi.1944, *Mendonça* (BM; LISC). LM: Sabié, fr. 28.xi.1944, *Mendonça* 3110 (BM; LISC; SRGH).
Known only from Mozambique. In low-altitude dry woodland, often with *Acacia* spp. or sometimes on termitaria.

The material from the Sul do Save and Lourenço Marques areas differs from the rest in having narrower, acute leaves but seems to differ in no other way and so can be considered no more than a slightly divergent form.

22. **Maerua salicifolia** Wild in Bol. Soc. Brot., Sér. 2, **32**: 46 (1958). Type: S. Rhodesia, Sebungwe Distr., 80 km. N. of Gokwe, *Vincent* 30 (K, holotype; SRGH).

Glaucous shrub up to 1·3 m. tall, glabrous except for the sepals; branches glaucous-green, striate and erect. Leaves simple; lamina 6–10 × 1·5–2·5 cm., narrowly elliptic, obtuse or subacute at the apex, mucronate, base narrowly cuneate, margin subcartilaginous, nerves in 4–6 pairs, subprominent on both sides; petiole up to 1 cm. long. Flowers single in the axils of the upper leaves; pedicels up to 3 cm. long. Receptacle 4·5 mm. long, 3 mm. in diam. at the mouth, cylindric, widened at the mouth, margin within with 4 subquadrate, irregularly dentate lobes 1·5–2 mm. long; sepals c. 2 × 0·6 cm., narrowly oblong-elliptic, apex acute, glabrous outside, sparsely pubescent inside, margins very shortly lanate. Petals whitish, c. 9 × 3·5 mm., narrowly elliptic, apex subacute. Androgynophore exserted from the receptacle, 7–8 mm. long. Stamens c. 30; filaments c. 2·5 cm. long; anthers c. 3 × 1·25 mm., oblong-elliptic. Ovary on a subterete or slightly sulcate gynophore c. 2·2 cm. long, oblong-ovoid, slightly rugose or subverrucose; ovules c. 8 on 2 placentas; stigma sessile, capitate. Fruit c. 2·5 × 2 cm., ovoid or ellipsoid, coarsely verrucose. Seeds several, c. 5 mm. in diam., irregularly ovoid; testa pale brown, rugose.

S. Rhodesia. N: Sebungwe, Lusulu R., Matabola Flats, fr. 27.v.1947, *West* 2341 (K; SRGH).
Known only from S. Rhodesia. Invariably recorded so far as occurring in *Colophospermum* woodland.

23. **Maerua angolensis** DC., Prodr. **1**: 254 (1824).—Sim, For. Fl. Port. E. Afr.: 10 (1909).—Gilg & Bened. in Engl., Bot. Jahrb. **53**: 257 (1915).—Burtt Davy, F.P.F.T. **1**: 122 (1926).—Steedman, Trees etc. S. Rhod. 9 (1933).—Arwidss. in Bot. Notis. **1935**: 360 (1936).—Exell & Mendonça, C.F.A. **1**: 59 (1937).—Hauman & Wilczek, F.C.B. **2**: 487 (1951).—O. B. Mill. in Journ. S. Afr. Bot. **18**: 16 (1952).—Milne-Redh. in Mem. N.Y.Bot. Gard. **8**, 3: 217 (1953). TAB. **35** fig. F. Type from Angola.
   Maerua arenicola sensu Eyles in Trans. Roy. Soc. S. Afr. **5**: 356 (1916).
   Maerua sp. Eyles, loc. cit.
   Maerua schinzii sensu O. B. Mill., in Journ. S. Afr. Bot. **18**: 16 (1952).

Shrub or small tree up to c. 8 m. tall ; young branches pubescent or glabrescent, becoming densely lenticelled. Leaves simple ; lamina 2·5–7 × 1·3–5·5 cm., lanceolate, broadly elliptic or ovate to broadly ovate, apex rounded or emarginate, mucronulate, base broadly cuneate or rounded, glabrous or pubescent on both sides, nerves in 5–6 pairs, slightly prominent on both sides ; petioles up to 3 cm. long, often slightly swollen just below the lamina. Inflorescence of short, terminal, corymbose racemes on the main branches or on short side branches or flower solitary in the axils of the upper leaves ; pedicels up to 2·5 cm. long, glabrous or pubescent ; bracts subulate, trifid with the longer middle lobe up to 3 mm. long or similar to the leaves but smaller. Receptacle 8–15 mm. long, 3 mm. wide, cylindric, slightly wider at the mouth, glabrous or pubescent outside, with a fimbriately lobed disk, c. 2·5 mm. long at its margin within ; sepals 1–1·7 × 0·5–0·7 cm., narrowly oblong-obovate or elliptic, apex rounded to a shortly apiculate apex, pubescent on both sides or glabrous on both sides except at the margins. Petals 0. Androgynophore c. the same length as the receptacle. Stamens c. 50 ; filaments c. 4 cm. long ; anthers 2 × 0·75 mm., narrowly oblong, arcuate. Ovary on a gynophore up to 4 cm. long, narrowly oblong ; ovules c. 80 on 2 placentas. Fruit 2–16 × 1 cm., narrowly cylindric, torulose. Seeds from several to numerous, c. 6 mm. in diam., subglobose, with a smooth, pale brown testa.

Caprivi Strip. E. of Kwando R., fl. x.1945, *Curson* 1229 (PRE). **Bechuanaland Prot.** N : Kwebe Hills, fl. & fr. 5.x.1897, *Lugard* 28 (K). SW : Mabeleapudi Hills, st. 26.vii.1955, *Story* 5046 (PRE). SE : Kanye, fl. 14.xi.1948, *Hillary & Robertson* 538 (PRE ; SRGH). **N. Rhodesia.** B : Chavuma, fr. *White* 3511 (FHO). N : Chiengi, fl. 5.vi.1933, *Michelmore* 387 (K). C : Lusaka Distr., Kafue Bridge, fl. *Duff* 337 (FHO). S : Mazabuka, fr. 20.vii.1931, *Trapnell* 340 (K ; PRE). **S. Rhodesia.** N : Shamva, fl. 12.vi.1922, *Eyles* 3507 (K ; SRGH). W : Matopos, fl. 27.ix.1952, *Plowes* 1482 (K ; SRGH). C : Hartley, fl. 2.vi.1930, *Eyles* 7153 (K ; SRGH). E : Umtali, Dora R., fl. 7.x.1951, *Chase* 4044 (BM ; K ; SRGH). S : Gwanda, Shashi R., fl. x.1954, *Davies* 806 (K ; PRE ; SRGH). **Nyasaland.** C : Kasungu, fl. 27.viii.1946, *Brass* 17440 (BM ; K ; SRGH). S : Ft. Johnston, Chipoka, fr. 7.vii.1954, *Banda* 40 (BM ; K). **Mozambique.** N : Mutuali, fl. 6.vi.1948, *Pedro & Pedrógão* 4169 (LMJ ; SRGH). Z : Maganja da Costa, fl. 31.vii.1943, *Torre* 5728 (BM ; LISC). T : Mutarara, fl. & fr. x.1944, *Mendonça* (BM ; LISC ; SRGH). MS : Baruè, Mungari, fl. 6.vi.1941, *Torre* 2818 (BM ; LISC). SS : Vilanculos, fl. & fr. 2.ix.1944, *Mendonça* 1973 (BM ; LISC). LM : Santaca, fl. 14.viii.1948, *Gomes e Sousa* 3789 (COI ; K ; PRE).

Widely distributed through tropical Africa and the more northerly parts of S. Africa. In low-altitude woodlands and thickets, also often on termitaria including those in *Brachystegia* woodland up to c. 1500 m.

Like *Maerua pubescens* this species is very variable as regards pubescence. Specimens may be entirely glabrous or pubescent on stems and leaves in any part of the range. Coarsely pubescent forms are particularly common in Matabeleland and Bechuanaland however, and have often been called *M. schinzii* Pax. Here this form is included in *M. angolensis*, and *M. schinzii* sensu strict. is considered, as far as the available material is concerned, to be confined to SW. Africa. The type of *M. schinzii* has rather thick, subfleshy leaves and may be distinct from *M. angolensis*.

8. BOSCIA Lam.

Boscia Lam., Tabl. Encycl. Méth. Bot. : t. 395 (1797) *nom. conserv.*

Shrubs or trees ; leaves petiolate, alternate or fasciculate, often coriaceous, simple ; stipules setaceous. Inflorescence of racemes, fascicles or corymbs, terminal or axillary. Sepals 4, free or nearly so, valvate. Petals 0. Receptacle disk-like, entire, granular or fimbriate, sometimes 4-sided. Stamens 6–∞ on a very short androgynophore. Ovary ovoid or globose with a gynophore, 1-locular with 6–12 or occasionally numerous ovules on 1–2 placentas ; style short or almost absent ; stigma capitate or subcapitate, rarely 2-lobed. Fruit globose, indehiscent, shortly stalked or subsessile, crustaceous or leathery. Seeds 1–several, embedded in pulp ; testa crustaceous, often rugose ; cotyledons folded, radicle curved.

Flowers in rather elongated racemes more than 5 cm. long ; sepals c. 5 mm. long or more
                                                  1. *mossambicensis*
Flowers in short racemes less than 5 cm. long or in compound corymbs or fascicles ; sepals less than 5 mm. long :
  Leaves reticulately veined at maturity ; nerves raised and prominent at least below :
    Inflorescence a terminal compound corymb -   -   -   -   3. *corymbosa*

Inflorescence of short axillary racemes :
  Stamens c. 12 ; gynophore pubescent ; fruit densely brownish-pubescent
                                                                  5. *matabelensis*
  Stamens c. 6 ; gynophore glabrous ; fruit glabrous -   -   -   2. *cauliflora*
Leaves not reticulately veined at maturity ; nerves and veins scarcely visible at maturity :
  Stamens 6–14 ; leaves minutely roughened or punctulate and usually pubescent on
    both sides, 3–15 cm. long :
    Leaves 4–15 cm. long, usually tapering to an acuminate apex or acute ; fruit
      c. 1·7 cm. in diam. ; bark dark grey, rough and flaking   -   4. *salicifolia*
    Leaves 3–5 cm. long, apex usually rounded or retuse ; fruit c. 1 cm. in diam. ;
      bark smooth, whitish -   -   -   -   -   -   -   -   7. *albitrunca*
  Stamens 6–8 ; leaves smooth on both sides, rather shiny above, dull below, glabrous
    on both sides or very sparsely puberulous (visible only at × 20 or above), 0·5–3·4
    cm. long :
    Gynophore pubescent ; inflorescence usually shortly racemose ; leaves 1·2–3·4 cm.
      long -   -   -   -   -   -   -   -   -   -   -   6. *filipes*
    Gynophore glabrous ; inflorescence always fasciculate ; leaves always fasciculate,
      0·5–1·4 cm. long   -   -   -   -   -   -   -   -   8. *rehmanniana*

1   **Boscia mossambicensis** Klotzsch in Peters, Reise Mossamb. Bot. **1** : 164 (1861).—
      Gilg & Bened. in Engl., Bot. Jahrb. **53** : 207 (1915). Type : Mozambique, Tete,
      Boror, *Peters* (B, holotype).
      Boscia angustifolia sensu Oliv., F.T.A. **1** : 92 (1868) pro parte.—Sim, For. Fl.
      Port. E. Afr : 10 (1909).
      Boscia carsonii Bak. in Kew Bull. **1895** : 288 (1895).—Gilg & Bened. in Engl.,
      Bot. Jahrb. **53** : 207 (1915).—Brenan, T.T.C.L. : 108 (1949).—Hauman & Wilczek,
      F.C.B. **2** : 507 (1951). Type : N. Rhodesia, Mweru Plateau, *Carson 37* (K,
      holotype).

Much branched shrub or small tree up to 6 m. tall ; young branches glabrous
and yellowish, older bark grey-brown. Leaf-lamina 3–8 × 1·5–3·5 cm., coriaceous,
oblong, oblanceolate or obovate, apex obtuse and apiculate, cuneate or abruptly
and narrowly rounded at the base, margin cartilaginous and sometimes minutely
scaberulous, glabrous on both sides, midrib impressed above, prominent below,
secondary nerves slightly raised below in adult leaves ; petiole up to 8 mm. long,
minutely pubescent on the upper side. Inflorescences in terminal, moderately
dense, c. 20-flowered racemes on short axillary branches with usually 1–2 leaves
below the inflorescence on each branch ; bracteoles 3–5 mm. long, subulate, trifid
with the middle lobe longest ; pedicels up to 1·2 cm. long, glabrous. Sepals
greenish, 5–7·5 × 3–4 mm., finally reflexed, narrowly ovate, acute at the apex,
pubescent within and densely ciliate at the margins ; receptacle granular. Stamens
15–40 ; filaments up to 1 cm. long, glabrous. Gynophore up to 9 mm. long,
glabrous ; ovary ovoid, glabrous ; stigma capitate, subsessile ; ovules 12–15 on 2
placentas. Fruit globose, up to 2 cm. in diam., yellowish, glabrous, shortly apiculate
at the apex. Seed usually 1, 7–8 mm. in diam., globose ; testa shallowly rugose.

Bechuanaland Prot. N : Chobe R., Ngoma, fl. 23.v.1954, *Munro 4* (K ; PRE ;
SRGH). **N. Rhodesia.** N : Mpulungu, fl. 6.vii.1941, *Greenway 6192* (K ; PRE).
W : Nkana, fl. 30.vi.1952, *Holmes 733* (FHO). S : Gwembe valley, fl. 5.iv.1952, *White
2610* (FHO ; K). **S. Rhodesia.** N : Sebungwe, Kariangwe, fl. 11.vi.1951, *Lovemore 56*
(K ; SRGH). W : 27 km. S. of Wankie, fr. 12.vii.1952, *Codd 7061* (K ; PRE ; SRGH).
S : Sabi-Lundi Junction, fr. 7.vi.1950, *Wild 3436* (K ; SRGH). **Mozambique.** N :
Imala, between Muíte and Mecuburi, fr. 25.x.1948, *Barbosa 2572* (BM ; LISC). Z :
Boror, fl. *Peters* (B). T : Boroma, fr. 8.vii.1950, *Chase 2654* (BM ; K ; SRGH). MS :
Mungari, st. 29.viii.1949, *Pedro & Pedrógão 8138* (LMJ ; PRE). SS : Guijá, Caniçado,
fr. 3.vii.1947, *Pedro & Pedrógão 1213* (K ; LMJ ; SRGH). LM : between Uanetze and
Magude, fl. 4.v.1944, *Torre 6557* (BM ; LISC ; PRE).
    Also in the Transvaal, Swaziland, Belgian Congo and Tanganyika. In lower-altitude
and drier types of woodland.
    The branches apparently have an unpleasant smell when broken.

2. **Boscia cauliflora** Wild in Bol. Soc. Brot., Sér. 2, **32** : 40 (1958). Type : N. Rhodesia,
      Mwinilunga, *Milne-Redhead 4454* (B ; BM ; BR ; COI ; K, holotype ; LISC ;
      P ; PRE ; SRGH).

Small tree up to 5 m. tall with the branches pendulous ; young branches pubescent but soon glabrescent ; bark grey-brown. Leaf-lamina 3–6·5 × 0·8–1·7 cm.,
oblanceolate, rounded or acute at the apex and apiculate or mucronate, cuneate at the
base, minutely and sparsely puberulous on both sides at least when young, midrib

more densely so, margin slightly revolute, nerves slightly raised on both sides, venation reticulate; petiole up to 1 cm. long, pubescent. Inflorescence of short, dense, axillary racemes, usually borne on the second year wood below the young leaves, sometimes on branches up to 6 years old; peduncles up to 7 mm. long, pubescent; bracteoles 2 mm. long, filamentous, pubescent, trifid; pedicels up to 1 cm. long, pubescent. Sepals greenish, c. 4 × 1·5 mm., lanceolate, subacute at the apex, pubescent on both sides, densely ciliate at the margins; receptacle 3 mm. in diam., concave, fimbriate. Stamens 6; filaments 4 mm. long, glabrous. Gynophore up to 1·5 mm. long, glabrous; ovary ovoid, glabrous; ovules 8–10 on 2 placentas. Fruit c. 1·3 cm. in diam., globose, brownish. Seeds 1–8, up to 8 mm. in diam., pale brown, subglobose, strongly rugose.

N. Rhodesia. W: Mwinilunga, Matonchi Farm, 6.ii.1938, *Milne-Redhead* 4454 (B; BM; BR; COI; K; LISC; P; PRE; SRGH).

So far known only from the Western Province of N. Rhodesia between Bwana Mkubwa and Mwinilunga but very likely occurring in neighbouring parts of Angola. Usually found on termite mounds in *Brachystegia* woodland and apparently deciduous for about 3 weeks only, towards the end of December. This species is closely related to *B. angustifolia* A. Rich. of W. and NE. Africa, but this latter species has a much denser reticulation of its veins similar to that of *B. corymbosa* Gilg.

3. **Boscia corymbosa** Gilg in Engl., Pflanzenw. Ost-Afr. **C**: 186 (1895).—Gilg & Bened. in Engl., Bot. Jahrb. **53**: 208 (1915).—O. B. Mill. in Journ. S. Afr. Bot. **18**: 14 (1952).—Milne-Redh. in Mem. N.Y.Bot. Gard. **8**, 3: 217 (1953). TAB. **36** fig. A. Type: Mozambique, *Peters* (B, holotype).

Boscia caloneura Gilg in Engl., Bot. Jahrb. **28**: 390 (1900).—Gilg & Bened., op. cit. **53**: 210 (1915).—Hauman & Wilczek, F.C.B. **2**: 509 (1951). Type from Tanganyika (Iringa).

Boscia homblei De Wild. in Fedde, Repert. **11**: 511 (1913).—R.E.Fr., Wiss. Ergebn. Schwed. Rhod.-Kongo-Exped. **1**: 52 (1914).—Gilg. & Bened. in Engl., Bot. Jahrb. **53**: 210 (1915). Type from the Belgian Congo (Elizabethville).

Small, evergreen, round-topped tree up to 8 m. tall; bark pale grey, deeply longitudinally and sinuously fluted; young branches pubescent, with raised ridges decurrent from below the petioles. Leaves alternate and single on the branches or often 2–4 together in fascicles on very abbreviated lateral branches; lamina 2–7 × 0·7–2 cm., coriaceous, somewhat glaucous, elliptic to oblanceolate, apex rounded or acute with a pungent apiculus, cuneate at the base, veins finely but strongly reticulate below, minutely puberulous and eventually glabrescent above, densely puberulous below; petiole up to 4 mm. long, pubescent. Inflorescence of terminal, crowded, compound corymbs c. 4 cm. in diam., often borne on short lateral branches; peduncles and pedicels densely pubescent; pedicels 2–8 mm. long; bracteoles 1–2 mm. long, filiform, trifid. Sepals 2–3 mm. long, lanceolate, subobtuse at the apex, pubescent on both sides and ciliate on the margins; receptacle disk-like, of 10–15 fimbriately divided segments. Stamens 8–10 with glabrous filaments c. 3 mm. long. Gynophore 1·5 mm. long, glabrous; ovary ovoid; ovules c. 8 on 2 placentas; stigma subsessile, capitate. Fruit yellowish, 1–1·3 cm. in diam., globose, slightly pitted or granular. Seeds 1–2, up to 8 mm. in diam., subglobose, slightly rugose, embedded in a sticky pulp.

Bechuanaland Prot. N: Kazungula, fl. iii.1936, *Miller* 123 (PRE). **N. Rhodesia.** N: Lufu R., fr. 3.xi.1911, *Fries* 1191 (UPS). W: Chingola, fr. 26.viii.1954, *Fanshawe* 1494 (K). E: Fort Jameson, fl. & fr. 3.vi.1958, *Fanshawe* 4513 (K). S: Gwembe valley, fr. 7.iv.1952, *White* 2622 (FHO; K). **S. Rhodesia.** N: Sebungwe, fl. ix.1955, *Davies* 1548 (K; SRGH). W: Bulawayo, Hope Fountain, fr. 19.vi.1947, *Keay* in FHI 21334 (BM; FHO; SRGH). C: Gwelo, fl. v.1927, *Eyles* 1372 (K; SRGH). E: Odzi R., st. 12.iv.1948, *Chase* 713 (BM; SRGH). S: Sabi-Lundi Junction, fr. 6.vi.1950, *Chase* 2281 (BM; K; SRGH). **Nyasaland.** N: Likoma I., fl. & fr. *Johnson* 39 (K). C: Kasungu, fr. 28.viii.1946, *Brass* 17457 (BM; K; SRGH). S: Lake Nyasa, Boadzulu I., fr. 18.v.1954, *Jackson* 1310 (FHO; K). **Mozambique.** N: Namapa, fl. & fr. 16.viii.1948, *Andrada* 1270 (BM; COI; LISC). Z: Nacala, Posto de Netia, fr. 30.x.1942, *Mendonça* 1156 (BM; LISC). T: between Chicôa and Fingoè, fr. 26.vi.1949, *Barbosa & Carvalho* 3291 (LM; SRGH). MS: R. Save, Maringas, fl. 27.vi.1950, *Chase* 2234 (BM; K; SRGH).

Also in the Belgian Congo and Tanganyika. In drier types of woodland i.e. *Acacia*, *Colophospermum mopane* or *Adansonia* etc. woodlands; if occurring in *Brachystegia* woodlands then confined to termite mounds.

The freshly cut wood is evil-smelling.

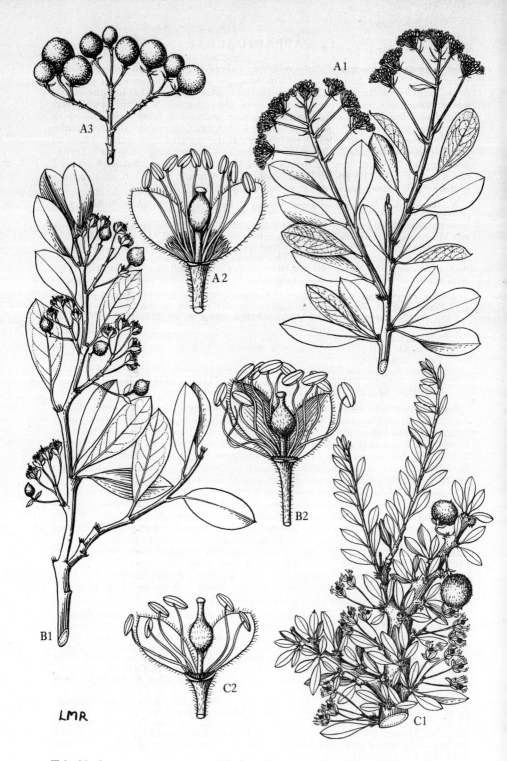

Tab. 36. A.—BOSCIA CORYMBOSA. A1, flowering branch (× ⅔) *Arnold* 859 ; A2, flower, front sepals removed (× 7) *Orpen* 12/53 ; A3, fruits (× ⅔) *Jackson* 1310. B.—BOSCIA MATABELENSIS. B1, flowering branch (× ⅔) ; B2, flower, front sepals removed (× 7), all from *Whellan* 446. C.—BOSCIA REHMANNIANA. C1, flowering and fruiting branch (× ⅘) ; C2, flower, front sepals removed (× 7), all from *Davies* 815.

4. **Boscia salicifolia** Oliv., F.T.A. **1** : 93 (1868) excl. specim. Welw.—Gilg & Bened. in Engl., Bot. Jahrb. **53** : 211 (1915).—Brenan, T.T.C.L. : 110 (1949).—Hauman & Wilczek, F.C.B. **2** : 510 (1952).—Milne-Redh. in Mem. N.Y.Bot. Gard. **8,** 3 : 217 (1953).—Keay, F.W.T.A. ed. 2, **1,** 1 : 93 (1954). Syntypes from Uganda and Nyasaland, Shire R., *Kirk* (K).

Boscia powellii Sprague & Green in Kew Bull. **1913** : 178 (1913).—Brenan, T.T.C.L. : 109 (1949). Type from Tanganyika.

Deciduous tree up to 15 m. tall, but often much less ; bark dark grey, rough and flaking ; young branches densely pubescent. Leaf-lamina 4–15 × 0·8–4·5 cm., linear, linear-lanceolate, narrowly elliptic, oblanceolate or oblong, apiculate, long-acuminate, acute or more rarely blunt at the apex, cuneate at the base, pubescent or scaberulous above, often glabrescent later, more persistently pubescent below ; midrib prominent below; petiole up to 1·5 cm. long, pubescent. Inflorescence of axillary racemes, 2–4 cm. long, often flowering at about the time of leaf-fall or sometimes a little later ; peduncles very short, 2–5 mm. long, densely pubescent ; pedicels 2–5 mm. long, densely pubescent ; bracteoles c. 1 mm. long, subulate, densely pubescent, entire or with two side teeth. Sepals greenish-yellow, c. 3 × 1·7 mm., reflexed at maturity, oblong-lanceolate to narrowly ovate, acute at the apex, pubescent within, densely velvety-pubescent outside, densely ciliate at the margins ; receptacle 4-sided, papillose-fimbriate. Stamens 6–14 with glabrous filaments 6–7 mm. long. Gynophore 4 mm. long, glabrous ; ovary ovoid, glabrous, with about 10 ovules on 2 placentas ; style c. 0·6 mm. long ; stigma small, not much wider than the style. Fruit c. 1·7 cm. in diam., globose, almost smooth, yellowish, with a minute apiculus, 1–3-seeded. Seeds brown, c. 1 cm. in diam., globose or somewhat compressed, rugose, embedded in a fleshy pulp.

Bechuanaland Prot. N : Tsotsoroga Pan, Goghe Hills, st. 8.vii.1937, *Erens* 355 (K ; PRE). **N. Rhodesia.** B : Balovale, fr. 12.x.1952, *Angus* 624 (FHO ; K). N : 26 km. S. of Samfya, fl. *Angus* 231 (FHO ; K). W : Chingola, fl. 26.viii.1954, *Fanshawe* 1496 (K). E : 483 km. E. of Lusaka, fl. viii.1927, *Burtt Davy* 965 (FHO ; K). C : 15 km. N. of Lusaka, fl., *Miller* 46 (FHO). S : Kafue, fr. 1932, *Stevenson* 435 (FHO ; K). **S. Rhodesia.** N : Mazoe, fr. i.1915, *Eyles* 608 (BM ; K ; SRGH). C : near Salisbury, fl. 20.ix.1936, *Eyles* 8792 (K ; SRGH). E : Umtali, fl. 4.vii.1947, *Chase* 552 (BM ; K ; SRGH). **Nyasaland.** S : Chikwawa, fr. 4.x.1946, *Brass* 17941 (K ; SRGH). **Mozambique.** N : between Metangula and Vila Cabral, fr. 11.x.1942, *Mendonça* 772 (BM ; LISC). Z : between Ile and Nhamarroi, fr. 25.ix.1941, *Torre* 3507 (BM ; SRGH). T : Mutarara, Sinjal, fl. 15.viii.1947, *Simão* 1480 (LM ; PRE ; SRGH). MS : Inhamitanga, fr. 14.x.1949, *Pedro & Pedrógão* 8656 (LMJ ; SRGH).

Also in Ghana, Nigeria, Cameroons, Uganda, Kenya and Tanganyika. In drier types of woodland or occurring on termite mounds in *Brachystegia* woodland.

The root of this species is eaten by natives in times of famine and the flowers produce a copious, scented nectar.

5. **Boscia matabelensis** Pest. in Bull. Herb. Boiss. **6,** Append. 3 : 115, t. 10 fig. 1, t. 11 fig. 3 (1898).—Gilg & Bened. in Engl., Bot. Jahrb. **53** : 211 (1915). TAB. **36** fig. B. Type : S. Rhodesia, between Sebenani and Nata, *Holub* 1265 (K, holotype).

Boscia hexamitocarpa Gilg-Bened. in Notizbl. Bot. Gart. Berl. **14** : 187 (1938). Type : Bechuanaland Prot., Maun-Mababe Road fork, *Erens* 341 (K ; PRE, holotype).

Shrub up to 2 m. tall, young branches pubescent. Leaf-lamina 1·7–7·5 × 0·6–2·5 cm., oblong, oblong-lanceolate to obovate, apex rounded or acute, apiculate, base rounded or subcuneate, scaberulous-pubescent on both sides particularly on the mid-nerve, glabrescent with age, nerves slightly raised on both sides but especially below, venation somewhat reticulate ; petiole up to 5 mm. long, pubescent. Inflorescence of short, dense, axillary racemes ; peduncle up to 1·2 cm. long, pubescent ; bracteoles c. 1 mm. long, filamentous, pubescent ; pedicels up to 5 mm. long, pubescent. Sepals greenish-yellow, 4 × 2·3 mm., lanceolate to ovate-oblong, apex blunt or subacute, 3-nerved, pubescent outside, pilose inside ; receptacle 4-sided, papillose-fimbriate. Stamens c. 12 on glabrous filaments 3–4 mm. long. Gynophore up to 2 mm. long, pubescent ; ovary ovoid, glabrous when young, papillose-puberulous later ; style 0·3 mm. long, glabrous ; stigma subcapitate, scarcely wider than the style ; ovules 12–14 on 2 placentas. Fruit 1·7 cm. in diam., globose, with a dense, brown, velvety pubescence. Seeds up to 0·9 cm. in diam., globose, smooth, usually single.

Bechuanaland Prot. N : S. Rhodesia-Bechuanaland border, Hunter's Rd., fr. 22.iv.1931, *Pole-Evans* 3326 (K ; PRE ; SRGH). SE : Bamangwato Reserve, Simanwana Stream, st. x.1949, *Miller* 951 (PRE). **S. Rhodesia.** N : Sebungwe, fr. ix.1955, *Davies* 1443 (K ; SRGH). W : Wankie, Siatshilaba's Kraal, fl. v. 1956, *Plowes* 1969 (K ; SRGH).
Confined to our area. Usually in *Colophospermum mopane* woodland.

The leaves are burnt by the Tonga people to drive away flies and mosquitoes.

6. **Boscia filipes** Gilg in Engl., Bot. Jahrb. **33** : 221 (1903).—Gilg & Bened., op. cit. **53** : 211 (1915). Type ; Mozambique, Lourenço Marques, *Schlechter* 11707 (B, holotype† ; BM ; COI ; K).
 Capparis albitrunca var. *parvifolia* Sim, For. Fl. Port. E. Afr. : 11, t. 3 fig. 4 (1909). Type : Mozambique, Lourenço Marques, *Sim* 5157 (PRE, holotype).

Shrub up to 4 m. tall ; young branches minutely puberulous, soon glabrous ; bark smooth, greyish or yellowish-green. Leaves alternate or in fascicles of 2–4 ; lamina 1·2–3·4 × 0·5–0·8 cm., oblanceolate to narrowly obovate, apex rounded, sometimes emarginate, apiculate, narrowed to a cuneate or narrowly rounded base, glabrous on both sides, somewhat shining above, dull below, midrib prominent, veins and nerves scarcely visible ; petiole 1–2 mm. long, puberulous or glabrescent, Inflorescence of very short axillary racemes or rarely reduced to axillary fascicles ; peduncle up to 4 mm. long, puberulous ; bracteoles 1 mm. long, filamentous, puberulous ; pedicels up to 7 mm. long, puberulous. Sepals 2·5–3 × 1·3 mm., oblong-lanceolate, apex subacute, puberulous outside and with longer hairs within ; receptacle divided into c. 12 digitately divided segments. Stamens 6–8 ; filaments 4 mm. long, glabrous. Gynophore c. 1·3 mm. long, pubescent ; ovary ovoid, glabrous ; ovules c. 8 on 2 placentas ; style 0·3 mm. long, glabrous ; stigma subcapitate, scarcely wider than the style. Fruit up to 1 cm. in diam., globose, very closely and densely velvety-brown-pubescent. Seeds brown, usually single, subglobose, 5–6 mm. in diam. ; testa rugose.

Mozambique. SS : Inhambane, fl. 2.ix.1944, *Mendonça* 1949 (BM ; LISC). LM ; Porto Henrique, fl. 25.viii.1948, *Gomes e Sousa* 3807 (COI ; K ; LM ; PRE ; SRGH). Also in the Transvaal. In low-altitude, usually coastal scrub or woodland.

7. **Boscia albitrunca** (Burch.) Gilg & Bened. in Engl., Bot. Jahrb. **53** : 212 (1915).— Burtt Davy, F.P.F.T. **1** : 123 (1926).—O. B. Mill. in Journ. S. Afr. Bot. **18** : 14 (1952). Type from Cape Province (Griqualand West).
 Capparis albitrunca Burch., Trav. Int. S. Afr. **1** : 343 (1822).—Harv. in Harv. & Sond., F.C. **1** : 63 (1860).—Sim, For. Fl. Port. E. Afr. : 10 (1909). Type as above.
 Boscia pechuelii Kuntze in Jahrb. Berl. Bot. Gart. **4** : 261 (1886).—Passarge, Die Kalahari : 789 (1904). Type from SW. Africa (Hereroland).

Much-branched tree up to 10 m. tall ; crown dense and rounded ; bark smooth and whitish ; trunk stout. Leaves alternate or fascicled 2–4 together on very reduced side-shoots ; lamina 3–5 × 0·7–1·2 cm., coriaceous, oblanceolate or narrowly oblong, apex rounded or retuse or very rarely acute, mucronate, narrowed to a cuneate or narrowly rounded base, puberulous on both sides or glabrescent, midrib prominent beneath, nervation scarcely visible ; petiole 1–3 mm. long, puberulous. Inflorescence of very short, dense, axillary racemes ; peduncle almost 0, rachis pubescent ; bracteoles 1·5 mm. long, filamentous, pubescent, trifid with the middle lobe longest ; pedicels up to 5 mm. long, pubescent. Sepals 4 × 2·5–3 mm., ovate-oblong, apex subacute, very minutely puberulous outside, pubescent within, densely so at the margins ; receptacle disk-like, papillose-fimbriate. Stamens 6–14, on glabrous filaments c. 5 mm. long. Gynophore c. 3 mm. long, glabrous ; ovary ovoid, glabrous, with c. 10 ovules on 2 placentas ; style 0·5 mm. long, glabrous ; stigma subcapitate, scarcely wider than the style. Fruit up to 1 cm. in diam., globose, yellowish, glabrous, smooth. Seed usually single, c. 0·7 cm. in diam., subglobose, rugose.

Caprivi Strip. Katima Molilo, fr. 23.x.1954, *West* 3251 (K ; SRGH). **Bechuanaland Prot.** N : Ngamiland, fl. ix.1949, *Pole Evans* 4626 (PRE). SW : 435 km. NW. of Molepole, 23° E, 22° 20′ S., st. 3.vii.1955, *Story* 4998 (PRE). SE : Lobatsi, fl. x.1913, *Rogers* 6211 (PRE ; SRGH). **N. Rhodesia.** B : Sesheke, Malabwe Forest, fl. *Brenan &* *Keay* 7669 (FHO). S : Livingstone Distr., fl. 24.viii.1947, *Brenan* 7742 (FHO ; K). **S.**

Rhodesia. W : Gwaai Res., fr. 29.ix.1947, *West* 2426 (SRGH). E : Melsetter Distr., Birchenough Bridge, fl. 12.ix.1949, *Chase* 1751 (SRGH). S : Gwanda, Beit Bridge, fr. xi.1952, *Davies* 378 (SRGH). **Mozambique.** LM : between Santaca and Catuane, fr 13.iv.1949, *Myre & Balsinhas* 593 (LM ; SRGH) ; Goba, fl. 23.viii.1944, *Mendonça* 1822 (BM ; LISC).

Also in the Cape, Transvaal, Natal, Orange Free State, and SW. Africa. In drier woodlands and bush, sometimes on termite mounds.

The leaves and young branches of this species are often browsed and the root is used as a food by Africans in time of famine.

8. **Boscia rehmanniana** Pest. in Bull. Herb. Boiss. **6,** App. 3 : 95, t. 5 fig. 3 (1898).—Gilg & Bened. in Engl., Bot. Jahrb. **53** : 213 (1915).—Burtt Davy in F.P.F.T. **1** : 123 (1926).—O. B. Mill. in Journ. S. Afr. Bot. **18** : 15 (1952). TAB. **36** fig. C. Syntypes from Transvaal and SW. Africa (23° S).

    Boscia microphylla Oliv., F.T.A. **1** : 93 (1868) pro parte quoad specim. Baines.—Passarge, Die Kalahari : 791 (1904).

    Boscia kalachariensis Pest. in Bull. Herb. Boiss. **6,** App. 3 : 98 (1898).—Gilg & Bened. in Engl., Bot. Jahrb. **53** : 213 (1915). Type : Bechuanaland, Lake Ngami, *Fleck* 247 (Z, holotype).

    Boscia seineri Gilg ex Engl., Pflanzenw. Afr. **3,** 1 : 242, fig. 158 d–f (1915) *nom. nud.*

Shrub or small tree up to 5 m. tall ; branches stout, glabrous ; bark smooth, grey. Leaves in fascicles ; lamina 0·5–1·3 × 0·2- 0·5 cm., oblanceolate to narrowly obovate, rounded or emarginate at the apex, mucronulate, broadly cuneate at the base, glabrous on both sides or very minutely and sparsely puberulous (hairs visible only at magnifications above × 20); petiole c. 1 mm. long, glabrous. Inflorescence of axillary fascicles (reduced racemes) ; pedicels up to 6 mm. long, slender, minutely and sparsely puberulous or glabrescent ; bracteoles 1 mm. long, filamentous, puberulous. Sepals yellowish-green, c. 2·3 × 1·2 mm., narrowly obovate, obtuse at the apex, puberulous on both sides ; receptacle disk-like, papillose-fimbriate. Stamens c. 6–8, on glabrous filaments up to 3·5 mm. long. Gynophore 1·5 mm. long, glabrous ; ovary ovoid, glabrous ; ovules c. 6 on 2 placentas ; style up to 0·4 mm. long, glabrous ; stigma subcapitate, scarcely wider than the style. Fruit up to 1 cm. in diam., globose, yellowish-brown, bluntly apiculate, minutely but densely pubescent. Seed usually single, c. 7 mm. in diam., globose, rugose.

Bechuanaland Prot. N : Ngamiland, Kwebe Hills, fl. & fr. 29.ix.1897, *Lugard* 27 (K). SE : Lobatsi, st. 6.iii.1930, *van Son* (BM ; L ; PRE ; SRGH). **S. Rhodesia.** S : Gwanda Distr., Shashi R., fl. x.1954, *Davies* 815 (K ; SRGH).

Also in the Transvaal and SW. Africa. Confined to our driest types of woodland and bushland.

The flowers have a rather strong unpleasant scent.

9. CAPPARIS L.

Capparis L., Sp. Pl. **1** : 503 (1753) ; Gen. Pl. ed. 5 : 222 (1754).

Shrubs or small trees ; branches often sarmentose, sometimes climbing. Stipules often transformed into spines. Leaves simple and entire. Inflorescences of terminal or axillary racemes sometimes condensed into corymbs or umbels, or occasionally of single flowers. Sepals 4, free or very nearly so, the two outer often overlapping the two inner, equal, or the two inner exceeding the outer. Petals usually 4, exceeding, equalling or much smaller than the calyx-segments. Stamens 6–∞, free, inserted on a very short androgynophore. Ovary with a gynophore, with 1–several loculi and with 2–several placentas ; stigma sessile or subsessile. Fruit globose, ovoid or elongated, usually indehiscent. Seeds 1–∞, embedded in a fleshy pulp, embryo convolute.

Calyx with sepals equal or subequal ; fruit not ridged longitudinally :
  Sepals glabrous or glabrescent outside, glabrous, ciliate or bearded at the margins :
    Flowers fascicled 2–5-together in the axils or solitary :
      Spines recurved ; leaves not reticulate -   -   -   -   -   7. *lilacina*
      Spines straight ; tertiary venation reticulate in adult leaves - 8. *orthacantha*
    Flowers in subumbellate racemes on short side-shoots or in leafless panicles :
      Leaves narrowly oblong-lanceolate, 2·5–6 × 0·4–1·7 cm. ; flowers polygamous with
        males predominating -   -   -   -   -   -   -   2. *brassii*

Leaves lanceolate, oblong, obovate or broadly elliptic, 1·2–5 × 0·4–2·8 cm. ;
flowers normally bisexual :
Sepals 6–7 mm. long ; leaves coriaceous, quite glabrous or sparsely patent-
pilose, shining above - - - - - - - 3. *citrifolia*
Sepals 4–5 mm. long ; leaves with a fine appressed pubescence, submembranous
in the flowering stage, dull on both sides - - - - - 4. *sepiaria*
Sepals tomentose or densely pubescent outside :
Flowers large, c. 2 cm. in diam., aggregated into terminal corymbs or racemes or
occasionally axillary - - - - - - - - 1. *tomentosa*
Flowers small, c. 1 cm. in diam., in axillary fascicles on very abbreviated shoots or in
axillary racemes :
Branches ferruginously hairy when young ; flowers all in axillary fascicles ; petals
sparsely pubescent on both sides - - - - - 5. *elaeagnoides*
Branches fulvous-hairy ; flowers in terminal or axillary racemes ; petals villous
outside - - - - - - - - - - 6. *rudatisii*
Calyx with the inner sepals petaloid and much longer than the outer sepals ; fruit
longitudinally ridged :
Leaves usually acute or acuminate at the apex ; ovary glabrous - 10. *erythrocarpos*
Leaves rounded or blunt at the apex ; ovary densely pilose - - - 9. *rosea*

1. **Capparis tomentosa** Lam., Encycl. Méth., Bot. **1** : 606 (1785).—Gilg & Bened. in
Engl., Bot. Jahrb. **53** : 189 (1915).—Exell & Mendonça, C.F.A. **1, 1** : 62 (1937).—
Hauman & Wilczek, F.C.B. **2** : 460, t. 46 (1951). Type from Senegal.
　　Capparis tomentosa var. β Oliv., F.T.A. **1** : 96 (1868). Syntypes from Sene-
gambia, Ethiopia, Angola and Bechuanaland Prot., Lake Ngami, *McCabe* 27 (K).
　　Boscia tomentosa sensu O. B. Mill. in Journ. S. Afr., Bot. **18** : 15 (1952).

A spiny, erect or scrambling shrub 1–3 m. tall or climbing over other vegetation
to 6 m. or perhaps more ; young branches covered with a dense, greyish or brownish pubescence; shorter side-branches often at right angles to the main stems.
Leaf-lamina 3–6 × 1·5–3 cm., oblong or elliptic, rounded and mucronulate or
slightly emarginate at the apex, rounded or broadly cuneate at the base, tomentose,
thinly hairy or glabrescent above, ± tomentose below ; petioles 5–10 mm. long,
densely pubescent. Flowers scented, in terminal corymbs of 3–15 flowers, the
lower pedicels solitary in the axils of the upper leaves, the upper with linear
tomentose bracts c. 3 mm. long ; pedicels up to 3 cm. long, densely pubescent.
Buds globose. Sepals 8–10 mm. long, orbicular, with a caducous greyish or
brownish tomentum outside. Petals pale yellowish-green, spathulate or cuneiform-
truncate at the apex, with white hairs at the base within. Stamens c. 80 with white
or pinkish slender filaments up to 4 cm. long. Gynophore up to 4 cm. long,
slender ; ovary ovoid, bluntly apiculate at the apex. Fruit up to 5 cm. in diam.,
globose, reddish-orange, on a very stout gynophore. Seeds c. 1·4 × 0·8 cm. when
mature, ovoid, smooth, embedded in a pinkish flesh.

Caprivi Strip. E. of Kwando R., fr. x.1945, *Curson* 1181 (PRE). **Bechuanaland
Prot.** N : Ngamiland, Lake R., fl. ix.1896, *Lugard* 18 (K). **N. Rhodesia.** B : Senanga,
fl. 5.viii.1952, *Codd* 7406 (BM ; K ; PRE). N : Lake Bangweulu, Chiluwi I., fl.
31.viii.1933, *Michelmore* 561 (K). W : Ndola, fl. 1953, *Duff* 165/33 (FHO ; K). C :
Chilanga, fl. 31.viii.1929, *Sandwith* 2 (K). S : Katombora, fl. 21.ix.1955, *Gilges* 435 (K ;
PRE ; SRGH). **S. Rhodesia.** N : Mtoko, fl. 18.x.1955, *Lovemore* 455 (K ; SRGH).
W : Wankie, fr. iii.1931, *Pardy* in GHS 4705 (SRGH). C : Umsweswe, fl. x.1919, *Eyles*
1834 (K ; PRE ; SRGH). E : Melsetter, Hot Springs, fl. & fr. 21.x.1948, *Chase* 1203
(BM ; K ; SRGH). S : between Lundi R., and Fort Victoria, fl. 19.x.1930, *F.N. & W.*,
2115 (BM ; LD ; PRE ; SRGH). **Nyasaland.** C : Kota Kota, Chia, fl. 4.ix.1946,
Brass 17523 (BM ; K ; SRGH). S : Upper Shire, Mikena, fl. 28.ix.1859, *Kirk* (K).
Mozambique. N : Macondes, between Muêda and Nangade, fr. 19.x.1942, *Mendonça*
979 (BM ; LISC). Z : Guruè, Lioma, fl. 13.ix.1949, *Barbosa & Carvalho* 4071 (LM ;
SRGH). T : Chicôa, Posto de Chiôco, fl. & fr. 26.ix.1942, *Mendonça* 436 (BM ; LISC).
MS : Gorongosa, fl. 23.x.1956, *Gomes e Sousa* 4314 (BM ; LISC). SS : Guijá, between Mapai and
Mabelane, fr. 3.xi.1944, *Mendonça* 2753 (BM ; LISC). LM : Magude, Mavabaze, fr.
29.x.1944, *Mendonça* 3133 (LISC).
Widespread throughout tropical Africa and also in the Transvaal and Natal. Common
in the drier types of woodland and bushland at lower altitudes, sometimes becoming a
climber or even a liane in riverine fringes and also occurring at higher altitudes, up to
2000 m., on termite mounds.

It is reported that the fruit is not edible but that the leaves make a palatable browse for
cattle.

2. **Capparis brassii** DC., Prodr. **1** : 248 (1824). TAB. **37** fig. D. Type from Ghana.
   Capparis thonningii Schumach. in Kongel. Dansk. Vid. Selsk. Naturvid. Math. Afh.
   4 : 236 (1829).—Oliv., F.T.A. **1** : 97 (1868).—Gilg & Bened. in Engl., Bot. Jahrb.
   53 !: 190 (1915).—Keay, F.W.T.A. ed. 2, **1**, 1 : 90 (1954) Type from Ghana.
   Capparis gueinzii Sond. in Harv. & Sond., F.C. **1** : 62 (1860).—Gilg & Bened. in
   Engl., Bot. Jahrb. **53** : 191 (1951). Type from Natal.

A thorny climber or trailing bush with the young branches flexuous and pubescent. Leaf-lamina 2·5–6·0 × 0·4–1·7 cm., narrowly oblong-lanceolate, blunt and minutely emarginate at the apex, rounded at the base, glabrous on both sides or the midrib downy ; midrib immersed above, prominent below, lateral nerves inconspicuous ; petiole up to 5 mm. long, somewhat channelled and pubescent on the upper side. Flowers sweet-scented, in axillary 4–10-flowered racemes as long as the leaves or shorter, or the uppermost in leafless panicles ; pedicels up to 8 mm. long, slender, pubescent ; bracts up to 1·5 mm. long, pubescent, linear. Sepals up to 5 mm. long, rotund-ovate, rounded at the apex, glabrous. Petals yellow, as long as the sepals, broadly oblong, rounded at the apex, sparsely pubescent towards the base within. Stamens up to c. 30 with filaments c. 6 mm. long ; plants often with staminate flowers only, fertile flowers usually with fewer stamens and these caducous. Gynophore up to 6 mm. long ; ovary ovoid ; stigma capitate on a very short style. Fruit orange or red, up to 1·7 cm. in diam., globose, several-seeded.

Mozambique. SS : Gaza Distr., Manjacaze, fl. 25.vii.1944, *Torre* 6804 (LISC). LM : Maputo, Santaca, fl. 14.viii.1948, *Gomes e Sousa* 3790 (COI ; K ; PRE ; SRGH). Also in Natal, Sierra Leone, Ghana, Togoland, Dahomey and Nigeria. In coastal bushland, fringing forest and forest edges.

The discontinuity in the distribution of this species is very striking, but material from our area Natal and West Africa is very uniform morphologically. This type of distribution is not, however, unique.

3. **Capparis citrifolia** Lam., Encycl. Méth. Bot. **1** : 606 (1785).—Gilg & Bened. in
   Engl., Bot. Jahrb. **53** : 191 (1915). Type from Cape of Good Hope.
   Capparis corymbosa var. *sansibarensis* Pax in Engl., Bot. Jahrb. **14** : 297 (1892).
   Type from Tanganyika (Dar es Salaam).
   Capparis sansibarensis (Pax) Gilg in Engl., Bot. Jahrb. **33** : 213 (1903).—Gilg &
   Bened., tom. cit. 192 (1915).—Brenan, T.T.C.L. : 113 (1949). Type as above.

Thorny shrub 1–3 m. tall with glabrous branches or the youngest branches pubescent. Leaf-lamina 1·8–4·5 × 0·8–2·5 cm., coriaceous, oblong, obovate or broadly elliptic, rounded, mucronulate and often emarginate at the apex, rounded or broadly cuneate at the base, glabrous on both sides or sparsely pubescent with patent hairs, more densely so below or on the midrib only, margin cartilaginous and sometimes recurved, midrib prominent below ; petiole up to 5 mm. long, glabrous or minutely pubescent. Flowers in terminal, subumbellate racemes or the lower pedicels single in the axils of the upper leaves ; pedicels up to 2·5 cm. long, slender, glabrous or sparsely patent-pilose. Buds globose. Sepals 6–7 mm. long, orbicular, concave, glabrous except for the ciliate margins. Petals c. 8 × 4 mm., oblong or oblong-spathulate, rounded at the apex, ciliate at the margins, villous at the base and within. Stamens 30–40, on slender filaments up to 1·8 cm. long. Gynophore up to 1·4 cm. long, slender ; ovary ovoid, the stigma forming an apiculus at the tip. Fruit c. 1 cm. in diam., spherical, somewhat fleshy, 1–few-seeded. Seeds 5–7 mm. in diam., smooth, brown, compressed-orbicular.

Mozambique. N : Porto Amelia, fl. 26.x.1942, *Mendonça* 1079 (BM ; LISC). SS : Inharrime, fl. 16.x.1957, *Barbosa & Lemos* 8068 (K ; LMJ). LM : Santaca, fl. 17.x.1947, *Gomes e Sousa* 3636 (COI ; K).
Also in the Cape, Natal and Tanganyika. In coastal bush and woodland.

One specimen, collected by an unknown collector at Boane in the Lourenço Marques Province (fl. 9.x.1907, No. 127 (COI)), probably belongs here but has patently pubescent leaves and stems. It may be a distinct variety but more material is needed to decide this, for variously pubescent forms of *C. citrifolia* occur also in the Cape Province.

4. **Capparis sepiaria** L., Syst. Nat. ed. 10, **2** : 1071 (1759). TAB. **37** fig. C. Type from
   India.
   Capparis corymbosa sensu Oliv., F.T.A. **1** : 97 (1868).
   Capparis citrifolia sensu Arwidss. in Bot. Notis, **1935** : 358 (1935).

Thorny, much branched or subscandent shrub to 6 m. tall or sometimes climbing over tall trees; young branches green and smooth, glabrous or with an appressed pubescence on the young parts. Leaf-lamina 1·2–5 × 0·4–2·8 cm., lanceolate, oblong-elliptic or elliptic, rounded or emarginate at the apex, rounded or slightly cordate at the base, appressed-pubescent on both sides; petiole up to 5 mm. long, appressed-pubescent. Flowers 4–6 which are together in condensed racemes, terminal or often on short side branches 2–4 cm. long, or single in the axils of the upper leaves; pedicels up to 2 cm. long, slender, glabrescent. Buds globose. Sepals 4–6 mm. long, concave, orbicular, membranous-margined, glabrous or almost so, sometimes minutely ciliolate on the margins. Petals up to 8 mm. long, spathulate, rounded at the apex, pilose within and at the margins, more densely so towards the base. Stamens 40–50, with slender filaments c. 1·3 cm. long. Gynophore about the same length as the stamens; ovary ovoid, sometimes oblique, with an apiculate stigma. Fruit yellowish, 1–1·3 cm. in diam., globose. Seed usually single, c. 6 mm. in diam., discoid, brown, smooth.

N. Rhodesia. N : Mpika, Munyamadzi R., fl. 24.x.1957, *Savory* 232 (SRGH). **S. Rhodesia.** N : Urungwe, fl. 21.xi.1953, *Wild* 4234 (K ; PRE ; SRGH). E : Melsetter, Hot Springs, fl. 21.x.1930, *F.N. & W.* 2213 (BM ; LD ; PRE ; SRGH). S : Nuanetsi, fl. 2.xi.1955, *Wild* 4693 (K ; SRGH). **Nyasaland.** S : Fort Johnston, Chipoka, fl. 21.x.1954, *Jackson* 1377 (FHO ; K). **Mozambique.** N : Porto Amelia, fl. 1.ix.1948, *Barbosa* 2339 (BM ; LISC). T : between Mungari and Changara, fl. 26.x.1943, *Torre* 6085 (BM ; LISC ; PRE). MS : Lower R. Buzi, fr. 20.xii.1906, *Swynnerton* (BM).
Also in Angola, Transvaal, India and Ceylon. In drier types of woodland and bushland or a climber in low-altitude riverine fringes.

This species has a coastal distribution in Angola and Mozambique but penetrates the Transvaal, Rhodesia and Nyasaland up the valleys of the Zambezi, Shire, Sabi and Limpopo.

5. **Capparis elaeagnoides** Gilg in Engl., Bot. Jahrb. **33** : 215 (1903).—Gilg & Bened. in Engl., Bot. Jahrb. **53** : 196 (1915).—Brenan, T.T.C.L. : 112 (1949).—Hauman & Wilczek, F.C.B. 2 : 465 (1951). Type from Tanganyika.
   Capparis bangweolensis R.E.Fr., Wiss. Ergebn. Schwed. Rhod.-Kongo-Exped. **1** : 50 (1914).—Gilg & Bened., loc. cit. Type : N. Rhodesia, Lake Bangweulu, *Fries* 1018 (B ; K ; UPS, holotype).
   Capparis sp. cfr. *C. oligantha.*—O. B. Mill. in Journ. S. Afr. Bot. **18**, 16 (1952).

Thorny, scrambling shrub up to 5 m. tall or a climber; branches rather zigzag, densely ferruginously hairy when young, remaining green for several years. Leaf-lamina 3–6 × 0·8–3 cm., oblong, elliptic, obovate or somewhat rhomboid, obtuse or emarginate at the apex, broadly cuneate or obtuse at the base, pubescent above but soon glabrescent, rather more densely pubescent below but also becoming glabrous with age; petiole up to 6 mm. long, densely pubescent. Flowers in few- to many-flowered axillary fascicles; pedicels up to 1·2 cm. long, ferruginously hairy. Buds globose. Sepals c. 4 × 2·5 mm., the inner pair somewhat narrower, broadly elliptic, obtuse at the apex, densely ferruginously appressed-hairy outside. Petals cream, 5–6 × 1·5–2·5 mm., oblanceolate, obtuse at the apex, sparsely pilose on both sides. Stamens (5) 8 (10) with filaments c. 6 mm. long. Gynophore c. 6 mm. long, sparsely pilose or glabrescent, inserted on a convex receptacle; ovary ovoid, attenuated into a short style c. 1 mm. long; stigma not wider than the style. Fruit orange, up to 1·2 cm. in diam., globose, slightly fleshy, apiculate. Seeds 2–10, 3–4 mm. in diam., globose.

Bechuanaland Prot. N : near Serondela, Chobe R., fl. 7.viii.1950, *Robertson & Elffers* 108 (K ; PRE ; SRGH). **N. Rhodesia.** N : between Panta and Mokawe, fr. 16.x.1911, *Fries* 1018 (K ; UPS). S : Nambala Mt., fl. 18.ix.1947, *Brenan & Greenway* 7878 (FHO ; K).
Also in Kenya, Uganda, Tanganyika and the Belgian Congo. Thicket-forming species or climber in riverine fringes or on termite mounds, rather rare in our area but more common in *Acacia* thornbush to the north in Tanganyika.

This species is very near *C. rothii* Oliv. of NE. and W. Africa but the latter has (15) 18 (22) stamens. There appear to be no intermediates, and *C. elaeagnoides* is evidently a good species.

6. **Capparis rudatisii** Gilg & Bened. in Engl., Bot. Jahrb. **53** : 198 (1915).—Burtt Davy, F.P.F.T. **1** : 122 (1926). Type from Natal.

Thorny, scrambling shrub up to 4 m. tall, often climbing over other shrubs and trees; branches rather zigzag, fulvous-pubescent at least when young; bark green or greenish-black ; spines sometimes absent on the young branches. Leaf-lamina 2·2–6·0 × 0·7–2·2 cm., coriaceous, narrowly oblong to elliptic-oblong, rounded or subacute and often emarginate at the apex, broadly cuneate or rounded at the base, sparsely pubescent especially near the midrib or glabrescent; petiole up to 6 mm. long, pubescent or glabrescent. Flowers in aphyllous racemes borne terminally on side branches or in the upper leaf-axils, or the lower flowers in the axils of reduced leaves ; pedicels up to 8 mm. long, densely fulvous-pubescent. Sepals 4–6 mm. long, broadly elliptic to rotund, concave, rounded at the apex, densely fulvous-pilose outside. Petals whitish, 5–8 × 2–3·5 mm., obovate-oblong to elliptic, rounded at the apex, villous outside and towards the base inside. Stamens c. 8 on filaments up to 1·0 cm. long. Gynophore c. 1 cm. long ; ovary ovoid, narrowing into a short conical style ; stigma minutely capitate, not wider than the style. Ripe fruit not known.

Mozambique. SS : Guijá, Caniçado, fl. 3.vii.1947, *Pedro & Pedrógão* 1231 (K ; LMJ ; SRGH). LM : Maputo, Bela Vista, fl. 30.vi.1944, *Torre* 6668 (BM ; LISC). Also in Natal and the Eastern Cape. In coastal and low-altitude bushland.

7. **Capparis lilacina** Gilg in Engl., Bot. Jahrb. **33** : 215 (1903).—Gilg & Bened. in Engl., Bot. Jahrb. **53** : 199 (1915).—Brenan, T.T.C.L. : 112 (1949). Type from Tanganyika (Uzaramo).

Scandent thorny shrub to 3 m. tall ; spines recurved ; branches smooth, green and glabrous. Leaf-lamina 3–5 × 1–2·5 cm., membranous, ovate-oblong, rhomboid or elliptic, apex blunt or subacute, mucronulate, rounded or broadly cuneate at the base, glabrous on both sides, shining above, dull below; petiole up to 5 mm. long, glabrous or very sparsely pubescent. Flowers 1–3(4) together in the upper axils ; pedicels up to 1 cm. long, slender, glabrous. Sepals lilac or green, 4–6 × 2–3 mm., oblong-elliptic, subacute at the apex, glabrous except for the bearded margins. Petals cream, c. 1·0 × 0·3 cm., spathulate-oblong, rounded at the apex, greyish-tomentose outside, glabrous inside. Stamens c. 6, 2 cm. long or more on very slender filaments. Gynophore 3 cm. long or more, very slender ; ovary ovoid, narrowing into a very short style and punctiform stigma. Ripe fruit not known.

Mozambique. N : Imala, between Muíte and Mecubúri, fl. 25.x.1948, *Barbosa* 2573 (LISC). MS : between Lacerdónia and Inhamitanga, fl. 10.x.1949, *Pedro & Pedrógão* 8630 (LMJ ; SRGH). Also in Tanganyika. In coastal thickets and bushland.

Barbosa 2573 is rather unusual in having a very caducous, white tomentum on its stems, petioles and the undersides of the young leaves but otherwise it agrees well with the type.

8. **Capparis orthacantha** Gilg-Bened. in Notizbl. Bot. Gart. Berl. **12** : 504 (1936). TAB **37** fig. B. Type from Tanganyika (Lindi Distr.).

Low shrub 1–2 m. tall, very similar to *C. lilacina* but readily distinguishable by its straight, erecto-patent, not recurved, spines. The rather coriaceous leaves have their tertiary veins prominently reticulate except when young, they are more acuminate at the apex than in *C. lilacina* and the petals are somewhat larger, attaining 1·4 × 0·6 cm. and sparsely pilose rather than tomentose outside. The stamens are also more numerous (18–22).

Mozambique. N : Macondes, between Muêda and Nangade, fl. 19.x.1942, *Mendonça* 958 (BM ; LISC). Also in Tanganyika. Little is known of the ecology of this species but it is probably a constituent of low-altitude and coastal thickets.

Mendonça 958 is a most unusual specimen as all the ovaries are monstrosities and consist of small " flowers " complete in all parts. Their " ovaries " in turn are abnormal and consist of even smaller sepals and petals. The other floral parts have not developed.

9. **Capparis rosea** (Klotzsch) Oliv., F.T.A. **1** : 99 (1868).—Bak. f. in Journ. Linn. Soc., Bot. **40** : 21 (1911).—Gilg & Bened. in Engl., Bot. Jahrb. **53** : 202 (1915).—Milne-

Redh. in Mem. N.Y.Bot. Gard. **8**, 3 : 216 (1953). TAB. **37** fig. A. Type : Mozambique, Sena, *Peters* (B, holotype).
Petersia rosea Klotzsch in Peters, Reise Mossamb. Bot. **1** : 168, t. 30 (1861). Type as above.
Capparis carvalhoana Gilg in Engl., Pflanzenw. Ost-Afr. **C** : 185 (1895).—Gilg & Bened. in Engl., Bot. Jahrb. **53** : 203 (1915). Type : Mozambique, Sena, *Carvalho* (B, holotype ; COI).

Much branched, thorny shrub or climber over trees up to 10 m. tall ; branches slightly zigzag, greyish-pubescent or tomentose when young. Leaf-lamina 2·5–5 × 1·2–3 cm., coriaceous, elliptic, rhomboid or ovate, obtuse at the apex, minutely mucronulate, rounded at the base, margin often somewhat undulate, greyish-tomentose on both sides, pubescent or glabrate when older ; petiole up to 5 mm. long, pubescent. Flowers single in the axils on pubescent or tomentose pedicels up to 1·3 cm. long. Buds campanulate-globose. Outer sepals 4–6 mm. long, spreading and shorter than the bud before expansion, ovate, concave, acute or mucronulate at the apex with a caducous, stellate tomentum on both sides, inner sepals greenish-white, petaloid, 3–4 times the length of the outer sepals, oblong, rounded at the apex, stellately greyish-tomentose on both sides. Petals similar to the inner sepals, up to 2·5 × 0·8 cm. Stamens 40–50, up to 2·8 cm. long, glabrous or sparsely stellate-pubescent. Gynophore c. 2 cm. long, stellately pubescent towards base and apex ; ovary oblong-ovoid with about 8 longitudinal ridges, densely grey-tomentose ; stigma sessile, truncate, almost as broad as the ovary. Fruit rose-pink, c. 4 × 2·5 cm., ellipsoid, shortly narrowed at each end, strongly 6–8-ridged, surface minutely verrucose. Seeds c. 4 mm. in diam., shining, dark brown, irregularly reniform-discoid.

N. Rhodesia. E : Beit Bridge, Luangwa R., fl. 5.ix.1947, *Brenan & Greenway* 7803 (FHO ; K). **S. Rhodesia.** N : Chirundu, fr. 17.xii.1947, *Whellan* in GHS 18464 (K ; SRGH). **Nyasaland.** S : Chikwawa, fl. 3.x.1946, *Brass* 17910 (BM ; K ; PRE ; SRGH). **Mozambique.** N : Metangula, Lago, fl. & fr. 11.x.1942, *Mendonça* 762 (BM ; LISC). Z : between Aguas Quentes and Metalola, fl. 12.viii.1942, *Torre* 4558 (BM ; LISC). T : between Mungari and Tambara, fl. 21.ix.1943, *Torre* 5823 (BM ; LISC). MS : Gorongosa, fl. 27.ix.1953, *Chase* 5092 (BM ; K ; SRGH). SS : Inhambane, between Macovane and Mambone, fl. 1.ix.1942, *Mendonça* 95 (BM ; LISC).
Also in Tanganyika. In low-altitude, dry bushland or woodland and often a constituent of thickets near river banks on alluvial soils.

. **Capparis erythrocarpos** Isert in Schr. Ges. Nat. Fr. Berl. **9** : 334, t. 9 fig. 3 (1789).—Oliv., F.T.A. **1** : 98 (1868).—Gilg & Bened. in Engl., Bot. Jahrb. **53** : 200 (1915).—Exell & Mendonça, C.F.A. **1**, 1 : 62 (1937).—Brenan, T.T.C.L. : 112 (1949).—Hauman & Wilczek, F.C.B. **2** : 468 (1951). Type from W. Africa.

A species very like *C. rosea* but with its leaves as a whole rather larger and often having acute or acuminate apices. The ovary is always glabrous, not tomentose. The whole plant is more glabrescent as a rule.

N. Rhodesia. N : Kafulwe, fr. *Angus* 710 (FHO ; K).
Also in Angola, Belgian Congo, W. Africa, Sudan, Uganda, Kenya and Tanganyika. On termite mounds and in semi-evergreen thickets.

C. erythrocarpos and *C. rosea* are very close but the presence or absence of tomentum on the ovary seems a very clear-cut character and the two species appear to be fairly well geographically separated, approaching each other only in Tanganyika.

10. CRATEVA L.

Crateva L., Sp. Pl. **1** : 444 (1753) ; Gen. Pl. ed. 5 : 203 (1754).

Shrubs or trees with 3-foliolate, petiolate leaves. Inflorescence of axillary or terminal, corymbose racemes. Flowers bisexual or unisexual by abortion. Sepals 4, open in aestivation, arising from a shallow receptacle. Petals 4 (5), rather large, clawed. Androgynophore short, dilated. Stamens 8–50. Ovary ovoid or globose on a long gynophore, 1-locular or 2-locular by intrusion of the 2 placentas, multi-ovulate ; stigma subsessile. Fruit globose or ovoid with a coriaceous pericarp, borne on a stout stipe. Seeds many, reniform with a coriaceous testa.

Crateva religiosa Forst. f., Pl. Escul. : 45 (1786) (" *Crataeva* ")—Oliv., F.T.A. **1** : 99 (1868).—Gilg & Bened. in Engl., Bot. Jahrb. **53** : 169 (1915),—Hauman & Wilczek,

Tab. 37. A.—CAPPARIS ROSEA. A1, flowering branch (× ⅔) *Dawe* 437 ; A2, fruit (× ⅔)
Kirk s.n. B.—CAPPARIS ORTHACANTHA, flowering branch (× ⅔) *Eggeling* 6750.
C.—CAPPARIS SEPIARIA. C1, flowering branch (× ⅔) *Wild* 4693 ; C2, flower (× 2)
Wild 4693. D.—CAPPARIS BRASSII. D1, flowering branch (× ⅔) *Gomes e Sousa*
3790 ; D2, male flower (× 4) *Gomes e Sousa* 3790 ; D3, bisexual flower (× 4) *Irving* 66,

F.C.B. **2** : 478 (1951).—Keay, F.W.T.A. ed. 2, **1**, 1 : 90 (1954). TAB. **33** fig. B. Type from the Society Islands.

Glabrous, deciduous shrub or smallish tree up to 9 m. tall with smooth, brown bark and prominent, yellowish lenticels on the young branches. Leaves 3-foliolate, appearing just before or just after the flowers ; leaflets membranous, 4–12 × 1–3·5 cm., elliptic, elliptic-lanceolate or ovate-lanceolate, tapering to an acuminate apex, cuneate at the base, somewhat discolorous; petiole up to 9 cm. long ; petiolules 2–9 mm. long. Flowers in terminal or axillary corymbose racemes with 10–15 flowers ; pedicels slender, up to 5 cm. long; bracts up to 1·2 cm. long, linear, often trifid in the upper half or the lower ones similar to the leaves but smaller. Sepals c. 6 × 3 mm., oblong or ovate-oblong, apex acute or subacute ; receptacle 5–6 mm. in diam., saucer-shaped, with a thickened (? nectariferous) rim opposite the sepal-bases. Petals white, c. 2·3 × 1·2 cm. in male or bisexual flowers, often much reduced or absent in female flowers, with a broadly elliptic or ovate lamina, apex obtuse, margin somewhat undulate, claw ⅓–¼ of the length of the lamina. Androgynophore very short, dilated. Stamens 16–20 ; filaments c. 3 cm. long ; anthers c. 2 × 1·5 mm., broadly elliptic. Ovary ellipsoid or globose on a gynophore c. 3·5 cm. long, rudimentary in male flowers ; ovules many on 2 placentas ; stigma subsessile, capitate or slightly bilobed. Fruit c. 6 cm. in diam., globose with a hard smooth pericarp, on a thickened woody gynophore c. 4 cm. long. Seeds numerous, 8–9 mm. in diam., subreniform.

N. Rhodesia. N : Mporokoso, Chisi Lake, fl. 24.ix.1956, *Richards* 6281 (K).
Widely distributed in tropical Africa north of our area and in Madagascar and through tropical Asia to the Pacific Islands. In riverine forest, swamp forest and on forest edges. Also found on termite mounds.

Most African material of this species has globose ovaries whilst that from the Pacific Islands has ellipsoid ovaries. There are plenty of intermediates in Madagascar and India, however, so this character has little significance at least at the species level.

11. RITCHIEA R. Br. ex G. Don

Ritchiea R. Br. [in Denham & Clapperton, Trav. : 225 (1826) *nom. nud.*] ex G. Don, Gen. Syst. **1** : 276 (1831) (" Richiea ").

Shrubs, trees or lianes with alternate or more rarely subverticillate or subopposite leaves ; petioles often rather long ; blade 3–5-foliolate or by reduction 1-foliolate or simple. Inflorescence in terminal or axillary racemes. Flowers bisexual with 4 valvate sepals in a single series arising from a cupular receptacle of which the margin is slightly thickened and annular. Petals 4 or many, often longer than the sepals. Androgynophore very short. Stamens 12–∞; filaments slender, long, the outer ones sometimes sterile and petaloid. Ovary borne on a gynophore. 1-locular with 2–4 multi-ovulate placentas ; stigma sessile or on a short style. Fruit cylindric, ellipsoid or ovoid, with 3–8 coriaceous valves or indehiscent. Seeds numerous, embedded in a fleshy pulp.

Shrubs or trees ; fruit ellipsoid or cylindric :
  Fruit cylindric ; leaflets leathery - - - - - - - - - 1. *insignis*
  Fruit ellipsoid or ovoid, 2·5–4·5 × 1·5–2·5 cm. ; leaflets membranous - 2. *albersii*
Liane ; fruit narrowly ellipsoid - - - - - - - - 3. *gossweileri*

1. **Ritchiea insignis** (Pax) Gilg in Engl., Bot. Jahrb. **33** : 209 (1903).—Gilg & Bened. in Engl., Bot. Jahrb. **53** : 179 (1915).—Brenan, T.T.C.L. : 120 (1949). TAB. **38** fig. A. Type from Tanganyika (Usambara Mts.).
    Maerua insignis Pax in Engl., Pflanzenw. Ost-Afr. **C** : 187 (1895). Type as above.
    Ritchiea bussei Gilg in Engl., Bot. Jahrb. **33** : 209 (1903).—R.E.Fr. in Wiss. Ergebn. Schwed. Rhod.-Kongo-Exped. **1** : 50 (1914).—Gilg & Bened. in Engl., Bot. Jahrb. **53** : 179 (1915). Type from Tanganyika (Kilwa Distr.).

Shrub or small tree 2–4 m. tall with somewhat pendulous, brown, densely lenticelled branches. Leaves 3–5-foliolate ; terminal leaflet 7–18 × 3–8 cm., lateral leaflets 5–15 × 3–6 cm., oblong-elliptic, oblong or obovate-oblong, rounded or shortly acuminate, often recurved at the apex, rounded or shortly cuneate at the base, glabrous on both sides, nerves raised below and sometimes above ; petiole up to 7 cm. long ; petiolules 5–7 mm. long, channelled above and articulated at the

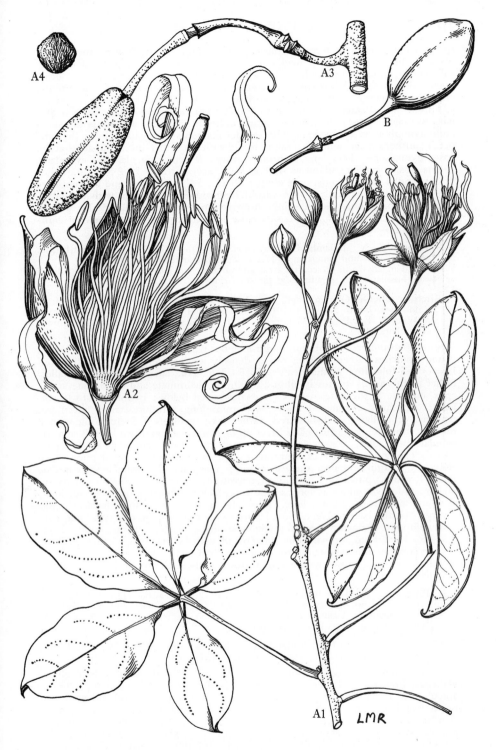

Tab. 38. A.—RITCHIEA INSIGNIS. A1, flowering branch (× ⅔) *Richards* 6145 ; A2, flower, one sepal removed (× 2) *Richards* 6145 ; A3, fruit (× ⅔) *Busse* 588 ; A4, seed (× ⅘) *Busse* 588; B.—RITCHIEA ALBERSII, fruit (× ⅔) *Geesteranus* 5424.

base. Inflorescence of terminal racemes with about 12–30 flowers, moderately dense at first ; pedicels up to 4 cm. long, glabrous, with three basal minute, subspinous, subulate bracteoles c. 0·5 mm. long, the middle one very early deciduous. Sepals 2–3 × 1–1·6 cm., ovate-acuminate, glabrous except for a minute, crisped pubescence on the margins, with about 7 longitudinal nerves ; receptacle c. 1 mm. tall. Petals 4, white, up to 4·5 × 0·5 cm., narrowly oblanceolate with a narrower claw almost as long as the lamina, or linear. Stamens numerous on a short androgynophore c. 1 mm. long, with slender filaments somewhat shorter than the petals ; anthers linear, somewhat sagittate and with the anther-thecae subacute at the base ; gynophore c. 4 cm. long, glabrous ; ovary c. 0·8 cm. long, glabrous, cylindric ; stigma sessile, capitate, about as wide as the ovary. Fruit 5–7 × 2·5–3 cm., cylindric, rounded at both ends, on a stout pedicel 2·5–4·5 cm. long and a gynophore 2·5–3 cm. long, 4- or rarely 3- or 6-valved, longitudinally dehiscent with the valves thick and coriaceous, rather shiny, brown-verrucose, somewhat fleshy within, many-seeded. Seeds 0·6 × 1·0 cm., rugose with a brown testa, somewhat compressed and 3-sided.

N. Rhodesia. N : Kalambo Falls, fr. 28.xi.1911, *Fries* 1372 (UPS). **Mozambique.** N : between Palma and Nangade, fl. 17.x.1948, *Andrada* 1360 (BM ; LISC).
Also in Tanganyika in the coastal areas and on Zanzibar Island. In coastal woodland or coastal bushland and sometimes at forest margins.

2. **Ritchiea albersii** Gilg in Engl., Bot. Jahrb. **33** : 208 (1902).—Gilg & Bened. in Engl., Bot. Jahrb. **53** : 178 (1915).—Brenan, T.T.C.L. : 120 (1949).—Hauman & Wilczek, F.C.B. **2** : 476 (1951). TAB. **38** fig. B. Type from Tanganyika (Usambara Mts.).

Small glabrous tree 4–5 m. tall in our area but reaching 10 m. tall farther north ; young branches brown, striate, rather densely lenticellate, somewhat drooping. Leaves 3–5-foliolate ; median leaflets 8–15 × 3–6·5 cm., laterals somewhat smaller, rather membranous, narrowly elliptic to elliptic, acuminate and mucronate at the apex, cuneate at the base, nerves somewhat raised below ; petioles up to 10 cm. long ; petiolules channelled above, 4–8 mm. long. Inflorescences racemose, terminal or axillary, 3–10-flowered, rather lax ; pedicels 2–5 cm. long with a minute, subulate, caducous, basal bract up to 2 mm. long and two lateral, smaller, subulate bracteoles. Sepals 1·2–2·5 × 0·6–1·2 cm., ovate, acuminate at the apex, with 7–9 longitudinal nerves, glabrous except for a minute, crisped pubescence along the margin and near the tip within ; receptacle 2–3 mm. long, cupular. Petals 4, white, up to 3 × 0·2 cm., linear, narrowed to both ends. Stamens many, on a short androgynophore as long as the receptacle ; filaments up to 4 cm. long, slender ; anthers c. 2 mm. long, linear-oblong with the anther-thecae slightly divergent at the base. Ovary c. 3 × 1·5 mm., on a gynophore c. 3 cm. long, narrowly ellipsoid or narrowly ovoid ; stigma sessile, capitate, c. 1 mm. wide. Fruit 2·5–4·5 × 1·5–2·5 cm., brown, ellipsoid or ovoid, with 4 coriaceous, verrucose valves, on a gynophore stouter but not much longer than in the flowering stage. Seeds 8–9 mm. in diam., brown with a smooth testa, embedded in a fleshy pulp, somewhat compressed.

N. Rhodesia. N : Lake Bangweulu, Samfya, fl. 12.vii.1955, *Fanshawe* 2393 (K). **S. Rhodesia.** E : Umtali, Vumba Mts., fr. 28.vi.1955, *Chase* 5464 (BM ; K ; SRGH).
Also in Tanganyika, Uganda, Kenya, Belgian Congo, Cameroons and S. Nigeria. At forest margins and in forest patches up to 1700 m.

3. **Ritchiea gossweileri** Exell & Mendonça in Bol. Soc. Brot., Sér. 2, **25** : 102 (1951) ; C.F.A. **1**, 1 : 360 (1951). Type from Angola (Moxico).

Slender-stemmed woody climber up to 3 m. tall with densely lenticellate, glabrous branches. Leaves 3-foliolate ; leaflets 6–9 × 2·5–3·5 cm., elliptic, apiculate-acuminate at the apex, acute at the base, glabrous on both sides, midrib prominent below, nerves 5–7, rather prominent above, more so below ; petiole slender, up to 5 cm. long ; petiolules c. 3 mm. long. Inflorescence of axillary racemes with 3–6 flowers ; pedicels 2·5–3 cm. long, glabrous ; basal bracteoles minute, c. 1 mm. long, subulate. Sepals 3·0 × 0·7–0·9 cm., narrowly and sometimes obliquely lanceolate, apex acuminate, margin minutely pubescent. Petals 4, greenish-yellow, up to 3·0 × 0·2 cm., narrowly linear-lanceolate. Stamens numerous ; filaments up to 2·5 cm. long. glabrous ; anthers 4 × 0·8 mm., arcuate.

linear. Ovary 5 × 1·2 mm. on a glabrous gynophore up to 2·5 cm. long, narrowly ellipsoid, glabrous. Fruit up to 4 × 1·0 cm. (? not quite mature), green and glossy, narrowly ellipsoid, papillose-verrucose, narrowed to a blunt apex, 4-valved. Seeds c. 5 mm. in diam., subglobose, embedded in a somewhat fleshy pulp.

N. Rhodesia. W : Mwinilunga, fr. 29.ix.1952, *Angus* 505 (FHO ; K).
Known only from Mwinilunga Distr. and Angola (Moxico). In evergreen riverine forest.

This species is very near *R. youngii* Exell also from Angola and *R. wittei* Wilczek from the Katanga Province of the Belgian Congo. Both these species, however, have much longer petals. All three are known from very few gatherings and more material is badly needed.

NOTE

Ritchiea englerana Busc. & Muschl. in Engl., Bot. Jahrb. **49** : 466 (1913), of which the type was said by the authors to have been collected by the Duchess of Aosta (No. 329) between Broken Hill and Bwana Mkubwa, was, according to Gilg and Benedict (in Engl., Bot. Jahrb. **53** : 183 (1915)), certainly not collected by the Duchess of Aosta and probably came from the Usambara Mts. in Tanganyika. It does not, therefore, concern our flora.

15. RESEDACEAE

By P. Taylor

Annual or perennial herbs, rarely shrubs. Leaves alternate or fasciculate, simple to pinnatipartite, stipules small. Flowers in terminal spikes or racemes, bisexual or rarely unisexual, usually zygomorphic. Calyx persistent or deciduous, 4–8-partite. Petals 4–7, free or slightly coherent, usually unequal, simple or laciniate. Disk usually present, often dilated on the adaxial side. Stamens 3–40, perigynous or inserted on the disk, often declinate, free or monadelphous at the base, not covered by the petals in bud ; anthers 2-thecous, introrse. Ovary superior, sessile or stipitate, of 2–6 free or connate carpels. Ovules 1–∞, inserted on parietal placentas or at the base of the ovary. Fruit a closed or open capsule, rarely baccate or of as many follicles as carpels. Seeds mostly numerous, reniform or hippocrepiform, without endosperm and with a curved embryo.

Caylusea abyssinica (Fresen.) Fisch. & Mey. occurs at the northern end of Lake Nyasa in southern Tanganyika and should be looked for in northern Nyasaland. *Oligomeris linifolia* (Vahl) J. F. Macbr. occurs on the eastern borders of SW. Africa and may possibly extend into Bechuanaland.

RESEDA L.

Reseda L., Sp. Pl. **1** : 448 (1753) ; Gen. Pl. ed. 5 : 207 (1754).

Characters of the family.

Reseda odorata L., Sp. Pl. ed. 2, **1** : 646 (1762). Type from Egypt.

Decumbent or erect annual herb up to 30 cm. tall ; stems ribbed, glabrous or with a few scattered hairs on the ribs. Leaves up to 7 × 2 cm., mostly simple, entire, oblanceolate, the uppermost cauline leaves sometimes 3-lobed and relatively shorter and broader, glabrous or with a few scattered hairs on the midrib. Flowers fragrant, in terminal many-flowered racemes up to 15 cm. long ; bracts 2–4 mm. long, subulate, persistent ; pedicels 2–3 mm. long in flower, elongating up to 8 mm. in fruit. Sepals 6, about 2·5 mm. long, linear, obtuse, persistent and slightly accrescent in fruit. Petals 6, yellow, 1·5–2·5 mm. long, with numerous small linear-spathulate lobes : anthers pinkish-brown or orange ; ovary narrowed at the base and shortly stipitate, apex with 3 conspicuous stigmatiferous lobes. Fruit about 10 × 7 mm., pendulous, oblong, slightly angled, rugose with bulging seeds, pericarp scaberulous. Seeds 1·5–2 mm. long, dull brown, reniform.

S. Rhodesia. W : Shangani Distr., Gwampa Forest Reserve, fl. & immature fr. iii.1957, *Goldsmith* 20/58 (K ; SRGH).
A native of Egypt and Cyrenaica, commonly cultivated for its fragrant flowers and often becoming naturalized.

16. VIOLACEAE

By N. K. B. Robson

Shrubs or perennial or annual herbs (more rarely trees). Leaves alternate (rarely opposite or whorled), simple, entire or serrate to dentate (rarely ± dissected), usually with 2 stipules. Flowers actinomorphic or, more often, ± zygomorphic, bisexual (rarely polygamous or dioecious). Sepals 5, free or shortly united, quincuncial or open in bud, usually persistent. Petals 5, free, equal or ± unequal, the anterior one frequently ± spurred, imbricate, usually deciduous, alternating with the sepals. Stamens 5, antisepalous, similar or ± dissimilar, the anterior pair in zygomorphic flowers with appendages which project into the spur, filaments free or ± united, often forming a cylinder round the ovary ; anthers usually introrse, free or ± united, usually with a prolongation of the connective. Ovary free, sessile, usually ± ovoid, 1-locular, with (2) 3 (4–5) parietal placentas each bearing 1–∞ ovules ; styles completely united, usually thickened above, often ± S-shaped in zygomorphic flowers, stigma usually undivided. Fruit a loculicidal capsule usually with contractile carinate valves, rarely a berry or nut. Seeds sometimes with a small aril, usually with abundant endosperm.

A family of over 800 species, practically confined to the tropical regions of both hemispheres, apart from the large predominantly temperate genus *Viola*.

Flowers actinomorphic or slightly zygomorphic ; anterior petal not saccate or spurred ;
  stipules deciduous ; shrubs or small trees - - - - - **1. Rinorea**
Flowers zygomorphic ; anterior petal nearly always saccate or spurred ; stipules per-
  sistent ; annual or perennial herbs, rarely shrubs or small trees :
  Anterior petal unguiculate ; stipules entire, subulate, glandular ; herbs or shrubs with
    cuneate leaf-bases - - - - - - - - **2. Hybanthus**
  Anterior petal sessile ; stipules laciniate, foliaceous, eglandular ; herb with cordate
    leaf-bases - - - - - - - - - - **3. Viola**

1. RINOREA Aubl.

Rinorea Aubl., Hist. Pl. Guian. Fr. **1** : 235, t. 93 (1775).
Alsodeia Thou., Hist. Vég. Isl. Austr. Afr. : 55 (1806).

Shrubs or small trees. Leaves alternate or, more rarely, opposite or whorled, petiolate, rarely sessile, entire or ± dentate ; stipules small, enclosing the terminal bud, deciduous. Inflorescence simple or compound, cymose or ± racemose, rarely reduced to a single flower, terminal or axillary ; pedicels articulated. Flowers actinomorphic or slightly zygomorphic. Sepals ± equal, margin ciliolate. Petals ± equal, often ± reflexed, white or greenish-white to yellow or red. Stamens with filaments simple or bearing an erect dorsal appendage, free or more frequently ± united, when the fused dorsal appendages form a " free margin " to the stamen-tube ; anthers free, with a thin dorsal prolongation of the connective (" connective-appendage ") and frequently also two free or ± united ventral (" thecal ") appendages. Ovary with 3 placentas, each bearing 1–3 or more ovules ; style usually erect ; stigma terminal. Fruit a loculicidal capsule with 3 contractile valves, rarely semi-succulent and indehiscent. Seeds few or one per capsule, usually glabrous and smooth, with abundant fleshy endosperm.

A medium-sized genus (200–300 species) characteristic of the shrub-layer of tropical forest in both hemispheres.

Inflorescence axillary, simple (rarely with a basal branch in *R. convallarioides*) :
  Stamen-filaments ± free ; connective-appendages orbicular ; petals reflexed at the
    apex only - - - - - - - - - **1. *convallarioides***
  Stamen-filaments completely united to form a tube ; connective-appendages lanceo-
    late ; petals eventually ± completely reflexed :
    Stamen-tube without free margin ; capsule rather succulent, glabrous, 1-seeded ;
      petals 6–10 mm. long, eventually ± reflexed - - - - **2. *elliptica***
    Stamen-tube with free margin ; capsule coriaceous, glabrous or ± pilose, 3-seeded
      (in *R. holtzii*) ; petals 5–6 (7) mm. long, soon reflexed :
      Young shoots and petioles shortly pubescent ; leaves ± oblong, slightly cordate to

broadly cuneate at the base ; capsule yellow-green to red-brown, glabrous
or ± pilose   -   -   -   -   -   -   -   -   -   3. *holtzii*
Young shoots and petioles glabrous ; leaves elliptic, broadly to narrowly cuneate
at the base ; capsule chocolate-brown, pilose   -   -   -   -   9. *sp. A.*
Inflorescence terminal (sometimes also axillary in *R. ilicifolia*), compound :
Young shoots and petioles glabrous ; leaves eglandular ; ovary glabrous :
Capsule rugose, dark reddish- to yellow-brown ; sepals frequently ± ribbed, midrib
not prominent, glabrous or shortly pubescent ; leaves frequently spinose-
serrate :
Sepals always markedly ribbed, usually glabrous, cucullate ; petals rather thick,
rounded at the apex ; leaves ± serrate with spines 0·5–3 mm. long ; petiole
and lamina ± coplanar   -   -   -   -   -   -   -   4. *ilicifolia*
Sepals markedly ribbed only in fruit, usually shortly pubescent, not cucullate ;
petals thin, acute ; leaves almost entire, or serrate with spines up to 1 mm. long ;
petiole and lamina forming an angle   -   -   -   -   -   7. *arborea*
Capsule smooth, bluish-green to brown ; sepals not ribbed, midrib frequently
prominent, ± pilose (rarely glabrous) ; leaves bluntly serrate   -   -   5. *poggei*
Young shoots and petioles pubescent ; leaves eglandular or glandular (if spinose-
serrate cf. *R. ilicifolia*) ; ovary usually pilose :
Leaves with glands on the lower surface ; petals thick, recurved above ; capsule
smooth, pilose   -   -   -   -   -   -   -   -   6. *welwitschii*
Leaves eglandular ; petals thin, erect ; capsule rugose, usually ± pilose   8. *gazensis*

1. **Rinorea convallarioides** (Bak. f.) Eyles in Trans. Roy. Soc. S. Afr. **5** : 421 (1916).—
Melchior in Engl. & Prantl., Nat. Pflanzenfam. ed. 2, **21** : 350 (1925).   TAB. **39**
fig. B.   Syntypes : S. Rhodesia, Chirinda Forest, *Swynnerton* 2119 (BM ; K)
Swynnerton 2119a (BM ; K).
   Alsodeia convallarioides Bak. f. in Journ. Linn. Soc., Bot. **40** : 21 (1911).   Syntypes
as above.
   Rinorea burtt-davyi Dunkley in Kew Bull. **1937** : 466 (1937).—Milne-Redh. in
Mem. N.Y.Bot. Gard. **8**, 3 : 218 (1953).   Type : Nyasaland, Cholo Mt., *Burtt Davy*
22182 (K, holotype).

Shrub or small tree, 1·5–7 m. high ; bark smooth. Branches slender, terete,
shortly pubescent at first, becoming glabrous. Leaves petiolate ; lamina (3)
4–7 (9) × 1·5–4 cm., oblong or elliptic, acute or shortly acuminate at the apex,
cuneate to rounded at the base, ± acutely serrate, rarely almost entire, subcoriaceous, olive- to greyish-green, glabrous except for the ± hirsute main veins
below, with venation prominent on both surfaces ; petiole 2–4 mm. long, channelled above, pubescent or rarely glabrous ; stipules 3–4 mm. long, triangular-
subulate, pubescent, longitudinally ribbed. Inflorescence 5–10-flowered, racemose, simple or occasionally with a short branch at the base, single or rarely ±
clustered, axillary, elongating before anthesis ; peduncle stout, greenish, shortly
brownish-pubescent. Bracts and bracteoles ovate, cucullate, longitudinally ribbed,
persistent. Flower-buds sessile, green, ovoid. Flowers erect or secund, pedicellate ; pedicels 4–6 (8) mm. long, orange, pubescent. Sepals 1–1·5 mm. long,
ovate, obtuse, cucullate, longitudinally ribbed. Petals 5–7 mm. long, c. 5 times as
long as the sepals, cream- or greenish-white, oblong to ovate-lanceolate, rounded
at the apex, ± reflexed at the apex only, glabrous. Stamens free or filaments sometimes slightly fused together at the base ; anthers oblong, almost as long as the
filament, with a ± orbicular connective-appendage broader than the anther, not
decurrent, and usually 2 thecal appendages or one bifid appendage. Ovary ±
ovoid, obscurely 3-lobed, glabrous, style 2–5 times as long as the ovary. Capsule
c. 13 mm. long, green, 3-sided, glabrous, prominently reticulate, coriaceous, few-
seeded. Seeds c. 5 mm. long.

S. Rhodesia. E : Chirinda Forest, fr. x.1947, *Chase* 495 (BM ; K ; SRGH). **Nyasaland.** C : Nchisi Mt., fl. 20.ii.1959, *Robson* 1690 (BM ; K ; SRGH). S : Cholo Mt.,
fl. 20.ix.1946, *Brass* 17687 (BM ; K ; PRE ; SRGH). **Mozambique.** MS : Vila de
Manica, Penhalonga waterfall, fl. 27.vii.1941, *Torre* 3197 (BM ; LISC).
   In Nyasaland, Mozambique and S. Rhodesia. A lower storey shrub or small tree of
evergreen forest.

   R. convallarioides differs from *R. convallariiflora* Brandt (Kenya, Cameroons and Ghana)
and *R. affinis* Robyns & Lawalrée (E. Belgian Congo, Uganda and Tanganyika) by its
sharply serrate, scarcely acuminate leaves and stout greenish peduncles. In the last two
species the leaves are entire to undulate or bluntly serrate, acuminate, and the peduncles
are slender and yellowish or orange at anthesis ; while, in addition, the petal-margins of
R. affinis are ciliolate and the flowers are clustered.

2. **Rinorea elliptica** (Oliv.) Kuntze, Rev. Gen. Pl. **1** : 42 (1891).—Sim, For. Fl. Port. E. Afr. : 11 (1909).—M. Brandt in Engl., Bot. Jahrb. **50**, Suppl. : 411 (1914).—De Wild. in Bull. Jard. Bot. Brux. **6** : 161 (1920).—Melchior in Engl. & Prantl, Nat. Pflanzenfam. ed. 2, **21** : 350, t. 151 fig. D (1925).—Burtt Davy & Hoyle, N.C.L. : 75 (1936).—Brenan, T.T.C.L. : 645 (1949). Type : Mozambique (or Tanganyika?), R. Rovuma, 32 km. from the mouth, *Kirk* (K, holotype).
Alsodeia elliptica Oliv., F.T.A. **1** : 108 (1868). Type as above.

Shrub or small tree, 1·5–12 m. high, much branched, with spreading crown, evergreen ; bark smooth, greyish- or brownish-green. Branches slender, terete, glabrous, with prominent lenticels. Leaves petiolate ; lamina 4·5–9 (11) × 2·5–5 (7) cm., broadly elliptic to oblong, acute or obtuse to rounded or slightly emarginate at the apex, cordate (less frequently rounded or broadly cuneate) at the base, serrate or serrulate, subcoriaceous to membranous, bright green, glabrous, with venation prominent on both surfaces ; petiole 6–10 mm. long, channelled above, minutely pubescent or glabrous ; stipules 6-11 mm. long, subulate, glabrous. Inflorescence up to 10-flowered, racemose, simple, axillary ; peduncle orange, pubescent. Bracts and bracteoles oblong-elliptic, cucullate, vertically ribbed, with a ciliolate margin, deciduous. Flower-buds conical, pedicellate. Flowers erect or spreading, pedicellate ; pedicels 5–10 (14) mm. long, orange, pubescent, rarely almost glabrous. Sepals c. 2 mm. long, ovate, obtuse, vertically ribbed in the upper part, glabrous or almost so apart from the margin. Petals 6–10 mm. long, 3–4 × the sepals, cream-white or pink, lanceolate, obtuse or acute at the apex, erect or ± spreading at first, becoming ± completely reflexed, margin ciliolate. Stamen-filaments completely fused to form a tube, glabrous ; anthers sessile on the margin of the tube, ovoid, pubescent, with a lanceolate, acute, decurrent connective-appendage, and two lanceolate-subulate thecal appendages. Ovary ± globose, glabrous ; style about twice as long as the ovary. Capsule 6–7 mm. long, red, globose or slightly 3-sided, glabrous, smooth, rather succulent, dehiscent at the apex, 1-seeded. Seeds c. 5 mm. long, pale brown.

Nyasaland. S : Malindi Stream, Port Herald, fl. *Topham* 541 (FHO). **Mozambique.** N : Mocímboa do Rovuma, fl. 23.ix.1948, *Pedro & Pedrógão* 5325 (SRGH). Z : Quelimane Distr., Namagoa, fl. x., *Faulkner* K 97 (COI ; K). MS : Moribane, between Muxamba and R. Revuè, fl. 3.x.1953, *Pedro* 4203 (K ; PRE). SS : Guijá, between Caniçado and Regulado Ximuzi, R. Limpopo, fr. 13.xii.1940, *Torre* 2396 (BM ; LISC).
In tropical E. Africa from Kenya and Tanganyika to Mozambique and Nyasaland. A lower storey shrub or small tree of damp evergreen lowland forest. Also in fringing forest, where it may form part of the tree layer.

R. elliptica is closely related to the Comoro Islands species *R. comorensis* Engl. (*R. hildebrandtiana* Perrier) in which the leaf-apices are shortly acuminate.

3. **Rinorea holtzii** Engl., Bot. Jahrb. **34** : 316 (1904).—M. Brandt in Engl., Bot. Jahrb. **50**, Suppl. : 411 (1914).—De Wild. in Bull. Jard. Bot. Brux. **6** : 166 (1920).— Brenan, T.T.C.L. : 645 (1949). TAB. **39** fig. A. Type from Tanganyika (Dar es Salaam).
Rinorea myrsinifolia Dunkley in [N.C.L. : 75 (1936) *nom. nud.*] Kew Bull. **1937** : 466 (1937). Type : Nyasaland, Mangoche Mt., *Clements* 465 (FHO ; K, holotype).

Shrub or small tree 1–5 m. high (up to 9 m. in Tanganyika) ; bark smooth, grey-brown. Branches slender, terete, brownish-pubescent, eventually glabrous. Leaves shortly petiolate ; lamina 2–7 (9) × 1–2·5 (3) cm., narrowly elliptic to oblong or oblong-ovate, obtuse to shortly acuminate at the apex, ± rounded to subcordate at the base, bluntly serrate, subcoriaceous, dark green, wholly glabrous (or pubescent along the midrib below) with venation prominent on both surfaces ; petiole 2–3 mm. long, channelled above, pubescent ; stipules narrowly lanceolate, 2–3 mm. long, pubescent. Inflorescence 1–8 (15)-flowered, racemose, simple, axillary ; peduncle brown, densely pubescent. Bracts ovate-oblong, cucullate, vertically ribbed, persistent. Flower-buds conical, pedicellate. Flowers erect to pendulous, pedicellate, with pedicels 4–8 mm. long (rarely to 14 mm. in fruit), brown, densely pubescent. Sepals c. 2 mm. long, ovate to oblong, obtuse, vertically ribbed in the midrib region, ± glabrous elsewhere. Petals 5–6 mm. long, white, lanceolate, acute at the apex, becoming ± completely reflexed, glabrous. Stamen-filaments completely fused to form a tube with a free ciliate margin ; anthers sessile, inserted on the inside of the tube, oblong, glabrous, with

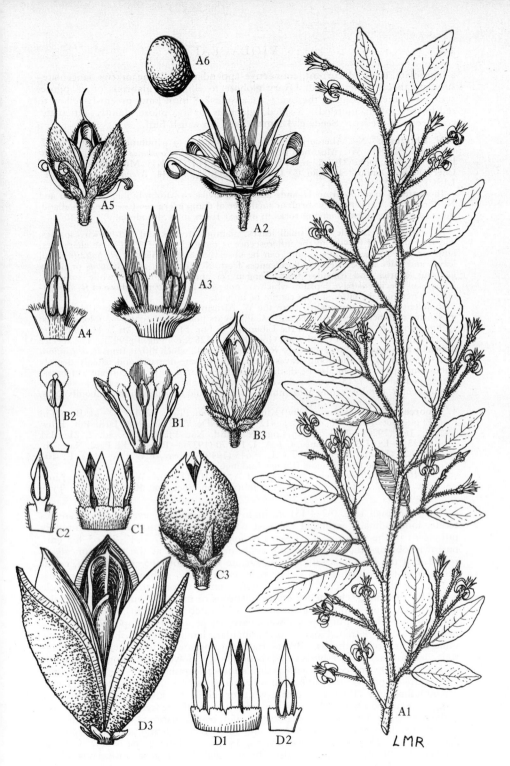

Tab. 39. A.—RINOREA HOLTZII. A1, fl. branch (× ⅔) *Fanshawe* 2671 ; A2, fl., 1 sepal, 2 petals and 1 stamen removed (×4) ; A3, andr. (×6) ; A4, anther (×6), A2–4 *Richards* 2252 ; A5, fr. (×2) ; A6, seed (×2), A5–6 *Fanshawe* 2948. B.—RINOREA CONVALLARIOIDES. B1, andr. (×6) ; B2, anther (×6), B1–2 *Kleinschmidt* 6 ; B3, fr. (×2) *Chase* 495. C.—RINOREA GAZENSIS. C1, andr. (×6) ; C2, anther (×6), C1–2 *Steedman* 5141 ; C3, fr. (×2) *Greenway* 3323. D.—RINOREA ARBOREA. D1, andr. (×6) ; D2, anther (×6), D1–2 *Dawe* 529 ; D3, fr. (×2) *Verdcourt* 1077.

a lanceolate, acute, decurrent, connective-appendage, and one or two lanceolate-subulate thecal appendages.   Ovary globose to ellipsoid, glabrous or $\pm$ pilose, style c. 4 times as long as the ovary.   Capsule 7–8 mm. long, greenish-yellow to red-brown (often mottled), 3-sided, glabrous or $\pm$ pilose, faintly reticulate, coriaceous, 3-seeded.   Seeds c. 4·5 mm. long, brownish-buff.

N. Rhodesia. N : Abercorn Distr., Inono R. close to Mpulungu Rd., fl. 1951–2. *Richards* 2252 (K). W : Mutundu R., Mufulira, fr. 22.i.57, *Fanshawe* 2948 (K). **Nyasaland.** S : Mangoche Mt., fl. xi.1935, *Clements* 573 (FHO ; K). **Mozambique.** MS : Cheringoma, Dondo, Serração C. Lopes, fl. & fr. 17.x.1944, *Mendonça* 2491 (BM ; LISC).

In scattered localities from Uganda and Tanganyika to Mozambique, Nyasaland and N. Rhodesia. A lower storey shrub or small tree of damp evergreen forest and woodland, but also forming thickets among rocks in or near rivers in N. Rhodesia.

R. myrsinifolia Dunkley has small leaves, glabrous ovaries and red-brown capsules, and the number of flowers in the inflorescence is reduced to one. All these characters, however, except the capsule colour, can be observed in specimens of *R. holtzii*—small leaves and few- to 1-flowered inflorescences from N. Rhodesia, and nearly or nearly glabrous ovaries and capsules from Tanganyika and Mozambique—while some N. Rhodesian specimens have capsules which are not paler in colour than those of *R. myrsinifolia*. Therefore the latter species cannot be upheld.

R. holtzii can be distinguished from the W. African *R. ardisiiflora* (Welw.) Kuntze by its shorter peduncles and pedicels, and usually by its pilose ovaries and $\pm$ oblong leaves.

Sim records *R. ardisiiflora* from the Libombos in For. Fl. Port. E. Afr. : 11 (1909) and cites his own For. & For. Fl. Col. Cap. Good Hope : 125 & t. 15 fig. 1 (1907). The species drawn by Sim appears, however, to be *R. natalensis* Engl., which differs from *R. ardisiiflora* and *R. holtzii* by its rhomboid leaves (frequently sinuate-dentate) and usually densely pubescent sepals. It can also be distinguished from *R. holtzii* by its cuneate leaf-bases, longer pedicels (c. 7–15 mm. in flower) and usually glabrous ovary. I have seen no specimen of *R. natalensis* from Mozambique, but it should be sought in the Libombo Mts.

4. **Rinorea ilicifolia** (Welw. ex Oliv.) Kuntze, Rev. Gen. Pl. **1** : 42 (1891).—M. Brandt in Engl., Bot. Jahrb. **50**, Suppl. : 412 (1914).—De Wild. in Bull. Jard. Bot. Brux. **6** : 166 (1920).—Melchior in Engl. & Prantl, Nat. Pflanzenfam. ed. 2, **21** : 351 (1925).—Exell & Mendonça, C.F.A. **1**, 1 : 70 (1937).—Robyns, Fl. Parc. Nat. Alb. **1** : 628 (1948).—Brenan, T.T.C.L. : 645 (1949).—Keay, F.W.T.A. ed. 2, **1** : 101 (1954). Type from Angola (Pungo Andongo).

Alsodeia ilicifolia Welw. ex Oliv., F.T.A. **1** : 108 (1868).—Welw. in Trans. Linn. Soc. **27** : 12, t. 2 (1869).—Bak. f. in Journ. Linn. Soc., Bot. **40** : 21 (1911). Type as above.

Shrub or small tree, 0·3–4 (5) m. high, much branched, evergreen. Branches terete, longitudinally wrinkled (at least when dry), glabrous or rarely minutely pubescent in the upper part. Leaves petiolate ; lamina (7·5) 12–22 × 3–8·5 cm., narrowly oblong-elliptic to obovate, acute or acuminate at the apex, cuneate to rounded at the base, coarsely spinose-serrate with spines 0·5–3 mm. long, coriaceous, bright- or bluish-green above, pale green below, glabrous or rarely pubescent along the main veins, with venation prominent on both surfaces ; petiole 5–25 (30) mm. long, channelled above, glabrous or minutely pubescent ; stipules c. 5–10 mm. long, subulate, glabrous (? or minutely pubescent), soon caducous. Inflorescence with $\pm$ numerous flowers, narrowly paniculate, compound, terminal ; peduncle stout, flattened or $\pm$ quadrangular, longitudinally ribbed, minutely pubescent or rarely glabrous. Bracts triangular, minutely pubescent or rarely glabrous, deciduous. Flower-buds ovoid, pendulous. Flowers becoming spreading or $\pm$ erect, shortly pedicellate, with pedicels minutely pubescent. Sepals 2–3 mm. long, ovate-elliptic, obtuse, ribbed fanwise, glabrous or minutely pubescent. Petals c. 5 mm. long, c. 2 × the sepals, cream-white or yellow, thick, oblong-lanceolate, obtuse, $\pm$ recurved above, glabrous. Stamen-tube without free margin, glabrous ; anthers on short filaments (? or almost sessile), ovoid, glabrous with an ovate, obtuse, decurrent connective-appendage, and one entire or emarginate thecal appendage. Ovary $\pm$ globose, glabrous ; style c. 1½ times as long as the ovary. Capsule 3-lobed, 10–20 mm. long, purplish to brown, glabrous, rugose, coriaceous, (2) 3-seeded. Seeds c. 7 mm. long.

Nyasaland. S : Malawi Mt., Port Herald, fl. ix.1937, *Topham* 1019 (FHO). **Mozambique.** MS : Chimoio, Serra do Garuso, fr. 10.xii.1943, *Torre* 6290 (BM ; LISC). From French Guinea to French Cameroons and the Sudan, and south to Tanganyika,

Nyasaland, Mozambique, Natal (Zululand) and Angola (Cuanza Norte). A lower storey shrub or small tree of damp evergreen forest and fringing forest. Occurs up to 1500 m. in E. Africa.

The stems, leaves and inflorescences of *R. ilicifolia* specimens from Mozambique and Tanganyika tend to be more pubescent than those from other parts of its range. However, the pubescence is not so long as on *R. angolensis* Exell from Angola, a very closely allied species which appears to grow in drier regions than *R. ilicifolia*. Other closely related species with spinose-dentate leaves are *R. khutuensis* Engl. (Tanganyika), in which the leaves are ± sessile and cordate-auriculate, and *R. spinosa* Baill. (Madagascar and the Comoros), which has narrowly oblong leaves. Sterile specimens of *R. ilicifolia* have been confused with *Rawsonia lucida* Harv. & Sond. (*Flacourtiaceae*).

5. **Rinorea poggei** Engl., Bot. Jahrb. **33** : 137 (1902).—M. Brandt in Engl., Bot. Jahrb. **50**, Suppl. : 414 (1914).—De Wild. in Bull. Jard. Bot. Brux. **6** : 181 (1920).—Brenan, T.T.C.L. : 646 (1949).—Keay, F.W.T.A. ed. 2, **1** : 104 (1954). Syntypes from the Belgian Congo (R. Lulua and Tondoa).
 Alsodeia poggei (Engl.) Th. & H. Dur., Syll. Fl. Cong. : 35 (1909). Type as above.
 Rinorea brachypetala sensu Robyns, Fl. Parc. Nat. Alb. **1** : 628, t. 63 (1948).

Shrub or small tree, 1·8–6 (7·5) m. high ; bark smooth, pale grey-brown. Branches terete, vertically ribbed when young, always glabrous. Leaves petiolate ; lamina 8–15 (19) × (3) 4–7·5 (9) cm., oblong-elliptic to oblanceolate, acute or usually acuminate at the apex, narrowly cuneate to rounded at the base, ± bluntly serrate, sometimes almost entire near the base, coriaceous, wholly glabrous, eglandular ; petiole 5–30 (42) mm. long, channelled above, glabrous ; stipules c. 3 mm. long (? or longer), subulate, pubescent, soon caducous. Inflorescence compound, terminal, narrowly paniculate, with few-flowered cymose clusters at the ends of short lateral branches ; peduncle stout, longitudinally ribbed, glabrous or ± pilose. Bracts ovate-triangular to lanceolate, ciliate round the margin, glabrous or ± pilose elsewhere, persistent. Flower-buds ovoid or cylindric, erect or pendulous. Flowers ± pendulous, shortly pedicellate, pedicels shortly pilose. Sepals c. 2 mm. long, orbicular-elliptic, rounded at the apex, not ribbed, frequently ± carinate, wholly glabrous or ± shortly pilose (especially near the midrib). Petals c. 4 mm. long, 1½–2 × the sepals, pale yellow or cream, thick, oblong, obtuse or rounded, recurved above, slightly unequal, glabrous, or ± ciliate round the margin. Stamen-tube without a free margin, glabrous ; anthers ± sessile or on very short filaments (i.e. with small spaces between the bases of the anthers), oblong, pilose at the base and apex and along the ventral midline, with an ovate obtuse decurrent connective-appendage, but without thecal appendages. Ovary ± globose, glabrous ; style about twice as long as the ovary. Capsule 11–16 mm. long, bluish-green to brown, ± erect, 3-lobed, smooth, coriaceous, 3- to several-seeded. Seeds c. 6 mm. long, pale brown.

N. Rhodesia. N : Kawambwa Distr., 16 km. W. of Kawambwa Boma, fl. & fr. 31.x.1952, *White* 3562 (FHO).
From Nigeria and French Cameroons to the Sudan, Uganda, Kenya, Tanganyika, Belgian Congo and N. Rhodesia. A lower storey shrub or small tree of damp evergreen forest and fringing forest, 1050–1530 m.
 R. poggei can be distinguished immediately from *R. welwitschii* by the absence (i) of glands on the lower surface of the leaves and (ii) of hairs on the young shoots, petioles, leaf-laminae and ovaries.
 R. dawei (Sprague) Brandt, from Uganda, is very closely allied to *R. poggei* but differs in having ± pubescent young shoots and (usually) petioles, brick-red capsules and a short free margin to the stamen-tube.

6. **Rinorea welwitschii** (Oliv.) Kuntze, Rev. Gen. Pl. **1** : 42 (1891).—De Wild. in Bull. Jard. Bot. Brux. **6** : 191 (1920).—Melchior in Engl. & Prantl, Nat. Pflanzenfam. ed. 2, **21** : 351, t. 151 fig. F (1925).—Exell & Mendonça, C.F.A. **1**, 1 : 73 (1937). Type from Angola (Cuanza Norte).
 Alsodeia welwitschii Oliv., F.T.A. **1** : 110 (1868) pro parte excl. specim. Guin. alt. Type as above.
 Rinorea elliottii Engl., Bot. Jahrb. **33** : 141 (1902).—M. Brandt in Engl. Bot. Jahrb. **50**, Suppl. : 414 (1914).—De Wild., tom. cit. : 161 (1920).—Melchior, loc. cit.— Keay, F.W.T.A. ed. 2, **1** : 103 (1954). Type from Sierra Leone.

Shrub or small tree, 1·5–9 m. high ; bark smooth, grey-brown. Branches terete, longitudinally ribbed and ± densely brownish-pubescent when young, eventually

± glabrous. Leaves petiolate ; lamina (6) 8·5–15 (19) × 3–7 cm., obovate to oblanceolate, abruptly acuminate at the apex, cuneate at the base, ± bluntly serrate, sometimes almost entire near the base, coriaceous, glabrous above, the main veins pubescent below, with venation prominent on both surfaces and sessile brown glands on the lower one ; petiole 5–20 (30) mm. long, channelled above, pubescent ; stipules c. 6–8 mm. long, subulate, pubescent, soon caducous. Inflorescence compound, terminal, narrowly paniculate to triangular, with few-flowered cymose clusters at the ends of short lateral branches ; peduncle stout, flattened, vertically ribbed, shortly brown-pilose. Bracts triangular to lanceolate, shortly brown-pilose, persistent. Flower-buds ovoid or cylindric, pendulous. Flowers ± pendulous, shortly pedicellate ; pedicels shortly pilose. Sepals c. 2 mm. long, ovate-elliptic, obtuse, not ribbed, pubescent or shortly pilose. Petals 3–4 mm. long, 1½–2 times as long as the sepals, yellow, thick, oblong, obtuse, recurved above, slightly unequal, sparsely pubescent or almost glabrous. Stamen-tube with free margin produced to form narrow lobes behind at least some of the anthers in each flower, glabrous ; anthers ± sessile, ovoid, pubescent or glabrous, with an ovate-obtuse, decurrent connective-appendage, and entire, bifid or paired thecal appendages. Ovary ± globose, densely pilose ; style 1½–2 times as long as the ovary. Capsule 10–13 mm. long, dark red-brown, ± erect, 3-lobed, pubescent, ± smooth, coriaceous, several-seeded. Seeds c. 4–5 mm. long.

N. Rhodesia. W : Mwinilunga, Zambezi R., 6 km. N. of Kalene Hill, fl. 20.ix.1952, *Angus* 503 (FHO).
From Sierra Leone to the Belgian Congo, N. Rhodesia and Angola. A lower storey shrub or small tree of damp evergreen forest and fringing forest.

R. welwitschii is very closely related to other species from the Belgian Congo, Tanganyika and West Africa with glandular leaves ; but it can be distinguished from all these by the following combination of characters : pubescent petioles and leaf nerves, stamen-tube with at least a partially free margin, ± sessile anthers and a pilose ovary. The anthers of *R. elliottii* are usually ± pubescent while those of *R. welwitschii* are glabrous, but this distinction breaks down sufficiently frequently to allow them to be considered conspecific.

7. **Rinorea arborea** (Thou.) Baill. in Bull. Soc. Linn. Par. **1** : 583 (1886).—Perrier in Mém. Inst. Sci. Madag., Sér. B, **2** : 329 (1949) ; in Humbert, Fl. Madag. Violacées : 42 (1955). TAB. **39** fig. D. Type from Madagascar.
  Alsodeia arborea Thou., Hist. Vég. Isl. Austr. Afr.: 57 (1805). Type as above.
  Rinorea orientalis Engl., Pflanzenw. Afr. **1**, 1 : 290 (1910) *nom. nud.* Type from Tanganyika.

Shrub or small tree, 4–9 m. high (? sometimes much taller) ; bark grey. Branches spreading, ± terete, somewhat angular when young, glabrous. Leaves petiolate ; lamina 9–24 × 4–9 cm., ± deflexed, oblong to elliptic or oblanceolate, acute (rarely acuminate) to rounded at the apex, cuneate at the base, subentire or sinuate to serrate or shortly spinose, rather coriaceous, glabrous, with prominent venation on both surfaces, eglandular ; petiole (10) 20–55 (60) mm. long, channelled above, glabrous ; stipules soon caducous. Inflorescence up to c. 20 cm. long, compound, terminal, paniculate, broadly cylindrical to pyramidal ; peduncle stout, angular, pubescent. Bracts triangular, pubescent, persistent. Flower-buds ovoid or cylindric, pendulous. Flowers becoming erect, shortly pedicellate ; pedicels pubescent. Sepals 2–3 mm. long, ovate to oblong, rounded, obscurely longitudinally ribbed or smooth, ± pubescent. Petals 4–5 mm. long, 1½–2 times as long as the sepals, white to greenish-yellow, ovate-lanceolate, acute or obtuse, scarcely recurved above, equal, ± pubescent. Stamen-tube with or without a short free 5-lobed margin, glabrous or occasionally pubescent ; anthers sessile, inserted opposite the lobes of the stamen-tube, ovate-truncate, glabrous or occasionally pubescent, with a triangular-ovate, acute, not decurrent, connective-appendage, with or without a single, short, lanceolate, thecal appendage. Ovary globose to ovoid or ellipsoid, glabrous ; style c. 3 times as long as the ovary. Capsule 20–25 mm. long, greenish or brown, ± erect, obovoid or ellipsoid, glabrous, rugose, fibrous, woody, several-seeded. Seeds c. 11 mm. long, dark brown.

Mozambique. MS : Cheringoma, Dondo, Serração C. Lopes, fl. 17.x.1944, *Mendonça* 2486 (BM ; LISC). SS : between Caniçado and Chibuto, fl. 14.xii.1942, *Mendonça* 1447 (BM ; LISC).
In eastern districts of Kenya (Mombasa), Tanganyika and Mozambique ; also in

Zanzibar and Madagascar. In evergreen rain-forest and open deciduous forest. Sea-level to 810 m., but never far from the coast.

The fruit of *R. arborea* distinguishes it at once from any related species except *R. ilicifolia* and the Madagascar species *R. longipes* (Tul.) Baill. and (?) *R. viridiflora* (Tul.) Baill. ; but these Madagascar species have smaller leaves and flowers, and shorter petioles than *R. arborea*, while *R. ilicifolia* differs by several characters (see key).

Most of the specimens of *R. arborea* are described as shrubs or small trees from 4–9 m. high ; but *Semsei* 925 (Tanganyika, Morogoro) is a " tall straight tree up to 190 ft." (=57 m.), and *Honey* 747 (Mozambique, Siluvu Hills) is a " large tree . . . buttressed at the base of the trunk." The specimens from Mozambique tend to have more markedly spinose-serrate leaves than those from E. Africa or Madagascar.

8. **Rinorea gazensis** (Bak. f.) M. Brandt in Engl., Bot. Jahrb. **50,** Suppl. : 416 (1914) " gazana ".—Eyles in Trans. Roy. Soc. S. Afr. **5** : 422 (1916).—De Wild. in Bull. Jard. Bot. Brux. **6** : 162 (1920).—Steedman, Trees etc. S. Rhod. : 52 (1933). TAB. **39** fig. C. Syntypes : S. Rhodesia, Chirinda Forest, *Swynnerton* 132 (BM ; K) ; *Swynnerton* 6500 (BM).

    Alsodeia usambarensis Engl. in Phys. Abh. Königl. Akad. Wiss. Berl. **1894** : 36 (1894) *nom. nud.*

    Alsodeia gazensis Bak. f. in Journ. Linn. Soc., Bot. **40** : 22 (1911). Syntypes as above.

    Rinorea usambarensis Engl. apud M. Brandt in Engl., Bot. Jahrb. **51** : 126 (1913).—Brandt, op. cit. **50,** Suppl. : 417 (1914).—De Wild., tom. cit. : 189 (1920).—Melchior in Engl. & Prantl, Nat. Pflanzenfam. ed. 2, **21** : 351 (1925).—Brenan, T.T.C.L. : 647 (1949). Type from Tanganyika (Usambara).

    Rinorea zimmermannii Engl. apud M. Brandt in Engl., Bot. Jahrb. **51** : 121 (1913).—Brandt, op. cit. **50,** Suppl. : 416 (1914).—De Wild., tom. cit. : 193 (1920).—Melchior, loc. cit. (1925). Type from Tanganyika (Usambara).

Shrub or small tree, c. 1–5 m. high ; bark smooth. Branches terete, longitud-inally ribbed and brownish-pubescent when young, eventually glabrous. Leaves petiolate ; lamina (8) 10–20 (26) × 4–8 cm., obovate to oblanceolate, abruptly acuminate at the apex, cuneate to subrotund at the base, serrate to dentate, sub-entire near the base, membranous, glabrous above except sometimes along the midrib, main veins ± pubescent and prominent on the lower surface only, eglandular ; petiole 5–40 (55) mm. long, channelled above, ± pubescent, curved upward at junction with lamina ; stipules subulate, pubescent, soon caducous. Inflorescence compound, terminal, paniculate, loosely cylindric to pyramidal or subcorymbose ; peduncle stout, flattened, longitudinally ribbed, pubescent. Bracts and bracteoles triangular to subulate, pubescent, persistent. Flower-buds ovoid or cylindric, pendulous. Flowers becoming erect, shortly pedicellate ; pedicels pubescent. Sepals (1) 2–3 (4) mm. long, ovate (rarely ± orbicular), obtuse to rounded, obscurely longitudinally ribbed or smooth, pubescent at least along the midrib region and margin. Petals (3) 4–5 mm. long, 1½–3 times as long as the sepals, yellow or yellowish-green, ovate-lanceolate, obtuse, not recurved above, ± equal, ± pubescent along the midrib region and margin. Stamen-tube with free irregularly 5-lobed margin, ± pubescent ; anthers on slender filaments, inserted between the lobes of the stamen-tube, oblong, ± pubescent or glabrous, with an ovate-obtuse, decurrent connective-appendage, and paired lanceolate thecal appendages. Ovary ± globose, pubescent or almost glabrous ; style 2–3 times as long as the ovary. Capsule 11–15 mm. long, blackish-brown, ± erect, 3-lobed, sparsely pubescent to shortly pilose, rugose, coriaceous, several-seeded. Seeds c. 5 mm. long.

S. Rhodesia. E : Chirinda Forest, fl. 26.x.1947, *Wild* 2219 (K ; SRGH). **Mozam-bique.** N : Muêda, Chomba, bud 25.ix.1948, *Pedro & Pedrógão* (SRGH). MS : Chimoio, Amatongas, fl. & fr. 10.xii.1943, *Torre* 6288 (BM ; LISC).

From SE. Kenya (Kwale) to Mozambique and S. Rhodesia. A lower-storey shrub or small tree of evergreen rain-forest.

The specimens of *R. gazensis* from S. Rhodesia and Mozambique all have large flowers and the stems and floral parts are always pubescent ; but the plants from the northern part of the range of the species tend to be less pubescent in general. The ovary is rarely, if ever, completely glabrous.

Engler contrasts the large flowers and paniculate inflorescence of *R. zimmermannii* with the small flowers and cymose inflorescence of *R. usambarensis*. These distinctions, how-ever, cannot be upheld as (i) flowers of intermediate size occur frequently, and (ii) the inflorescence is always cymose but varies from pyramidal or loosely cylindric to sub-corymbose.

9. Rinorea sp. A.

Branches slender, green and longitudinally ribbed when young, eventually terete, always glabrous. Leaves elliptic, petiolate ; lamina 5·5–9 × 1·8–3·8 cm., elliptic, shortly acuminate at the apex, cuneate at the base, bluntly and shallowly serrate, bluish-green above and paler green below, with venation prominent below and often darker in colour ; petiole 4–7 mm. long, channelled above, glabrous or sparsely pubescent ; stipules 3–4 mm. long, narrowly lanceolate, glabrous or sparsely pubescent. Inflorescence 8–10-flowered, racemose, simple, axillary ; peduncle brown, densely pubescent. Bracts ovate-oblong, cucullate, vertically ribbed, persistent. Flower-buds and young flowers not seen. Pedicels (in fruit) 5–6 mm. long. Sepals (in fruit) c. 2 mm. long, oblong, obtuse, longitudinally ribbed, glabrous (except round the margin). Petals (in fruit) c. 5 mm. long, lanceolate, acute, glabrous. Stamen-filaments completely fused to form a tube with a free ciliate margin ; anthers sessile, inserted on the inside of the tube, oblong, glabrous, with a lanceolate-acute, decurrent connective-appendage and two (?) lanceolate-subulate thecal appendages. Ovary not seen. Capsule 9–13 mm. long, dark chocolate-brown, 3-sided, pilose, faintly reticulate, coriaceous. Seeds not seen.

Nyasaland. N : Mugesse, Misuku, fr., *Lewis* 19 (FHO). In rain-forest.

This specimen is very similar to the Tanganyikan species *R. albersii* Engl. which, in turn, does not seem to be distinct from *R. gracilipes* Engl. from W. Africa. The inflorescences of the Nyasaland plant, however, are much shorter and bear fewer flowers than those of *R. albersii*, and the capsule is larger and darker in colour. These differences may not be of specific value ; but, until more material from Nyasaland is available, its status must remain in doubt.

2. HYBANTHUS Jacq.

Hybanthus Jacq., Enum. Syst. Pl. Ins. Carib. : 2, 17 (1760).
Ionidium Vent., Jard. Malm. **1** : sub t. 27 (1803).

Herbs, shrublets or shrubs. Leaves alternate or opposite, petiolate or sessile, entire or ± serrate ; stipules usually small and persistent, rarely foliaceous or deciduous. Inflorescence of single flowers in the leaf axils or, more rarely, terminal racemes or axillary racemes or clusters ; pedicels articulated. Flowers zygomorphic. Sepals almost equal, without a basal appendage. Petals unequal, the anterior one smaller, or usually much larger, than the others, unguiculate, with a short basal spur. Stamens with filaments of varying length, free or ± coherent, the two anterior ones (and rarely also the lateral ones) with free or coherent appendages which extend into the spur and may secrete nectar (" spur-appendages ") ; anthers free or coherent, with a thin dorsal prolongation of the connective (" connective-appendage "). Ovary with 3 placentae each bearing (2) 3–∞ ovules ; style bent downwards and ± thickened towards the apex, entire ; stigma terminal. Fruit a loculicidal capsule with 3 elastic valves, keeled and thickened along the suture. Seeds spheroidal-ovoid, smooth or ribbed or pitted, with abundant endosperm and a small aril.

A genus of nearly 100 species, distributed throughout the tropical regions but most abundant in South and Central America.

Capsule glabrous ; seeds longitudinally ribbed, rarely almost smooth ; perennial (rarely annual) herbs or shrubs - - - - - - - - 1. *enneaspermus*
Capsule shortly white-pubescent ; seeds pitted in lines or lightly ribbed ; annual herb
2. *densifolius*

1. **Hybanthus enneaspermus** (L.) F. Muell., Fragm. **10** : 81 (1876).—Eyles in Trans. Roy. Soc. S. Afr. **5** : 422 (1916).—Gomes e Sousa, Bol. Soc. Estud. Col. Moçamb. **26** : 42 (1935).—Exell & Mendonça, C.F.A. **1**, 1 : 76 (1937).—Robyns, Fl. Parc. Nat. Alb. **1** : 630 (1948).—Martineau, Rhod. Wild Fl. : 54 (1954). TAB. **40** fig. A. Type from Ceylon.
   Viola enneasperma L., Sp. Pl. **2** : 937 (1753). Type as above.
   Ionidium enneaspermum (L.) Vent., Jard. Malm. **1** : sub t. 27 (1803). Type as above.
   Calceolaria enneasperma (L.) Kuntze, Rev. Gen. Pl. **1** : 41 (1891). Type as above.

Perennial herb (occasionally annual or ± shrubby), flowering the first year, up to 60 cm. high (rarely to 3 m.). Stems slender, branching from near the base, erect or ± spreading, terete or flattened or ± ridged, with raised lines decurrent from the stipules and the leaf-bases, glabrous to pubescent or densely hirsute. Leaves not crowded, subsessile or narrowing to a short petiole ; lamina (1) 2–6·5 × 0·2–2·2 cm., linear to elliptic or obovate, acute to obtuse or rounded, cuneate at the base, entire or ± crenate to serrate-dentate, membranous or ± coriaceous, glabrous to pubescent or densely hirsute, venation prominent or not ; stipules c. 1·5–4 mm. long, subulate, glabrous or pubescent, terminating in a gland. Flowers solitary in the axils of foliage leaves. Pedicels 4–18 mm. long, slender, glabrous or ± pubescent, with two triangular or subulate bracteoles in the upper half. Sepals 3- 4 mm. long, subequal, lanceolate, acute, ± prominently carinate, glabrous or ciliate-pubescent (especially the keel). Anterior petal (5) 10–15 (20) mm. long, including a short obtuse spur, unguiculate, the claw exceeding the calyx, the lamina varying in shape from suborbicular or obcordate to subquadrate and in colour from rose-pink to violet or bluish, glabrous or ciliate ; lateral petals asymmetric, ± triangular-ascending, rounded and often darker in colour at the apex ; dorsal petal symmetric, triangular-acuminate, often expanded and darker at the apex. Androecium shorter than the lateral petals, the 3 lateral stamens with short filaments, the 2 anterior ones with longer filaments bearing pilose spur-appendages ; anthers oblong-elliptic, glabrous or pilose, with a rounded connective-appendage. Ovary ovoid-globose, glabrous ; style about twice as long as the ovary, expanded above, ascending. Capsule 5–8 mm. long, 3-sided, pale yellow, glabrous, coriaceous, smooth or slightly reticulate. Seeds c. 1·5 mm. long, c. 9–12 per capsule, ovoid-ellipsoid, longitudinally ribbed, glabrous.

A very variable plant, usually occurring in open habitats from Senegal and the Sudan to Angola, Bechuanaland, Transvaal and Natal. Also in Madagascar, the Comoro Islands, Socotra, Arabia, tropical Asia from Ceylon to Hainan, Malaysia and Australia.

Extreme forms of *H. enneaspermus* appear to be very distinct, but it does not seem possible to designate any of these as species because they all intergrade with one another. However, they are sufficiently distinct to be recognized as varieties, of which the following occur in our area.

Leaves with reticulate venation prominent on upper surface, oblanceolate, entire or slightly crenate-serrate, often obtuse or rounded at the apex, rather coriaceous ; stems little-branched, erect, herbaceous - - - - - var. *nyassensis*
Leaves without prominent venation on upper surface (or only the midrib and laterals slightly prominent) :
Leaves all linear or narrowly elliptic to oblanceolate (or the upper ones linear and the lower ones lanceolate-elliptic), glabrous to densely hirsute, usually ± entire ; stems usually spreading or ascending, herbaceous - - var. *enneaspermus*
Leaves elliptic to obovate, rarely the uppermost ones ± linear, but then plant bushy, erect :
Leaves strongly serrate with acute, spreading or ascending teeth ; stems branched above the base, becoming rather woody below ; plant ± bushy var. *serratus*
Leaves entire to crenate or shortly serrate, with teeth usually ± obtuse or rounded :
Leaves rather coriaceous, yellowish-green above, margins usually incurved ; stem and leaves ± pubescent, sometimes hoary ; herb up to 60 cm. high
var. *caffer*
Leaves thin, deep green above, margins plane ; stem and leaves glabrous or sparsely pubescent, not hoary ; woody herb or shrub up to 3 m. high
var. *latifolius*

Var. enneaspermus

Ionidium heterophyllum Vent., Jard. Malm. **1** : sub t. 27 (1803). Type from Madagascar.

Ionidium hirtum Klotzsch in Peters, Reise Mossamb. Bot. **1** : 148 (1861). Type : Mozambique, Rios de Sena, *Peters* (B, holotype †).

Ionidium enneaspermum var. *hirtum* (Klotzsch) Oliv., F.T.A. **1** : 106 (1868), ("hirta").—Engl., Pflanzenw. Ost-Afr. **C** : 277 (1895).—Bak. f. in Journ. Linn. Soc., Bot. **40** : 21 (1911). Type as for *I. hirtum*.

Hybanthus heterophyllus (Vent.) Baill., Bot. Méd. **2** : 841 (1884).—Perrier, Fl. Madag. Violacées : 8 (1955). Type as for *Ionidium heterophyllum*.

Hybanthus hirtus (Klotzsch) Engl., Bot. Jahrb. **55** : 399 (1919). Type as for *I. hirtum*.

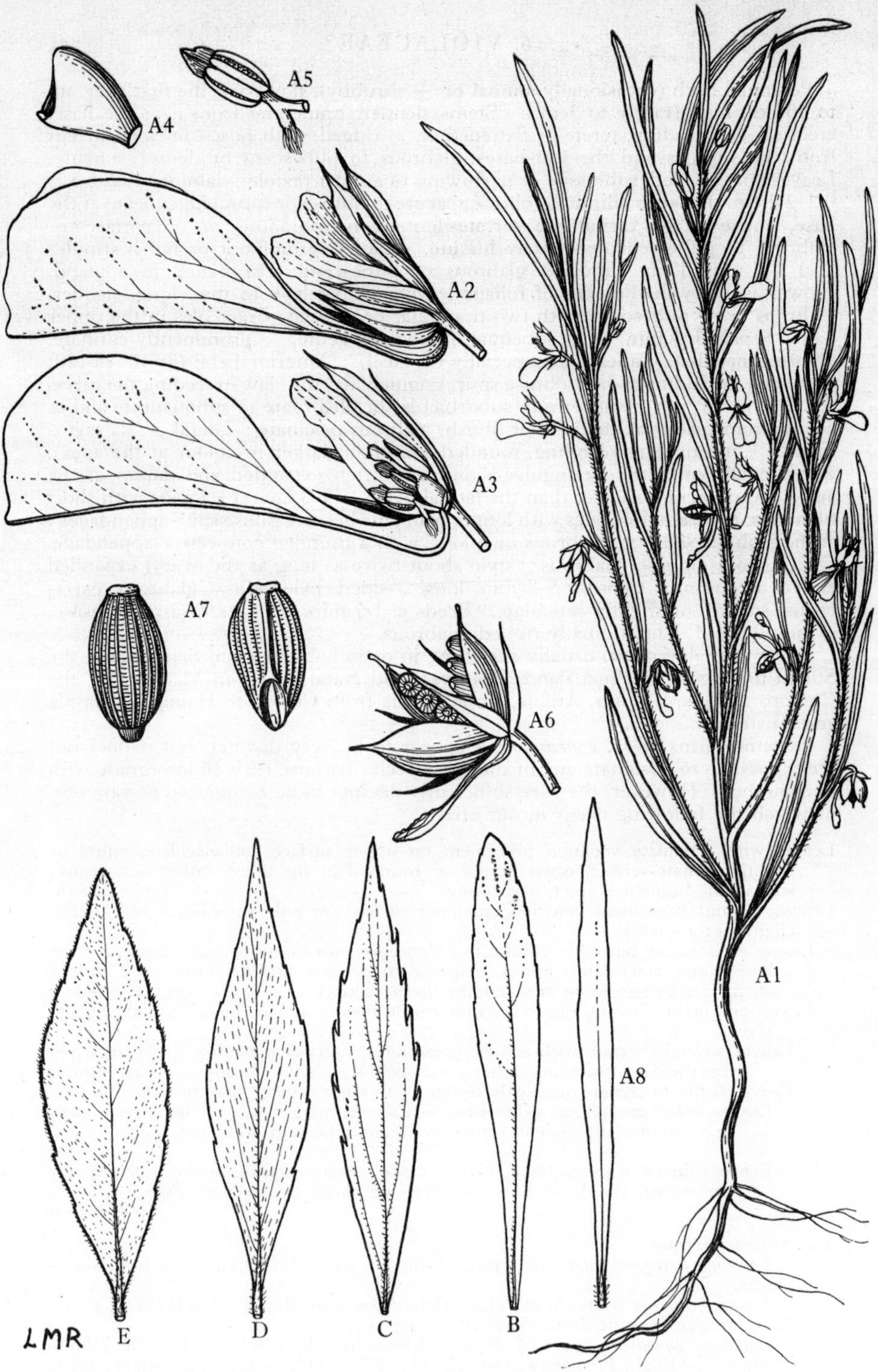

Tab. 40. A.—HYBANTHUS ENNEASPERMUS VAR. ENNEASPERMUS. A1, whole plant (× ⅔) *Kirk* s.n.; A2, flower (× 3); A3, flower in longitudinal sect. (× 3); A4, style apex (× 14); A5, anterior stamen (× 6), A2–A5 *Schlechter* 12172; A6, fruit (× 3); A7, seed (2 views) (× 10); A8, leaf (× $\frac{4}{3}$), A6–A8 *Kirk* s.n. B.—VAR. NYASSENSIS, leaf (× $\frac{4}{3}$) *Johnson* 469. C.—VAR. SERRATUS, leaf (× $\frac{4}{3}$) *Wild* 3726. D.—VAR. LATIFOLIUS, leaf (× $\frac{4}{3}$) *Gomes e Sousa* 4306. E.—VAR. CAFFER, leaf (× $\frac{4}{3}$) *Faulkner* 112.

Hybanthus hirtus var. *klotzschii* Engl., loc. cit. Type as for *I. hirtum*.
Hybanthus hirtus var. *glabrescens* Engl., loc. cit.—Burtt Davy, F.P.F.T. **1**: 131 (1926). Syntypes from Somaliland, Kenya, Tanganyika, Belgian Congo, Transvaal and Mozambique : Beira, *Schlechter* 12171 (B †).

Erect or spreading, perennial or rarely annual herb up to c. 30 cm. high ; branches arising from a ± woody rootstock. Stems and leaves densely hirsute to pubescent or glabrescent. Leaves linear to narrowly elliptic, sometimes narrowing progressively up the stem (" *heterophyllus* "), entire or rarely remotely serrate, with margins often ± reflexed.

N. Rhodesia. S : Mazabuka, fl. & fr. 4.iii.1921, *Woods* 39 (BM). **S. Rhodesia.** E : Umtali, Tsungwesi R., fl. & fr. 1.xii.1954, *Wild* 4651 (K ; SRGH). S : S. of Lundi R. on Beitbridge Rd., fl. & fr. 15.ii.1955, *E. M. & W.* 377 (BM ; SRGH). **Mozambique.** Z : Namagoa, Mocuba, fl. & fr. xi-xii.1943, *Faulkner* 260 (K ; PRE). T : Boroma, fr. i.1891, *Menyhart* 550 (K). MS : Beira, fl. 2.iv.1898, *Schlechter* 12172 (BM ; COI ; K ; PRE). SS : Muchopes, Chicomo, fl. & fr. 8.xi.1944, *Mendonça* 3326 (BM ; LISC). LM : Maputo, Goba, fl. & fr. 19.xii.1947, *Barbosa* 761 (BM ; LISC).
Distribution as for the species. In roadsides, grassland, rock clefts and other exposed dry places, often on sandy soils. From the coast to c. 1000 m. in our area.

Var. **nyassensis** (Engl.) N. Robson in Bol. Soc. Brot., Sér. 2, **32** : 168 (1958). TAB. **40** fig. B. Lectotype : Nyasaland, Shire Highlands, Blantyre, *Last* (K).
Ionidium nyassense Engl., Pflanzenw. Ost-Afr. **C** : 277 (1895). Syntypes: Nyasaland, Shire Highlands, *Last* (B† ; K) and *Buchanan* 1325 (B† ; BM).
Hybanthus nyassensis (Engl.) Engl., Bot. Jahrb. **55** : 400 (1919). Syntypes as above.

Perennial herb up to c. 30 cm. high, with erect branches arising from a woody rootstock. Stems and leaves glabrous to scabrid or shortly pubescent. Leaves oblanceolate, rather coriaceous, obtuse or rounded at the apex, increasing in size up the stem, entire or shallowly and remotely serrate, with prominent reticulate venation on the upper surface.

N. Rhodesia. N : Mpika, fr. 10.ii.1955, *Fanshawe* 2047 (K). E : Chadiza, fl. 28.xi.1958, *Robson* 755 (BM ; K ; SRGH). **Nyasaland.** C : Kota Kota Distr., fl. & fr. 3.iii.1944, *Benson* 93 (PRE). S : Likwenu, fl. i.1947, *Seddon* 1 (K). **Mozambique.** N : Vila Cabral, fl. & fr. 1.xi.1934, *Torre* 36 (BM ; COI ; LISC).
Apparently confined to S. Tanganyika, Nyasaland and the northern parts of N. Rhodesia and Mozambique. In woodland or grassland from 480–1800 m.

Var. **serratus** Engl., Bot. Jahrb. **55** : 398 (1919).—Burtt Davy, F.P.F.T. **1** : 130 (1926). TAB. **40** fig. C. Syntypes from the Transvaal and Mozambique : Ungulubi, *Schlechter* 12140 (B† ; BM ; K ; PRE).
Hybanthus sp. sensu Eyles in Trans. Roy. Soc. S. Afr. **5** : 422 (1916).

Perennial (? or annual) herb, usually erect, up to c. 45 cm. high, often branched from above the base and hence bushy in appearance. Stems and leaves ± scabrid or pubescent. Leaves lanceolate or narrowly elliptic, acute at the apex, usually densely serrate with acute, spreading or ascending teeth, rarely almost entire.

Bechuanaland Prot. N : Francistown, fl. 7.i.1926, *Rand* 44 (BM). **S. Rhodesia.** W : Bulawayo, fl. & fr. xi.1902, *Eyles* 1214 (BM ; K ; SRGH). E : Chipinga Distr., Giriwayo, fl. & fr. 19.i.1957, *Phipps* 43 (K). S : Ndanga, Triangle, fl. 22.xii.1951, *Wild* 3726 (K ; SRGH). **Mozambique.** SS : Ungulubi, fl. & fr. 14.ii.1898, *Schlechter* 12140 (BM ; K ; PRE).
Confined to the Transvaal, NE. Bechuanaland, and the southern parts of S. Rhodesia and Mozambique.
In open woodland, grassland and rocky places from 150 m. to 1350 m.

Var. **caffer** (Sond.) N. Robson in Bol. Soc. Brot., Sér. 2, **32** : 169 (1958). TAB. **40** fig. E. Type from Natal.
Ionidium caffrum Sond. in Linnaea, **23** : 13 (1850) ; in Harv. & Sond., F.C. **1** : 74 (1860). Type as above.
Hybanthus caffer (Sond.) Engl., Bot. Jahrb. **55** : 400 (1919).

Erect or spreading perennial herb up to c. 60 cm. high, rather bushy. Stems and leaves pubescent, rarely glabrous, sometimes hoary. Leaves obovate below, increasing in size upwards and becoming oblanceolate to narrowly elliptic, the uppermost ones smaller and narrower, usually rather coriaceous, acute or obtuse at the apex, entire or remotely crenate-serrate.

Mozambique. Z : Lugela Distr., Muobede, fl. & fr. 12.i.1948, *Faulkner* 112 (COI ; K).

In four isolated regions : the Natal coast round Durban, the Cape, Madagascar, and Mozambique (Namagoa and Muobede). A lowland plant of forest margins, roadside banks and sandy places.

Var. **latifolius** (De Wild.) Engl., Bot. Jahrb. **55** : 398 (1919). TAB. **40** fig. D.
Type from the Belgian Congo.
Ionidium enneaspermum var. *latifolium* De Wild., Pl. Thonn. Congol., Sér. **2** : 239, t. 17 (1911) ; in Fedde, Repert. **10** : 523 (1912). Type as above.

Erect woody herb or less frequently a spindly shrub, usually up to 50 cm. high but can attain 1·5–3 m. Stem and leaves glabrous or sparsely pubescent, especially along the main veins of the leaves below. Leaves broadly to narrowly elliptic, sometimes with a distinct petiole, membranous, acute or shortly acuminate at the apex, shallowly serrate or more rarely almost entire.

Mozambique. MS : Beira, Corone, fr. 20.iv.1956, *Gomes e Sousa* 4306 (K).
The predominant variety in W. Africa and the Congo region ; also occurs in S. Tanganyika and Mozambique.
In the ground layer of forest or woodland, and in ditches and similar damp places. From the coast to 2400 m. (in Tanganyika).

2. **Hybanthus densifolius** Engl., Bot. Jahrb. **55** : 398 (1919).—Melchior in Engl. & Prantl, Nat. Pflanzenfam. ed. 2, **21** : 359 (1925). Type from SW. Africa (Otjihua).

Annual herb, 10–15 (20) cm. high. Stems erect, unbranched or branching from the base, glabrous or scabrid to shortly pubescent. Leaves crowded, narrowing into a short petiole ; lamina linear, 3–8 × 0·1–0·35 cm., acute, entire or rarely remotely serrate, membranous, glabrous or minutely scabrid, sometimes with reflexed margins. Flowers solitary, axillary, ± hidden by the leaves. Pedicels 3–4 mm. long, scabrid. Sepals ± scabrid-pubescent. Anterior petal c. 6 mm. long, the lamina ± obovate, pink or purple, glabrous. Ovary and capsule shortly white-pubescent. Seeds with lines of pits, or lightly ribbed. Otherwise similar to *H. enneaspermus*.

Bechuanaland Prot. N : Kwebe Hills, Ngamiland, fl. & fr. 29.xii.1897, *Lugard* 74 (K).
Confined to SW. Africa and Bechuanaland Prot. Granitic and micaceous hills at c. 1200–1500 m.

In habit and leaf-shape *H. densifolius* sometimes approaches some glabrescent, narrow-leaved forms of *H. enneaspermus* var. *enneaspermus*, but the pubescent capsule and (usually) the pattern on the testa are distinctive.

3. VIOLA L.

Viola L., Sp. Pl. **2** : 933 (1753) ; Gen. Pl. ed. 5 : 402 (1754).

Herbs, rarely shrublets. Leaves alternate, petiolate, usually ± serrate ; stipules sometimes foliaceous, persistent. Inflorescence almost always of single flowers in the leaf axils ; pedicels not articulated. Flowers zygomorphic. Sepals almost equal, usually with a basal appendage (± absent in *V. abyssinica*). Petals unequal, the anterior usually larger than the others with a basal spur of varying length, blue or purple to yellow or white but always yellow at the base. Stamens with very short free filaments ; anthers free or slightly coherent, with a ± thick dorsal prolongation of the connective (" connective-appendage "), and the ventral anthers with two appendages which extend into the spur and secrete nectar (" spur-appendages "). Ovary with 3 placentas, each bearing numerous ovules ; style usually bent downwards and ± thickened towards the apex, entire or with various appendages ; stigma terminal, or apparently lateral when the style apex is reflexed. Fruit a loculicidal capsule with 3 contractile valves keeled and thickened along the sutures. Seeds globose-ovoid, usually smooth, with abundant endosperm and with or without a thickened aril.

A large cosmopolitan genus (400–500 species), most abundant in temperate regions of the Northern Hemisphere.

Viola abyssinica Steud. ex Oliv., F.T.A. **1** : 105 (1868).—Bak. f. in Journ. Linn. Soc., Bot. **40** : 21 (1911).—R.E.Fr., Wiss. Ergebn. Schwed. Rhod.-Kongo-Exped. **1** : 155

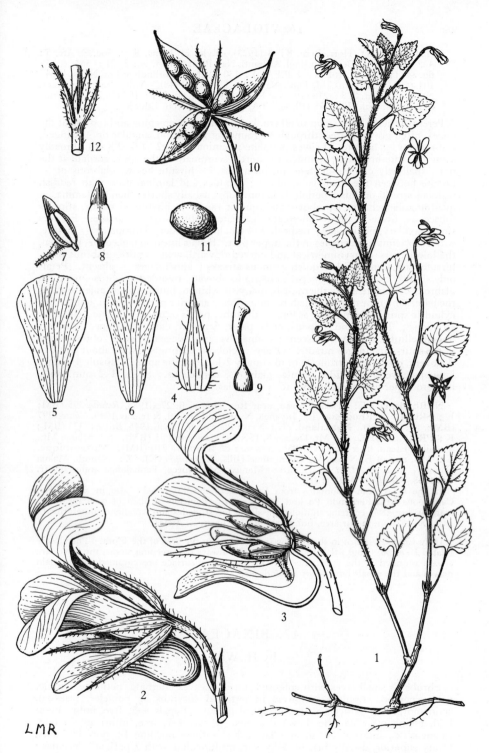

LMR

Tab. 41. VIOLA ABYSSINICA. 1, flowering and fruiting shoot (×⅔) *Eyles* in GHS 1143 ;
2, flower (×5) ; 3, flower in longitudinal sect. (×5) ; 4, sepal (×5) ; 5, upper petal (×5) ;
6, lateral petal (×5) ; 7, stamen with spur appendage (×5) ; 8, stamen without spur
appendage (dorsal view) (×5) ; 9, ovary and style (×5) ; 2–9 *Whyte* 113 ;
10, fruit (×4) and 11, seed (×6) *Sturgeon* in GHS 16948 ; 12, stipules (×3) *Eyles*
in GHS 1143.

(1914) ; in Act. Hort. Berg. **8** : 3 (1923).—Eyles in Trans. Roy. Soc. S. Afr. **5** : 422 (1916).—W. Becker in Engl. & Prantl, Nat. Pflanzenfam. ed. 2, **21** : 364, t. 159 fig. 34 (1925).—Burtt Davy, F.P.F.T. **1** : 130 (1926).—Norlindh in Bot. Notis. **1934** : 104 (1934).—Robyns, Fl. Parc. Nat. Alb. **1** : 631 (1948).—Milne-Redh. in Mem. N.Y.Bot. Gard. **8,** 3 : 218 (1953).—Keay, F.W.T.A. ed. 2, **1** : 107 (1954).—Martineau, Rhod. Wild Fl. : 55 (1954). TAB. **41.** Type from Ethiopia.

Perennial herb. Stems up to 60 cm. long, trailing or creeping and rooting at the nodes, often growing erect through surrounding vegetation, angular or ± winged, glabrous or ± hirsute. Leaves petiolate ; lamina 0·9–3·2 × 0·7–2·8 cm., broadly ovate to ovate-orbicular, rounded to shortly acuminate at the apex, cordate at the base, shallowly crenate-dentate, membranous, ± hirsute below, glabrous or ± hirsute between the main veins above ; stipules and lamina often with reddish resinous streaks or dots ; petiole 4–32 mm. long, usually shorter than the lamina, ± quadrangular, channelled above, glabrous or rarely ± hirsute, slender ; stipules foliaceous, ovate to linear-lanceolate, laciniate, glabrous or ± hirsute. Flowers single, axillary, pedicellate ; pedicels 1–4 cm. long, slender, glabrous or ± hirsute, with two laciniate bracteoles in the upper part. Sepals linear to lanceolate, unequal, the lower pairs ± asymmetrical and curved upward, acute, entire, glabrous or ± hirsute, sometimes with reddish resinous streaks ; appendages ± absent. Petals only slightly exceeding the sepals, oblong to obovate, rounded or obtuse, unequal, glabrous or hirsute along the midrib, mauvish-white to bluish-violet, paler towards the base, sometimes with reddish resinous dots ; anterior petal 5–8 mm. long with a cylindric spur 2–3 mm. long, as long as the other petals or up to ⅓ shorter, with dark striations ; lateral and upper petals paler, without dark striations. Anthers oblong-elliptic, shortly pubescent ; connective-appendage orange, ± obtuse ; tips of the spur-appendages hirsute. Ovary ± globose, glabrous ; style about equal to the ovary, obconic, ± truncate or unequally 2-lobed at the apex. Capsule 4–6 mm. long, pale yellow, 3-sided, acute, glabrous, coriaceous. Seeds 1–2 mm. long, numerous per capsule, ellipsoid to ± globular.

N. Rhodesia. E : Nyika Plateau, near Rest House, fl. 27.x.1958, *Robson* 395 (BM ; K ; SRGH). **S. Rhodesia.** E : Inyanga, Nyangani Farm, fl. & fr. 2.iv.1949, *Chase* 1297 (BM ; K ; SRGH). **Nyasaland.** N : Nyika Plateau, fl. 18.viii.1946, *Brass* 17323 (BM ; K ; PRE ; SRGH). C : Mt. Dedza, fl. 15.x.1937, *Longfield* 31 (BM). S : Mlanje Mt., Tuchila Plateau, fl. & fr. 26.vii.1956, *Newman & Whitmore* 210 (BM). **Mozambique.** N : Mts. east of L. Nyasa, W. Livingstone Hills, fl., *Johnson* (K). Z : Namuli, Makua Country, fl. & fr., *Last* (K). MS : Manica, Macequece, Penhalonga waterfall, fr. 9.vi.1948, *Mendonça* 4472 (LISC).

In the E. African mountains from Ethiopia to the Transvaal ; also in S. Nigeria, Cameroons Mt., Fernando Po and Madagascar. Typically a plant of montane forest margins, but also occurring in moist grassland, forest glades and woodland undergrowth. At 1200–2520 m. in our area, but reaching 3350 m. in E. Africa.

Only *Viola abyssinica* in the narrow sense occurs in the area of the Flora. The closely related *V. eminii* (Engl.) R.E.Fr. from E. Africa always has creeping stems, the leaves are always rounded at the apex, the hairs on the upper surface are confined to the main nerves and the sepals have short appendages.

17. BIXACEAE

By H. Wild

Shrubs or small trees with coloured juice. Leaves alternate, petiolate, simple, palmatinerved, stipulate. Flowers in terminal corymbs or panicles, bisexual, actinomorphic ; pedicels with 5–6 apical glands. Sepals 4–5, free, imbricate in bud, falling off when the flower expands. Petals 4–7, free, imbricate in bud. Stamens ∞, inserted on an annular disk ; anthers opening by pore-like slits. Ovary superior, usually bristly, 1-locular, multiovulate with 2 parietal placentas ; style 1, stigma 2-lobed. Fruit a densely echinate-setose or smooth capsule, 2-valved ; valves thick with the placentas in the middle. Seeds numerous ; testa fleshy, densely studded with red glands ; endosperm copious ; embryo large.

BIXA L.

Bixa L., Sp. Pl. **1** : 512 (1753) ; Gen. Pl. ed. 5 : 228 (1754).

Characters of the family.

Bixa orellana L., Sp. Pl. **1** : 512 (1753).—Oliv., F.T.A. **1** : 114 (1868).—Ficalho, Pl. Ut. Afr. Port. : 92 (1884).—Sim, For. Fl. Port. E. Afr. : 11, t. 2 fig. A (1909).— Exell & Mendonça, C.F.A. **1**, 1 : 77 (1937).—Brenan, T.T.C.L. : 74 (1949).—Keay, F.W.T.A. ed. 2, **1**, 1 : 183 (1954). Type from tropical America.

Shrub or small tree up to 8 m. tall ; branches and rhachis of the inflorescence covered with rusty scales. Leaf-lamina 7·5–24 × 4–16 cm., ovate, apex long-acuminate, base cordate or truncate, densely scaly beneath at first, later glabrescent, 5-nerved from the base, petiole up to 12 cm. long, thickened at the base and apex. Flowers in terminal 8–50-flowered panicles or corymbs ; bracts 5–10 mm. long, caducous ; pedicels 8–10 mm. long, red-scaly, apex thickened and with 5–6 sessile glands. Sepals 10–12 mm. long, purplish, scaly, obovate, concave, blunt. Petals 5–7, unequal, obovate, obtuse or retuse, pinkish-white. Anthers violet. Ovary globose, with dense, red-blotched bristles ; style 1·2–1·5 cm. long ; stigma-lobes very short. Capsule narrowly ovoid, acute, densely covered with stiff, slender bristles, opening to the base by two valves. Seeds c. 4 mm. long, numerous, obovoid-angular ; testa covered with minute, red, sessile glands.

S. Rhodesia. E : Umtali, fr. 14.vi.1951, *Chase* 3899 (BM ; SRGH). **Mozambique,** N : Nametil, fr. 12.vii.1948, *Pedro & Pedrógão* 4431 (LMJ ; SRGH). Z : Mocuba, fr. 6.vi.1949, *Barbosa & Carvalho* 2987 (LM ; SRGH). T : Zobuè, fl. & fr. 9.v.1948, *Mendonça* 4146 (BM ; LISC). SS : Inharrime, fl. & fr. 9.x.1945, *Pedro* 258 (BM ; PRE).

An ornamental shrub and hedge plant native of tropical America. Widely cultivated in the warmer parts of Africa. The seeds are used to make a yellow dye.

18. FLACOURTIACEAE
(incl. *SAMYDACEAE*)
By H. Wild

Trees or shrubs. Leaves petiolate, alternate, simple, entire, crenate or serrate, crenations often glandular. Stipules caducous or persistent, small or large and foliaceous or wanting. Inflorescences subterminal or more usually axillary, of racemes, panicles or cymes or reduced to fascicles or glomerules, or flowers solitary. Flowers bisexual or unisexual, monoecious or dioecious, occasionally polygamous, actinomorphic, sepals and petals dissimilar or more rarely spirally arranged and ± undifferentiated ; pedicels often articulated. Sepals 3–6 or more, often persistent, sometimes accrescent, valvate or imbricate, free or connate below into a calyx-tube (or receptacle ?). Petals 3–12 or rarely more, sometimes accrescent, free, valvate or imbricate, often inserted on the margin of a hypogynous or perigynous disk, or absent. Disk extrastaminal or with free glands between the stamens or of stami-node-like scales inserted at the base of the petals (or of threads or corona-like outside our area), often adnate to the receptacle or developed as appendages to the receptacle. Stamens 5–∞, free or rarely connate into a cylinder, sometimes alternating with staminodes ; anthers 2-thecous, dehiscing longitudinally, or rarely by terminal pores, connective sometimes appendaged. Ovary superior or more rarely semi-inferior, 1-locular with 2–8 parietal placentas or incompletely 2–8-locular ; ovules 2–∞, anatropous ; styles 1–10, free or connate. Fruit a more or less fleshy berry or a capsule. Seeds sometimes arillate, with copious endosperm ; embryo straight ; cotyledons usually broad, foliaceous.

Petals or inner perianth-segments with a fleshy gland on the inner face or staminodes
    petaloid with fleshy glands :
    Stamens numerous ; perianth-segments spirally arranged :
        Racemes subspicate, up to 2 cm. long ; leaf-margins spinulose-serrate **1. Rawsonia**
        Racemes slender, 3–4 cm. long ; leaf-margins entire or remotely and minutely
          serrate - - - - - - - - - - **2. Dasylepis**

Stamens 8–10 ; perianth-lobes not spirally arranged ; sepals valvate   **3. Kiggelaria**
Petals or petaloid staminodes without fleshy glands, petals sometimes absent :
  Petals present :
    Petals more than the sepals :
      Sepals valvate ; fruit c. 4 cm. in diam., echinate with puberulous branching
        bristles or spines ; a shrublet less than 1 m. tall   **4. Buchnerodendron**
      Sepals imbricate ; fruit smooth or rarely with simple bristles and c. 2 cm. in diam. :
        Fruit bristly :
          Flowers 1·5 cm. in diam., in racemes or solitary in the upper axils
            5. Lindackeria
          Flowers up to 10 cm. in diam., in fascicles on old or 2nd year wood
            6. Caloncoba
        · Fruit not bristly ; flowers large and showy ; hairs not stellate :
          Stems unbranched from a woody rootstock ; young leaves shining and
            glutinous with minute tubercular glands ; petioles jointed
              6. Caloncoba
          Stems branching ; shrubs or small trees ; leaves without tubercular glands ;
            petioles not jointed :
              Branches without spines ; fruit ovoid, beaked, dehiscing into 4–8 valves
                7. Xylotheca
              Branches spiny ; fruit globose, indehiscent - - -   **8. Oncoba**
    Petals as many as the sepals :
      Stamens numerous, not collected into bundles ; fruit a berry -   **9. Scolopia**
      Stamens equal in number to the petals or in bundles opposite the petals ; fruit
        usually capsular :
          Leaves with 5–9 nerves from the base ; flowers dioecious ; seeds with a red aril,
            testa tessellate - - - - - - -   **17. Trimeria**
          Leaves penninerved ; flowers bisexual ; seeds without an aril :
            Style simple with a capitate or minutely bilobed stigma ; flowers in peduncu-
              late cymes ; stipules deltoid - - - - - - **13. Gerrardina**
            Styles 2–6 or 2–6-cleft ; flowers in racemes or panicles ; stipules absent or
              large and orbicular or reniform - - - - - **14. Homalium**
  Petals absent :
    Leaves with linear or circular pellucid glands - - - -   **15. Casearia**
    Leaves without pellucid glands :
      Stamens in bundles of 3–10 alternating with the disk glands :
        Seeds cottony ; leaves penninerved - - - - -   **16. Bivinia**
        Seeds not cottony, testa tessellate-pitted and with a red aril ; leaves 5–9-nerved
          from the base - - - - - - - -   **17. Trimeria**
      Stamens not in bundles :
        Flowers bisexual ; style very short ; stigma peltate ; leaves narrowly elliptic
          to oblanceolate - - - - - - -   **10. Aphloia**
        Flowers unisexual or more rarely bisexual ; styles 2–8 ; leaves from elliptic to
          obovate or orbicular :
            Stamens intermingled with the disk glands and alternating with them ; ovules
              1–6 per placenta - - - - - - -   **11. Dovyalis**
            Stamens surrounded by a ring of disk glands ; ovary incompletely 4–8-locular
              with 2 ovules per loculus one above the other ; fruit with seeds in pairs one
              above the other - - - - - - -   **12. Flacourtia**

1. RAWSONIA Harv. & Sond.

Rawsonia Harv. & Sond., F.C. **1** : 67 (1860).

Evergreen shrubs or small trees. Leaves petiolate ; lamina often rather leathery,
usually oblong or elliptic-oblong, margins spinulose-serrate. Flowers in axillary
spike-like or lax racemes, bisexual or male. Sepals 4–5, free, imbricate, concave,
unequal. Petals similar to the sepals but rather larger, with petaloid scales larger
than the petals, opposite each petal, and with a fleshy gland at their base. Stamens
numerous, the inner row arising from the receptacle, the outer attached at the base
to the petaloid scales. Ovary on a convex receptacle, 1-locular with 4–5 parietal
multiovulate placentas ; style very short ; stigmas 4–5. Fruit berry-like when
first ripe but tardily longitudinally dehiscent when dry into 4–5 sections. Seeds
few, subglobose.

Rawsonia lucida Harv. & Sond., F.C. **1** : 67 (1860).—Gilg in Engl., Bot. Jahrb. **40** :
449 (1908) ; in Engl., & Prantl, Nat. Pflanzenfam. ed. 2, **21** : 394 (1925).—Sim, For.
Fl. Port. E. Afr. : 12 (1909).—Bak. f. in Journ. Linn. Soc., Bot. **40** : 23 (1911).—
Eyles in Trans. Roy. Soc. S. Afr. **5** : 422 (1916).—Engl., Pflanzenw. Afr., **3**, 2 :

559 (1921).—Burtt Davy, F.P.F.T. **1**: 215 (1926). TAB. **42** fig. A Type from Natal.
Rawsonia schlechteri Gilg in Engl., Bot. Jahrb. **40**: 449 (1908) ; in Engl. & Prantl, Nat. Pflanzenfam. ed. 2, **21**: 394 (1925).—Engl., loc. cit.—Brenan, T.T.C.L.: 235 (1949). Syntypes from Tanganyika and Nyasaland, Blantyre, *Buchanan* in Herb. Medley Wood 6886 (B† ; K).

Evergreen shrub or small tree up to c. 7 m. tall ; vegetative parts quite glabrous except for the stipule margins. Leaf-lamina 7–16 × 2·5–6 cm., oblanceolate-oblong to lanceolate-oblong, apex acuminate, base cuneate, margin spinulose-serrate, lateral nerves in c. 7 pairs, slightly raised on both sides, venation laxly reticulate ; petioles up to 1·2 cm. ; stipules 5–7 mm. long, very caducous, narrowly oblong, ciliate at the margins. Flowers in short, axillary, spike-like racemes usually 0·5–2 cm. long ; peduncles 2–3 mm. long ; pedicels stout, c. 1 mm. long, bracteoles 1 mm. long, very caducous, ovate, acute, with ciliate margins. Sepals 4–5, 1·5–3·5 mm. in diam., unequal, orbicular, imbricate, very concave, margin ciliolate otherwise glabrous. Petals whitish-green, similar to the sepals but c. twice the size, also unequal ; petaloid scales c. 0·7 × 0·35 mm., oblong or elliptic-oblong, rounded at the apex, margin ± ciliolate with a fleshy, pubescent, elliptic or more usually bilobed gland at the base. Stamens numerous ; filaments white, 8 mm. long, glabrous ; anthers oblong, anther-thecae c. 2 × 0·75 mm., slightly divaricate at the base. Ovary pinkish, narrowly ovoid, glabrous ; style very short ; stigmas 3–5, 1 mm. long, spreading. Fruit yellow, c. 2·5 cm. in diam., globose, smooth, style persistent as a short point, slightly fleshy and berry-like at first, tardily dehiscent into 4–5 longitudinal sections when dry. Seeds few, c. 1 cm. in diam., subglobose.

N. Rhodesia. N : Kasama, fl. 26.xi.1952, *White* 3759 (FHO). W : Chingola, fl. 14.xi.1955, *Fanshawe* 2604 (K). E : Lundazi, Nyika, fl. 7.v.1952, *White* 2768 (FHO). **S. Rhodesia.** E : Umtali, fl. 21.xi.1954, *Chase* 5340 (K ; SRGH). S : Victoria, fl. iv.1921, *Eyles* 3786 (SRGH). **Nyasaland.** N : Vipya, Chamambo Forest, fr. 22.i.1956, *Chapman* 270 (FHO). C : Nchisi Mt., st. 3.ix.1929, *Burtt Davy* 21139 (FHO ; K ; PRE). S : *Buchanan* 293 (BM). **Mozambique.** MS : Baruè, Chôa Mts., fl. 8.ix.1943, *Torre* 5881 (BM ; LISC). SS : Chongoene, fr. 11.ii.1942, *Torre* 3925 (BM ; LISC).
Also in Kenya, Tanganyika, Swaziland, Transvaal and Natal. A species of evergreen forest or forest relics, common as an understorey species.

2. DASYLEPIS Oliv.

Dasylepis Oliv. in Journ. Linn. Soc., Bot. **9** : 170 (1865).

Bushes or tall trees with the vegetative parts quite glabrous. Leaves petiolate with caducous stipules ; lamina entire, serrulate or dentate. Flowers bisexual or male by abortion, in lax racemes or crowded and more or less spicate. Sepals 4–5, almost free, the outer scarcely smaller, usually orbicular. Petals 4–7, imbricate, like the inner sepals but larger, with thick hairy scales adnate at the base within. Stamens indefinite, free ; anthers linear to narrowly lanceolate, dehiscing longitudinally. Ovary free, glabrous or hairy, 1-locular with 2–4 multiovulate placentas ; style short or long, simple or divided in the upper half into 2–4 branches, stigmas 2–4, short. Fruit a globose capsule with a hard leathery pericarp, style persistent, splitting into 2–4 longitudinal sections. Seeds few.

Dasylepis burtt-davyi Edlin in Kew Bull. **1935** : 255 (1935). TAB. **42** fig. B. Type : Nyasaland, Mt. Mlanje, *Burtt Davy* 22043 (FHO ; K, holotype).

Tree with slender, dark-brown, striate branches. Leaf-lamina 10–15 × 3–5 cm., coriaceous, oblong-lanceolate, apex acute, base broadly cuneate, margin carti-laginous, entire or remotely and minutely serrulate, often somewhat undulate, lateral nerves in 6–9 pairs, looping within the margin, slightly raised on both sides, venation laxly reticulate ; petioles up to 1 cm. long. Flowers in slender, c. 6-flowered axillary racemes 3–4 cm. long ; pedicels c. 2 mm. long, with 2–3 suborbicular, ciliate-margined bracteoles 0·5–1 mm. in diam. Sepals 5, the three outer ones similar to the bracteoles but up to 3·5 mm. in diam., the two inner c. 4 × 3 mm., less coriaceous, ciliate at the margins, broadly elliptic or rotund. Petals 5, 4–5 × 3–4 mm., broadly elliptic to obovate with a tomentose scale 1·5 mm. long adnate to the base inside. Stamens c. 20 ; filaments (young) 2 mm. long ; anthers

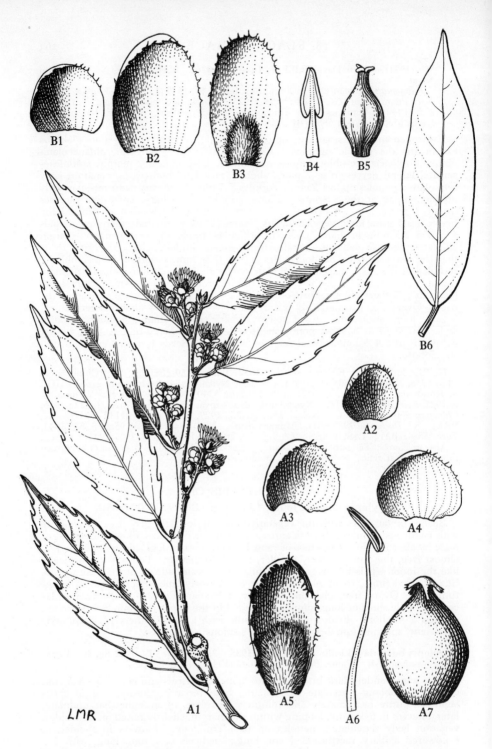

LMR

Tab. 42. A.—RAWSONIA LUCIDA. A1, flowering branch (× ⅔) *Wild* 2037 ; A2, outer sepal
(×6) *Dawe* 478 ; A3, inner sepal (×6) *Dawe* 478 ; A4, petal (×6) *Dawe* 478 ; A5,
petaloid scale (×6) *Dawe* 478 ; A6, stamen (×6) *Dawe* 478 ; A7, gynoecium (×6)
Dawe 478. B.—DASYLEPIS BURTT-DAVYI. B1, outer sepal (×6); B2, inner sepal
(×6) ; B3, petal with adnate scale (×6) ; B4, stamen (×6) ; B5, gynoecium (×6) ;
B6, leaf (× ⅔), all from *Burtt Davy* 22043.

3 × 1 mm., narrowly lanceolate, with the anther-thecae slightly divergent below. Ovary sulcate, glabrous, very shortly stipitate; styles 3, connate half-way, the stigmatic apices 1·3 mm. long, spreading, (? immature). Fruit at present not known.

Nyasaland. S : Mlanje Mt., Luchenya Plateau, fl. ix.1929, *Burtt Davy* 22043 (FHO ; K).
Endemic so far as is known to Mt. Mlanje. In montane, evergreen forest.

Known so far only from the type gathering and more material would be welcome. It is closely related to *Dasylepis integra* Warb. from Kenya and Tanganyika but the pedicels in that species are much longer. It must be noted that all the floral measurements recorded above are from rather immature flowers.

3. KIGGELARIA L.

Kiggelaria L., Sp. Pl. **2** : 1037 (1753) ; Gen. Pl. ed. 5 : 459 (1754).

Unarmed shrubs or trees with minute stellate hairs. Leaves petiolate, simple, entire or serrulate, exstipulate. Flowers dioecious. Male flowers in axillary cymes or racemes, female flowers usually solitary in the axils. Sepals 5, almost free, more or less valvate in bud. Petals 5, free with a more or less fleshy scale adnate to the base within. Male flowers : stamens 8–12, free ; filaments shorter than the anthers ; anthers 2-thecous, opening by terminal pores. Female flowers : ovary sessile, 1-locular with 2–5 parietal placentas, ovules ∞ ; styles 2–5, short, free or connate at the base. Fruit a globose, woody or slightly fleshy capsule, with 2–5 valves dehiscing from above. Seeds few or fairly numerous, testa surrounded by a viscid rather fleshy coating ; endosperm copious.

Kiggelaria africana L., Sp. Pl. **2** : 1037 (1753).—Harv. in Harv. & Sond., F.C. **1** : 71 (1860).—Bak. f. in Journ. Linn. Soc., Bot. **40** : 24 (1911).—Eyles in Trans. Roy. Soc. S. Afr. **5** : 422 (1916).—Engl., Pflanzenw. Afr. **3**, 2 : 571 (1921).—Gilg in Engl. & Prantl, Nat. Pflanzenfam. ed. 2, **21** : 413 t. 179 fig. F–H (1925).—Burtt Davy, F.P.F.T. **1** : 217 (1926).—Brenan, T.T.C.L. : 233 (1949).—Milne-Redh. in Mem. N.Y.Bot. Gard. **8**, 3 : 219 (1953). TAB. **43**. Type from Cape Province.
Kiggelaria grandiflora Warb. in Engl., Pflanzenw. Ost-Afr. **C** : 278 (1895).—Gilg in Engl., Bot. Jahrb. **40** : 468 (1908) ; in Engl. & Prantl, Nat. Pflanzenfam. ed. 2, **21** : 413 (1925).—Engl., Pflanzenw. Afr. **3**, 2 : 571 (1921). Type : Nyasaland, *Buchanan* 1469 (B, holotype † ; BM ; K).

Bush or tree up to 13 m. tall, probably evergreen ; bark pale grey, smooth ; young branches yellowish-brown, stellately tomentellous, the youngest parts often with tufted ferruginous hairs also. Leaf-lamina 3·5–9 × 2–5 cm., ovate-oblong, elliptic or lanceolate, apex acute or sometimes rounded, base broadly cuneate or rounded, margin entire, slightly undulate or shallowly and distantly toothed, finely yellowish-stellate-tomentellous above when young, later glabrescent, densely yellowish-stellate-tomentellous below although some forms are eventually glabrescent on both sides ; nerves in c. 9 pairs, rather prominent below, veins slightly raised, ± parallel between the nerves ; petiole up to 1·4 cm. long, tomentellous. Male flowers yellowish-green, in short 3–7-flowered axillary cymes ; peduncle c. 5 mm. long, tomentellous ; pedicels similar, 2–7 mm. long with minute, caducous, tomentellous bracts at the base. Sepals c. 5 × 2·5 mm. narrowly ovate, slightly keeled, subacute at the apex, densely yellowish-puberulous on both sides, free almost to the base. Petals 2 mm. long, broadly obovate, slightly keeled, obtuse or subacute at the apex, densely puberulous on both sides, basal scale subfleshy, oblong. Stamens c. 10 ; filaments up to 1 mm. long ; anthers oblong, 2·2 × 1·3 mm., stellately puberulous. Female flowers solitary in the upper axils ; peduncles c. 5 mm. long ; pedicels up to 1·5 cm. long with 1–2 minute bracts at the base. Sepals narrower as a rule than in male flowers, c. 7 × 2·5–3 mm., narrowly oblong. Petals c. 8 × 3·5 mm., narrowly oblong, subacute at the apex, otherwise as in the male. Ovary ovoid or obovoid, densely puberulous and minutely tuberculate ; styles 5, c. 3 mm. long, glabrous or very sparsely puberulous, divergent, free for c. 2 mm., connate below. Fruit 2 cm. in diam., a hard, woody, globose capsule, splitting from the apex into 5 valves, yellowish-green, densely tomentellous, warted. Seeds c. 10, 7 mm. in diam., bright orange-red, viscid, subglobose ; testa smooth.

Tab. 43. KIGGELARIA AFRICANA. 1, branch with male flowers (× ⅔) *Greenway* 6361 ; 2, male flower, front sepals and petal removed (×4) *Greenway* 6361 ; 3, female flower, front sepal and petal removed (×4) *Codd* 6658 ; 4, fruits (*Worsdell* × ⅔) s.n. ; 5, seed, part of fleshy covering removed (×2) *Worsdell* s.n. ; 6, petal of male flower (×4) *Greenway* 6361 ; 7, petal of female flower (×4) *Codd* 6658.

S. Rhodesia. E : Melsetter, fr. 25.iv.1947, *Wild* 1959 (K ; SRGH). **Nyasaland.** N : Nyika Plateau, fr. 16.viii.1946, *Brass* 17238 (K ; PRE ; SRGH). S : Zomba Plateau, fl. 23.x.1941, *Greenway* 6361 (K). **Mozambique.** T : Angónia, Mt. Domuè, fl. 18.x.1943, *Torre* 6053 (BM ; LISC).

Distributed along the mountains of eastern Africa from Mt. Kilimanjaro to the Cape Province.

A species of submontane forest often becoming a bush of about 3 m. at forest margins.

4. BUCHNERODENDRON Gürke

Buchnerodendron Gürke in Engl., Bot. Jahrb. **18** : 151, t. 6 (1893).

Unarmed shrubs or small trees. Leaves petiolate with more or less caducous stipules ; lamina serrate and frequently cordate. Flowers in axillary, cymose panicles, racemes or fascicles, bisexual or unisexual, rarely dioecious. Sepals 3, subvalvate, free to the base. Petals 6–12, imbricate, larger than the sepals. Stamens ∞, in two series, the outer somewhat longer than the inner ; filaments slender, rather short ; anthers linear, dehiscing by slits. Ovary 1-locular, with 3–5 parietal multiovulate placentas ; style simple, apex subentire. Fruit a globose, tardily dehiscent, echinate capsule splitting into 3–5 longitudinal valves (? or indehiscent). Seeds moderately numerous, ovoid or compressed, with a crustaceous testa, arillate at the base, sometimes pubescent ; embryo straight, cotyledons foliaceous, ovate.

Buchnerodendron lasiocalyx (Oliv.) Gilg in Engl., Bot. Jahrb. **40** : 467 (1908) ; in Engl. & Prantl, Nat. Pflanzenfam. ed. 2, **21** : 406 (1925).—Engl., Pflanzenw. Afr., **3, 2** : 571 (1921).—Brenan, T.T.C.L. : 230 (1949). TAB. 44 fig. A. Type from Tanganyika.

Oncoba lasiocalyx Oliv., in Hook., Ic. Pl. : t. 1485 (1885). Type as above.

Small bush usually from 0·3–1 m. tall ; branches stiffly erect, golden-tomentose at first, later glabrescent and brownish-purple. Leaf-lamina 5–17 × 3·5–10 cm., ovate-oblong, broadly ovate or obovate, apex acute or obtuse, base cordate, margins serrate, pilose above and tomentose below, at least when young ; lateral nerves in 6–9 pairs with 5–7 basal nerves, slightly prominent above, very prominent below, venation reticulate and prominent below ; petioles up to 7 cm. long, golden-tomentose ; stipules c. 1 cm. long, lanceolate to linear-lanceolate, acuminate at the apex, yellowish-tomentose. Flowers bisexual, white, in 3–8-flowered, axillary cymose panicles ; peduncle 1–3 cm. long, golden-tomentose ; pedicels similar, c. 1 cm. long ; bracts at the base of the pedicels c. 7 mm. long, narrowly elliptic, cucullate, apex acuminate, the margins serrate, tomentellous. Sepals 1–1·5 × 0·5–0·7 cm., elliptic, blunt at the apex, with many puberulous, setose processes at the back. Petals 8–12, c. 2 × 0·7 cm., obovate-oblong, blunt at the apex, narrowed at the base, margin somewhat undulate, tomentose on the back. Stamens very numerous ; filaments 3–5 mm. long, slender, puberulous ; anthers c. 5 mm. long, linear, puberulous. Ovary ellipsoid to globose, covered with soft, puberulous, setose processes ; style slender, c. 5 mm. long, puberulous except near the tip ; stigmatic apex not wider than the style. Fruit a globose capsule, echinate with many puberulous, branching bristles or spines, up to 4·5 cm. in diam. with the bristles, dehiscing into 3 valves, bristles 1·5–2 cm. long. Seeds several, c. 4 × 4 mm., compressed ; testa pubescent ; aril dark brown when dry.

Mozambique. N : Cabo Delgado, fl. & fr. 27.x.1942, *Mendonça* 1101 (BM ; LISC). Z : Mocuba, Muobede Rd., fr. 16.xi.1948, *Faulkner* 343 (COI ; K).

Also in Tanganyika. A handsome species of dense woodland or evergreen bush, usually in shady places.

5. LINDACKERIA C. Presl

Lindackeria C. Presl, Reliq. Haenk. **2** : 89, t. 65 (1835).

Unarmed shrubs or trees. Leaves petiolate, petioles sometimes elongate ; lamina usually rather large, glabrous or hairy, hairs stellate or simple, margins usually toothed. Stipules present. Flowers bisexual or male by abortion, in racemes or solitary in the axils. Sepals 3, imbricate, concave. Petals 6–12, imbricate, not much longer than the sepals. Stamens ∞, filaments slender, free or

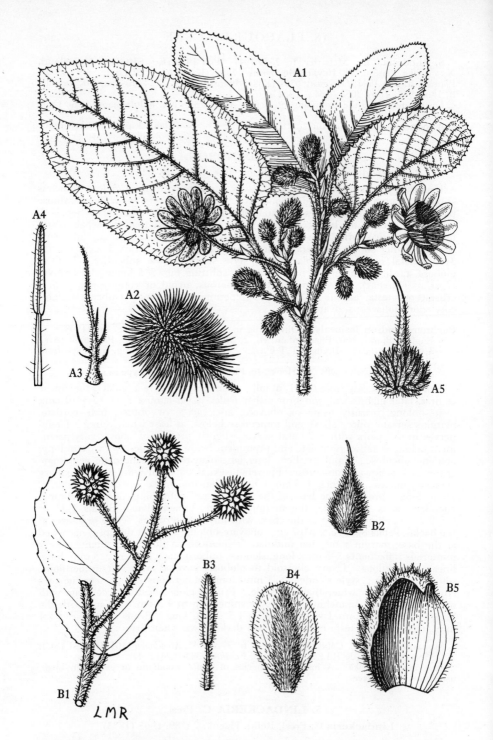

LMR

Tab. 44. A.—BUCHNERODENDRON LASIOCALYX. A1, flowering stem (× ⅔) *Faulkner* 343 ; A2, fruit (× ⅔) *Allen* 147 ; A3, fruit bristle (×2) *Allen* 147 ; A4, stamen (×6) *Faulkner* 343 ; A5, gynoecium (×4) *Faulkner* 343. B.—LINDACKERIA BUKOBENSIS. B1, leaf and fruits (× ⅔) *Richards* 5251 ; B2, gynoecium (×4) *Gillman* 460 ; B3, stamen (×6) *Gillman* 460 ; B4, petal (×4) *Gillman* 460 ; B5, sepal (×4) *Gillman* 460.

rarely connate in a tube ; anthers linear. Ovary shortly stalked, smooth, tuberculate or shortly echinate, usually hairy, 1-locular with 3 parietal placentas ; placentas multiovulate or with relatively few ovules ; style simple with inconspicuous stigmas. Fruit a globose, woody, echinate or warted capsule dehiscing tardily into 3 longitudinal valves. Seeds 1–3, with copious endosperm and a large embryo with flat, cordate cotyledons.

Leaves tapering to an acuminate apex, with somewhat appressed silky hairs on the midrib, nerves and veins ; flowers solitary in the axils - - - - - - 1. *fragrans*
Leaves abruptly acuminate, shortly pubescent or glabrescent ; flowers in 3–many-flowered axillary racemes - - - - - - - - - - 2. *bukobensis*

1. **Lindackeria fragrans** (Gilg) Gilg in Engl., Bot. Jahrb. **40**: 465 (1908); in Engl., Pflanzenw. Afr. **3**, 2 : 569 (1921).—R.E.Fr., Wiss. Ergebn. Schwed. Rhod.-Kongo-Exped. **1** : 155 (1914).—Brenan, T.T.C.L. : 233 (1949). Type from Tanganyika.
   Oncoba fragrans Gilg in Engl., Bot. Jahrb. **30** : 357 (1902). Type as above.

Shrub or small tree to 6 m. tall ; young branches silky-pubescent, later glabrescent and densely lenticelled. Leaf-lamina obovate-oblong, 2·5–8·0 × 1·2–3 cm., apex acuminate, cuneate or rounded at the base, margin finely and regularly serrulate, silky hairy on both sides on the midrib, nerves and veins but less so above ; petiole up to 1 cm. long. Flowers solitary in the upper axils ; peduncles c. 2 cm. long, densely pubescent, articulated near the base ; bracts very caducous, not seen. Sepals c. 1 × 0·5–0·6 cm., concave, broadly elliptic, apex rounded, margins membranous, silky pubescent on the back. Petals 6–8, up to 2·3 × 1·3 cm., obovate, narrowed to the base. Stamens numerous, filaments 3–5 mm. long, puberulous, anthers 3 mm. long, linear, puberulous, slightly twisted and dilated near the apex. Ovary ovoid, from muricate to softly echinate, puberulous ; style simple, 8 mm. long, slender, puberulous near the base. Mature fruit not so far known.

N. Rhodesia. N : Kalambo Falls. fl. xi.1911, *Fries* 1386 (UPS).
Also in Tanganyika. In deciduous bush or woodland.

2. **Lindackeria bukobensis** Gilg in Engl., Bot. Jahrb. **40** : 465 (1908) ; in Engl., Pflanzenw. Afr. **3**, 2 : 569 (1921).—Brenan, T.T.C.L. : 233 (1949). TAB. **44** fig. B. Syntypes from Tanganyika.

Shrub c. 5 m. tall (in Tanganyika up to 13 m. tall) with the young branches brownish-pubescent, later glabrescent and purple-brown. Leaf-lamina 5–14 × 3·5–7·5 cm., obovate or obovate-oblong, apex acute, shortly acuminate or obtuse, base rounded, margin coarsely serrate-dentate, the teeth subspinose, sparsely pilose above, densely so below, some of the hairs stellate ; lateral nerves in c. 8 pairs, prominent below, veins laxly reticulate, slightly raised below ; petiole pubescent, c. 2 cm. long. Flowers in 3–many-flowered, axillary racemes ; peduncle up to 4 cm. long in fruit, densely yellowish-pubescent ; pedicels similar, up to 2·5 cm. long ; bracts 2 mm. long, pubescent, lanceolate, very caducous. Sepals 10 × 7·5 mm., obovate, rounded at the apex, very concave, densely golden-pilose outside. Petals c. 10, 7 × 5 mm. (probably not fully developed), obovate-elliptic, rounded at the apex, densely pilose outside. Stamens numerous ; filaments 2–3 mm. long, pilose ; anthers linear, 3·5 mm. long, pilose. Ovary ovoid, densely pilose ; style simple, up to 8 mm. long, slender, pilose except towards the narrow apex. Fruit a globose capsule c. 2 cm. in diam. (including bristles), covered with pilose bristles swollen at the base. Seeds 1–3, c. 7 × 4·5 mm., irregularly compressed, with a smooth, sparsely pilose testa.

Nyasaland. N : Misuku Hills, fl. *Lewis* 37 (FHO). **N. Rhodesia.** N : Kasama, Luombe R., fr. 31.iii.1944, *E.M. & W.* 1372 (BM ; K ; LISC ; SRGH).
Also in Uganda and Tanganyika. In fringing forest and forest patches.

This species is not well known and good mature flowers have so far not been collected. It is possible that *Fanshawe* 2966 from Nchelenge on Lake Mweru is this species but the leaves are more cuneate and the whole specimen is more nearly glabrate than usual. It is, moreover, a fruiting specimen and it will be necessary to see a range of flowering material from this locality before making a final decision on this plant.

6. CALONCOBA Gilg

Caloncoba Gilg in Engl., Bot. Jahrb. **40** : 458 (1908).
Ventenatia Beauv., Fl. Owar. & Ben. **1** : 29, t. 17(1805)
non Koel. (1802).

Shrubs or trees with unarmed branches. Leaves on long petioles or almost sessile ; lamina glabrous, scaly or hairy, sometimes glandular-punctate. Stipules caducous. Flowers often large, in axillary fascicles or solitary, bisexual and male, often appearing before the leaves. Sepals 3, imbricate, concave. Petals 8–12, larger than the sepals. Stamens ∞, with linear or sagittate-linear anthers dehiscing by slits or pores. Ovary 1-locular with 5–8 multiovulate placentas ; style simple with 5–8 distinct stigmas or the stigmatic apex only slightly lobed and somewhat peltate. Fruit an echinate or smooth, dehiscent, ovoid, globose or ellipsoid capsule splitting into 5–8 valves, many-seeded and sometimes with a fleshy or gelatinous pulp.

Petiole 4–15 cm. long ; leaves broadly ovate ; fruit spiny, spines 1·5–2 cm. long
1. *welwitschii*
Petiole c. 2 mm. long ; leaves broadly oblong or ovate-oblong ; fruit smooth
2. *suffruticosa*

1. **Caloncoba welwitschii** (Oliv.), Gilg in Engl., Bot. Jahrb. **40** : 462 (1908).—Sleumer in Exell & Mendonça, C.F.A. **1**, 1 : 82 (1937). Type from Angola.
   Oncoba welwitschii Oliv., F.T.A. **1** : 117 (1868). Type as above.

Small or medium sized tree up to c. 14 m. tall, branches glabrous or puberulous. Leaves collected towards the ends of the branches ; lamina up to 25 × 18 cm., membranous, ovate, apex acuminate, base rounded or slightly cordate, 5-nerved from the base; petiole up to 15 cm. long ; stipules up to 2·5 cm. long, subulate-aristate, caducous. Flowers up to 10 cm. in diam., scented, borne on the previous year's branches or on older wood, appearing with the young leaves, in fascicles of 2–5 ; pedicels up to c. 2·5 cm. long, sparingly glandular. Sepals 2 × 1·3 cm., imbricate, very concave, glandular on exposed parts outside, oblong. Petals white, c. 10, about twice the size of the sepals, spathulate-oblong, tapering to a short basal claw, strongly veined towards the base. Stamens very numerous with slender filaments up to 2 cm. long ; anthers linear, 4 mm. long, dehiscing by apical slits. Ovary tuberculate ; placentas 5–6 ; style slender, c. 1 cm. long, stigma-lobes 5–6, linear, obtuse or capitate. Fruit a densely echinate capsule c. 8 cm. in diam. (including spines), with slender spines 1·5–2 cm. long, splitting into 5–6 recurved valves when ripe ; style persistent. Seeds numerous, 6–7 mm. in diam., globose, puberulous.

Nyasaland. N : Mugesse Forest, Misuku Hills, fl. & fr. x.1955, *Chapman* 165 (FHO). **Mozambique.** N : Muatua, Nametil, fl. 21.vii.1948, *Pedro & Pedrógão* 4611 (LMJ ; SRGH).
Also in Angola, Nigeria, Cameroons, Gaboon, Belgian Congo and Tanganyika. In the lower storey of evergreen forests and forest patches.

2. **Caloncoba suffruticosa** (Milne-Redh.) Exell & Sleumer in Fedde, Repert. **39** : 274 (1936).—Sleumer in Exell & Mendonça, C.F.A. **1**, 1 : 82 (1937). TAB. **45**. Type : N. Rhodesia, Solwezi Distr., *Milne-Redhead* 1133 (K, holotype).
   Paraphyadanthe suffruticosa Milne-Redh. in Hook., Ic. Pl. **32** : t. 3168 (1932). Type as above.

Small deciduous shrub up to c. 1 m. tall with many erect, ± unbranched stems arising from a thick woody rootstock ; branches brown, longitudinally striate. Leaf-lamina 13–26 × 7·5–16 cm., elliptic or ovate-oblong, sometimes slightly un-equal-sided, apex obtuse, bases lightly cordate, margins entire or slightly undulate, lateral nerves in 6–10 pairs, prominent on both sides, looping within the margin, venation laxly reticulate, young leaves shining and glutinous on both surfaces, minutely glandular-tubercular, minute gland-dots can also be seen on the under-sides of all but very old leaves ; petiole c. 2 mm. long, thick ; stipules c. 2 mm. long, subulate, very caducous. Flowers often, but not invariably, appearing with the very young leaves, male or bisexual, in axillary fascicles of (1) 2–6 ; pedicels 1·5–6 cm. long, densely purple-glandular, with a minute, triangular bracteole at the base. Sepals 1·2 × 0·9 cm., imbricate, very concave, broadly elliptic, glandular

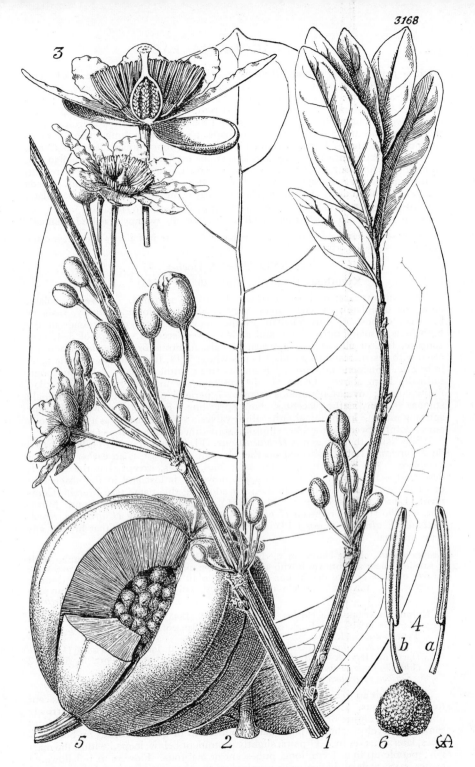

Tab. 45. CALONCOBA SUFFRUTICOSA. 1, part of flowering stem showing young leaves
(×1) *Milne-Redhead* 1133 ; 2, leaf (×1) *Milne-Redhead* 470 ; 3, longitudinal section of bisexual flower (×⅔) *Milne-Redhead* 1133 ; 4a, b anthers (×4) *Milne-Redhead*
1133 ; 5, fruit with part of pericarp removed (×⅔) *Milne-Redhead* 1133 A ; 6, seed,
(×2) *Milne-Redhead* 1133 A. From *Hooker's Icones Plantarum* with acknowledgements to the Bentham-Moxon Trustees,

on those parts exposed in bud. Petals 8, c. 2 × 1 cm. in male flowers, c. 3 × 1·3 cm. in bisexual flowers, white with golden-yellow veins towards the base, oblong-elliptic, rounded at the apex, margin irregularly undulate. Stamens c. 50; filaments 2–4 mm. long ; anthers linear, 6–8 mm. long, dehiscing by apical pores. Ovary ovoid, longitudinally striate, slightly tuberculate, slightly puberulous near the apex ; placentas 5–7 with numerous ovules ; style 4 mm. long ; stigma somewhat peltate with 5–7 folds. Fruit a subglobose or obovoid capsule up to 12 cm. in diam. splitting into 5–7 longitudinal sections each with a shallow median rib, subfleshy at first but drying out as it splits and the sections tending to remain connate at the base and apex ; pedicels elongating to 5 cm. long in fruit. Seeds many, c. 7 mm. in diam., subglobose or angularly compressed, densely tuberculate.

N. Rhodesia. W : Solwezi, fr. 10.vi.1930, *Milne-Redhead* 470 (K).
Also in Angola. In *Brachystegia* woodland or seasonally swampy grassland (dambos).

This species often flowers in a more or less leafless state just after a burn before the beginning of the dry season.

7. XYLOTHECA Hochst.

Xylotheca Hochst. in Flora, **26** : 69 (1843).

Small or moderately large shrubs ; branches unarmed. Leaves petiolate, entire or undulate ; stipules small, very caducous. Flowers solitary, male or bisexual, cymose or subumbellate in the upper leaf-axils or terminal on the branchlets, usually rather large and showy, sweet-scented. Calyx of 3–4, very concave, free or almost free imbricate sepals, glabrous or variously pubescent, often with sessile resinous glands. Petals white, 7–14, free, narrowed to the base, imbricate. Stamens numerous ; filaments free ; anthers linear, dehiscing longitudinally from above. Ovary sessile (rudimentary in male flowers), 1-locular, multiovulate ; ovules pendulous from c. 7 parietal placentas ; style terminal ; stigmas as many as the placentas, short, spreading. Fruit a tough woody capsule splitting into c. 8 longitudinal rather thick valves ; style persistent as a hard, woody apical point. Seeds numerous, ellipsoid, sometimes with a resinous aril.

It is possible that the genus *Heptaca* Lour. (Fl. Cochinch. : 657 (1790)), based on *H. africana* Lour., collected on the Mozambique coast, is an earlier name for *Xylotheca*. The description, as far as it goes, agrees, except that the ovary is described as 7-locular. Loureiro's type, however, is lost and in the absence of proof there is insufficient justification for adopting this name (see Merrill, Comment. Lour. Fl. Cochinch. : 271 (1935)). Planchon (in Hook., Lond. Journ. Bot. **6** : 296 (1847)) reduced *Heptaca* to *Oncoba*, transferring *H. africana*, which became *O. africana* (Lour.) Planch., but there is no evidence that he saw any type material.

Leaves elliptic, oblong-elliptic or rarely obovate ; flowers up to 7 cm. in diam. ; seeds
   smooth, pilose and brownish with a red aril down one side - - 1. *kraussiana*
Leaves obovate or obovate-oblong ; flowers up to 10 cm. in diam. ; seeds smooth or
   rugose, pale brown or buff, glabrous, not obviously arillate but embedded in a thin pulp
                                          2. *tettensis*

1. **Xylotheca kraussiana** Hochst. in Flora **26** : 69 (1843).—Gilg in Engl., Bot. Jahrb.
   40 : 455 (1908) ; in Engl. & Prantl, Nat. Pflanzenfam. ed. 2, **21** : 402 (1925). Type from Natal.
   Oncoba kraussiana (Hochst.) Planch. in Hook., Lond. Journ. Bot. **6** : 296 (1847).— Harv. in Harv. & Sond., F.C. **1** : 66 (1860).—Sim, For. Fl. Port. E. Afr. : 12 (1909). Type as above.
   Oncoba petersiana sensu Sim, loc. cit. (1909).

Small shrub up to 1·6 m. tall, young branches from densely yellowish-pubescent to quite glabrous. Leaf-lamina 6–10 × 2–5 cm., elliptic, oblong-elliptic or narrowly obovate, apex obtuse or acute, base rounded or broadly cuneate, margin sometimes somewhat revolute, slightly paler below, pubescent or quite glabrous on both surfaces, lateral nerves in 7–11 pairs, slightly prominent below, looped within the margin ; petiole up to c. 1 cm. long, pubescent or glabrous. Flowers in 1–3-flowered cymes in the axils of the leaves or terminal on the branches ; peduncle up to 2 cm. long, pubescent or glabrous ; pedicels up to 3·5 cm. long, pubescent or glabrous, with minute, caducous, deltoid bracts at their base. Sepals usually 3, c. 1·5 × 1 cm., broadly obovate, very concave, apex rounded, margins membranous, ± densely

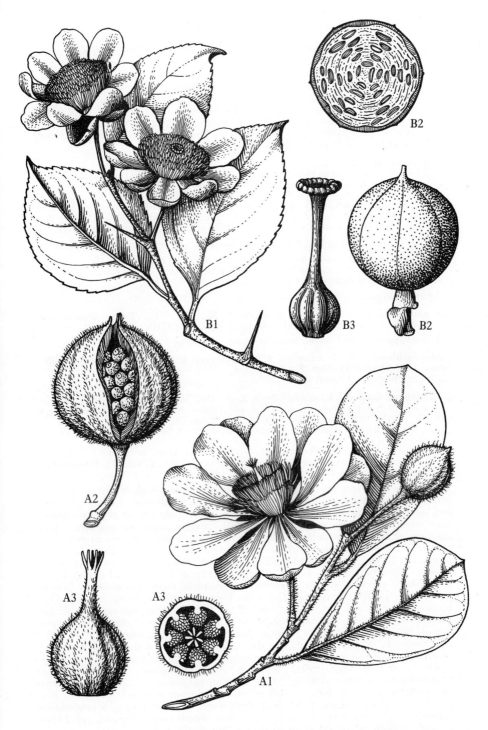

Tab. 46. A.—XYLOTHECA TETTENSIS VAR. MACROPHYLLA. A1, flowering stem (× ⅔) *Scott* s.n. ; A2, fruit (× ⅔) *Faulkner* P24 ; A3, ovary and style with cross section (× 2) *Scott* s.n. B.—ONCOBA SPINOSA. B1, flower (× ⅔) *Robinson* 256 ; B2, fruit and cross section (× ⅔) *Robinson* 256 ; B3, ovary and style (× 2⅔) *Purves* 225.

pubescent outside. Petals white, 7–12, up to 3·5 × 1·5 cm., oblanceolate to obovate, tapering to the base, with scattered woolly hairs. Stamens very numerous; filaments slender, c. 6 mm. long, glabrous or sparsely pubescent; anthers c. 6 × 1 mm., linear, glabrous, or sparsely pubescent. Ovary ellipsoid, densely hairy, sulcate; style up to 5 mm. long, columnar, somewhat sulcate, glabrous or somewhat hairy at the base; stigmas 4–6, spreading, up to 1 mm. long. Fruit woody, c. 4 × 2·5 cm., ovoid or ellipsoid, with c. 8 longitudinal ridges, splitting into c. 8 longitudinal segments; style persistent as a hard point. Seeds many, 9 × 5 mm., ellipsoid, brownish, pilose with a red sticky aril down one side.

Var. **kraussiana**

> *Oncoba tettensis* sensu Hook. f. ex. Harv., F.C. **2** : 584 (1862) quoad specim. Forbes. excl. specim. Kirk.—Schinz in Mém. Herb. Boiss. **10** : 52 (1900).
> *Oncoba macrophylla* sensu Schinz, loc. cit.
> *Xylotheca lasiopetala* Gilg in Engl., Bot. Jahrb. **40** : 457 (1908) ; in Engl. & Prantl, Nat. Pflanzenfam. ed. 2, **21** : 402 (1925). Syntypes : Mozambique, Delagoa Bay, *Monteiro* 12 (K) ; *Junod* (Z) ; *Schlechter* 11578 (COI ; K) ; *Quintas* 68 (COI).

Pubescent on the vegetative parts, varying from sparsely to densely pubescent.

Mozambique. SS : Inharrime, fr. 27.ii.1955, *E.M. & W.* 681 (BM ; LISC ; SRGH). LM : Costa do Sol, fl. 7.xii.1947, *Barbosa* 679 (BM ; LISC).
Also in Natal and Zululand. Coastal areas.

A handsome plant with sweetly scented flowers.

Var. **glabrifolia** Wild in Bol. Soc. Brot., Sér. 2, **32** : 53 (1958). Type : Mozambique, Quissico, *E.M. & W.* 698 (BM, holotype ; LISC ; SRGH).

Leaves, petioles and young branches quite glabrous.

Mozambique. SS : R. Limpopo, fl. & fr. 6.iii.1941, *Torre* 2618 (BM ; LISC). LM : Inhaca I., fl. 11.xii.1955, *Noel* (K ; PRE).
Also in Zululand. Coastal areas.

2. **Xylotheca tettensis** (Klotzsch) Gilg in Engl., Bot. Jahrb. **40** : 456 (1908) ; in Engl. & Prantl, Nat. Pflanzenfam. ed. 2, **21** : 402 (1925).—Wild in Bol. Soc. Brot., Sér. 2. **32** : 53 (1958). Type : Mozambique, Sena, *Peters* (B, holotype †).

> *Chlanis tettensis* Klotzsch in Peters, Reise Mossamb. Bot. **1** : 145 (1861). Type as above.
> *Oncoba tettensis* (Klotzsch) Hook. f. ex Harv. in Harv. & Sond., F.C. **2** : 584 (1862) quoad specim. Kirk. excl. specim. Forbes.—Oliv., F.T.A. **1** : 116 (1868).— Sim, For. Fl. Port. E. Afr. : 12 (1909). Type as above.

Shrub or small tree up to c. 5 m. tall, young branches densely yellowish-pilose with rather long hairs, puberulous with short hairs, or glabrous. Leaves appearing just before the flowers ; lamina 1·5–10 × 0·8–7·5 cm., obovate or obovate-oblong, apex rounded, base rounded or subcordate, margin entire or undulate, quite glabrous or sparsely or densely pilose on both sides, lateral nerves in 4–8 pairs, prominent below, looping within the margin, venation reticulate and subprominent below; petiole up to 1·5 cm. long or rarely up to 3 cm. long on old leaves. Flowers usually solitary but occasionally in 1–3-flowered cymes in the leaf-axils or terminal on the branchlets ; peduncle short, glabrous to densely pubescent ; pedicels similar, up to 3·5 cm. long with deltoid, caducous, pubescent bracts at the base c. 2 mm. long. Sepals 3 (4), 2 × 1·5–2 cm., rotund to orbicular, rounded at the apex, margins membranous, from glabrous to densely pubescent outside but always with some resinous glands. Petals 7–12, white, up to 4·5 × 2·4 cm., narrowly obovate, rounded at the apex, cuneate to a shortly clawed base, glabrous. Stamens very numerous on glabrous filaments up to 10 mm. long ; anthers c. 6 mm. long, linear. Ovary urceolate, ± densely pilose, smooth or longitudinally sulcate ; style up to 1 cm. long, columnar, pilose below, stigmas recurved, up to 2 mm. long. Fruit woody, subglobose or ovoid, smooth or with c. 14 longitudinal ridges, glabrous or pubescent, splitting into c. 8 longitudinal segments, style-base persisting as a woody point. Seeds many, ellipsoid, pale brown or buff, smooth or rugose, embedded in a thin scarlet pulp.

Var. **tettensis**

Leaves only sparsely pubescent, usually less than 3 cm. long; fruit usually sulcate.

Nyasaland. N : lat. 12° S, W. of Lake Nyasa, fl. *Livingstone* (K). S : Port Herald, Malawe Hills, fl. 16.xi.1933, *Lawrence* 122 (K). **Mozambique.** Z : Pebane, fl. viii.1950, *Munch* 258 (SRGH). T : Tete, fl. & fr. xi.1858, *Kirk* (K). MS : Chemba, Maringuè, fr. 19.vii.1941, *Torre* 3114 (BM ; LISC).
Also in Tanganyika. In low-altitude woodland or bushland.

Var. **macrophylla** (Klotzsch) Wild in Bol. Soc. Brot., Sér. 2, **32** : 54 (1958). Type : Mozambique, Sena, fl. *Peters* (B, holotype †). TAB. **46** fig. A.
    Chlanis macrophylla Klotzsch in Peters, loc. cit. Type as for var. *macrophylla*.
    Oncoba macrophylla (Klotzsch) Warb. in Engl. & Prantl, Nat. Pflanzenfam. **3**, 6a : 18 (1893) ; in Engl., Pflanzenw. Ost-Afr. **C** : 277 (1895). Type as for var. *macrophylla*.
    Oncoba petersiana Oliv., F.T.A. **1** : 116 (1868) *nom. illegit.* Type as for var. *macrophylla*.
    Oncoba stuhlmannii Gürke in Engl., Bot. Jahrb. **18** : 164 (1894). Type : Mozambique, Quelimane, *Stuhlmann* Ser. 1 no. 707 (B, holotype †).
    Oncoba angustipetala De Wild., Pl. Nov. Herb. Hort. Then. **1** : 13, t. 4 (1904). Type : Mozambique, Morrumbala, *Luja* 395 (BR, holotype).
    Xylotheca stuhlmannii (Gürke) Gilg in Engl., Bot. Jahrb. **40** : 456 (1908) ; in Engl. & Prantl, Nat. Pflanzenfam. ed. 2, **21** : 402 (1925). Type as for *Oncoba stuhlmannii*.
    Xylotheca macrophylla (Klotzsch) Sleumer in Fedde, Repert. **45** : 20 (1938).— Brenan, T.T.C.L. 237 (1949). Type as for var. *macrophylla*.

Mature leaves always more than 3 cm. long, densely pubescent; fruit usually smooth.

Nyasaland. S : Chiromo, fl. i. *Scott-Elliot* 8810 (BM ; K). **Mozambique.** N : Mucojo, Macomia, fl. & fr. 30.ix.1948, *Barbosa* 2294 (BM ; LISC ; LM). Z : Lugela, Mocuba, fl. & fr. ix. *Faulkner* 24 (BM ; COI ; K ; PRE). T : Boror, fl. 26.iii.1943, *Torre* 5006 (BM ; LISC). MS : Inhamitanga, fl. 13.x.1949, *Pedro & Pedrógão* 8632 (LMJ ; SRGH).
Also in Tanganyika. Ecology similar to that of var. *tettensis*.

Var. **kirkii** (Oliv.) Wild in Bol. Soc. Brot., Sér. 2, **32** : 55 (1958). Type : Mozambique, Rovuma Bay, *Kirk* (K, holotype).
    Oncoba kirkii Oliv., F.T.A. **1** : 116 (1868). Type as for var. *kirkii*.
    Xylotheca kirkii (Oliv.) Gilg in Engl., Bot. Jahrb. **40** : 455 (1908) ; in Engl. & Prantl, Nat. Pflanzenfam. ed. 2, **21** : 402 (1925).—Brenan, T.T.C.L. : 236 (1949). Type as for var. *kirkii*.

Very like var. *macrophylla* but the vegetative parts are quite glabrous. The young parts are often noticeably glutinous with resin glands.

Mozambique. N : Rovuma Bay, fl. & fr. iii.1861, *Kirk* (K).
Also in Tanganyika and near Mombasa in Kenya. Ecology similar to that of var. *tettensis*.

Kirk's gathering from Rovuma Bay may be on the Tanganyika side of the Bay but this variety is almost certain to occur on both sides. No other Mozambique gatherings are known so far.

8. ONCOBA Forsk.

Oncoba Forsk., Fl. Aegypt.-Arab. : CXIII, 103 (1775).

Trees or shrubs ; branches often spinose. Leaves alternate, petiolate. Flowers terminal or axillary, solitary, often on short side-shoots, often large and showy. Sepals 3 or 4, free or united at the base, imbricate. Petals 5–10 or more, exceeding the sepals, imbricate. Stamens numerous, free, anthers linear. Ovary 1-locular with 2–10 multiovulate parietal placentas ; style simple. Fruit indehiscent, smooth or slightly ridged with a hard, shell-like pericarp, 1-locular, many seeded. Seeds with a smooth, horny testa ; embryo with leafy cotyledons.

Oncoba spinosa Forsk., Fl. Aegypt.-Arab. : CXIII, 103 (1775).—Oliv., F.T.A. **1** : 115 (1868).—Gibbs in Journ. Linn. Soc., Bot. **37** : 429 (1906).—Sim, For. Fl. Port. E. Afr. : 12, t. 2 fig. B (1909).—Bak. f. in Journ. Linn. Soc., Bot. **40** : 23 (1911).— R.E.Fr., Wiss. Ergebn. Schwed. Rhod.-Kongo-Exped. **1** : 155 (1914).—Eyles in Trans. Roy. Soc. S. Afr. **5** : 422 (1916).—Gilg in Engl. & Prantl, Nat. Pflanzenfam. ed. 2, **21** : 402, t. 172 (1925).—Burtt Davy. F.P.F.T. **1** : 214, t. 30 (1926).—Exell & Mendonça, C.F.A. **1, 1** : 81 (1937).—Brenan, T.T.C.L.: 234 (1949).—Evrard in Bull. Soc. Roy. Bot. Belg. **86** : 11 (1953).—Keay, F.W.T.A. ed 2, **1, 1** : 188, t. 71 (1954).—

Pardy in Rhod. Agr. Journ. **53** : 62, cum tab. (1956). TAB. **46** fig. B. Type from Arabia.

Glabrous shrub or small tree up to c. 5 m. tall ; branches with pale lenticels and very sharp, spreading, axillary spines up to 5 cm. long. Leaf-lamina 3·5–12 × 2–10 cm., usually rather membranous, elliptic, ovate-elliptic or oblong-elliptic, acuminate at the apex, abruptly cuneate at the base, serrate or serrate-crenate except near the base or serrations almost obsolete, with 7–11 pairs of lateral nerves, venation reticulate but not very conspicuous ; petiole up to 8 mm. long. Flowers scented, solitary and terminal or lateral on short, axillary shoots. Calyx globose in bud with 4 very concave, imbricate, persistent lobes shortly united at the base ; lobes c. 1·5 × 1·2 cm., rounded at the apex and rather membranous at the margins. Petals white, 8–10, c. twice the size of the calyx-lobes, imbricate, oblong-spathulate, rounded at the apex. Anthers yellow, very numerous, 1·75 mm. long, linear with the connective produced as a minute narrowly triangular tip ; filaments up to 6 mm. long. Ovary globose with a style as long as the stamens ; stigma 4 mm. in diam., peltate or somewhat funnel-shaped and lobulate at the margins. Fruit dark green or finally brown, c. 5 cm. in diam., globose, smooth, with a hard shell marked with c. 8 rather faint longitudinal lines, the old calyx persistent at the base, many-seeded. Seeds c. 6 × 4 mm., the shape of a somewhat flattened apple pip, with a shiny, rich brown testa.

Bechuanaland Prot. N : Serondela, Chobe R., fr. v.1952, *Miller* B 1317 (PRE). **N. Rhodesia.** B : near Chavuma, st. 13.x.1952, *White* 3495 (FHO). N : Abercorn, Kaka R. gorge, fl. 14.xi.1956, *Richards* 6979 (K). W : Kabompo R., fl. 23.xi.1952, *Holmes* 1018 (FHO). C : Chilanga, fl. 3.x.1929, *Sandwith* 36 (K). S : Mapanza, fl. & fr. 22.v.1953, *Robinson* 256 (K). **S. Rhodesia.** N : Urungwe, fr. 18.iv.1954, *Lovemore* 395 (K ; SRGH). W : Victoria Falls, fl. i.1904, *Allen* 110 (K ; SRGH). C : Makoni, fl. 29.xi.1955, *Drummond* 5075 (SRGH). **Nyasaland.** S : Ncheu Distr., Livulezi valley, fl. 15.x.1950, *Jackson* 295 (K). **Mozambique.** N : Mandimba, fl. 15.xii.1941, *Hornby* 2450 (K). Z : Quelimane, Namagoa, fl. xi.*Faulkner* 111 (K). MS : Chimoio, Bandula, fl. 8.i.1948, *Barbosa* 826 (LISC ; LM). SS : Inhambane, fl. ii.1938, *Gomes e Sousa* 2088 (COI ; K). LM : Maputo, fl. 23.x.1948, *Gomes e Sousa* 3868 (COI ; K ; PRE).

Widely distributed throughout tropical Africa and also in the Transvaal and Arabia. Usually found in the hotter and drier types of woodland and sometimes thicket forming, particularly in river valleys.

The fruits are made by natives into snuff boxes or into rattles for the ankles of dancers.

9. SCOLOPIA Schreb.

Scolopia Schreb. in L., Gen. Pl. ed. 8, **1** : 335 (1789) *nom. conserv.*

Shrubs or trees ; branches often spiny. Leaves glabrous, usually leathery, entire or toothed. Flowers bisexual in axillary racemes or solitary in the axils. Receptacle usually funnel-shaped. Sepals c. 5, imbricate. Petals c. 5, similar to the sepals but slightly smaller. Disk fleshy, filling and sometimes elevated above the receptacle, usually with numerous glandular lobules at its margin and sometimes between the stamens. Stamens numerous, borne on the surface of the disk ; filaments very slender ; anthers 2-thecous, the connective produced as a short horn or apiculus. Ovary sessile, 1-locular with 2-6 parietal placentas ; ovules few or numerous ; style simple, stigmas 3-6, short. Fruit a more or less fleshy berry with several seeds.

Flowers in axillary racemes ; leaves discolorous ; fruit up to 1 cm. in diam.   1. *zeyheri*
Flowers solitary in the axils ; leaves not discolorous ; fruit c. 2 cm. in diam. 2. *stolzii*

1. **Scolopia zeyheri** (Nees) Szyszyl., Polypet. Thalam. Rehm.: 111 (1887).—Sim, For. & For. Fl. Col. Cap. Good Hope : 126, t. 2 (1907).—Gilg in Engl., Bot. Jahrb. **40** : 481 (1908) ; in Engl. & Prantl, Nat. Pflanzenfam. ed. 2, **21** : 418 (1925).—Phillips in Bothalia **1**, 2 : 84 (1922). Type from S. Africa, Cape Province.
Eriudaphus zeyheri Nees in Eckl. & Zeyh., Enum. Pl. Afr. Austr. Extratrop. **2** : 272 (1836). Type as above.
Eriudaphus ecklonii Nees, loc. cit. Type from S. Africa, Cape Province.
Phoberos zeyheri (Nees) Arn. in Hook., Journ. Bot. **3** : 150 (1841).—Harv. in Harv. & Sond., F.C. **1** : 68 (1860). Type as for *Scolopia zeyheri*.
Phoberos ecklonii (Nees) Arn. ex Harv. in Harv. & Sond., loc. cit. Type as for *Eriudaphus ecklonii*.
Scolopia ecklonii (Nees) Szyszyl., loc. cit.—Gilg in Engl., Bot. Jahrb. **40** : 481

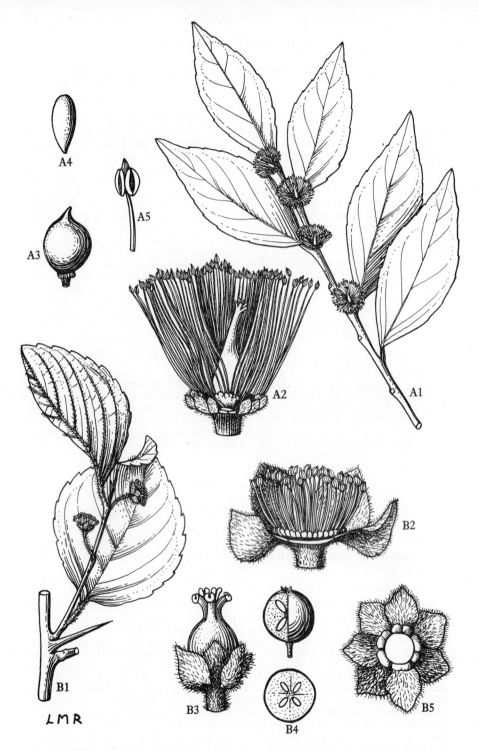

LMR

Tab. 47. A.—SCOLOPIA STOLZII. A1, flowering branch (× ⅔) *Stolz* 1742 ; A2, flower, front sepals and petal removed to show disk (× 4) *Stolz* 1742 ; A3, fruit (× ⅔) *Greenway* 2751 ; A4, seed (× 2) *Greenway* 2751 ; A5, anther (× 16) *Stolz* 1742. B.—FLACOURTIA INDICA. B1, shoot with male flowers (× ⅔) *Faulkner* 96 ; B2, male flower, front sepals removed to show disk (× 6) *Faulkner* 96 ; B3, female flower (× 6) *Faulkner* 374 ; B4, fruit in longitudinal and transverse section (× ⅔) *Milne-Redhead* 2790a ; B5, female flower, gynoecium removed to show disk (× 6) *Faulkner* 374.

(1908).—Phillips, tom. cit.: 86 (1922).—Gilg in Engl. & Prantl, Nat. Pflanzenfam. ed. 2, **21** : 420 (1925).—Burtt Davy, F.P.F.T. **1** : 215 (1926). Type as for *Eriudaphus ecklonii*.

    Scolopia mundtii sensu O.B. Mill. in Journ. S. Afr. Bot. **18** : 59 (1952) (*"mundii"*).

Shrub or small tree up to about 8 m. tall (in the Cape Province of S. Africa it can attain heights of 23 m.), glabrous except for the inflorescence ; branches unarmed or with straight spines up to 9 cm. long ; bark dark grey and rather thin. Leaf-lamina 2-8 × 1-3·5 cm., coriaceous, very variable, ovate, obovate, lanceolate, oblanceolate or rhomboid-lanceolate, apex obtuse or rounded, rarely acute, cuneate at the base, margin entire, repand or bluntly toothed, dark green above, paler below, often reddish when young, nerves in 4–6 pairs, at a narrow angle with the midrib, slightly prominent on both sides, venation laxly reticulate ; petiole up to 1 cm. long. Flowers in axillary racemes 1–3 cm. long ; pedicels 2–5 mm. long with basal, caducous, triangular bracts c. 0·75 mm. long, rhachis and pedicels glabrous or puberulous. Receptacle puberulous outside or glabrous, broadly funnel-shaped. Sepals 5–6, c. 1·5 × 1 mm., ovate, acute at the apex, glabrous or puberulous outside and minutely ciliate at the margins. Petals 5–6, similar to the sepals but somewhat smaller. Disk annular with many small fleshy lobes at the margin, surface densely villous. Stamens c. 40 with slender filaments c. 3·5 mm. long ; anthers oblong, somewhat arcuate, connective produced as a broadly triangular apiculus. Ovary subglobose, glabrous ; style c. 2·5 mm. long, somewhat sulcate, rather stout, glabrous ; stigma bifid, subsessile. Fruit a globose, 2–3-seeded, fleshy berry, c. 0·8 cm. in diam., crowned with the persistent style.

Bechuanaland Prot. SE : Kanye, fl. xi.1944, *Miller* 338 (PRE). **N. Rhodesia.** S : Mazabuka, st. vii.1952, *White* 3864 (FHO). **S. Rhodesia.** C : Salisbury, st. 9.iv.1929, *Eyles* 6326 (K ; SRGH). E : Umtali, Inyamatshira Mts., 9.ix.1951, *Chase* 3938 (K ; SRGH). **Mozambique.** LM : Muchopes, fl. 17.iii.1948, *Torre* 7509 (BM ; LISC).

    Also in Kenya, Uganda, Tanganyika, Angola, Swaziland, Transvaal, Natal and the Cape Province. A very variable species with forms (? ecotypes) in submontane evergreen forest, open woodland or bushland down to sea-level.

    As long ago as 1907, Sim (loc. cit.) concluded that *S. zeyheri* and *S. ecklonii* represented forms of the same polymorphic species. Unfortunately Phillips (loc. cit.) in his later revision of the S. African species decided to treat them once more as distinct on the basis of differences in leaf-shape and the absence or presence of pubescence on the inflorescence. There are so many intermediates however, that Sim's view is followed here. Furthermore, if this concept of the species is accepted, it is apparent that *S. rigida* R.E.Fr. of Kenya, Uganda and Tanganyika and *S. gossweileri* Sleumer and *S. dekindtiana* Gilg of Angola must be considered as forms of the same species. For some reason the species seems to be very rare in our area, where the specimens cited above are all that are known.

    S. zeyheri is the " thorn-pear " of S. Africa : its wood is excessively hard and difficult to work.

2. **Scolopia stolzii** Gilg [in Pflanzenw. Afr. **3**, 2 : 577 (1921) ; in Engl. & Prantl, Nat. Pflanzenfam. ed. 2, **21** : 420 (1925) *nom. nud.*] ex Sleumer in Notizbl. Bot. Gart. Berl. **12** : 142 (1936).—Brenan, T.T.C.L. : 236 (1949). TAB. **47** fig. A. Type from Tanganyika.

Small tree up to c. 10 m. tall, glabrous except for the flowers or occasionally with the young stems and petioles minutely pubescent ; bark grey, smooth. Leaf-lamina 5–12 × 2–5·5 cm., coriaceous, elliptic, elliptic-oblong or ovate-elliptic, bluntly acuminate or subobtuse at the apex, cuneate at the base, margin entire or irregularly and remotely subserrate, somewhat shining above, dull below, nerves in 4–6 pairs, slightly raised on both sides, venation laxly reticulate ; petiole up to 7 mm. Flowers solitary and sessile in the leaf-axils. Receptacle very small, funnel-shaped. Sepals 5–7, c. 1·5 × 1·5 mm., overlapping considerably, broadly ovate, subacute at the apex, margin shortly ciliate, sparsely puberulous outside. Petals similar to the sepals but more membranous. Disk annular, irregularly lobed, elevated and granular between the stamens. Stamens very numerous on slender filaments c. 7 mm. long ; anthers cordate-oblong with a prominent and somewhat recurved apiculus at the apex. Ovary urceolate, glabrous or sparsely puberulous ; style c. 5 mm. long, stout, glabrous ; stigmas c. 6. Fruit a 10–12-seeded, globose, fleshy berry, c. 2 cm. in diam., crowned with the persistent style.

N. Rhodesia. N : Mpika, fr. 11.ii.1955, *Fanshawe* 2054 (K). W : Mufulira, fl. 27.xi.1954, *Fanshawe* 1696 (K). **Nyasaland.** N : Matipa Forest, fl. ix.1954, *Chapman*

236 (FHO). C : Nchisi Mt., st. 7.ix.1929, *Burtt Davy* 21379 (FHO). S : Zomba, fr. 1936, *Clements* 592 (FHO). **Mozambique.** MS : Vila de Manica, fl. 5.xi.1950, *Chase* 3074 (K ; SRGH).
Also in Tanganyika. In fringing forest.

The N. Rhodesian material differs somewhat from the type. *Fanshawe* 1696, for instance, has puberulous young stems and petioles but is not otherwise abnormal. Another specimen, *Fanshawe* 1534 from the same locality, has glabrous stems and petioles but the apiculate connective is lacking in all but a very few anthers. The flowers are young and so this is not due to the caducous nature of the apiculum. It was at first thought that this latter specimen must be an undescribed species but the presence of these two variants in the same area points rather to the conclusion that both are forms of *S. stolzii*.

10. APHLOIA (DC.) Benn.

Aphloia (DC.) Benn. in Benn. & Br., Pl. Jav. Rar. **2** : 192 (1840).
Prockia sect. *Aphloia* DC., Prodr. **1** : 261 (1824).

Shrubs or trees, entirely glabrous. Leaves petiolate, simple, more or less dentate-serrulate at the margins ; stipules minute. Flowers bisexual, axillary, solitary, in few-flowered fascicles or racemes ; bracts minute, scale-like ; pedicels with 1–3 minute, scaly bracteoles in the lower half. Sepals 4–6, free except at the base, imbricate. Petals 0. Stamens very numerous, free, inserted towards the margin of a slightly concave receptacle. Ovary sessile or very shortly stipitate, 1-locular with one parietal placenta ; ovules rather few, in 2 rows, ± horizontal ; stigma subsessile, large and peltate with a median furrow. Fruit a fleshy berry with c. 6 seeds. Seeds discoid with a smooth crustaceous testa ; endosperm fleshy.

Aphloia theiformis (Vahl) Benn. in Benn. & Br., Pl. Jav. Rar. **2** : 192 (1840).—
Bak. f. in Journ. Linn. Soc., Bot. **40** : 23 (1911).—Perrier, Fl. Madag., Flacourt. : 13 (1946). TAB. **48**. Type from Réunion I.
Lightfootia theiformis Vahl, Symb. Bot. **3** : 69 (1794). Type as above.
Prockia theiformis (Vahl) Willd. in L., Sp. Pl. ed. 4, **2** : 1214 (1800). Type as above.
Neumannia theiformis (Vahl) A. Rich., Ess. Fl. Cub. : 97 (1845).—Dur. & Schinz, Consp. Fl. Afr. **1**, 2 : 218 (1898).—Gilg in Engl., Bot. Jahrb. **40** : 503 (1908) ; in Engl. & Prantl, Nat. Pflanzenfam. ed. 2, **21** : 437, t. 200 (1925).—Eyles in Trans. Roy. Soc. S. Afr. **5** : 422 (1916).—Engl., Pflanzenw. Afr. **3**, 2 : 584, t. 260 (1921).—Steedman, Trees etc. S. Rhod. : 53 (1933). Type as above.
Aphloia myrtiflora Galpin in Kew Bull. **1895** : 142 (1895).—Burtt Davy, F.P.F.T. **1** : 215 (1926).—Brenan, T.T.C.L. : 229 (1949).—Pardy in Rhod. Agr. Journ. **53** : 953, cum tab. (1956). Type from the Transvaal.
Neumannia myrtiflora (Galpin) Th. Dur. in Dur. & Schinz, Consp. Fl. Afr. **1**, 2 : 218 (1898). Type as for *Aphloia myrtiflora*.

Glabrous, evergreen tree up to 14 m. tall ; branchlets brown, longitudinally striate with a stronger line decurrent from a stipular cushion. Leaf-lamina 3–8 × 1·2–2·75 cm., narrowly elliptic to elliptic or oblanceolate, apex obtuse or subacute, base cuneate, margins serrate or serrulate, often entire towards the base, lateral nerves in c. 10 pairs, inconspicuous ; petiole up to c. 3 mm. long. Flowers solitary, or in fascicles of 1–3 in the leaf axils ; pedicels greenish, c. 1 cm. long with 1–3 minute, deltoid bracteoles c. 1 mm. long towards their bases. Sepals white turning yellowish, c. 5 × 5 mm., imbricate, orbicular, connate for 1–1·5 mm. at the base, the inner more membranous and petaloid, margins membranous and entire or denticulate. Stamens very numerous ; filaments slender, 4–5 mm. long ; anthers orbicular, c. 0·7 mm. in diam. Ovary ellipsoid, and sometimes very shortly stipitate ; stigma subsessile with a median furrow, irregularly sublobed, as wide or wider than the young ovary. Fruit a white, fleshy berry with the stigma persistent at the apex. Seeds c. 6, c. 2 mm. in diam., discoid with a glossy whitish testa.

S. Rhodesia. E : Inyanga, source of Mtenderere R., fl. 22.x.1946, *Wild* 1455 (K ; SRGH). **Nyasaland.** N : Nymkowa Hill, fl. ix.1903, *McClounie* 181 (K). S : Zomba, fl., *Clements* 68 (FHO ; K). **Mozambique.** Z : Milange, Tumbine Mt., fl. 12.xi.1942, *Mendonça* 1392 (BM ; LISC). MS : Gorongosa, fl. 15.ix.1946, *Simão* 1100 (LM ; SRGH).
Also in Tanganyika, Transvaal, Madagascar, Comoro Is., Mascarene Is. and the Seychelles. In submontane forest.
The flowers are very sweet-scented.

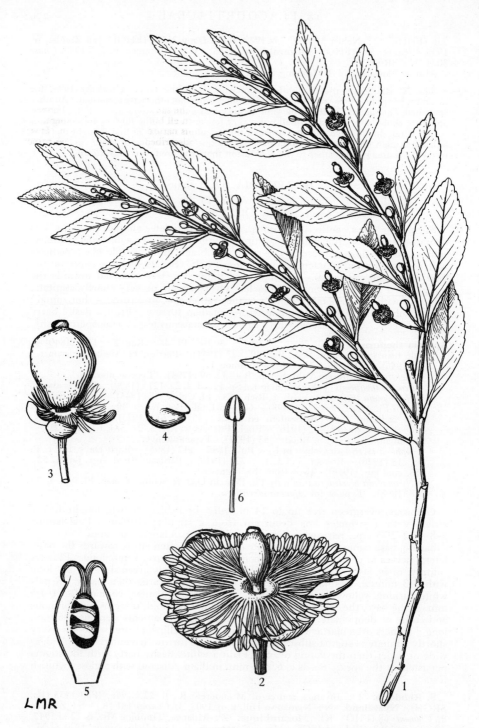

LMR

Tab. 48. APHLOIA THEIFORMIS. 1, flowering branch (× ⅔) *Johnston* 37 ; 2, flower (×4) *Clements* 68 ; 3, fruit (×2) *Carmichael* 326 ; 4, seed (×4) *Carmichael* 326 ; 5, longitudinal section of ovary (×8) *Clements* 68 ; 6, stamen (×6) *Clements* 68.

In Africa this species is fairly uniform morphologically but in the remainder of its range it is excessively variable. Our material is matched very well by *A. theiformis* subsp. *madagascariensis* var. *closii* Tul. (see Perrier, tom. cit. : 18 (1946)) which occurs on the mainland of Madagascar and in the Seychelles. There is no need to resort to this cumbersome collection of epithets for our plant, however, unless it is necessary to relate it to the extra-African forms. I have, moreover, compared our material with Vahl's type collected by Commerson in Réunion (*Commerson* in Herb. Thouin (C, holotype)) and apart from its rather longer pedicels it does not differ from our plant.

11. DOVYALIS E. Mey. ex Arn.

Dovyalis E. Mey. ex Arn. in Hook., Journ. Bot. **3** : 251 (1841).

Shrubs or trees, unarmed or spiny. Leaves alternate, simple, entire, denticulate or crenate, exstipulate. Flowers dioecious. Male flowers fasciculate or in short racemes ; calyx (3) 4–7-lobed almost to the base, lobes valvate or almost so, variously pubescent or tomentose ; petals 0 ; stamens 12–∞, inserted on a sub-fleshy receptacle and alternating with minute, often pubescent glands. Female flowers often solitary in the axils or in fascicles of 2–4 ; calyx 4–8-lobed almost to the base, persistent and sometimes accrescent in fruit, sometimes bordered with stalked glands ; petals 0 ; disk undulate or lobed or subentire ; ovary sessile on the disk, 1–2-locular or occasionally imperfectly 2–8-locular with up to 8 placentas ; each placenta with 1–6 ovules ; styles 2–8, divergent. Fruit fleshy. Seeds few or occasionally up to about 12, glabrous or hairy.

Ovary with 2–3 (rarely 4) placentas ; styles 2–3 (rarely 4) ; male flowers in lax fascicles of
    1–4 (rarely up to 7) :
  Calyx much enlarged in fruit, calyx-lobes fimbriate-ciliate at the margins
                                                       1. *macrocalyx*
  Calyx not enlarged in fruit :
    Calyx-lobes densely pubescent, tomentose or tomentellous on both sides :
      Male calyx-lobes c. 2 mm. long ; stamens c. 15 -   -   -   -   2. *lucida*
      Male calyx-lobes 5–6 mm. long ; stamens c. 40 :
        Leaves broadly elliptic, nearly glabrous, 3–9·5 × 1·75–4·75 cm. ; male sepals
          minutely dentate or fimbriate near the apex -   -   -   3. *spinosissima*
        Leaves obovate or rarely broadly elliptic, often more or less pubescent,
          1·5–6 × 1–3·5 cm. ; male sepals entire -   -   -   -   4. *zeyheri*
    Calyx-lobes sparsely hispidulous outside -   -   -   -   5. *hispidula*
  Ovary with 4–8 placentas ; styles 4–8 ; male flowers in dense fascicles -   6. *caffra*

1. **Dovyalis macrocalyx** (Oliv.) Warb. in Engl. & Prantl, Nat. Pflanzenfam. **3**, 6a :
    44 (1893).—Gilg in Engl., Bot. Jahrb. **40** : 506 (1908).—Eyles in Trans. Roy. Soc.
    S. Afr. **5** : 423 (1916).—Gilg in Engl. & Prantl, Nat. Pflanzenfam. ed. 2, **21** : 441
    (1925).—Steedman, Trees etc. S. Rhod. : 52, t. 50 & 51 (1933).—Sleumer in Exell &
    Mendonça, C.F.A. **1**, 1 : 86 (1937).—Brenan, T.T.C.L. : 231 (1949) ; in Mem.
    N Y Bot. Gard. **8**, 3 : 218 (1953). TAB. 49 fig. A. Type from Angola.
    Aberia ? macrocalyx Oliv., F.T.A. **1** : 122 (1868).—Bak. f. in Journ. Linn. Soc.,
    Bot. **40** : 23 (1911). Type as above.
    Dovyalis salicifolia Gilg in Engl., Bot. Jahrb. **40** : 505 (1908). Syntypes : Nyasa-
    land, Blantyre, *Buchanan* in Herb. Wood 6833 (B†) ; *Buchanan* 347 (B† ; BM ; E).
    Dovyalis chirindensis Engl., Pflanzenw. Afr. **3**, 2 : 587 (1921).—Gilg in Engl. &
    Prantl, Nat. Pflanzenfam. ed. 2, **21** : 441 (1925) *nom. nud.*

Bush or small tree up to c. 7 m. tall ; young branches slender, pubescent at first, becoming glabrous, usually armed with slender, straight spines up to c. 5 cm. long. Leaf-lamina 2·5–9 × 1·4–4·3 cm., membranous or thinly coriaceous, somewhat shining above, narrowly ovate, elliptic or ovate, apex obtuse or subacute, broadly cuneate, obtuse or rarely slightly cordate at the base, margin remotely crenulate or entire, glabrous on both sides, with 5–7 basal nerves, venation laxly reticulate and subprominent on both sides ; petiole c. 2 mm. long, puberulous. Male flowers in 1–4-flowered axillary fascicles ; pedicels 2–3 mm. long, shortly and densely pubescent ; calyx lobed almost to the base, lobes 5–6, 3–4 × 1–2·5 mm., lanceolate to ovate-lanceolate, apex acute, margin entire or occasionally with a few coarse teeth and sometimes with one or two glandular hairs ; stamens c. 20 ; filaments slender, 5 mm. long, inter-staminal glands minute, puberulous. Female flowers 1–2 per axil ; calyx of 6–10 lobes divided almost to the base, c. 6·5 × 2 mm., narrowly lanceolate, often incurved at the acuminate apex, margin with dense stalked glands, both sides pubescent ; annular disk pubescent, undulate and segmented ; ovary ovoid, pubescent with two pubescent divergent styles c. 3 mm. long ; stigmas narrow, ± bilobed. Fruiting calyx accrescent and pinkish-green or red, the marginal glands are also much enlarged and become fimbriate-ciliate, the calyx-lobes equal or slightly exceed the fruit. Fruit fleshy, red, ellipsoid, puberulous, 2-seeded. Seeds c. 8 × 6 mm., obovoid ; testa with a dense brown wool.

N. Rhodesia. N : Isoka, fl. 20.xi.1952, *White* 3731 (FHO). W : Solwezi Falls, fl. 23.ix.1930, *Milne-Redhead* 1177 (K ; PRE). **S. Rhodesia.** E : Chipinga, Chipete Forest, fl. 22.x.1947, *Wild* 2129 (K ; SRGH). **Nyasaland.** S : Cholo Distr., Nswadzi R., fl. 29.ix.1946, *Brass* 17862 (BM ; K ; PRE ; SRGH). **Mozambique.** Z : Milange, Tumbine Mts., fl. 12.x.1942, *Torre* 4580 (BM ; LISC). MS : Gorongosa, Vunduzi, fl. 5.x.1944, *Mendonça* 2369 (BM ; LISC). LM : Delagoa Bay, fl. 28.ix.1906, *Swynnerton* (BM).

Also in Angola, Sudan, Uganda, Kenya and Tanganyika. At evergreen forest edges and in fringing forest.

The fruit is edible.

Swynnerton's specimen from Delagoa Bay has oblanceolate leaves and approaches *D. longispina* (Harv.) Warb. from Natal. A number of specimens of *D. macrocalyx* exist, however, which have a proportion of their leaves of this shape and so for the present, at least, Swynnerton's specimen is regarded as a variant of *D. macrocalyx* at the southern extremity of the range of this species. Swynnerton's specimen is a male ; it would be useful to have female material from the same area.

2. **Dovyalis lucida** Sim, For. & For. Fl. Col. Cap. Good Hope : 131, t. 6 (1907). TAB. 49 fig. C. Syntypes from Cape Province.

Shrub or small tree up to c. 7 m. tall, quite glabrous except for the inflorescences ; branches unarmed (or armed in some S. African material), grey with pale lenticels. Leaf-lamina 2·5–8 × 2–5 cm., coriaceous, broadly elliptic to rotund or obovate-oblong, apex obtuse or subacute, cuneate at the base, margins entire or rarely with a few irregular teeth, shining and dark green above, paler and dull below, lateral nerves in 4–6 pairs, slightly prominent on both sides, venation very laxly reticulate ; petiole 2–3 mm. long. Male flowers in axillary fascicles of 2–7 ; pedicels up to 3 mm. long, densely and shortly pubescent ; calyx lobed almost to the base, lobes usually 4, c. 2 × 1·5 mm., oblong, apex obtuse, pubescent on both sides ; stamens c. 15, filaments c. 2 mm. long, interstaminal glands minute, pubescent. Female flowers in axillary fascicles, 1–4 together ; pedicels up to 4 mm. long ; calyx-lobes (3) 4–6, larger than in the males, c. 3 × 2 mm., annular disk slightly undulate, minutely puberulous or tomentellous ; ovary ovoid, tomentellous ; styles 2–3 (4), 1 mm. long, glabrous. Ripe fruit 1·5–2 × 1·3 cm., fleshy, orange-red, ellipsoid, densely puberulous. Seeds 1–2, c. 0·9 × 0·6 cm., ellipsoid, woolly.

S. Rhodesia. E : Chimanimani Mts., fl. 16.x.1950, *Wild* 3633 (K ; SRGH). Also in the Cape Province and the mountains of the Transvaal. In S. Rhodesia a species of submontane forest known only from the Chimanimani and forests near Umtali and apparently rather rare throughout its known range.

3. **Dovyalis spinosissima** Gilg in Engl., Bot. Jahrb. **40** : 509 (1908) ; in Engl. & Prantl, Nat. Pflanzenfam. ed. 2, **21** : 441 (1925). Syntypes : Nyasaland, Blantyre, *Buchanan* 338 (B† ; BM), 6808 and 7018 in Herb. Wood (PRE).

Shrub or small tree glabrous except for the inflorescences or with a very few scattered setulose hairs on the leaves ; branches greyish with prominent pale lenticels and axillary spines up to 4·5 cm. long. Leaf-lamina 3–9·5 × 1·75–4·75 cm., broadly elliptic, apex obtuse or bluntly acuminate, base broadly cuneate or rounded, margins entire, slightly revolute, nerves in 4–5 pairs, slightly prominent on both sides, venation very laxly reticulate below, petiole up to 5 mm. long. Male flowers 2–3-fasciculate in the leaf-axils on densely and shortly pubescent pedicels up to 6 mm. long ; calyx 5–7-lobed almost to the base, lobes c. 5 × 2–2·5 mm., lanceolate-oblong, acuminate at the apex, margin minutely laciniately lobed near the apex, entire below, densely pubescent on both sides ; stamens c. 40 on filaments up to 3·5 mm. long ; inter-staminal glands ciliate at the apex. Female flowers and fruit so far unknown.

Nyasaland. S : Blantyre, fl. 1895, *Buchanan* in Herb. Wood 7018 (PRE). Endemic so far as is known to Nyasaland. Ecology unknown.

LMR

Tab. 49. A.—DOVYALIS MACROCALYX. A1, branch with male flowers (×⅔) ; A2, fruit
with accrescent calyx (×1) ; A3, male flower, front sepals removed (×6), all from
Brass 17812. B.—DOVYALIS CAFFRA. B1, male flower, front sepals removed (×6)
Adlam 563 ; B2, female flower, front sepals removed (×6) *Mogg* 14114. C.—
DOVYALIS LUCIDA. C1, male flower, front sepal removed (×6) *Wild* 3634 ; C2, female
flower, front sepals removed (×6) *Chase* 1615. D.—DOVYALIS ZEYHERI. D1, male
flower, front sepals removed (×2) *Eyles* 882 ; D2, female flower (×2) *Chase* 6238.

D. spinosissima is near *D. abyssinica* (A. Rich.) Warb. which is found in Ethiopia and southwards to Tanganyika. It is possible that it is synonymous with this latter species but the minute laciniae on the male calyx lobes are not found on any specimens of *D. abyssinica* so far examined so it is preferable at present to consider it distinct. Further material is badly needed of this obscure species.

4. **Dovyalis zeyheri** (Sond.) Warb. in Engl. & Prantl, Nat. Pflanzenfam. **3**, 6a : 44 (1893).—Gilg in Engl., Bot. Jahrb. **40** : 506 (1908) ; in Engl. & Prantl, Nat. Pflanz-enfam. ed. 2, **21** : 441 (1925).—Eyles in Trans. Roy. Soc. S. Afr. **5** : 423 (1916).—Burtt Davy, F.P.F.T. **1** : 216 (1926). TAB. **49** fig. D. Type from the Transvaal.
    Aberia zeyheri Sond. in Linnaea **23** ; 10 (1850).—Harv. in Harv. & Sond., F.C. **1** : 70 (1860). Type as above.
    Aberia tristis Sond., tom. cit. : 9 (1850).—Harv. in Harv. & Sond., loc. cit. Type from Cape Province.
    Dovyalis tristis (Sond.) Warb., loc. cit. (1893).—Gilg in Engl., Bot. Jahrb. **40** : 506 (1908) ; in Engl. & Prantl, Nat. Pflanzenfam. ed. 2, **21** : 441 (1925).—Burtt Davy, loc. cit.—Suesseng. in Proc. & Trans. Rhod. Sci. Ass. **43** : 87 (1951). Type as for *Aberia tristis*.

Shrub or tree up to about 10 m. tall, sometimes armed with straight spines up to about 2·5 cm. long or rarely more, branches with a pale grey bark, often rough with pale lenticels, young branchlets with patent yellowish hairs or occasionally practi-cally glabrous. Leaf-lamina 1·5–6 × 1–3·5 cm., obovate to broadly elliptic, apex rounded, mucronate or occasionally retuse, cuneate at the 3-nerved base, margin remotely crenulate or subentire, often somewhat revolute, both sides pubescent or more rarely glabrous, the margins or at least the crenatures often ciliate-pubescent, with a pair of lateral nerves arising from the midrib some 5 mm. above the base noticeably stronger and more prominent than the remainder and subparallel with the margins, petiole up to 8 mm., ± pubescent. Male flowers in axillary fascicles of 2–4 ; pedicels 4–7 mm. long, densely pubescent ; calyx of 5 (6) lobes, divided almost to the base, lobes c. 6 × 2·5 mm., oblong-lanceolate to ovate-lanceolate, acute at the apex, densely pubescent to tomentose on both sides ; stamens c. 40 on glabrous filaments c. 4 mm. long ; inter-staminal glands 0·5 mm. long, obovoid, densely setulose-ciliate. Female flowers solitary ; calyx lobes 5–7 (8), broader than in the male, ovate-lanceolate ; annular disk with short, rounded, villous lobes alternating with the sepals ; ovary narrowly ovoid, densely but shortly tomentose ; styles 2–3, 1·5–2 mm. long, divergent, pubescent, stigmas broadly bilobed. Fruit c. 2 × 1·3 cm., densely and shortly golden-tomentose, ellipsoid, fleshy, 2–3-seeded.

S. Rhodesia. W : Matopos, fl. 23.xi.1951, *Plowes* 1329 (SRGH). C : Salisbury, fl. xii.1917, *Eyles* 882 (BM ; K ; SRGH). E : Penhalonga, fl. 11.xi.1956, *Chase* 6238 (K ; SRGH). S : Zimbabwe, fl. 4.x.1949, *Wild* 2990 (K ; SRGH).
    Also in the Cape, Transvaal and Natal. In open *Brachystegia* or mixed woodland usually above about 1200 m.

This is a very variable species in its pubescence. The forms with glabrous leaves have in the past been named *D. tristis* but intermediates are so common that this species is not worth keeping up. In addition, *D. zeyheri* in the strict sense is described as being spiny and *D. tristis* as being unarmed but here again there is variation which is not correlated apparently with any other character including the degree or kind of pubescence.

5. **Dovyalis hispidula** Wild in Bol. Soc. Brot., Sér. 2, **32** : 51 (1958). Type : S. Rhodesia, Lundi R., Chipinda Gorge, *Davies* 2197 (SRGH, holotype).

Shrub or small tree with grey or greyish-brown spiny branches with scattered pale lenticels, pubescent at first, glabrescent later ; spines axillary, numerous, slender, straight and very acute, pubescent at first, glabrescent later, up to 4 cm. long. Leaf-lamina 0·8–3·5 × 0·6–2·3 cm., broadly elliptic to obovate, apex obtuse, rounded or retuse, base obtuse or cuneate, margin shallowly crenate or subentire, sparsely hispidulous and dull on both sides, nerves inconspicuous above, visible below but fading towards the leaf margins, 3-nerved from the base, 3–4 pairs of nerves above the base anastomosing well within the margins ; petiole up to 0·8 cm. long, hispidulous. Male flowers in axillary fascicles of 1–4 ; pedicels slender, 2–3 mm. long, puberulous ; calyx of 4–6 subequal lobes divided almost to the base, lobes broadly ovate, c. 2 × 1·3 mm., apex obtuse, margin entire or minutely dentate towards the apex, sparsely hispidulous outside ; stamens numerous on filaments (? immature) c. 0·5 mm. long ; inter-staminal glands very sparsely hispidulous at the apex. Female flowers solitary, pedicels slender, glabrescent, up to 1·2 cm.

long ; calyx lobes 5–6, c. 2×2 mm., persistent, but not accrescent in fruit, broadly ovate to orbicular, margins minutely dentate or very shortly fimbriate, hispidulous outside ; annular disk scarcely undulate, subentire, sparsely hispidulous ; ovary pilose or glabrescent, ovoid ; styles 2–3, 1 mm. long, pubescent or glabrous. Fruit 1–2 cm. in diam., fleshy, hispidulous. Seeds 2–3, c. 7×5 mm., ellipsoid with a woolly testa.

S. Rhodesia. S : Lundi R., Chipinda Gorge, fl. xi.1956, *Davies* 2197 (SRGH). **Mozambique.** N : between Namapa and Lumbo, fr. 30.x.1942, *Mendonça* 1146 (BM ; LISC). MS : Lower R. Buzi, Boka, fr. 28.xii.1906, *Swynnerton* (BM). Not known outside our area. In low-altitude dry woodlands or bush.

6. **Dovyalis caffra** (Hook f. & Harv.) Warb. in Engl. & Prantl, Nat. Pflanzenfam. **3**, 6a : 44 (1893).—Gilg in Engl., Bot. Jahrb. **40** : 508 (1908) ; in Engl. & Prantl, Nat. Pflanzenfam. ed. 2, **21** : 441 (1925).—Sim, For. Fl. Port. E. Afr. : 13 (1909).—Eyles in Trans. Roy. Soc. S. Afr. **5** : 423 (1916).—Burtt Davy, F.P.F.T. **1** : 216 (1926). TAB. **49** fig. B. Syntypes from Cape Province.
    Aberia caffra Hook. f. & Harv. in Harv. & Sond., F.C. **2** : 584 (1862). Syntypes as above.
    Dovyalis sp.—Eyles in Trans. Roy. Soc. S. Afr. **5** : 423 (1916).

Shrub up to about 3·5 m. tall, glabrous in all its vegetative parts ; branches smooth, grey and armed with stout spines up to about 6 cm. long. Leaves often fasciculate on short side-branches ; lamina 2–5·5 ×0·5–2·7 cm., narrowly obovate, obovate or obovate-elliptic, apex rounded or retuse, cuneate or narrowly rounded and 3-nerved at the base, margins entire, slightly revolute, venation slightly raised on both sides and very laxly reticulate ; petiole up to 5 mm. long. Male flowers densely fascicled in the axils or commonly on abbreviated side-shoots ; pedicels 1–3 mm. long, puberulous ; calyx with 5 lobes divided almost to the base, lobes c. 2·5 ×1·5 mm., narrowly ovate, apex acute, shortly pubescent on both sides ; stamens c. 15 ; filaments c. 5 mm. long ; inter-staminal glands very minutely pubescent, at least at their apices. Female flowers solitary on pedicels up to 8 mm. long ; calyx lobes 5–8, slightly larger and wider than the males ; annular disk undulately lobed, rather thick, minutely and sparsely puberulous or glabrous ; ovary globose-ovoid, glabrous, with 5–8 divergent, puberulous styles c. 2·5 mm. long ; stigmas somewhat spreading. Fruit fleshy, apricot-coloured, up to 4 cm. in diam., globose, glabrous, crowned with the persistent styles. Seeds c. 12, woolly.

S. Rhodesia. W : Bulalima-Mangwe, Empandini, fl. 27.i.1942, *Feiertag* in GHS 45400 (K ; SRGH). E : Umtali, st. 15.iii.1955, *Chase* 5494 (BM ; K ; SRGH). S : Belingwe, Bankwe vlei, fl. 16.ix.1948, *West* 2797 (SRGH). **Nyasaland.** S : Blantyre, fr. 22.i.1944, *Hornby* 2949 (PRE) cult.
Also in Natal and the Cape. Often found near rivers.
A rather rare species in S. Rhodesia.

This is the " kei apple " of S. Africa. The fruit is very acid but suitable for making jelly or preserves. The plant is also used for hedges. It has been introduced into Nyasaland, East Africa, California, Australia and even England. In the case of S. Rhodesian specimens it is not always possible, from herbarium material, to tell whether the plants were indigenous or introduced but all those quoted above were probably indigenous. Sim (loc. cit.) also records this species from the Libombo Mts., presumably on the Mozambique side but he quotes no specimen number. He is very likely correct, however, although no other collector seems to have produced Mozambique material as yet. In addition Sim (loc. cit. : t. 17, fig. B) also records *Dovyalis celastroides* Sond. (=*D. rotundifolia* (Thunb.) Harv.) from the Delagoa Bay area. Once again, however, he cites no specimen and his drawing looks more like *D. caffra* than *D. rotundifolia*.

12. FLACOURTIA L'Hérit.

Flacourtia L'Hérit., Stirp. Nov. **3** : 59, t. 30, 31 (1786).

Shrubs or trees ; branches often spinose, trunk occasionally spinose. Leaves petiolate, mostly crenate. Flowers dioecious, rarely bisexual, small, in short axillary racemes or solitary. Sepals 4–7, slightly connate at the base, imbricate. Petals 0. Male flowers with an extrastaminal disk usually broken into free glands ; stamens 15–∞, anthers dorsifixed, rudiment of ovary 0. Female flowers with a usually entire or crenulate disk, ovary incompletely (2) 4–6 (10)-locular by false septa ; ovules 2 per loculus one above the other ; styles as many as the loculi, free or connate, persistent. Fruit a fleshy berry with 4–16 seeds usually in pairs one above the other. Seeds obovoid ; cotyledons ± orbicular.

Flacourtia indica (Burm. f.) Merr., Interpr. Rumph. Herb. Amboin. : 377 (1917).— Gilg in Engl. & Prantl, Nat. Pflanzenfam. ed. 2, **21** : 440, t. 201 (1925).—Brenan, T.T.C.L. : 231 (1949).—Palgrave, Trees of Central Africa : 189 cum photogr. et tab. (1957). TAB. **47** fig. B. Type from Java.
  Gmelina indica Burm. f., Fl. Ind. : 132, t. 39 fig. 5 (1768). Type as above.
  Flacourtia ramontchi L'Hérit., Stirp. Nov. **3** : 59, t. 30, 31 (1786).—Oliv., F.T.A. **1** : 120 (1868).—Sim, For. Fl. Port. E. Afr. : 13 (1909).—Bak. f., Journ. Linn. Soc., Bot. **40** : 23 (1911).—R.E.Fr., Wiss. Ergebn. Schwed. Rhod.-Kongo-Exped. **1** : 157 (1914).—Eyles in Trans. Roy. Soc. S. Afr. **5** : 422 (1916).—Perrier, Fl. Madagasc., Flacourt. : 9 (1946).—O. B. Mill. in Journ. S. Afr. Bot. **18** : 59 (1952). Type from Madagascar.
  Flacourtia hirtiuscula Oliv., F.T.A. **1** : 121 (1868).—Sim, loc. cit. ("hirtinscula").— Bak. f., loc. cit.—Eyles, loc. cit.—Gilg, loc. cit.—Burtt Davy, F.P.F.T. **1** : 215 (1926).—Steedman, Trees, etc. S. Rhod. : 52 (1933).—O. B. Mill., loc. cit. Type : Mozambique, near Sena, *Kirk* (K, holotype).

Shrub or small tree up to 10 m. tall, with the bark rough and yellowish or orange-brown, occasionally silvery on young branches ; axillary straight spines present or absent on the branches, sometimes with fearsome branching spines up to 12 cm. long on the trunk near the base or with very spiny coppice shoots. All the vegetative parts except the older branches vary from quite glabrous to densely pubescent. Leaf-lamina 2·5–12 × 1·3–7·5 cm., very variable, membranous or coriaceous, suborbicular, ovate, elliptic, obovate or ovate-elliptic, apex rounded, obtusely or rarely obtusely acuminate at the apex, base usually cuneate, occasionally rounded, margins crenate, crenate-serrulate or subentire, nerves in 4–7 pairs, slightly prominent above and below, venation laxly reticulate ; petiole up to 1·3 cm. long. Flowers dioecious or occasionally bisexual in short axillary racemes or occasionally solitary in the axils ; peduncles very short ; rhachis up to 2 cm. long, ± pubescent ; pedicels up to 1 cm. long, ± pubescent, with caducous, deltoid, pubescent bracts at the base. Sepals 1·5–2·5 × 1·5–2·5 mm., imbricate, united for about 1 mm. at the base, broadly ovate, acute or rounded at the apex, pubescent on both sides. Male flower with very numerous stamens on filaments c. 2·5 mm. long, often with lobulate, glabrous glands forming a disk around the outer stamens. Female flowers with a lobulate, fleshy, glabrous disk clasping the base of the ovary ; ovary ovoid, glabrous ; styles 4–8, 0·5–1·5 mm. long, spreading, longitudinally grooved above ; stigmas truncate. Bisexual flowers similar to the female but with c. 5 stamens. Fruit reddish or reddish-black when ripe, up to 2·5 cm. in diam., fleshy, globose, becoming sulcate when dry, glabrous, with persistent styles, up to 10-seeded or thereabouts. Seeds c. 8 × 7 mm., obovoid and somewhat flattened ; testa pale brown, rugose.

Bechuanaland Prot. N : Chobe R., Kazane, fr. 13.vii.1937, *Erens* 422 (K ; PRE ; SRGH). **N. Rhodesia.** B : Balovale, fr. 16.v.1954, *Gilges* 364 (K ; SRGH). N : Abercorn, fr. 27.iii.1952, *Richards* 1416 (K). W : Mwinilunga, Matonchi Farm, fl. 2.xi.1937, *Milne-Redhead* 3063 (K). C : Chilanga, fl. 14.x.1929, *Sandwith* 54 (K). E : Katete, St. Francis Mission, fl. ix.1955, *Wright* 24 (K). S : Victoria Falls, fl. 19.ix.1949, *Wild* 3104 (K ; SRGH). **S. Rhodesia.** N : Urungwe, fl. 17.xi.1956, *Phelps* 162 (K ; SRGH). W : Inyati, fr. 18.iv.1947, *Keay* in FHI 21206 (FHO ; SRGH). C : Marandellas, fl. 11.x.1949, *Corby* 499 (K ; SRGH). E : Melsetter, Cashel, fl. 12.xi.1953, *Chase* 5120 (K ; SRGH). S : Sabi-Lundi Junction, fr. 7.vi.1950, *Wild* 3442 (K ; SRGH). **Nyasaland.** N : 29 km. N. of Mzimba, fr., *Pole Evans & Erens* 653 (K ; PRE). C : Salima Bay, fl. 23.ix.1935, *Galpin* 15034 (K ; PRE). S : Shire Highlands, fr. *Buchanan* (BM). **Mozambique.** N : Massangulo, fr. 15.v.1948, *Pedro & Pedrógão* 3528 (LMJ ; SRGH). Z : Quelimane Distr., Namagoa, fl. xi-xii, *Faulkner* 96 (COI ; K). T : Changara, fl. 28.ix.1948, *Wild* 2659 (K ; SRGH). MS : Manica, between R. Revuè and R. Ouro, fl. 18.iii.1948, *Barbosa* 1211 (BM ; LISC ; LM). SS : Inhambane, Massinga, fr. 21.v.1941, *Torre* 2724 (LISC).

Widely distributed in central and eastern tropical Africa and also in the Transvaal, Madagascar, India, Ceylon, Indo-China, Indonesia and China. Found in all types of woodland from the Mozambique coast up to about 1800 m. A form with fierce branching thorns at the base of its trunk is sometimes found in riverine fringes.

A species showing extreme variability in indumentum, leaf-shape and presence or absence of spines. There appear to be no clear-cut varieties and the species is equally variable in Asia. The fruit is edible but acid.

13. GERRARDINA Oliv.

Gerrardina Oliv. in Hook., Ic. Pl. : t. 1075 (1870).

Shrubs or small trees. Leaves alternate, petiolate, serrate or crenate, stipulate. Flowers bisexual in axillary, pedunculate cymes. Calyx campanulate with a cupular receptacle and 5 imbricate sepals. Petals 5 inserted on the margin of the disk. Disk cupular, ± adnate to the receptacle. Stamens opposite the petals, inserted on the disk margin. Ovary 1-locular ; placentas 2, each with 2 anatropous, pendulous ovules. Fruit dry or somewhat fleshy, 1–4-seeded, apparently indehiscent. Seeds obovoid or ellipsoid ; testa smooth or minutely reticulate, endosperm fleshy.

Gerrardina eylesiana Milne-Redh. in Hook., Ic. Pl. **34**: t. 3390 (1939). TAB. **50**. Type: S. Rhodesia, Umtali Distr., Stapleford Forest, *McGregor* 57/37 (BM ; FHO ; K, holotype).

Scrambling shrub to 6 m. tall, with slender pubescent or glabrescent branches. Leaf-lamina reddish when young, 2–5·5 × 1·8–3 cm., broadly ovate to lanceolate, acuminate to acute at the apex, ± cordate, truncate or broadly cuneate at the base ; margin crenate to crenate-serrate, teeth minutely glandular at the apex, glabrous on both sides or pubescent on the midrib below, lateral nerves in c. 7 pairs, prominent below, venation laxly reticulate and fairly prominent below ; petiole up to 5 mm. long ; stipules up to 3 × 2 cm., foliaceous, deltoid, acute to acuminate, truncate or slightly auriculate at the base, margin glandular-crenate. Flowers in axillary, pedunculate, few-flowered cymes ; peduncle up to 2 cm. long, slender, glabrous or pubescent near the base, secondary branches short, puberulous ; bracts minute, ovate, glabrous, pubescent ; pedicels 3–5 mm. long, glabrous. Receptacle c. 2 mm. in diam., cupular, glabrous ; sepals c. 2 × 1·5 mm., deltoid-ovate, minutely ciliate at the margins. Petals white, 1·5–2 × 1·5 mm., broadly ovate, rounded at the apex, margin ciliolate in the upper half. Stamens 5, opposite the petals ; filaments inserted on the margin of the disk, c. 0·5 mm. long ; anthers c. 0·6 × 0·5 mm., ovate, apiculate at the apex. Disk cupular, ± adnate to the receptacle, obscurely pentagonal. Ovary slightly immersed in the disk, ovoid, glabrous ; style c. 0·5 mm. long, glabrous ; stigma minutely bifid. Fruit c. 8 × 6 mm., pendulous, somewhat fleshy, elliptic-ovoid. Seeds 1–4, c. 5 × 3 mm., ellipsoid to plano-convex ; testa golden-brown, minutely reticulate.

S. Rhodesia. E : Chimanimani Mts., fl. 15.x.1945, *Wild* 3631 (K ; LISC ; SRGH).
Nyasaland. S : Mlanje Mt., Luchenya Plateau, fl. 26.vi.1946, *Brass* 16441 (K ; SRGH).
Also in Tanganyika. At forest edges and in forest undergrowth from 1500–2000 m.

14. HOMALIUM Jacq.

Homalium Jacq., Enum. Syst. Pl. Ins. Carib. : 5, 24 (1760).

Trees or shrubs. Leaves alternate (in African spp.), petiolate ; lamina entire or crenate-serrate with a gland near the apex of each tooth on the underside ; stipules absent, minute or large and orbicular, caducous or persistent. Flowers bisexual in axillary or terminal racemes or paniculate racemes, solitary or fasciculate on the inflorescence-branches, sessile or pedicelled, subtended by small caducous or persistent bracts. Receptacle (or calyx tube?) adnate to the ovary ; sepals (4) 5–8 (12), usually narrow, sometimes accrescent. Petals alternating with the sepals, persistent, sometimes accrescent. Stamens epipetalous, solitary or in fascicles of 2–3 (or up to 12 outside our area) ; filaments slender ; anthers small, extrorse, dorsifixed. Disk represented by a usually tomentose gland opposite each sepal. Ovary with its lower half sunk in the receptacle, 1-locular with 2–3 (8) placentas each with (1) 2–3 (7) ovules near the apex ; styles 2–4 (7), free or variously connate ; stigmas small. Capsule half inferior, woody or coriaceous, 2–3 (8)-valved from the apex or indehiscent. Seeds solitary or few.

Stamens solitary before each petal :
  Leaves ovate, oblong-ovate, broadly ovate, broadly obovate, oblong-ovate or elliptic, 4–10 (13) cm. long ; petals elliptic, not much larger than the sepals :
    Leaves glabrous, or minutely and sparsely pubescent, or with tufts of hairs in the nerve-axils on the underside ; flowers in axillary panicles ; style divided near the apex into short, stigmatic arms 0·5 mm. long or less   -   -   1. *dentatum*

3390

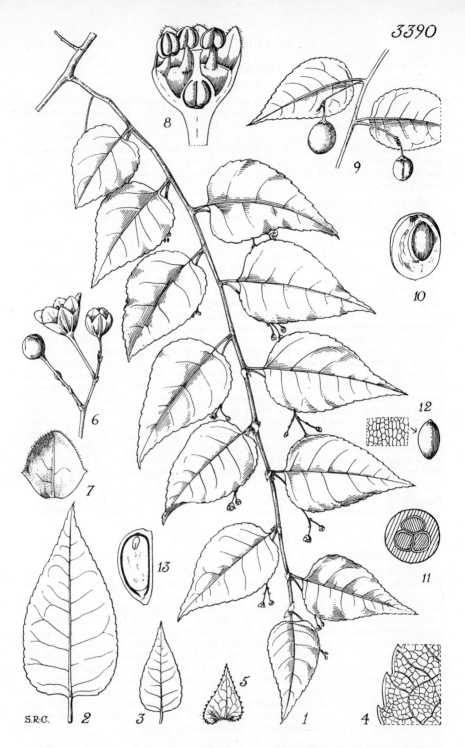

Tab. 50. GERRARDINA EYLESIANA. 1, flowering branch (×1); 2, 3, leaves (×1); 4, lower surface and margin of leaf (×4); 5, stipule (×4); 6, inflorescence (×4); 7, petal (×12); 8, longitudinal section of flower (petals removed) (×12), all from *McGregor* 57/37; 9, part of fruiting branch with mature fruits (×1); 10, longitudinal section of a 1-seeded fruit (×2); 11, diagrammatic transverse section of a 3-seeded fruit (×2); 12, seed (×2) and surface of testa (×24); 13, longitudinal section of seed (×4); 9–13 from *Pardy* s.n. From *Hooker's Icones Plantarum* with acknowledgements to the Bentham-Moxon Trustees.

Leaves densely yellowish-pubescent on both sides ; flowers in axillary racemes ;
  styles free almost to the base - - - - - - - 2. *chasei*
Leaves elliptic-oblong, often very large, 10–26 cm. long ; petals broadly spathulate or
  oblong-obovate, about twice the size of the sepals :
Leaves glabrous beneath ; vegetative branchlets quite glabrous, inflorescence
  branchlets sparsely and shortly pilose :
Stipules caducous ; leaves broadly cuneate or rounded at the base 3. *africanum*
Stipules persistent ; leaves cordate at the base - - - 4. *sarcopetalum*
Leaves permanently hairy on the midrib and nerves beneath ; branchlets densely
  hirsute - - - - - - - - 5. *molle* var. *rhodesicum*
Stamens in fascicles of 3 before each petal :
Leaves with the midrib and nerves setulose-pubescent below
                          6. *abdessammadii* subsp. *abdessammadii*
Leaves and petioles quite glabrous - 6. *abdessammadii* subsp. *wildemanianum*

1. **Homalium dentatum** (Harv.) Warb. in Engl. & Prantl, Nat. Pflanzenfam. **3,** 6a :
   36 (1893).—Engl., Pflanzenw. Afr. **3,** 2 : 581 (1921).—Gilg in Engl. & Prantl, Nat.
   Pflanzenfam. ed. 2, **21** : 427 (1925). Type from Natal.
   Blackwellia dentata Harv. in Harv. & Sond., F.C. **2** : 585 (1862). Type as above.

Tall tree to 17 m. tall or perhaps more with a straight unbranched bole, glabrous
in all its vegetative parts or sometimes the young leaves puberulous or with tufts of
hairs in the nerve axils on the underside of the leaves ; bark grey, smooth ; young
branches brown with pale lenticels. Leaf-lamina 4–13 × 3–7·5 cm., rotund,
broadly ovate or broadly obovate, apex usually abruptly acuminate, base broadly
cuneate or occasionally truncate, margins serrate-dentate or crenate, cartilaginous ;
nerves in 7-8 pairs, slightly prominent above, prominent below ; veins laxly
reticulate ; petiole up to 2·5 cm. long. Flowers yellowish-green, in lax, axillary
panicles up to 10 cm. long ; peduncles up to 5 cm. long, younger branches of
inflorescence puberulous ; pedicels c. 2 mm. long, in fascicles of 3–5, articulate
near the middle, puberulous, bracts very caducous. Sepals 7–9, 2 × 0·5 mm.,
oblong-linear, pubescent. Petals 7–9, c. 2·3 × 1 mm., elliptic, subacute at the apex,
densely pilose on both sides. Disk glands 0·5 mm. in diam., discoid-tomentellous.
Stamens 7–9 ; filaments c. 2·3 mm. long, slender, slightly widened at the base,
glabrous ; anthers 4 mm. in diam., rotund. Ovary conical, densely pilose both
outside and inside, ovules 9–12 ; style c. 1 mm. long, columnar, pilose, dividing
near the apex into three short stigmatic arms 0·5 mm. long or less. Fruit capsular,
not much larger than the ovary, up to 3 mm. in diam., probably tardily dehiscent
into 3 valves. Seeds c. 2 mm. in diam., biconical, often reduced to a single one
filling the capsule ; testa thin and brown.

N. Rhodesia. W : Nchanga, fl. i.1942, *Ferrar* (K ; SRGH). **S. Rhodesia.** W :
Matopos, fl. 27.v.1951, *Plowes* 1146 (SRGH). E : Umtali, fl. i.1947, *Chase* 305 (BM ;
SRGH).
Also in Natal and the Transvaal. A forest species or occasionally on rocky hillsides where
there is some fire protection.
The wood is hard and cream with brown streaks, taking a good polish (Chase).

2. **Homalium chasei** Wild in Bol. Soc. Brot., Sér. 2, **32** : 59 (1958). TAB. **51** fig. C.
   Type : S. Rhodesia, Umtali, *Chase* 6096 (K, holotype ; PRE ; SRGH).

Tree c. 8 m. tall, with a straight bole and smooth grey bark ; young branches
densely yellowish-pubescent. Leaf-lamina 4–7 × 2·5–4·2 cm., ovate, broadly ovate
or oblong-ovate, apex obtuse or subacute, base sometimes slightly asymmetric,
truncate or abruptly cuneate, pubescent on both sides, margin shallowly crenate,
entire towards the base ; nerves in 7–8 pairs, slightly prominent above, prominent
below ; veins laxly reticulate ; petiole up to 1 cm. long, densely pubescent.
Flowers in subsessile, axillary or terminal, dense racemes ; pedicels in fascicles
of 2–3, articulate in the middle, with a caducous, lanceolate, puberulous bract 1·5
mm. long at the base of each fascicle. Sepals 5–7, c. 2·2 × 0·5 mm., narrowly
linear-lanceolate, puberulous. Petals 5–7, c. 2·7–1·5 mm., rhomboid-elliptic,
obtuse at the apex, densely pilose. Disk glands c. 0·8 mm. in diam., discoid, tomentellous. Stamens 5–7, filaments c. 2·7 mm. long, slender, sparsely pilose ; anthers
0·6 mm. in diam., orbicular. Ovary conical, filled with long filamentous hairs
inside, shortly pilose outside ; ovules 6–12, compressed ; styles 3–4, pilose, up to
2·5 mm. long, free almost to the base. Mature fruit at present not known.

S. Rhodesia. E : Umtali Distr., Zimunya Reserve, fl. 6.v.1956, *Chase* 6096 (K ; SRGH).

Endemic in S. Rhodesia. Apparently confined to rocky hillsides where it probably receives some protection from fire and grazing.

3. **Homalium africanum** (Hook. f.) Benth. in Journ. Linn. Soc., Bot. **4** : 35 (1859).—Mast. in Oliv., F.T.A. **2** : 497 (1871).—Gilg in Engl., Bot. Jahrb. **40** : 488 (1908) ; in Engl. & Prantl, Nat. Pflanzenfam. ed. 2, **21** : 426 (1925).—Engl., Pflanzenw. Afr. **3**, 2 : 579 (1921).—Keay, F.W.T.A. ed. 2, **1**, 1 : 196 (1954). Type from Sierra Leone.
   Blackwellia africana Hook. f., in Hook., Niger Fl. : 361 (1849). Type as above.

Shrub or small tree, the vegetative parts quite glabrous. Leaf-lamina 12–26 × 5–12 cm., coriaceous, elliptic-oblong, bluntly acuminate or rounded at the apex, broadly cuneate or rounded at the base, margin repand or crenate ; nerves in 9–13 pairs, prominent below ; venation laxly reticulate, slightly prominent on both sides ; petiole up to 1·2 cm. long ; stipules c. 1 cm. long, foliaceous, subreniform to linear, caducous. Flowers in dense clusters along the branches of a terminal panicle c. 25 × 15 cm. ; inflorescence branches sparsely and shortly pilose ; pedicels very short, puberulous, articulate near the apex. Sepals 5, c. 1·2 mm. long, narrowly triangular, acute, densely puberulous. Petals 5, c. 2 × 1·3 mm., broadly spathulate or oblong-obovate, rounded or apiculate at the apex, densely pubescent on both sides. Disk glands 0·5 mm. in diam., discoid, tomentellous. Stamens 5 ; filaments 1 mm. long, glabrous, arching inwards ; anthers 0·4 mm. in diam., subglobose. Ovary a depressed cone deeply sunk in the receptacle, fairly long-pilose inside, shortly pilose outside ; ovules 6–9 ; style 0·75 mm. long, columnar, pilose, stigmatic arms spreading, 0·4 mm. long. Fruit (seen only in West African specimens) with the petals somewhat accrescent, the whole with a diameter of c. 7 mm. ; capsule not much larger than the unripe ovary. Seed usually single, c. 1·7 mm. in diam., with a thin brown testa, biconical.

N. Rhodesia. N : Kasama, fl. 18.x.1949, *Hoyle* 1296 (FHO). W : Mwinilunga, Luakera Falls, fl. 2.x.1937, *Milne-Redhead* 2524 (K). **Nyasaland.** S : Mt. Mlanje, fl. x.1891, *Whyte* (BM).
Widely distributed in West Africa and the Cameroons.
A species of evergreen fringing forest.

Apparently a rare species in our area but perhaps in the past more widely distributed as indicated by its present discontinuity of distribution.

4. **Homalium sarcopetalum** Pierre in Bull. Soc. Linn. Par., N. S. **14**: 119 (1899).—Keay, F.W.T.A. ed. 2, **1**, 1 : 196 (1954).—Fernandes & Diniz in Garcia de Orta, **5**, 2 : 252 (1957). TAB. **51** fig. B. Type from Gaboon.
   Homalium riparium Gilg in Engl., Bot. Jahrb. **40** : 494 (1908) ; in Engl. & Prantl, Nat. Pflanzenfam. ed. 2, **21** : 427 (1925).—Engl., Pflanzenw. Afr. **3**, 2 : 580 (1921).—Brenan, T.T.C.L. : 549 (1949).—Fernandes & Diniz, tom. cit. : 249 (1957). Syntypes from Tanganyika and Mozambique : Namuli, Makua, *Last* (K, syntype).
   Homalium stipulaceum sensu Garcia in Est. Ens. Docum. Junta Invest. Ultr. **12** : 161 (1954).

Bush or small tree, very similar to *H. africanum* but the leaves are cordate rather than cuneate or rounded at the base and the stipules, which are orbicular-reniform and c. 1·5 cm. in diam., are persistent. The petals are often, but not always, swollen and fleshy.

Nyasaland. S : Blantyre Distr., *Buchanan* 935 (K). **Mozambique.** Z : R. Lua, Guruè-Mocuba Rd., fl. 9.xi.1942, *Mendonça* 1346 (COI ; LISC).
Also in Tanganyika, Fernando Po, French Cameroons, Gaboon and Spanish Guinea. In evergreen fringing forest.

Like *H. africanum* this appears to be a rare species in our area.

5. **Homalium molle** Stapf in Journ. Linn. Soc., Bot. **37** : 100 (1905).—Gilg in Engl., Bot. Jahrb. **40** : 493 (1908) ; in Engl. & Prantl, Nat. Pflanzenfam. ed. 2, **21** : 427 (1925).—Keay, F.W.T.A. ed. 2, **1**, 1 : 196 (1954). Syntypes from Liberia and Sierra Leone.
   Homalium stipulaceum Welw. ex Mast. in Oliv., F.T.A. **2** : 498 (1871) pro parte quoad specim. Afzel. et Mann.

Spreading tree very similar to *H. africanum* and *H. sarcopetalum*. It has the persistent orbicular-reniform stipules and somewhat cordate leaves of *H. sarco-*

petalum but differs from both in the underside of the leaves being persistently softly pubescent.

Var. **rhodesicum** R.E.Fr., Wiss. Ergebn. Schwed. Rhod.-Kongo-Exped. **1** : 156 (1914).— Engl., Pflanzenw. Afr. **3**, 2 : 580 (1921). Type : N. Rhodesia, Northern Province, Mporokoso Distr., Kunkuta, *Fries* 1180 (UPS, holotype).

Differs from the type in having the hairs on the undersides of the leaves concentrated on the underside of the midrib, nerves and veins and also in the hairs on the young branches being densely and patently hirsute rather than shortly pubescent. The only two specimens collected so far appear to be shrubs or scramblers rather than tall trees.

N. Rhodesia. N : Pansa R., fl. 16.x.1949, *Bullock* 1144 (K).
The type variety occurs in tropical W. Africa whilst var. *rhodesicum* is endemic, so far as is known, to N. Rhodesia. In fringing forest or forest patches.

The characters separating *H. africanum*, *H. sarcopetalum* and *H. molle* are hardly perhaps of specific value but these species are so rare in our area that we shall have to await further collecting before a final evaluation of these taxa can be made.

6. **Homalium abdessammadii** Aschers. & Schweinf. in Sitz.-Ber. Ges. Nat. Fr. Berl. **1880** : 130 (1880).—Gilg in Engl., Bot. Jahrb. **40** : 494 (1908) ; in Engl. & Prantl, Nat. Pflanzenfam. ed. 2, **21** : 428 (1925).—Engl., Pflanzenw. Afr. **3**, 2 : 581 (1921).— Fernandes & Diniz in Garcia de Orta, **5**, 2 : 252 (1957). Type from the Sudan (Equatoria).

Small tree c. 10 m. tall ; young branches pubescent or glabrous, often purplish-brown and with pale lenticels. Leaf-lamina 5–10 × 3·5–5·5 cm., obovate or broadly elliptic, apex acuminate or rarely rounded, base broadly cuneate or truncate, margin coarsely and irregularly crenate-serrate, lateral nerves in c. 6 pairs, slightly prominent above, prominent below, venation laxly reticulate, glabrous on both sides or setulose-pubescent on the midrib and main nerves below, sometimes with tufts of hairs in the nerve-axils below ; petiole often purplish, up to 1·5 cm. long, pubescent or glabrous. Flowers in lax, terminal or axillary panicles up to 12 cm. long, the individual flowers 1–2 together and sessile or nearly sessile on the puberulous branches of the panicle. Sepals 5–7, 3·5–5 × 1·5–2·5 mm., elliptic or ovate, acute at the apex, densely pubescent on the back. Petals 5–7, 3·5–4 × 2–2·5 mm., broadly rhomboid-elliptic or ovate, subacute at the apex, densely pubescent on the back. Stamens in fascicles of 3 opposite the petals ; filaments 3·5 mm. long, slender, glabrous or sparsely pilose ; anthers 0·5 mm. in diam., globose, the thecae free in the lower half and the connective with a blunt glandular process at the apex. Disk glands 1–1·5 mm. in diam., discoid, tomentose. Ovary conical, pilose outside, full of long pilose hairs inside ; ovules 6–9, discoid ; style 2–3·5 mm. long, columnar, pilose at the base ; stigmatic arms divergent, very short. Fruit capsular, woody and perhaps indehiscent, surrounded by the persistent corolla, pilose, similar in shape to the ovary. Seed usually solitary, c. 3 × 2 mm., narrowly ovoid; testa thin and brown.

Subsp. **abdessammadii**
   Homalium macranthum Gilg in Engl., Bot. Jahrb. **40** : 496 (1908) ; in Engl. & Prantl, Nat. Pflanzenfam. ed. 2, **21** : 428 (1925).—Engl., loc. cit.—Brenan, T.T.C.L. : 549 (1949).—Fernandes & Diniz in Garcia de Orta, **5**, 2 : 250 (1957). Type : Mozambique, R. Rovuma, opposite Lissenga Mt., *Busse* 1049 (B, holotype † ; EA).

Young branchlets and petioles pubescent. Midrib of leaves and often the lateral nerves setose-pubescent below, the midrib usually puberulous above. Occasionally there are tufts of hairs in the nerve axils. The flowers have sparsely pilose filaments and the sepals (5 × 2·5 mm.) and petals (4 × 2·5 mm.) are rather larger than in subsp. *wildemanianum*.

Mozambique. N : R. Rovuma, fl. ii. *Busse* 1049 (B† ; EA). Z : Mocuba, Namagoa, fl. 12.i.1949, *Faulkner* 381 (K).
Also in the Sudan, Kenya and Tanganyika. In fringing forest.

Subsp. **wildemanianum** (Gilg) Wild in Bol. Soc. Brot., Sér. 2, **32** : 57 (1958). TAB. **51** fig. A. Type : Belgian Congo, Katanga.
   Homalium abdessammadii sensu De Wild., Études Fl. Katanga : 93 (1903).
   Homalium wildemanianum Gilg in Engl., Bot. Jahrb. **40** : 497 (1908) ; in Engl.

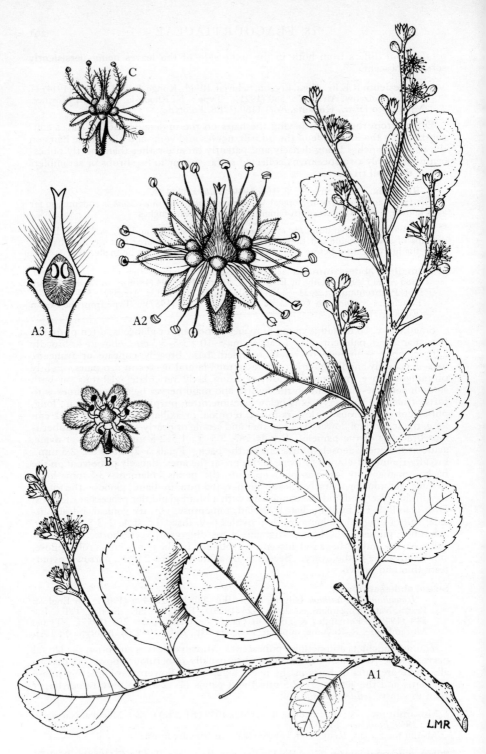

Tab. 51. A.—HOMALIUM ABDESSAMMADII SUBSP. WILDEMANIANUM. A1, flowering branch
(× ⅔) ; A2, flower (× 4) ; A3, longitudinal section of gynoecium (× 4), all from
Martin 66. B.—HOMALIUM SARCOPETALUM, flower (× 4) *Faulkner* 6. C.—HOMALIUM
CHASEI, flower (× 4) *Chase* 6096.

& Prantl, Nat. Pflanzenfam. ed. 2, **21** : 428 (1925).—Engl., loc. cit. (1921). Type as for subsp. *wildemanianum*.

Homalium rhodesicum Dunkley in Kew Bull. **1934** : 182, cum tab. (1934). Type : N. Rhodesia, Kafue, *Martin* 66 (FHO ; K, holotype).

In this subspecies, all the vegetative parts are quite glabrous and the filaments are also glabrous. The sepals and petals are slightly smaller than in subsp. *abdessammadii*.

Caprivi Strip. Katima Mulilo, fl. 23.x.1954, *West* 3245 (K ; SRGH). **N. Rhodesia.** B : near Senanga, fl. 30.vii.1952, *Codd* 7264 (BM ; K ; PRE). N : Ft. Rosebery, Mansa R., fl., *Lawton*, 131 (K). W : Chingola, fl. 28.ix.1955, *Fanshawe* 2463 (K ; SRGH). C : Kafue, fl. 26.vii.1956, *Clarke* 166 (PRE). S : Katombora, fr. 20.i.1956, *Gilges* 566 (SRGH). **S. Rhodesia.** W : Victoria Falls, Cataract I., fl. 23.xi.1949, *Wild* 3170 (K ; LISC ; SRGH).

Also in Tanganyika, Belgian Congo and Angola. Like subsp. *abdessammadii*, occurring in fringing forest.

Although I have not been able to trace *Curson* 521 and 732 from the northern division of Bechuanaland (Ngamiland), named *Homalium rufescens* by O.B.Miller (in Journ. S. Afr. Bot. **18** : 59 (1952)), they probably belong to this subspecies, as does *Curson* 969 (PRE) from that part of Ngamiland lying within the Caprivi Strip. They could hardly be *H. rufescens* (E. Mey.) Benth., which is a high-rainfall species from Natal.

15. CASEARIA Jacq.

Casearia Jacq., Enum. Syst. Pl. Ins. Carib. : 4, 21 (1760).

Shrubs or trees, sometimes very tall. Leaves alternate, manifestly distichous, entire, crenate or serrate, usually pellucid-punctate and/or -striate. Stipules caducous (in all our material) or persistent. Flowers axillary, in fascicles or glomerules, or single ; pedicels articulated near the base and surrounded there by many scale-like bracts which often form a cushion in the leaf-axil. Receptacle cupular or funnel-shaped ; sepals 5, imbricate, persistent. Petals 0. Stamens (5) 7–10 (12), equal or alternate ones longer. Staminodes alternating with the stamens, clavate or flattened, usually hairy above and united below with the stamens into a ± perigynous tube. Ovary free, ovoid or columnar ; style 0 or very short ; stigma capitate ; ovules few to many. Capsule coriaceous, hard or succulent, globose-ovoid or oblong-ovoid, 3-angled when fresh, often 6-ribbed when dry, 2–4-valved. Seeds few to numerous, ovoid or ellipsoid, enveloped in a soft, membranous, coloured and usually fimbriate aril ; testa crustaceous ; endosperm fleshy; cotyledons flat.

Shrub or small tree up to c. 8 m. ; leaves lanceolate or narrowly ovate ; stamens quite glabrous - - - - - - - - - - - - 1. *gladiiformis*
Tree up to about 40 m. ; leaves narrowly oblong to oblong ; stamens minutely pubescent 2. *battiscombei*

1. **Casearia gladiiformis** Mast. in Oliv., F.T.A. **2** : 493 (1871).—Gilg in Engl., Bot. Jahrb. **40** : 510 (1908) ; in Engl. & Prantl, Nat. Pflanzenfam. ed. 2, **21** : 454 (1925). —Engl., Pflanzenw. Afr. **3**, 2: 589 (1921).—Fernandes & Diniz in Garcia de Orta, **5**, 2 : 252 (1957). TAB. **52** fig. B. Type : Mozambique, Zambezi, Chupanga, *Kirk* (K, holotype).
    Casearia junodii Schinz in Mém. Herb. Boiss. **10** : 52 (1900).—Gilg in Engl., Bot. Jahrb. **40** : 513 (1908) ; in Engl. & Prantl, loc. cit.—Engl., tom. cit. : 590 (1921). Type : Mozambique, Delagoa Bay, *Junod* 351 (K ; Z, holotype).

Large shrub or small tree up to c. 8 m. tall ; young branches puberulous, soon becoming glabrous ; bark greyish. Leaf-lamina 5–17 × 2·8–6·5 cm., coriaceous, lanceolate or narrowly ovate, apex acuminate, acute or subobtuse, cuneate and usually oblique at the base, margins entire or undulate or when young sparsely and irregularly denticulate, glabrous on both sides or puberulous on the midrib when very young, with circular and linear pellucid dots, somewhat shining above, duller below, lateral nerves in c. 9 pairs, slightly prominent on both sides, venation laxly reticulate ; petiole up to 1·3 cm. long, channelled above, puberulous or glabrous. Flowers greenish-white, in dense axillary fascicles on a cushion of minute, brown, membranous, ovate bracteoles with ciliolate margins ; pedicels 1–2 mm. long, puberulous. Sepals 5, 2–3 × 2–2·5 mm., ovate, obtuse at the apex, margins membranous, puberulous or glabrescent on the back. Receptacle cupular, c. 2·5 mm. wide at the mouth, c. 1 mm. long, from puberulous to villous outside. Staminal tube united for about 0·5 mm. and then dividing into 10 fertile stamens and 10 alternating staminodes, glabrous ; filaments c. 0·8 mm. long, glabrous ; anthers 0·9 × 0·6 mm., narrowly ovate, connective bluntly apiculate ; staminodes up to 0·8 mm. long, oblong, villous. Ovary ovoid, hirsute at least towards the apex ; style very short ; stigma capitate. Fruit an angular, ellipsoid capsule splitting from above into 3–4 longitudinal valves. Seeds c. 10, c. 4 × 3 mm., ellipsoid with a pale smooth testa and clasped by a fleshy lobate aril hiding more than half the seed.

Nyasaland. S : Port Herald, fl. *Topham* 564 (FHO). **Mozambique.** Z : Maganja da Costa, fl. 28.ix.1949, *Barbosa & Carvalho* 4247 (LM ; SRGH). MS : Moribane, fl. 2.x.1953, *Pedro* 4194 (K ; LMJ ; PRE). SS : Bilene, fr. 9.xii.1940, *Torre* 2269 (BM ; LISC). LM : Santaca, fl. 6.x.1948, *Gomes e Sousa* 3855 (COI ; K ; PRE).

Endemic to our area although closely related species, which may even be conspecific, occur in Tanganyika and Natal. In coastal bushland or woodland.

The type of *C. junodii* has more glabrous flowers and smaller leaves than typical *C. gladiiformis* but a good range of intermediates exists, and there appears to be no geographical segregation of the two forms, so they must be treated as one species.

2. **Casearia battiscombei** R.E.Fr. in Notizbl. Bot. Gart. Berl. **9** : 326 (1925).   TAB. **52** fig. A.   Type from Kenya.

   Casearia chirindensis Engl., Pflanzenw. Afr. **3**, 2 : 590 (1921) *nom. nud.*

Tall tree up to about 40 m. tall with rough greyish bark ; young branches brown (or silvery in saplings), puberulous or glabrescent. Leaves on puberulous or glabrescent petioles up to 6 mm. long ; leaf-lamina 8–22 × 3–6 cm., papyraceous, narrowly oblong to oblong, shortly acuminate or obtuse at the apex, unequal-sided and rounded, slightly cordate or broadly cuneate at the base, margin entire, undulate or rarely shallowly and irregularly crenulate (the leaves on young saplings often serrate) glabrous on both sides, with circular and linear pellucid dots, lateral nerves in 12–20 pairs, prominent below ; venation laxly reticulate, not at all prominent ; petiole up to 6 mm. long, puberulous or glabrous ; stipules up to 2 mm. long, triangular, scarious, very caducous. Flowers numerous, greenish, in axillary fascicles borne on a cushion of minute, densely pubescent bracteoles ; pedicels 3–4 mm. long, increasing to 1 cm. in fruit, sparsely puberulous or glabrescent. Sepals c. 2 × 1·5 mm., broadly elliptic or rotund, obtuse at the apex, margins membranous, puberulous or glabrescent outside. Receptacle c. 2 mm. wide at the mouth, c. 1 mm. tall, funnel-shaped, puberulous or glabrescent outside. Staminal tube exceedingly short, stamens 6–10 on puberulous filaments 0·8–1 mm. long ; anthers c. 0·5 × 0·3 mm., ovate-lanceolate, minutely and sparsely puberulous or glabrescent, connective bluntly apiculate ; staminodes 0·5 mm. long, oblong or subquadrate, densely hirsute. Ovary ovoid, glabrous ; style very short, stigma capitate. Fruit an ellipsoid capsule, c. 1·3 × 0·8 cm., splitting from above into 2–4 longitudinal valves, glabrous. Seeds few, c. 6 × 4·5 mm., ellipsoid, with a smooth pale testa, almost enclosed in a fleshy aril.

S. Rhodesia. E : Chirinda Forest, fl. 22.x.1947, *Wild* 2113 (K ; SRGH). **Nyasaland.** S : Blantyre, fl. 1895, *Buchanan* 299 (B† ; BM).

Also in Kenya and Tanganyika. In evergreen forest at 1000–2000 m.

The heartwood is white and soft.

Gilg (in Engl., Bot. Jahrb. **40** : 510 (1908)) said that *Buchanan* 299 was very probably *C. macrodendron* Gilg, a species from Tanganyika closely related to *C. gladiiformis*. There is no doubt, however, that this gathering is perfectly good *C. battiscombei.*

16. BIVINIA Jaub. ex Tul.

Bivinia Jaub. ex Tul. in Ann. Sci. Nat., Sér. 4, Bot. **8** : 78 (1857).

Tree with alternate, petiolate, crenate-dentate leaves. Stipules absent. Flowers bisexual in axillary racemes ; bracts small and linear ; pedicels articulated near the middle. Receptacle very shallow. Sepals 5–6, valvate. Petals 0. Glands broad and truncate, pubescent, adnate to the base of each sepal. Stamens in fascicles alternating with the sepals ; filaments folded in bud ; anthers very small. Ovary 1-locular with 4–6 multiovulate placentas ; styles 4–6, filiform. Fruit a small dry capsule with 4–6 valves. Seeds few, with long hairs, endosperm fleshy, cotyledons thin and flat.

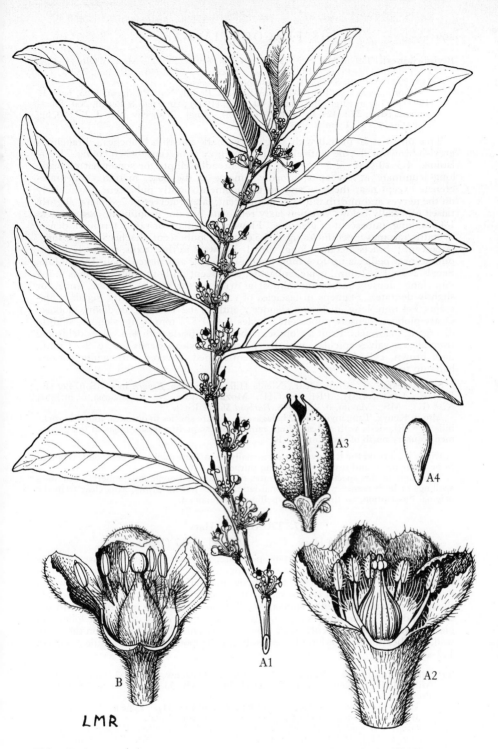

Tab. 52. A.—CASEARIA BATTISCOMBEI. A1, flowering branch (×⅔) *Battiscombe* 19 ;
A2, flower, one sepal, one stamen and one staminode removed (×14) *Wild* 2239 ;
A3, fruit (×2) *Battiscombe* 19 ;   A4, seed (×2) *Battiscombe* 19.   B.—CASEARIA GLADII-
FORMIS, flower, two sepals, five stamens and five staminodes removed (×6)
Pedro 4194.

Bivinia jalbertii Tul. in Ann. Sci. Nat., Sér. 4, Bot. **8**: 78 (1857).—Mast. in Oliv., F.T.A. **2**: 496 (1871).—Perrier, Fl. Madagasc., Flacourt.: 67 (1946).—Brenan, T.T.C.L.: 548 (1949). TAB. **53** fig. B. Type from Madagascar.

   Calantica jalbertii (Tul.) Warb. in Engl. & Prantl, Nat. Pflanzenfam. **3**, 6a: 37 (1893).—Gilg in Engl., Bot. Jahrb. **40**: 498 (1908); in Engl. & Prantl, Nat. Pflanzenfam. ed. 2, **21**: 429, fig. 191 F–G (1925).—Engl., Pflanzenw. Afr. **3**, 2: 580, fig. 25 F–G (1921). Type as above.

Tree up to c. 20 m. tall with a light grey bark and thin branches; branchlets greyish brown with pale lenticels, with a thin grey pubescence when young. Leaf-lamina, 4·5–10 (13) × 2·5–5·8 cm., ovate, broadly elliptic or somewhat obovate, long-acuminate at the apex, broadly cuneate at the base, margin ± dentate-crenate except near the base and on the acumen, slightly pubescent, particularly on the nerves and midrib but later glabrescent, lateral nerves in 7–8 pairs, slightly raised especially below, venation laxly reticulate but not prominent; petiole up to 1·4 cm. long, pubescent or glabrous. Flowers in axillary or rarely terminal, dense, cylindric racemes; peduncles up to 3·5 cm. long, pubescent, rhachis 5–12 cm. long, pubescent; bracts 2–3 mm. long, with 1–3 flowers in their axils, narrowly linear, pubescent; pedicels up to 4 mm. long, pubescent, articulated about the middle. Sepals 2–3 × 1·5–2 mm., ovate, subacute, pubescent on both sides. Glands c. 0·4 mm. long, densely pubescent, almost as wide as the sepals, truncate, entire or slightly dentate. Stamens in fascicles of c. 10 alternating with the sepals; filaments 3–4 mm. long, glabrous, slender; anthers 0·2–0·3 mm. in diam., globose. Ovary globose, densely pubescent; styles 4–6, 1·5–2 mm. long, slender, slightly divergent, glabrous. Fruit a globose, pubescent capsule with 4–6 valves dehiscing from above, valves apiculate with the persistent styles. Seeds few, c. 2 × 1 mm., dark brown, cylindric, somewhat crescent-shaped, covered with white cottony hairs up to c. 4 mm. long.

S. Rhodesia. S: Chibi Distr., Njenja Hills, fl. iii.1956, *Espach* in GHS 67499 (B; BR; K; LISC; LMJ; PRE; SRGH). **Mozambique.** N: R. Rovuma, fr. iii.1861, *Kirk* (K). MS: Mavita, fl. 22.i.1948, *Barbosa* 868 (LISC).

Also in Kenya, Tanganyika and Madagascar. A rare species in our area, found in rocky hills in S. Rhodesia with a locally high rainfall and some dry season mists in the escarpment country north of the Limpopo R.

Kirk's locality on the R. Rovuma may actually be in Tanganyika. The wood is reported to be borer proof and useful for building purposes.

The name of this species commemorates the botanists Boivin and Jaubert. Tulasne's spellings of both names were intentional latinizations not misprints, as is clear from the original description, so the original spellings are retained.

17. TRIMERIA Harv.

Trimeria Harv., Gen. S. Afr. Pl.: 417 (1838).

Shrubs or trees. Leaves alternate, petiolate, 5–9-nerved from the base, margins serrate or dentate. Stipules caducous, sometimes foliose. Flowers dioecious, with the males in axillary panicles and the females in spiciform racemes. Male flowers: perianth of 6–10 lobes in two rows, the inner slightly larger, disk glands opposite the outer perianth segments, stamens 9–12 in bundles of 3–4, alternating with the disk glands. Female flowers: perianth as in the male but with smaller lobes and glands, ovary sessile, 1-locular with 3 parietal placentas borne high up on the ovary wall, each bearing near its base 1–3 ascending anatropous ovules. Capsule 3-valved, 1–3-seeded.

Trimeria rotundifolia (Hochst.) Gilg in Engl., Pflanzenw. Afr. **3**, 2: 582 (1921).—Milne-Redh. in Kew Bull. **1939**: 34 (1939). TAB. **53** fig. A. Type from Cape Province.

   Monospora rotundifolia Hochst. in Flora, **24**: 661 (1841). Type as above.
   Monospora grandifolia Hochst., loc. cit. Type from Natal.
   Trimeria grandifolia (Hochst.) Warb. in Engl. & Prantl, Nat. Pflanzenfam. **3**, 6a: 37, fig. 13 H–J (1893).—Burtt Davy, F.P.F.T. **1**: 219 (1926). Type as for *Monospora grandifolia*.

Shrub or small tree up to 10 m. tall; young branches densely pubescent, becoming glabrous later; bark smooth, brown. Leaf-lamina 4–12 × 2·5–10 cm., orbicular, rotund or obovate, apex rounded, emarginate or occasionally apiculate,

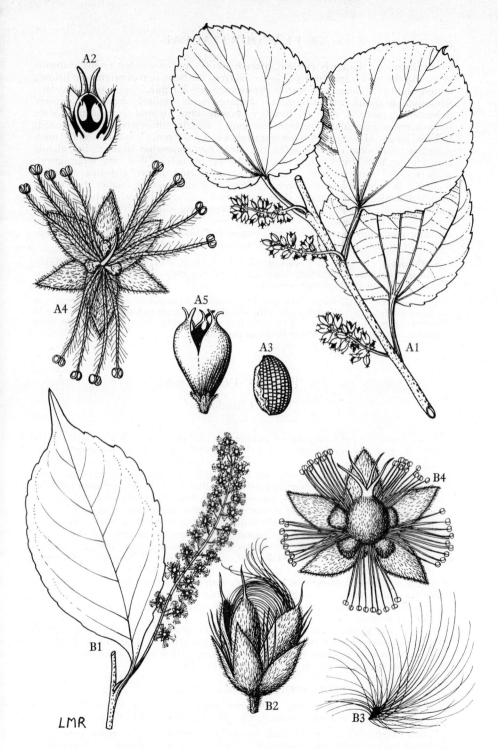

Tab. 53. A.—TRIMERIA ROTUNDIFOLIA. A1, branchlet with female flowers (× ⅔) *Chase* in GHS 16630 ; A2, female flower in longitudinal section (× 14) *Chase* in GHS 16630 ; A3, seed (× 6) *Chase* in GHS 16630 ; A4, male flower (× 10) *Wild* 4635 ; A5, fruit (× 4) *Chase* in GHS 16630. B.—BIVINIA JALBERTII. B1, flowering branchlet (× ⅔) *Espach* in GHS 67500 ; B2, fruit and persistent calyx (× 6) *Champion* in GHS 52429 ; B3, seed (× 6) *Champion* in GHS 52429 ; B4, flower (× 6) *Espach* in GHS 67500.

base truncate or shallowly cordate, margin serrate, pubescent but finally glabrous except on the nerves, palmately 5–7-nerved from the base, nerves prominent below ; petiole up to 3 cm. long, pubescent ; stipules c. 3–6 × 2–6 mm., caducous, foliaceous, lanceolate to reniform, pubescent. Inflorescences axillary, branches densely pubescent ; pedicels absent ; bracts c. 1 mm. long, deltoid, pubescent. Male inflorescences paniculate, up to 6 cm. long ; female inflorescences unbranched, 1–2 cm. long. Male flowers : perianth-lobes c. 1 × 0·5 mm., concave-oblong, apex subacute, pubescent on both sides, the inner row somewhat larger ; disk glands truncate, up to 0·4 mm. long ; stamens 10–12, filaments 2·5–3 mm. long, hairy ; anthers c. 0·3 mm. in diam., globose ; style aborted, linear, glabrous. Female flowers : perianth lobes sometimes reduced to 6 and slightly smaller than in the male ; disk glands also smaller than in the male ; ovary ellipsoid, glabrous, sometimes shallowly 3-furrowed ; styles 3, c. 0·4 mm. long, slightly divergent, glabrous. Fruits yellowish. Capsule c. 5 × 2·5 mm., narrowly obovoid, three-sided, glabrous, splitting from above into 3 valves. Seeds 1–2 (3), c. 2·5 × 1·5 mm., ellipsoid, dark grey, with a red aril attached on one side ; testa tessellate-pitted.

S. Rhodesia. E : Umtali, Engwa, fl. 1.ii.1955, *E.M. & W.* 51 (BM ; LISC ; SRGH). Also in Swaziland, Transvaal, Natal and Cape Province. Evergreen forest margins between 1000 and 2000 m. approximately.

The basionyms of *T. rotundifolia* and *T. grandifolia* were published simultaneously in the same publication. Warburg (loc. cit.) was the first author to transfer one of them to the correct genus, but Gilg (loc. cit.) was the first to unite them : his choice of epithet must therefore be followed. See Milne-Redhead (loc. cit.).

19. PITTOSPORACEAE

By G. Cufodontis

Woody plants, trees, shrubs or climbers with resin-ducts in the bark, rarely spiny. Leaves alternate, often crowded at the ends of the branches, simple, evergreen and ± leathery, entire or rarely dentate or lobed, without stipules. Inflorescences cymose or paniculate, terminal and/or axillary, rarely flowers in clusters on the old wood or solitary, terminal and/or axillary. Flowers actinomorphic or somewhat irregular, hypogynous, bisexual or functionally (rarely morphologically) unisexual. Sepals 5, free or ± connate, sometimes imbricate. Petals 5, free or with claws slightly connivent, usually with spreading or revolute blades, imbricate in bud. Stamens 5, free, with 2-thecous introrse anthers opening by slits, rarely by apical pores. Ovary sessile or shortly stipitate, paracarpous, with 2–5 carpels and parietal placentas, usually 1-locular, or, less often, 2–5-locular by central contact of the placentas ; style simple, stigma capitate or somewhat lobed ; ovules 2-several in 2 rows, anatropous, horizontal, with one integument. Fruit a berry or a capsule with entire, rarely split valves. Seeds without an aril but often covered by a viscid resin, rarely dry and winged, with a smooth testa, a hard endosperm and a very minute embryo.

In addition to the genus *Pittosporum* the genus *Hymenosporum* R. Br. ex F. Muell. is represented by the cultivated *H. flavum* (Hook.) R. Br. ex F. Muell., a tree or shrub with fragrant cream flowers turning yellow and winged seeds, introduced from Australia, which is grown in S. Rhodesia.

PITTOSPORUM Banks ex Soland.

Pittosporum Banks ex Soland. in Gaertn. Fruct. **1** : 286, t. 59 fig. 7 (1788).— Cufod. in Fedde, Repert. **55** : 27–112 (1952), 113 (1953).

Trees or shrubs, never climbing or spiny. Leaves usually entire, sometimes undulate, rarely subserrate or shallowly lobed, glabrous or pilose. Inflorescences usually variously paniculate or subracemose, terminal or terminal with axillary branches, rarely fascicled, or flowers solitary and axillary. Flowers actinomorphic, never more than 15 mm. long, of various colours (in our area only white, yellowish or greenish) sweet-scented, functionally unisexual ; male with long filaments, fertile anthers and slender sterile ovary; female with short filaments, reduced sterile anthers and stout fertile ovary. Usually all the flowers of one inflorescence alike (? monoecious or dioecious). Sepals free or ± connate, imbricate or not. Petals with claws sometimes connivent and blades usually spreading or revolute. Anthers opening by slits. Ovary 1-locular, with 2–5 (in our area always 2) carpels*; style short with a capitate or 2-lobed stigma, splitting in fruit according to the number of carpels; ovules ovoid. Fruit capsular with entire, leathery or woody, erect or spreading valves, sometimes reflexed when quite ripe. Seeds variously deformed by mutual pressure, 2–many in 2 rows on each parietal placenta, orange to red, covered with a sticky, slow-drying resin.

Plants with margins of leaves distinctly undulate; terminal inflorescences few-flowered, c. 12 mm. long; sepals more than 5 mm. long, connate in the lower half in a spathe, finally slit on one side, with subulate, hairy lobes bent outwards (cultivated)

5. *undulatum*

Plants not corresponding with every character above (indigenous):
Adult leaves with persistent spreading hairs on petiole and (fewer) on midrib; reticulation remaining conspicuous above for a long time; inflorescences small, few-flowered, with curly, persistent pubescence on axes and pedicels; petals sometimes (in *P. kapiriense*) up to 8 mm. long; sepals always free; capsules always more than 4-seeded:
    Adult leaves ± chartaceous, bright green with coloured reticulation remaining conspicuous for a long time on both surfaces, acutely acuminate, often cuspidate on account of the prolonged midrib; inflorescences terminal and axillary; petals up to 7 × 1·5 mm.; sepals c. 2·5 × 1 mm. - - - - 3. *rhodesicum*
    Adult leaves ± leathery, dull dark green, with a dense, regular tessellation above, glaucescent and with faint and incompletely visible reticulation beneath, shortly acuminate with very blunt tips; inflorescences terminal; petals up to 8 × 2 mm.; sepals up to 4 × 1·5 mm. - - - - - - - 4. *kapiriense*
Adult leaves usually quite glabrous, rarely with faint appressed pubescence on petiole and (less) on midrib; reticulation always coloured at first and entirely visible, soon turning pale beneath; inflorescences usually well developed with appressed or ± spreading pubescence on branches and pedicels (rarely almost glabrous); petals only exceptionally attaining 8 mm., usually 5–7 mm. long; sepals free or ± connate; capsules (in our area probably always) 4-seeded:
    Adult leaves sometimes with faint pubescence as above, usually obovate-spathulate to rounded-oblanceolate, glaucescent at least beneath, less often ± acuminate and then darker green, sometimes glossy above; midrib flat or in a shallow furrow above; veins thin, flat, merging into the reticulation, only exceptionally prominulous beneath, often conspicuous above; reticulation dense, uniform, ± impressed and often tessellate above when old; inflorescences sometimes reduced and subglabrous; sepals free or ± connate, petals mostly greenish-white; ripe valves plano-convex or dorsally gibbous, spreading or bent slightly upwards

1. *viridiflorum*

    Adult leaves always quite glabrous, usually ± acuminate, dull dark green above, ± yellowish-green beneath; midrib impressed above, often almost hidden in a deep furrow; veins evidently interconnected by arched junctions and ± prominent beneath, rarely slightly elevated above; reticulation lax, not uniform, usually laxer along the midrib, visible but never clearly tessellate above; inflorescences always well developed and pubescent; petals mostly creamy-white; sepals quite free or almost so, ripe valves dorsally shallow-concave, spreading or finally ± reflexed

2. *mannii* subsp. *ripicola*

1. **Pittosporum viridiflorum** Sims in Curt., Bot. Mag. **41**: t. 1684 (1814).—Putterl., Syn. Pittosp.: 11 (1839).—Sond. in Harv. & Sond., F.C. **1**: 443 (1860).—Eyles in Trans. Roy. Soc. S. Afr. **5**: 359 (1916).—Marloth, Fl. S. Afr. **2**: 30, t. 11 fig. E, fig. 15 (1925).—Burtt Davy, F.P.F.T. **1**: 213 (1926).—Exell & Mendonça, C.F.A. **1**, 1: 87 (1937).—Cufod. in Exell & Mendonça, C.F.A. **1**, 2: 362 (1951); in Fedde, Repert. **55**: 41 (1952).—Brenan in Mem. N.Y. Bot. Gard. **8**, 3: 219 (1953).—TAB. **54** fig. A. Type a specimen from the Cape of Good Hope cultivated at Malcolm and Sweet's Nursery, Stockwell Common, London.

    Pittosporum floribundum Wight & Arn., Prodr. Fl. Penins. Ind. Or. **1**: 154 (1834). —Gowda in Journ. Arn. Arb. **32**: 332 (1951). Type from S. India.

    Pittosporum commutatum Putterl., op. cit.: 10 (1839). Syntypes from S. Africa.

    Pittosporum abyssinicum var. *angolense* Oliv., F.T.A. **1**: 124 (1868). Syntypes from Angola (Huila).

---

* In the whole of Africa (not in our area) only two cases of abnormal trivalvate capsules have been observed, together with normal ones.

Pittosporum kruegeri Engl. in Notizbl. Bot. Gart. Berl. **2** : 26 (1897). Type from the Transvaal.

Pittosporum malosanum Bak. in Kew Bull. **1897** : 244 (1897). Type : Nyasaland, Zomba, near Mt. Malosa, *Whyte* 420 (K, holotype).

Pittosporum antunesii Engl., Bot. Jahrb. **32** : 130 (1902). Type from Angola (Huila).

Pittosporum vosseleri Engl., Bot. Jahrb. **43** : 371 (1909) ; in Pflanzenw. Afr. **3,** 1 : 851 (1915). Syntypes from Tanganyika (Usambara).

Pittosporum viridiflorum var. *commutatum* (Putterl.) Moeser and var. *kruegeri* (Engl.) Moeser in Engl., Pflanzenw. Afr. **3,** 1 : 850 (1915). Types as for *P. commutatum* and *P. kruegeri*.

Pittosporum quartinianum Cufod. in Oest. Bot. Zeitschr. **98** : 132, fig. 5b (1951). Type from Eritrea.

Pittosporum viridiflorum subsp *viridiflorum,* subsp. *angolense* (Oliv.) Cufod., subsp. *kruegeri* (Engl.) Cufod., subsp. *malosanum* (Bak.) Cufod., subsp. *quartinianum* (Cufod.) Cufod. in Fedde, Repert. **55** : 59, fig. 4a–c ; tom. cit. : 62, fig. 4i ; tom.· cit.: 68, fig. 4q–s ; tom. cit. : 71, fig. 4f–g, 5a ; tom. cit. : 69, fig. 4t–w (all 1952). Types as above.

Tree up to 10 m. high or large shrub. Adult leaves glabrous or faintly appressed-pubescent on petiole and midrib, very variable in size and shape, with lamina (including petiole) up to 11 × 4 cm., usually obovate to broadly oblanceolate, rounded or shortly acuminate, length-breadth ratio usually less than 3 : 1, rarely narrower and longer acuminate, in age usually glaucescent and ± tessellate above, rarely smoother, dull or somewhat glossy ; midrib flat or slightly impressed ; nerves generally thin, flat, rarely prominulous beneath, merging into the reticulation, often somewhat conspicuous above ; reticulation dense, uniform. Inflorescences rather dense, rarely reduced, with (in our area always) ± pubescent branches. Sepals free and up to 1·5 mm. long or rarely longer and variously connate. Petals usually greenish-white. Capsules (in our area probably always) 4-seeded ; valves of the ripe capsule plano-convex or dorsally gibbous, spreading or bent slightly upwards.

S. Rhodesia. N : Darwin, bank of Umsengedsi R. above gorge, fr. 10.v.1955, *Whellan* 884 (K ; SRGH). W : Matobo Distr., Besna Kobila Farm, on river bank, fr. i.1955, *Miller* 2627 (K ; SRGH). C : Salisbury, Twentydales, ♂ fl. 1.i.1946, *Wild* 617 (K ; SRGH). E : Chipinga, Woodstock Farm, 16 km. E. of Chipinga, ♂ fl. 23.x.1950, *Crook* M/220 (K ; SRGH). S : Victoria, Zimbabwe, fr. v.1951, *Gibson* 44/51 (K ; SRGH). **Nyasaland.** S : Zomba Mt., fr. 22.ix.1954, *Banda* 52 (FHO ; K). **Mozambique.** Z : Guruè, Serra do Guruè, 1600 m., fr. 20.ix.1944, *Mendonça* 2168 (BM ; LISC). MS : Serra da Gorongosa, Pico Gogogo, 1800 m., fr. 26.ix.1943, *Torre* 5962 (BM ; LISC).

Also in Cape Province, Natal, Basutoland, Orange Free State, Transvaal, Swaziland, Angola, Tanganyika, Kenya, Uganda, Sudan, Ethiopia, Somaliland Protectorate, tropical Arabia, Madagascar and S. India.

Riverine forests at lower altitudes and in evergreen mountain forests from 600–2500 m

P. viridiflorum as a whole is extremely variable in shape, size and colour of leaves, concrescence of sepals, number of seeds and degree of hairiness. The subspecies recognized by Cufodontis (1952) cannot be maintained because their geographical independence has proved to be too vague. Within our area variability is less pronounced. Typical features of the species are broadly obovate, rounded or only shortly acuminate leaves, ± glaucescent when old, with midrib flat above, pubescent inflorescences and free, small (c. 1·5 mm. long) sepals. Specimens from higher altitudes from the eastern border of Rhodesia, Nyasaland and adjacent territories of Mozambique are somewhat divergent. These show narrower, longer acuminate, not glaucescent, not tessellate, sometimes glossy adult leaves, more impressed midrib, reticulation less conspicuous beneath, longer calyx of ± connate unequal sepals and often larger capsules. This is the form represented by the type of *P. malosanum*.

Recently there has been evidence for the existence of *P. viridiflorum* in S. India. Gowda, loc. cit. (1951) referred to *P. floribundum* Wight & Arn. a specimen from S. Madagascar, *Humbert & Swingle* 5711, which undoubtedly belongs to *P. viridiflorum* which is widespread in Madagascar, as stated by Cufodontis, loc. cit. (1955).

2. **Pittosporum mannii** Hook. f. in Journ. Linn. Soc., Bot. **6** : 5 (1862). Type from Fernando Po.

Small tree up to 7 (12) m. tall or a shrub. Adult leaves glabrous, very variable in size but usually broad, up to 15 × 5 cm. including petioles, oblong-lanceolate to broadly oblanceolate, ± acuminate, length-breadth ratio mostly more than 3 : 1,

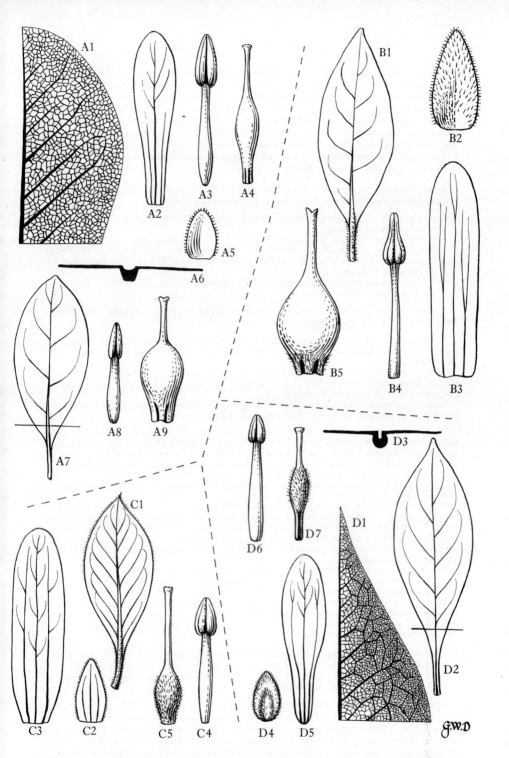

Tab. 54. A.—PITTOSPORUM VIRIDIFLORUM. A1, nervation of leaf (×3); A2, petal (×7½) *Eyles* 207; A3, stamen (×7½) *Eyles* 207; A4, sterile gynoecium (×7½) *Eyles* 207; A5, sepal (×7½) *Eyles* 207; A6, cross-section of leaf (×4); A7, leaf (×⅔) *Eyles* 206; A8, sterile stamen (×7½) *Eyles* 206; A9, gynoecium (×7½) *Eyles* 206. B.—PITTOSPORUM KAPIRIENSE. B1, leaf (×⅔); B2, sepal (×7½); B3, petal (×7½); B4, sterile stamen (×7½); B5, gynoecium (×7½). All from *Fanshawe* 3443. C.—PITTOSPORUM RHODESICUM. C1, leaf (×⅔); C2, sepal (×7½); C3, petal (×7½); C4, stamen (×7½); C5, sterile gynoecium (×7½). All from *Fanshawe* 1564. D.—PITTOSPORUM MANNII SUBSP. RIPICOLA. D1, nervation of leaf (×3); D2, leaf (×⅔); D3, cross-section of leaf (×3½); D4, sepal (×7½); D5, petal (×7½); D6, stamen (×7½); D7, sterile gynoecium (×7½). All from *Milne-Redhead* 3562.

fairly dark green and dull, rarely somewhat glossy above, paler and finally often yellowish-green beneath ; midrib considerably impressed, often almost hidden in a deep furrow above ; nerves arcuate, always ± prominent beneath, anastomosing, reticulation lax, not uniform, usually laxer along the midrib, ± impressed above but never clearly tessellate. Inflorescences well developed, many-flowered, always pubescent. Sepals up to 2 × 1 mm., free or nearly so or distinctly connate, not imbricate, ciliolate. Petals mostly creamy white, up to 7 × 1·5 mm. (not more than 5 mm. long in subsp. *mannii*). Capsule always 4-seeded, valves shallowly concave dorsally, spreading or often finally ± reflexed.

Subsp. *mannii* is restricted to Fernando Po, Cameroons Mt., the Bamenda Highlands and perhaps the Bambutos, at and above 2000 m.

Subsp. **ripicola** (J. Léonard) Cufod., comb. nov. TAB. **54** fig. D. Type from the Northern Belgian Congo.
　　Pittosporum ripicola J. Léonard in Bull. Jard. Bot. Brux. **20** : 47, fig. 7 (1950). ("ripicolum").—Cufod. in Oest. Bot. Zeitschr. **98** : 129 (1951) ; in Fedde, Repert. **55** : 77, fig. 5d–e, fig. 8 (1952).—Tisserant & Sillans in Not. Syst. Par. **15** : 92 (1954). Type as above.
　　Pittosporum ripicola subsp. *katangense* J. Léonard, loc. cit. 277 (1950). Type from south-eastern Belgian Congo.

Leaves rather larger and less stiff than in subsp. *mannii*. Sepals free. Petals always more than 5 mm. long.

N. Rhodesia. W : Mwinilunga Distr., Matonchi R., NE. of Kalenda Plain, ♂ fl. 8.xii.1937, *Milne-Redhead* 3562 (BM ; BR ; K) ; Kitwe, fr. 27.ii.1955, *Fanshawe* 2101 (K ; NDO).
　　Also in the Belgian Congo, Uganda, Kenya, Ethiopia, Sudan, Ubangi-Shari, French Congo, French Cameroons, British Cameroons, S. Nigeria and perhaps extending into Dahomey, Ivory Coast and French Guinea.
　　In riverine forest and less common in evergreen mountain forests from 650 m. upwards but commonest at middle altitudes from 1200–1800 m.

　　The differences between subsp. *mannii* and subsp. *ripicola* obviously cannot justify specific separation especially as in the French and the British Cameroons intermediates have been observed. Fruiting specimens are usually indeterminable as to the subspecies and as all specimens reported from regions W. of Nigeria are such, we cannot be sure to what subspecies they should be referred, but we may assume that they all belong to subsp. *ripicola*. The latter thus covers an immense area, descending (e.g. in the Nile basin) to unusually low altitudes for the genus in Africa.
　　The status of *P. ripicola* subsp. *katangense* is still uncertain, since developed flowers are as yet unknown. It was described as having pubescence on the very young leaves, sometimes ± rounded leaf-tips and smaller capsules. *Fanshawe* 3379 shows this last character well, but the shape of the leaves in all Rhodesian specimens does not differ from the average. Its reference to *P. viridiflorum* sensu lato by Cufodontis, loc. cit. (1952) must certainly be rejected.
　　Between the southernmost point of the main range of the subspecies in southern Uganda and its occurrences in Upper Katanga and the adjacent parts of N. Rhodesia is a gap of about 1300 km. without connecting records, which cannot easily be explained.
　　Just as *P. viridiflorum* has been shown to be identical with the S. Indian *P. floribundum*, *P. mannii* subsp. *ripicola* has been recognized as similar to and perhaps identical with the 4-seeded *P. nepaulense* var. *rawalpindiense* Gowda, from N. India, the Asiatic *Pittosporum* with the most westerly distribution.

3. **Pittosporum rhodesicum** Cufod. in Bol. Soc. Brot., Sér. 2, **25** : 102 (Apr. 1951) ; in Exell & Mendonça, C.F.A. **1**, 2 : 362 (Aug. 1951) ; in Fedde, Repert. **55** : 83, fig. 5i–j, fig. 8 (1952). TAB. **54** fig. C. Type : N. Rhodesia, Kyangozhi R., *Milne-Redhead* 1148 (BR ; K, holotype).

Small tree up to 4 m. high or shrub. Leaves up to 15·5 × 4·5 cm. (including petioles), length-breadth ratio 2·5 : 1, chartaceous, at first bright green on both sides, later a little darker above, with weak, persistent, spreading hairs on the petiole, midrib and sometimes also on the margin, broadly obovate-oblanceolate, usually shortly but acutely acuminate and often remarkably apiculate by the prolongation of the midrib ; midrib ± flat above, prominent beneath, veins very delicate on both sides, reticulation delicate, dense, ± uniform, pellucid, visible on both surfaces for a long time owing to its colouration. Inflorescences small, few-flowered, terminal and axillary, racemose-paniculate, with persistent, crisped pubescence on branches and pedicels. Sepals free, 2·5 × 1 mm., lanceolate, ciliolate. Petals creamy-yellowish, up to 7 × 1·5 mm., sublinear, ciliolate. Capsule up to 7 × 7 mm., shortly stipitate with valves thickish, convex-gibbous, suborbicular, spreading. Seeds 8, 4 on each placenta.

N. Rhodesia. N : Lake Bangweulu, near Samfya Mission, fr. 23.viii.1952, *White* 3140 (K). W : Kyangozhi R., W. of Solwezi, ♂ fl. 18.ix.1930, *Milne-Redhead* 1148 (BR; K) ; Mbulungu Stream, near Mutanda Bridge, fr. 8.vii.1930, *Milne-Redhead* 683/A (K).

Also in Angola (Lunda). Fringing forest.

P. rhodesicum is easily recognized by the unusually bright colour of the relatively thin leaves, whereby the coloured reticulation remains visible on both surfaces for a long time, almost until leaf-fall, and by the frequent, sometimes awn-like, prolongations of the midribs at the leaf tips.

4. **Pittosporum kapiriense** Cufod. in Bol. Soc. Brot., Sér. 2, **32** : 61 (1958). TAB. **54** fig. B. Kapiri Mposhi, *Fanshawe* 3443 (K, holotype : NDO : WU).

Tree up to 7 m. high, with trunk 10 cm. in diam. Leaves not crowded at the ends of the branches, subcoriaceous, dull and intense green above, paler and glaucescent beneath, with persistent spreading hairs on petiole and midrib ; lamina up to 14·5 × 4·5 cm., length-breadth ratio about 3 : 1, oblong-lanceolate, broadest at about the middle, cuneate at the base, shortly acuminate with blunt tips ; midrib narrow and scarcely impressed above, prominent beneath ; nerves thin, hardly prominulous beneath, almost invisible above ; reticulation thin and soon scarcely visible beneath, forming a regular, dense and minute tessellation above ; petiole 10–15 mm. long. Inflorescences always terminal, small, racemose-paniculate, with few- (often 1-) flowered branches and a persistent, crisped-villous pubescence ; bracts 5 mm. long, subulate, soon caducous. Sepals 4 × 1·5 mm., free, lanceolate, bluntish, puberulous. Petals at first greenish, later greenish-white, 8 × 2 mm., suberect, sublinear, somewhat broader in the lower half. Filaments 1 mm. broad at the base, subulate towards the apex, 4 mm. long, sterile anthers 2 mm. long. Fertile ovary 3·5 mm. long, glabrous, on a short, thick, pubescent stipe. Style 2 mm. long with a bilobed stigma, glabrous. Capsule (immature) dark green, with the valves granular outside and greenish inside and with 8 greenish ovules, 4 on each placenta, with scanty resin at this stage.

N. Rhodesia. C : Kapiri Mposhi, ♀ fl. & fr. 4.viii.1957, *Fanshawe* 3443 (K ; NDO ; WU).

Riverine forest.

Known only from the type-collection and characterized by the colour and tessellation of the leaves, persistent pilosity on the petioles and few-flowered inflorescences of relatively large flowers. It seems to be nearest to *P. cacondense* Exell & Mendonça.

5. **Pittosporum undulatum** Vent., Descr. Pl. Jard. Cels : t. 76 (1802).—Andr., Bot. Rep. **6** : t. 383 (1804).—Ker-Gawl. in Bot. Reg. **1** : t. 16 (1815).—Putterl., Syn. Pittosp.: 6 (1839).—Pritzel in Engl. & Prantl, Nat. Pflanzenfam. ed. 2, **18a** : 277 (1930).—Cufod. in Fedde, Repert. **55** : 104 (1952). Type from the Canary Islands, cultivated in Cels's Garden, Paris.

Tree (in its native country) up to 12 m. tall or shrub with wavy-margined leaves and umbellate-paniculate inflorescences of rather larger flowers than in any of the indigenous species.

S. Rhodesia. C : Salisbury, 1440 m., cultivated, ♂ fl. 15.viii.1931, *Brain* 6095 (K ; SRGH).

A native of SW. Australia. Introduced into the Canary Islands and British gardens. The species was introduced to Paris from the Canaries where it was for a long time believed to be indigenous.

20. POLYGALACEAE

By A. W. Exell

Small trees, woody climbers, shrublets or perennial or annual herbs. Leaves usually alternate, simple, entire, exstipulate. Sepals 5 ; two lateral (interior) ones often petaloid (wings) ; two anterior ones sometimes joined. Petals 3–5 ; two upper ones free or joined to the lower one (keel) ; two lateral ones free, often absent or vestigial. Stamens usually 5–8 (rarely 4) ; filaments usually united in a slit tube. Ovary superior, usually 2-locular (1–5), with 1 pendulous ovule in each loculus. Fruit a capsule, samara or drupe. Seeds often sericeous, usually carunculate ; endosperm usually present.

For a paper of general interest see M. R. Levyns " Some geographical features of the Family Polygalaceae in Southern Africa " in Trans. Roy. Soc. S. Afr. **34** : 379–386 (1955).

Petals 5, subequal ; stamens 5 ; fruit drupaceous   -   -   -   - **1. Carpolobia**
Petals 3 (others minute or absent) ; stamens usually 7–8 (4–8) ; fruit samaroid or capsular :
    Petals free ; fruit indehiscent, winged   -   -   -   -   -   - **2. Securidaca**
    Petals united, the two upper ones joined to the keel ; fruit capsular :
        Keel-appendage bilobed, expanded and leaflike ; flowers solitary (in our species) ;
            stamens 7   -   -   -   -   -   -   -   -   -   - **3. Muraltia***
        Keel-appendage (crest) fimbriate or at least plurilobed (rarely absent) ; flowers in
            racemes (some flowers solitary in *P. erioptera*) ; stamens 8 (or 6 fertile and 2
            sterile)   -   -   -   -   -   -   -   -   -   -   - **4. Polygala**

1. CARPOLOBIA G. Don

Carpolobia G. Don, Gen. Syst. **1** : 349, 370 (1831).

Shrubs or small trees. Flowers zygomorphic. Sepals 5, free, two inner slightly larger. Petals 5, joined at the base to the staminal sheath, keel with appendage. Stamens 5, monadelphous. Ovary 2–3-locular. Fruit drupaceous. Seeds ellipsoid, hairy.

Carpolobia conradsiana Engl., Pflanzenw. Afr. **3**, 1 : 839 (1915). Type from Tanganyika (Ukerewe).
    Carpolobia alba sensu R.E.Fr. in Wiss. Ergebn. Schwed. Rhod.-Kongo-Exped., **1** : 112 (1914).
    Carpolobia suaveolens Meikle in Kew Bull. **1950** : 337 (1951). Type : Mozambique, Lugela, Namagoa, *Faulkner* K 106 (K, holotype).

Shrub or small tree ; branchlets at first pubescent, becoming glabrous. Leaves shortly petiolate ; lamina 3–11 × 1–4 cm. (a small-leaved form with most of the leaves 2·5–3 cm. long occurs), ovate or obovate, lanceolate, elliptic or narrowly elliptic, reticulation prominent on both surfaces, glabrous above except for the midrib, sometimes sparsely pubescent below and always pilose along the midrib especially towards the base ; petiole 1·5–3 mm. long. Flowers white, cream or yellowish, sometimes scented, in very short axillary racemes with deltoid bracts 1·5–2 mm. long, and pubescent rhachis. Sepals 5–7 mm. long, broadly ovate-lanceolate, brownish-pubescent especially towards the apex. Petals oblong-elliptic, 2 upper ones 8–10 mm. long, 3 lower ones 11–15 mm., ciliate on the margins. Stamens 5 ; filaments 9 mm. long ; anthers 1 mm. long. Fruit 1 cm. in diam., subglobose, 2–3-locular. Seeds 8 × 6 mm., flattened, with silky hairs.

N. Rhodesia. N : Bulaya-Mwewe, fl. 22.x.1949, *Bullock* 1345 (K). W : Chingola, fl. 14.xi.1955, *Fanshawe* 2605 (BM ; K). C : Broken Hill, 21.ix.1947, *Brenan & Trapnell* 7904 (FHO). **Mozambique.** N : Tungue, Nangade, fl. 19.x.1942, *Mendonça* 980 (BM ; LISC). Z : Namagoa, fr. x.1944, *Faulkner* 110 (K).
Eastern tropical Africa from the Sudan to Mozambique. In dry bush and evergreen thickets.

2. SECURIDACA L.

Securidaca L., [Gen. Pl. ed. 5 : 316 (1754) pro parte] Syst. ed. 10, **2** : 1155 (1759) *nom. conserv.*

Small trees or shrubs, often scandent, or lianes with entire, alternate leaves. Flowers zygomorphic, in terminal and axillary racemes or panicles. Sepals 5, unequal, two lateral ones (wing-sepals) larger and petaloid, free. Petals 3 (occasionally 5, two being vestigial), lowest keel-shaped. Stamens 8, monadelphous. Ovary

---

* Our only species of *Muraltia* (*M. flanaganii*) approaches *Polygala* in several characters usually used to separate the two genera.

1-locular. Fruit a samara (a 1-locular, 1-seeded, winged nut) usually 1-winged, occasionally with an additional rudimentary wing. Seed without caruncle and without endosperm.

Plant a shrub or small tree ; young stems usually densely pubescent ; leaves rarely more
than about 1·5–2 cm. wide - - - - - - - 1. *longepedunculata*
Plant a scandent shrub or liane, sometimes climbing to considerable heights ; young stems
soon becoming glabrous ; leaves 1·5–5 cm. wide - - - - 2. *welwitschii*

1. **Securidaca longepedunculata** Fresen. in Mus. Senckenb. **2** : 275 (1837).—Oliv., F.T.A. **1** : 134 (1868).—Eyles in Trans. Roy. Soc. S. Afr. **5** : 392 (1916).—Burtt Davy, F.P.F.T. **1** : 132, fig. 13 J–K (1926).—Norlindh in Bot. Notis. **1935** : 368 (1935).—Exell & Mendonça, C.F.A. **1**, 1 : 89 (1937).—Milne-Redh. in Mem. N.Y. Bot. Gard. **8**, 3 : 219 (1953).—Codd in Fl. Pl. Afr. : t. 1191 (1955).—Williamson, Useful Pl. Nyasal. : 107 (1955).—Palgrave, Trees of Central Africa : 346, cum photogr. et tab. (1957).—E. Petit, F.C.B. **7** : 280 (1958).—TAB. **55** fig. A. Type from Ethiopia.
   Lophostylis pallida Klotzsch in Peters, Reise Mossamb. Bot., **1** : 115, t. 22 (1861). Type : Mozambique, Sena, *Peters* (B, holotype †, not seen).
   Securidaca longepedunculata var. *parvifolia* Oliv., loc. cit.—Gibbs in Journ. Linn. Soc., Bot. **37** : 430 (1906).—Eyles, loc. cit.—E. Petit, loc. cit. Syntypes from Nigeria and Angola.

Small tree up to about 6 m. or shrub, sometimes spiny ; stems usually pubescent at first, becoming glabrous. Leaves petiolate ; lamina 1–5 × 0·5–1·8 cm., very variable in size and shape, from broadly oblong to very narrowly elliptic, usually minutely pubescent when young, soon becoming glabrous ; petiole up to 5 mm. long. Flowers pink or purple sometimes variegated with white, sweet-scented, pedicels up to 14 mm. long, pubescent, in terminal or lateral racemes, 3–5 cm. long, often borne on short shoots which become spiny, with minute, deciduous bracts and bracteoles and pubescent rhachis. Posterior sepal up to 5 × 4 mm. ovate-acuminate, with ciliate margins ; wing sepals 5–11 × 4–9 mm., suborbicular, nearly glabrous ; anterior sepals up to 5 × 4·5 mm., broadly ovate. Upper petals up to 7·5 × 3·5 mm., narrowly elliptic, ciliate at the base ; carina up to 10 mm. long with a small lobed appendage about 1 mm. long near the apex. Staminal sheath ciliate on the upper margin. Fruit 3–5 × 0·8–2 cm., with an oblong or elliptic somewhat obliquely curved wing, sometimes with a second rudimentary wing. Nut containing the seed 8–10 mm. in diam., rugulose or smooth.

N. Rhodesia. B : Shangombo, 1030 m., fl. & fr. 16.viii.1952, *Codd* 7567 (BM). N : Abercorn, 1220–1830 m., fl. 4.x.1931, *Gamwell* 97 (BM). W : Mwinilunga Distr., Matonchi Farm, 1430 m., fl. 20.vii.1928, *Paterson* 3 (K). C : Chisamba, Golden Valley, fr. 17.iii.1933, *Michelmore* 660 (K). S : Mazabuka, fl. 7.x.1930, *Milne-Redhead* 1216 (K). **S. Rhodesia.** N : Sinoia, fl. 26.x.1926, *Rand* 273 (BM). W : Matopo Hills, 1520 m., fl. x.1905, *Gibbs* 244 (BM ; K). C : Salisbury, fl. x.1908, *Rand* 1365 (BM). E : Umtali, Commonage, fl. & fr. 7.ix.1948, *Chase* 1341 (BM ; SRGH). S : Fort Victoria, fl. 1909, *Monro* 570 (BM). **Nyasaland.** N : Upper Luangwa, fl. xi.1896, *Nicholson* (K). C : Lilongwe, fl. 6.xi.1951, *Jackson* 630 (BM ; K). S : Fort Johnston, Chilembwe Village, fr. 6.vii.1954, *Jackson* 1358 (BM ; K). **Mozambique.** N : Amaramba, Cuamba, fl. 22.x.1948, *Mendonça* 1440 (BM ; LISC). Z : between Mocuba and Guruè, fl. 5.iv.1943. *Torre* 5068 (BM ; LISC). T : Macanga, Furancungo, 1000 m., fl. 20.x.1943, *Torre* 6068 (BM ; LISC). MS : Madanda Forest, 120 m., fr. 4.xii.1907, *Swynnerton* 1032 (BM). SS : Inhachengo, fr. 26.ii.1955, *E.M. & W.* 627 (BM ; LISC ; SRGH). LM : Lourenço Marques, 30 m., 29.xi.1897, *Schlechter* 11523 (BM ; COI).
Widespread in tropical Africa. A sporadic constituent of many types of woodland and bush from near sea-level to about 1600 m. Rarely gregarious.

A very variable species in leaf-shape and flower-size as well as in habit and extent to which spines develop. It has been subdivided by various authors into a number of varieties. Specimens from Mozambique coastal regions tend to have longer, narrower leaves, while some of the smallest and roundest leaves occur on specimens from Angola. There is, however, great variation in most parts of the range.
This is known as the Violet Tree owing to the scent of the flowers. The bast produces a useful fibre which the Africans use as cotton and from which they used to weave a coarse cloth. An infusion of the roots is used as a remedy for snake-bite and also as a poison. For further uses etc. see Palgrave (loc. cit.).

2. **Securidaca welwitschii** Oliv., F.T.A. **1** : 135 (1868).—Exell & Mendonça, C.F.A. **1**, 1 : 89 (1937).—E. Petit, F.C.B. **7** : 279 (1958). Type from Angola (Cuanza Norte).

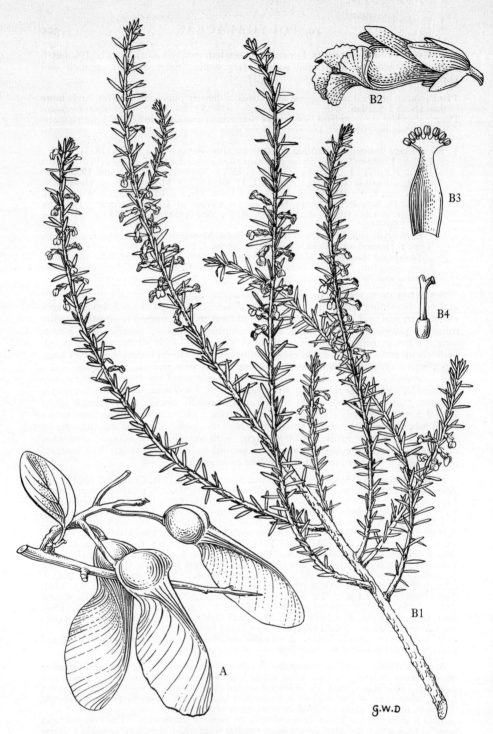

Tab. 55. A.—SECURIDACA LONGEPEDUNCULATA, fruiting branch (×1) *Phelps* in GHS
42556. B.—MURALTIA FLANAGANII. B1, flowering branch (×1); B2, flower (×6);
B3, androecium (×6); B4, ovary and style (×6). All from *Whyte* 1.

A spiny scandent shrub or liane with young stems at first minutely appressed-pubescent, rapidly glabrescent. Leaves petiolate ; lamina up to 10 × 5 cm., elliptic to oblong-elliptic or broadly oblong-elliptic, glabrous or nearly so, acuminate at the apex, cuneate to obtuse at the base ; petiole up to 8 mm. long. Inflorescences and flowers as in the preceding species but somewhat smaller than in the larger forms of *S. longepedunculata*, with the wing sepals up to about 8 mm. long. Fruit as in *S. longepedunculata* but wing relatively narrower and less oblique.

N. Rhodesia. W : Mwinilunga Boma, Lunga R., fl. 27.viii.1930, *Milne-Redhead* 977 (K).
Also in western tropical Africa, Uganda, Kenya, and Tanganyika. Evergreen forests.

Recorded only from the extreme NW. of our region.

3. MURALTIA DC.

Muraltia DC., Prodr. **1** : 335 (1824).

Ericoid shrubs or shrublets. Sepals 5, usually subequal (2 inner ones distinctly larger in *M. flanaganii*). Petals usually 3, two upper ones oblong and joined at the base to the keel. Keel-appendage expanded and leaflike. Stamens 7, monadelphous. Ovary 2-locular. Capsule usually 4-horned (hornless in *M. flanaganii*), 2-seeded.

Muraltia flanaganii Bolus in Journ. of Bot. **34** : 17 (1896).—Milne-Redh. in Mem. N.Y. Bot. Gard. **8**, 3 : 220 (1953). TAB. **55** fig. B. Type from S. Africa (Basutoland).
    Muraltia fernandi Chod. in Mitt. Bot. Mus. Univ. Zür. **76** : 612 (1916). Syntypes from S. Tanganyika and Nyasaland (Mt. Mlanje).

Shrublet up to 1 m. tall with slender, glabrous or minutely pubescent branches. Leaves subsessile or very shortly petiolate, 3–8 × 0·5–1 mm., linear, with needle-like points and margins scaberulous and often incurved. Flowers white and purple with pedicels 1 mm. long, solitary, axillary. Sepals 2–3 × 1·5–2 mm., ovate. Upper petals 4 mm. long, oblong, blunt ; carina 3·5 mm. long with bilobed leafy crest, 1·5 mm. long. Capsule 3 × 2·5 mm., broadly elliptic in outline, without horns. Seeds 1·5 × 0·8 mm. ellipsoid, pubescent ; caruncle 0·5 mm. long with very short caruncular appendages.

S. Rhodesia. E : Umtali Distr., Engwa, 2140 m., fl. & fr. 10.ii.1955, *E.M. & W.* 350 (BM ; LISC ; SRGH). **Nyasaland.** S : Mt. Mlanje, Chambe-Tuchila Pass, 2320 m., fl. 15.vii.1956, *Newman & Whitmore* 8 (BM ; BR ; SRGH ; WAG).
From S. Tanganyika to Natal and Basutoland (Drakensberg). Montane grasslands and among stunted *Widdringtonia* from 1600 m. upwards.

This species appears to be transitional in some respects between *Muraltia* and *Polygala*, having the two inner sepals considerably enlarged, coloured and almost wing-like and a hornless capsule indistinguishable from that of a *Polygala*. The seeds also entirely resemble *Polygala* seeds. The general appearance is very like that of *P. teretifolia* L.f. but the flowers are solitary instead of in racemes. The apparent close relationship with *Polygala* may not, however, be a real one (see Levyns in Trans. Roy. Soc. S. Afr. **34** : 379–386 (1955)).

4. POLYGALA L.

Polygala L., Sp. Pl. **2** : 701 (1753) ; Gen. Pl. ed. 5 : 315 (1754).

Shrubs, shrublets or perennial or annual herbs. Leaves alternate (in our species) and very shortly petiolate. Flowers zygomoric, papilionaceous, in terminal or lateral racemes or rarely solitary. Sepals unequal, the three outer ones usually sepaloid and the two inner ones (wings) larger and petaloid, all free or the two anterior ones united. Petals 3 (5) the lowest forming the carina (usually crested), the two lateral ones usually vestigial or absent, the two upper ones joined at the base to the carina. Stamens 8 (rarely 9, 4 or 5) sometimes only 6 fertile with 2 staminodes, monadelphous. Ovary 2-locular. Capsule flattened, 2-locular, 2-seeded. Seeds usually with a caruncle and with a silky indumentum (rarely glabrous or with glochidiate hairs).

A genus very well represented in this region. It has been suggested that the seeds are distributed by ants which drag them away to eat the caruncles. The long silky hairs allow the seeds to progress readily " head first " but not in the reverse direction.

The following synopsis may save time in identification before turning to the full key :

I. ANTERIOR SEPALS FREE

A. *Inflorescences lateral*

(a) *Species* 1–4. Inconspicuous greenish flowers. Ripe capsule broader than the wing sepals. Seeds with well-developed caruncular appendages. Mainly Mozambique.

(b) *Species* 5–7. Ericoid, montane species. S. Rhodesia (E), Nyasaland and Mozambique (mountains).

(c) *Species* 8–12. Remainder of the species with lateral inflorescences. A heterogeneous group. Two are widespread.

B. *Inflorescences terminal*

(d) *Species* 13–14. Small, montane species with an underground woody stem or rhizome. Mountain regions of N. Rhodesia (E), S. Rhodesia (E), Nyasaland and Mozambique.

(e) *Species* 15–18. Large, handsome-flowered species fairly widespread in the region.

(f) *Species* 19–23. Flowers medium to small, greenish-white, pinkish or pale purple. Mostly roadside weeds usually in sandy soil or in abandoned cultivation. Widespread in the area. No. 23 (*P. melilotoides*) is our only species with completely glabrous seeds.

(g) *Species* 24–38. Species mostly with grasslike leaves and elongated terminal inflorescences. Flowers medium to very small. Probably a heterogeneous group.

II. ANTERIOR SEPALS UNITED

(h) *Species* 39–49. Fertile stamens 8.

(i) *Species* 50–54. Fertile stamens 6. Mainly Kalahari Sand species of Angolan –SW. African affinity.

To use the key successfully it is necessary to note whether the anterior sepals (below the keel) are free or united and whether there are 8 or 6 fertile stamens, which needs dissection of dried specimens but is easily ascertained in fresh material. If the anterior sepals are free, the stamens are always 8 except in one rather rare species (*P. youngii*) so dissection may not be necessary.

Anterior sepals free or only slightly joined at the base :
  Flowers solitary or in lateral racemes :
    Wings greenish all over, width about half that of the ripe capsule :
      Capsule 2·5–3 × 3–3·5 mm. -   -   -   -   -   -   -   -   1. *limae*
      Capsule 4–5 × 4·5–5·5 mm. :
        Stems and leaves crisped-pubescent :
          Caruncular appendages about as long as the ripe seed   -   -   2. *torrei*
          Caruncular appendages only half as long as the ripe seed :
            Secondary nervation and reticulation of the leaves conspicuous   3. *goetzei*
            Secondary nervation and reticulation of the leaves inconspicuous
                                        4. *transvaalensis*
        Stems and leaves patent-pubescent   -   -   -   -   -   5. *francisci*
    Wings ± petaloid, as wide as or wider than the ripe capsule :
      Plant a shrub, shrublet or herb of ± ericoid appearance with rigid or semi-rigid leaves with ± revolute margins, standing perpendicular to the stem or reflexed :
        Leaves 10–30 mm. long, margins slightly revolute but under surface visible, pubescent : wings 8–14 mm. long   -   -   -   -   6. *gazensis*
        Leaves 5–12 mm. long, under surface glabrous ; wings up to 8 mm. long :
          Leaves 5–12 mm. long, margins usually strongly revolute, often concealing the lower surface completely or almost completely ; young stems usually densely minutely pubescent or tomentellous   -   -   -   7. *teretifolia*
          Leaves 5–7 mm. long, margins revolute but not entirely concealing the lower surface ; young stems crisped-pubescent   -   -   -   8. *adamsonii*

Plant a shrublet or herb, not ericoid in appearance ; leaves neither rigid nor with appreciably revolute margins :
  Wings suborbicular, 4–8 mm. broad :
    Caruncular appendage nearly as long as the seed ; inflorescences 1·5–3·5 cm. long (exceptionally up to 8 cm.) ; wings 6–10 mm. in diam. with ciliolate margins ; leaves blunt or retuse, often mucronulate   -   9. *senensis*
    Caruncular appendage up to 1 mm. long, not more than half the length of the seed ; inflorescences 3–14 (18) cm. long ; wings 4–7 mm. in diam. ; leaves usually pointed - - - - - - - - 10. *sphenoptera*
  Wings elliptic or broadly elliptic, up to 6 × 3 mm. :
    Wings up to 5 × 2·5 mm. with a longitudinal green stripe - 11. *erioptera*
    Wings up to 6 × 3·5 mm., concolorous - - - - 12. *sadebeckiana*
Flowers in terminal racemes, sometimes with additional lateral racemes :
  Plant with underground woody stem or rhizome producing annual shoots up to 5–15 cm. long, erect or prostrate :
    Wings 6–8 mm. long ; leaves narrowly elliptic or linear-oblong, glabrous or nearly so - - - - - - - - 13. *nyikensis*
    Wings 2·5–7 mm. long ; leaves pubescent or pilose at least on the margins :
      Wings 5–7 mm. long ; pedicels 2–4 mm. long - - 14. *ohlendorfiana*
      Wings 2·5–4 mm. long ; pedicels up to 1 mm. long - - 15. *wilmsii*
  Plant not as above, an erect or prostrate annual or perennial herb or shrub at least 20 cm. tall :
    *Wings 2 mm. long or longer :
      †Wings obliquely suborbicular or slightly longer than broad, 7–15 mm. broad ; plant a shrub or robust herb ; flowers usually purple or purple fading white :
        Bracts and bracteoles of inflorescence caducous ; plant a virgate shrub ; wings slightly longer than broad and up to 14–15 mm. long although sometimes shorter - - - - - - - 16. *virgata*
        Bracts and bracteoles of inflorescence persistent ; plant a robust herb or shrub ; wings obliquely suborbicular, up to 12 mm. long :
          Inflorescence elongated, up to 30–50 cm. long ; wings up to 11–12 mm. long ; bracts and bracteoles scarious, bract 5–6 mm. long with filiform tip ; stems and leaves usually nearly glabrous :
            Leaves elliptic to narrowly elliptic ; flowers bright violet or purple ; perennial - - - - - - - - 17. *gomesiana*
            Leaves linear to linear-elliptic ; flowers purple at first, fading to greenish-white ; annual - - - - - - - 18. *macrostigma*
          Inflorescences much shorter, usually up to 6 cm. long ; bracts and bracteoles opaque to subscarious, bract ovate-lanceolate, 4 mm. long
                                                                                          19. *exelliana*
      †Wings not more than 6 mm. broad ; plant usually a slender annual or perennial herb ; flowers usually greenish-white :
        Inflorescences both terminal and lateral rarely projecting much beyond the leaves ; leaves usually well developed :
          Wings 4·5–8 mm. long :
            Plant of sympodial growth, secondary flowering branches much exceeding the primary terminal racemes - - - 20. *arenaria*
            Plant unbranched or with secondary flowering branchlets usually shorter than the terminal primary raceme :
              Wings obliquely elliptic, hairy at the base - - - 21. *albida*
              Wings suborbicular, glabrous - - - 22. *persicariifolia*
          Wings less than 4 mm. long :
            Seeds sericeous or minutely pubescent - - - 23. *pygmaea*
            Seeds glabrous - - - - - - - 24. *melilotoides*
        Inflorescences terminal (though sometimes branched), projecting well beyond the leaves : leaves usually narrow and often grasslike, often rather poorly developed :
          Fertile stamens 8 :
            Wings 6–10 mm. long :
              Stem below the inflorescence with a dense indumentum of short, crisped hairs and longer patent ones ; flowers mauve, drying brownish-yellow - - - - - - 25. *mendoncae*
              Stem below the inflorescence with an indumentum of short, crisped hairs only or glabrous or nearly so ; flowers usually blue, drying cream or blue :
                Rhachis of inflorescence densely patent-pubescent ; margin of ovary ciliolate ; plant an annual :
                  Pedicels 2–3·5 mm. long ; wings 5–8 mm. long ; racemes up to 30 cm. long, usually little branched - - 26. *usafuensis*

Pedicels 4–5 mm. long ; wings 8–10 mm. long ; racemes 10–15
  cm. long, inflorescence usually much branched   -   27. *friesii*
Rhachis of inflorescence minutely appressed-pubescent, glabrous or
  with a few scattered hairs ; plant usually perennial :
  Margin of capsule ciliolate :
    Wings 5·5–6 mm. broad ; rhachis of inflorescence minutely ap-
      pressed-pubescent ; perennial -   -   -   28. *britteniana*
    Wings 4–5 mm. broad ; rhachis of inflorescence glabrous ;
      annual -   -   -   -   -   -   - 29. *nambalensis*
  Margin of capsule glabrous ; wings up to 5 mm. broad :
    Stems green ; carina 6 mm. long with crest 2–3 mm. long
                                                    30. *hottentotta*
    Stems greyish ; carina 3 mm. long with crest 1 mm. long
                                                    31. *seminuda*
Wings 2·5–5·5 mm. long :
  Wings broadly elliptic to suborbicular, up to 1½ times as long as broad :
    Wings 2–2·5 mm. long, glabrous :
      Flowers white or yellowish ; wings 1·3–1·5 mm. broad   - 32. *spicata*
      Flowers pink or purplish ; wings 0·6–0·9 mm. broad   - 33. *filicaulis*
    Wings 4–4·5 mm. long, pubescent :
      Stems and leaves densely patent-pilose -   -   - 34. *claessensii*
      Stems (below inflorescence) appressed-pubescent or nearly glabrous
                                                    35. *ukirensis*
  Wings narrowly elliptic to elliptic, 2–3 times as long as broad
                                                    36. *myriantha*
    Fertile stamens 6   -   -   -   -   -   -   -   - 37. *youngii*
*Wings 1–2 mm. long :
  Wings 1–1·5 mm. long, suborbicular ; seed 0·6 mm. long with glochidiate hairs ;
    plant usually 10–15 cm. tall (occasionally taller); flowers pink or mauve 38. *africana*
  Wings 1·5–2 mm. long, broadly elliptic ; seed 0·9–1 mm. long with appressed hairs
    (sometimes slightly glochidiate) ; plant 15–40 cm. tall (usually over 20 cm.) ;
    flowers usually white or purplish -   -   -   -   - 39. *capillaris*
Anterior sepals united for at least half their length, usually almost to the apex :
  Fertile stamens 8 :
    Crest of carina well developed :
      Wings 8–11 mm. long -   -   -   -   -   -   - 40. *homblei*
      Wings up to 8 mm. long :
        Stems and inflorescences densely pubescent :
          Wings 6–8 mm. long ; apical lobes of the ovary pointed -   41. *marensis*
          Wings 4–5 mm. long ; apical lobes of the ovary rounded - 42. *schinziana*
        Stems and inflorescences glabrous or minutely puberulous :
          Wings 4–8 mm. long :
            Wings 6–8 mm. long ; perennial, usually not exceeding 25–30 cm. in height
                                                    43. *rehmannii*
            Wings 4–6·5 mm. long :
              Leaves grasslike (sometimes very sparse) ; very gradually narrowed to a
                needle-like point -   -   -   -   -   - 44. *stenopetala*
              Leaves blunt, rounded or uncinate at the apex :
                Leaves uncinate, rounded at the tips with recurved apices ; plant
                  perennial 10–30 cm. tall -   -   -   -   - 45. *uncinata*
                Leaves rounded at the apex, but scarcely uncinate ; plant an annual
                  up to 150 cm. tall -   -   -   -   - 46. *producta*
            Wings 2·5–3 mm. long -   -   -   -   -   - 47. *rivularis*
    Crest of carina absent or very much reduced :
      Wings 4–6 × 2·5–3 mm., usually 5-nerved   -   -   -   - 48. *xanthina*
      Wings 2·5–5 × 1·5–2 mm., 3-nerved   -   -   -   - 49. *petitiana*
    [Insufficiently known species with anterior sepals partly connate and flowers
      in extra-axillary lateral racemes -   -   -   -   -   - 50. *westii*]
  Fertile stamens 6 :
    Stems pubescent ; flowers solitary or in few-flowered racemes   - 51. *kalaxariensis*
    Stems glabrous ; flowers in terminal or apparently terminal racemes :
      Wings 7–10 mm. long -   -   -   -   -   -   - 52. *baumii*
      Wings up to 6 mm. long :
        Pedicels 5–7 mm. long   -   -   -   -   - 53. *gossweileri*
        Pedicels up to 3 mm. long :
          Flowers in dense subcapitate racemes ; inflorescences typically sympodial
            with secondary branchlets much exceeding the primary terminal racemes
            (the latter sometimes appearing lateral)   -   -   - 54. *dewevrei*
          Flowers in laxer terminal racemes up to 3–4 cm. long ; growth not markedly
            sympodial -   -   -   -   -   -   -   - 55. *arenicola*

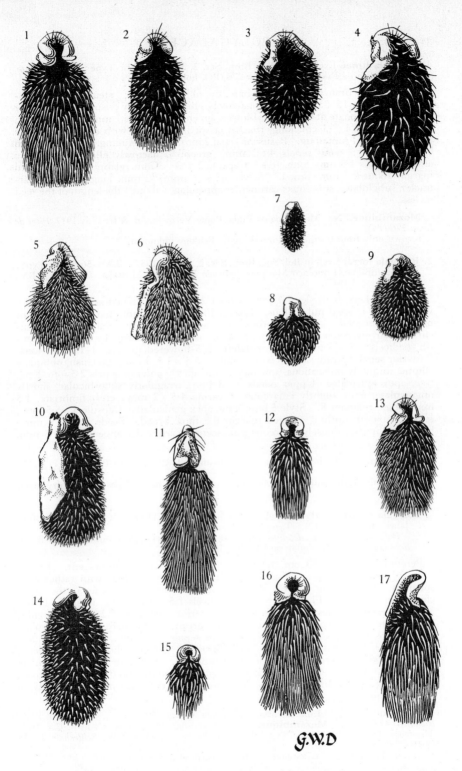

G.W.D

Tab. 56. Seeds of POLYGALA (all ×8). 1. P. MACROSTIGMA ; 2. P. EXELLIANA ; 3. P. ADAM-
SONII ; 4. P. GAZENSIS ; 5. P. FRANCISCI ; 6. P. TORREI ; 7. P. WILMSII ; 8. P. CLAESSENSII ;
9. P. SADEBECKIANA ; 10. P. SENENSIS ; 11. P. NAMBALENSIS ; 12. P. UKIRENSIS ; 13. P.
SPHENOPTERA ; 14. P. VIRGATA VAR. DECORA ; 15. P. USAFUENSIS ; 16. P. MENDONCAE ;
17. P. HOTTENTOTTA.

1. **Polygala limae** Exell in Bol. Soc. Brot., Sér. 2, **31** : 5 (1957). Type : Mozambique, Mocímboa da Praia, *Pires de Lima* 276 (PO, holotype).

Annual herb branched from the base, up to 18 cm. tall ; stems crisped-pubescent. Leaves 15–25 × 5–8 mm., narrowly elliptic, sparsely crisped-pubescent or glabrous, mucronate at the apex. Flowers greenish, pedicels 1 mm. long, in short, lateral racemes 1·5–3 cm. long, rhachis crisped-pubescent with deciduous bracts and bracteoles 1 mm. long. Posterior sepal 2 mm. long, including an apical mucro 0·7 mm. long ; wing sepals 4 × 2 mm., greenish, narrowly elliptic, glabrous ; anterior sepals 1·5 mm. long, free. Capsule 2·5–3 × 3 mm., glabrous, with winged margin about 0·3 mm. broad. Seeds 1·2 × 1·2 mm. (2 mm. long including caruncle), subglobose, sericeous ; caruncular appendages about ¼ the length of the seed or less.

Mozambique. N : Mocímboa da Praia, Ponta Vermelha, fl. & fr. 12.ix.1917, *Pires de Lima* 276 (PO).
Known only from northern Mozambique. Ecology unknown.

2. **Polygala torrei** Exell in Bol. Soc. Brot., Sér. 2, **31** : 6 (1957). TAB. **56** fig. 6. Type : Mozambique, Lourenço Marques, between Boane and Moamba, *Torre* 2171 (BM, holotype ; LISC).

Perennial herb with slender stems about 25 cm. tall, arched-ascending from a woody base, crisped-puberulous. Leaves 10–35 × 2·5–6 mm., narrowly elliptic or narrowly lanceolate, almost glabrous, apex acute, mucronate, margin slightly incurved, narrowed towards the base, base obtuse usually unequal. Flowers greenish, pedicels 2·5–3·5 mm. long, in lateral 2–5-flowered racemes 1–1·5 cm. long. Posterior sepal 2·5 mm. long ; wing sepals 5–5·5 × 2·5–3 mm., greenish, obliquely elliptic, minutely puberulous, apiculate at the apex ; anterior sepals 2·5 mm. long, free, apex apiculate. Upper petals 3 × 3 mm., irregularly suborbicular, shortly unguiculate, apex slightly emarginate ; carina 4·5 × 2 mm., crest fimbriate, 1·5 mm. long. Stamens 8. Style 5 mm. long, apex geniculate. Capsule 5 × 5·5 mm., suborbicular in outline, winged margin 0·6 mm. broad. Seeds 2·5 × 1·8 mm., broadly ellipsoid, densely sericeous-pubescent ; caruncular appendages 2·5 mm. long.

Mozambique. LM : between Boane and Moamba, fl. & fr. 2.xii.1940, *Torre* 2171 (BM ; LISC).
Known only from southern Mozambique. Perennial herb of pastures in dry, open bush.

3. **Polygala goetzei** Gürke in Engl., Bot. Jahrb. **28** : 417 (1900).—Chod. in Engl., Bot. Jahrb. **48** : 326 (1912). Type from Tanganyika.
  Polygala goetzei var. *depauperata* Chod., loc. cit. Type : Mozambique, near Beira, *Schlechter* 12248 (B, holotype † ; PRE, lectotype).

Perennial herb, 30–40 cm. tall ; stems slender, crisped-pubescent. Leaves 1–7 × 0·5–2·5 cm., elliptic or narrowly elliptic, almost glabrous with rather conspicuous anastomosing nerves, rounded or acute and acuminate at the apex, cuneate at the base. Flowers greenish, pedicels 1·5–2 mm. long, in short, lateral 3–15-flowered racemes 1·5–2 cm. long. Posterior sepal 2·5 × 1·2 mm. with ciliolate margin ; wing sepals 5–5·5 × 2·5–3 mm., greenish, obliquely elliptic, apiculate at the apex, glabrous or nearly so ; anterior sepals 2 × 1 mm., free, with ciliolate margins. Upper petals 3·5 × 3·1 mm., broadly obovate, shortly unguiculate ; carina 3 × 2·5 mm., crest 1·5 mm. long. Stamens 8. Style 5 mm. long, geniculate at the apex. Capsule 4 × 5 mm., glabrous with winged margin 0·6–0·8 mm. broad. Seeds 2·5 × 2 mm., broadly ovoid, sericeous-pubescent ; caruncular appendages 1·2 mm. long.

N. Rhodesia. N : Lunzua, above Kafakula Village, 940 m., fl. & fr., 5.iii.1955, *Richards* 4789 (K). **Mozambique.** N : Nampula, Murrupula road, fl. 3.x.1942, *Mendonça* 1213 (BM ; LISC). MS : near Beira, fl. 11.iv.1898, *Schlechter* 12248 (PRE).
Also in Tanganyika. Woodland and bush, sea-level to 1000 m.

The specimens cited agree fairly well with a small fragment of the type, *Goetze* 406 (BM), available for comparison.
In addition to the variety cited above in the synonymy, Chodat (loc. cit.) has described

P. goetzei var. *schlechteri* from Lourenço Marques based on *Schlechter* 11636 (B, holotype †, not seen). This may be either *P. torrei* Exell or *P. francisci* Exell.

4. **Polygala transvaalensis** Chod. in Mém. Soc. Phys. Hist. Nat. Genève, **31**, 2 : 374, t. 29 fig. 18 (1893). Type from the Transvaal.

> *Polygala amatymbica* sensu Burtt Davy, F.P.F.T. **1** : 134 (1926) pro parte.

Perennial herb 10–20 cm. tall ; stems slender, crisped-pubescent. Leaves 10–30 × 2·9 mm., narrowly elliptic to linear, glabrous or minutely pubescent, nerves almost inconspicuous, acute or rounded and often mucronulate at the apex, cuneate at the base. Flowers pink, crimson or purple, pedicels 2–3·5 mm. long, in short lateral 3–6-flowered racemes about 1 cm. long. Posterior sepal 2·5 × 1·2 mm., broadly ovate-acuminate, with ciliolate margin ; wing sepals 4 × 2·8 mm., greenish, obliquely elliptic, apiculate at the apex, glabrous or nearly so ; anterior sepals 2 × 1 mm., free, with ciliolate margins. Upper petals 4·5 × 2·5 mm., irregularly contorted-cuneate ; carina 4 × 1·5 mm., crest 1·5 mm. long. Stamens 8. Style 3·5 mm. long, sharply bent at the apex with a dorsal crest. Capsule 4·5–5 mm. in diam., suborbicular in outline, notched at the apex, and with a marginal wing 0·5 mm. broad. Seeds 3 × 1·3 mm., cylindric, sericeous-pubescent ; caruncular appendages 1–1·5 mm. long, extending about half the length of the seed.

S. Rhodesia. W : Matobo Distr., Besna Kobila Farm, fl. & fr. ii.1955, *Miller* 2688 (SRGH), fl. & fr. xii.1956, *Miller* 3989 (BM ; SRGH).
Also in the Transvaal. Grassland.

Chodat based *P. transvaalensis* on *Rehmann* 4198 from Pretoria, " Aaples poort " and *Rehmann* 6348 from Houtbos, neither of which I have seen ; but *Rehmann* 4196 (BM) from Aapies Poort seems to be the species and may indeed be the specimen Chodat intended to cite.
As Chodat says (loc. cit.), the " species " is larger-flowered and has larger capsules than *P. amatymbica* Eckl. & Zeyh. *C. E. Moss* 13320 (BM) from Haartebeestpoort near Pretoria is a connecting link with our Rhodesian material and there is a fairly complete series linking the two species in the Transvaal. The constantly repeated problem arises whether *P. transvaalensis* is no more than a large-flowered form of *P. amatymbica* or whether we have two species which meet and interbreed in the Transvaal. Burtt Davy (F.P.F.T. **1** : 134 (1926)) cites *P. transvaalensis* as a synonym of *P. amatymbica*.

5. **Polygala francisci** Exell in Bol. Soc. Brot., Sér. 2, **31** : 7 (1957). TAB. **56** fig. 5.
> Type : Mozambique, Inharrime, *Mendonça* 3356 (BM, holotype ; LISC, isotype).

Perennial herb or shrublet with slender patent-pubescent stems. Leaves 10–35 × 6–16 mm., elliptic or broadly elliptic, pubescent, nerves almost inconspicuous, apex somewhat acute, rounded or slightly emarginate, mucronulate, base acute or slightly rounded. Flowers violet, pedicels 1·5 mm. long, in lateral 1–5-flowered racemes 1·5 cm. long. Posterior sepal 2·5 mm. long ; wing sepals 6 × 3 mm., greenish, obliquely elliptic, apex apiculate, margin ciliolate ; anterior sepals 2 mm. long, free. Upper petals 3·5 × 3 mm., obovate ; carina 4 × 1·5 mm., crest 2 mm. long, purple. Stamens 8. Style 4 mm. long, apex geniculate. Capsule 5 × 5·5 mm., suborbicular in outline, glabrous, emarginate at the apex, margin narrowly winged, 0·5 mm. broad. Seeds 3 × 2 mm., broadly ovoid, sericeous-pubescent ; caruncular appendages 1 mm. long.

Mozambique. SS : Inharrime, between Jacobécua and Inharrime, fl. & fr. 9.xii.1944, *Mendonça* 3356 (BM ; LISC). LM : Maputo, between Umbeluzi and Bela Vista, fl. & fr. 20.xi.1940, *Torre* 2091 (BM ; LISC).
Southern Mozambique.

In open bush on white sand and on the edges of dense, mixed woodland.
Junod 80 (BR) from " Delagoa Bay ", with immature capsules, is very like *P. francisci* in appearance but has crisped pubescence.

6. **Polygala gazensis** Bak. f. in Journ. Linn. Soc., Bot. **40** : 24 (1911).—Eyles in Trans. Roy. Soc. S. Afr. **5** : 391 (1916). TAB. **56** fig. 4. Type : S. Rhodesia, Chimanimani, *Swynnerton* M. 32 (BM, holotype).
> *Polygala teretifolia* var. *gazensis* (Bak. f.) Norlindh in Bot. Notis. **1935** : 366, fig. 2, middle and right (1935). Type as above.
> *Polygala teretifolia* sensu Norlindh, loc. cit. quoad specim. 2398.

An ericoid shrub, erect or sprawling, up to 1–2·5 m. tall ; stems densely crisped- or patent-pubescent sometimes tomentose. Leaves 10–25 × 2–5 mm., narrowly elliptic, chartaceous to almost coriaceous, rather dark, shiny and usually nearly glabrous above, lighter in colour and somewhat pubescent or pilosulose below, apex pointed, margins usually revolute. Flowers mauve, purple or violet or occasionally white and mauve, pedicels 5–9 mm. long, solitary or in short lateral 2–5-flowered racemes up to about 1 cm. long. Posterior sepal 3 mm. long ; wing sepals 10–12 × 6–8 mm., obliquely ovate, unguiculate, glabrous ; anterior sepals 3 mm. long, free, ovate. Upper petals 6–7 × 2 mm., obliquely and irregularly oblong, forked at the apex nearly to the middle ; carina 8 × 5·5 mm., ciliate on its proximal margin. Stamens 8. Style 10 mm. long, ribbon-like, hooked at the apex and with incurved margins. Capsule 8 × 7·5 mm., suborbicular in outline, with a winged margin nearly 1 mm. broad. Seeds 3·5 × 2·5 mm., oblong-ellipsoid, somewhat sparsely pilose ; caruncular appendages about 1 mm. long.

S. Rhodesia. E : Inyanga, Pungwe R., c. 1400 m., fl. 18.xii.1930, *F. N. & W.* 3921 (BM). **Mozambique.** MS : Gorongosa, Pico Gogogo, 1800 m., fl. & fr. 10.x.1944, *Mendonça* 2417 (BM ; LISC).

Also in Pondoland according to Norlindh (loc. cit.) who has identified *P. esterae* Chod. (Type : *Bachmann* 745, not seen) with this. River banks and margins of evergreen forest at 1000–2000 m.

7. **Polygala teretifolia** L. f., Suppl. Pl.: 316 (1781).—Harv.& Sond., F.C. **1** : 83 (1860).—Norlindh in Bot. Notis. **1935** : 366, fig. 2 left (1935) excl. specim. 2398. Type from S. Africa (Cape Province).

Ericoid shrublet up to 1 m. tall ; stems densely crisped-pubescent or white-tomentellous or sometimes glabrous. Leaves up to 12 × 1·5 mm., linear becoming terete by revolution of the margins which usually entirely conceals the lower surface except for a narrow gap, pubescent or glabrous, often needle-like. Flowers purplish, pedicels 2–3 mm. long, in short terminal or lateral 1–5-flowered racemes up to about 1·5 cm. long. Posterior sepal 3 mm. long ; wing sepals 7–9 × 5–6 mm., obliquely ovate, unguiculate, glabrous; anterior sepals 2·5 × 2 mm., free, ovate. Upper petals, carina and stamens as in *P. gazensis*. Style 8 mm. long. Capsule and seeds only seen in an immature state ; caruncular appendages apparently about 1 mm. long.

S. Rhodesia. E : Inyanga, near Nyarawe R., c. 1700 m., fl. 4.xi.1930, *F.N. & W.* 2616 (BM). **Mozambique.** MS : Tsetsera, 1980 m., fl. 8.ii.1955, *E.M. & W.* 300 (BM ; LISC ; SRGH).

Also in S. Africa (Cape Province). Mountain grassland among rocks and beside streams, 1700–2000 m.

The apparent absence of this species from the Transvaal is curious.

The Cape Province specimens examined had slightly larger flowers than those of our specimens, though not so large as those of *P. gazensis*, and the caruncular appendages are longer.

F.N. & W. 2616 (BM) is a mixture of what I take to be the " normal " form and a glabrous form lacking the indumentum on the young branches and the cilia on the carina and staminal sheath but otherwise apparently identical in structure.

8. **Polygala adamsonii** Exell in Bol. Soc. Brot., Sér. 2, **31** : 9 (1957). TAB. **56** fig. 3. Type : Nyasaland, *Adamson* 340 (BM, holotype).

Annual ?, up to 30–40 cm. tall, with very slender crisped-pubescent stems and branchlets. Leaves 5–7 × 1·5–1·8 cm., linear-lanceolate, apex mucronate with a needle-like point, margins somewhat revolute, base truncate, glabrous. Flowers bright pink, pedicels 3·5–4 mm. long, in very slender 1–5-flowered lateral racemes 15–20 mm. long. Posterior sepal 2·5 mm. long ; wing sepals 5–6 × 4·5–5 mm., obliquely ovate, glabrous ; anterior sepals 2 × 1·3 mm., free, ovate. Upper petals 4 × 1 mm., contortedly narrow-oblong, forked at the apex for nearly 2 mm. ; carina 5 × 3 mm., minutely ciliolate on the proximal edge, crest 3 mm. long. Stamens 8, sheath minutely ciliate at the base. Style 5 mm. long, hooked at the upper end. Capsule 5·5 × 5 mm., suborbicular in outline, glabrous, margin with a narrow wing 0·4 mm. broad. Seeds 2·1 × 1·9 mm., subglobose, rather sparsely pubescent ; caruncular appendages c. 0·6 mm. long.

Nyasaland. S : Mt. Mlanje, *Adamson* 340 (BM ; BR ; K). **Mozambique.** Z : Serra de Guruè, Vale de Muchana, 1300 m., *Mendonça* 2151 (BM ; LISC).

S. Nyasaland and N. Mozambique. Ecology unknown, found at Guruè in cultivated fields.

Only known from the two gatherings cited ; Mendonça describes it as an annual herb.

9. **Polygala senensis** Klotzsch in Peters, Reise Mossamb. Bot. **1** : 113 (1861).—Oliv., F.T.A. **1** : 129 (1868) pro parte excl. syn. et specim. Schimper. ex Abyss.—Chod. in Mém. Soc. Phys. Hist. Nat. Genève, **31**, 2 : 323, t. 27 fig. 16 (1893). TAB. **56** fig. 10. Type : Mozambique, Sena, *Peters* (B, holotype †).

Polygala rogersiana Bak. f. in Journ. of Bot. **56** : 5 (1918). Type : Mozambique, Beira, Vila Machado, *Rogers* 4505 (BM, holotype ; K).

Perennial herb or shrublet (occasionally annual ?) about 0·6 m. high, often with annual shoots produced from a woody rhizome ; stems slender, densely crisped-pubescent. Leaves up to 3 × 1 cm., narrowly oblanceolate to oblong-oblanceolate, densely crisped-pubescent. Flowers yellowish, pedicels up to 6 mm. long, in lateral 3–12-flowered racemes 1·5–3·5 cm. long with persistent, ovate-acuminate bracts and bracteoles about 1 mm. long. Posterior sepal 4 × 2·8 mm., keel-shaped, pubescent ; wing sepals 8–10 × 7–8 mm., yellow, ovate-suborbicular, pubescent; anterior sepals 3·5 × 2·5 mm., free, keel-shaped, pubescent. Upper petals 6 × 2·5 mm., obliquely oblanceolate ; carina 6 × 2·5 mm., with crest 3 mm. long. Stamens 8. Capsule 6 × 5 mm., broadly oblong-elliptic in outline, pubescent, emarginate at the apex, margin very narrowly winged. Seeds 4 × 2 mm., at first densely sericeous, more sparsely so when ripe ; caruncular appendages broad, humped at the apex, and extending almost to the base of the seed.

S. Rhodesia. S : Nuanetsi R., 370 m., fl. 2.xi.1955, *Wild* 4682 (BM ; SRGH). **Mozambique.** MS : Sofala, between the R. Save and Nova Sofala, fl. & fr. 2.ix.1942, *Mendonça* 114 (BM ; LISC). LM : Sabiè, between Moamba and Ressano Garcia, fl. & fr. 3.xii.1940, *Torre* 2202 (BM ; LISC).

S. Rhodesia and Mozambique. Alkaline steppes, alluvial grasslands and dry open woodland up to 370 m.

Cufodontis (in Bull. Jard. Bot. Brux. **26** : 410 (1956)) and other authors have wrongly extended the distribution of this species to Ethiopia and Arabia. I have seen no material of it from north of the Zambezi.

10. **Polygala sphenoptera** Fresen. in Mus. Senckenb. **2** : 274 (1837).—Cufod. in Bull. Jard. Bot. Brux. **26** : 411 (1956).—E. Petit, F.C.B. **7** : 251, t. 27 (1958). TAB. **56** fig. 13. Type from Ethiopia.

Polygala quartiniana A. Rich. in Ann. Sci. Nat., Sér. 2, Bot. **14** : 263 (1840).— Norlindh in Bot. Notis. **1935** : 360 (1935).—Martineau, Rhod. Wild Fl. : 44, t. 14 fig. 3 (1953). Type from Ethiopia.

Polygala ukambica Chod. in Mém. Soc. Phys. & Nat. Hist. Genève, **31**, 2 : 329, t. 27 fig. 21 (1893).—Burtt Davy, F.P.F.T. **1** : 134 (1926).—Syntypes from Tanganyika and Natal.

? *Polygala tristis* Chod. in Bull. Herb. Boiss. **4** : 903 (1896). Type : Mozambique, Boroma, *Menyhart* 811 (Herb. Schinz, not seen). The type cannot be found in the Zurich Herbarium and has not been discovered elsewhere.

Polygala fischeri sensu R.E.Fr. in Wiss. Ergebn. Schwed. Rhod.-Kongo-Exped. **1** : 112 (1914).—Wild, Guide Fl. Vict. Falls : 143 (1953).

Polygala persicariaefolia sensu Eyles in Trans. Roy. Soc. S. Afr. **5** : 391 (1916) non DC.

Annual or perennial herb or shrublet up to 0·6–1 m. tall with slender stems usually crisped-pubescent sometimes patent-pubescent or nearly glabrous. Leaves 2–8 × 0·3–2·5 cm., very narrowly to narrowly elliptic, crisped-pubescent. Flowers pink or purple, pedicels slender, up to 6 mm. long, in slender lateral racemes usually about 5–8 cm. long but occasionally up to 18 cm. long; bracts and bracteoles 1·5–2 mm. long, persistent. Posterior sepal 3 × 1·8 mm. ; wing sepals 4–7 mm. in diam., suborbicular in outline, usually glabrous, frequently and characteristically tinged with purple round the extreme margin ; anterior sepals 3 mm. long, free. Upper petals 4 mm. long and 2 mm. broad near the apex, obliquely oblanceolate-oblong with a semi-circular incision at one side of the waist, puberulous from the waist downwards ; carina 4·5 × 2 mm., crest 2–3 mm. long. Capsule 4–6 × 3–4 mm., broadly oblong in outline, slightly broader at the apex or almost suborbicular, with a narrowly winged margin, 0·2–0·3 mm. broad. Seeds 2 × 1·5 mm., sericeous ; caruncular appendages 0·5–1 mm. long. (Note : the wings vary from 4 to 7 mm. in diam. and the rest of the flower parts vary more or less in proportion).

Bechuanaland Prot. N : Francistown, fl. ii.1926, *Rand* 52 (BM). SE : 18 km. N. of Lobatsi, fl. 15.xi.1948, *Hillary & Robertson* 556 (PRE). **N. Rhodesia.** B : Balovale Distr., near Mombezi-Kabompo confluence, fl. & fr., 8–10.v.1953, *Holmes* 1094 (FHO).

N : Mulungushi R., fl. xii.1907, *Kassner* 2074 (BM ; K). S : Livingstone, 910 m., fl. & fr. 1909, *Rogers* 7092 (K). **S. Rhodesia.** N : Mazoe, 1310 m., fl. & fr. iv.1906, *Eyles* 302 (BM). W : Matopos, fl. iii.1918, *Eyles* 972 (BM). C : Umsweswe, fl. 20.iv.1921, *Borle* 186 (PRE). E : Inyanga, near Cheshire, 1300 m., fl. 5.ii.1931, *Norlindh & Weimarck* 4881 (BM). S : Fort Victoria, *Monro* 1885 C (BM). **Nyasaland.** N : Rumpi, Njakwa Gorge, fl. & fr. 13.v.1952, *White* 2848 (FHO). S : Shire R., near Liwonde Ferry, 475 m., fl. 13.iii.1955, *E.M. & W.* 847 (BM ; LISC ; SRGH). **Mozambique.** N : Ribauè, fl. 24.iv.1937, *Torre* 1463 (COI ; LISC). Z : Lugela, Mocuba Distr., Namagoa, fl. 4.v.1948, *Faulkner* 262 (COI ; K : PRE). T : Tete, fl. & fr. ii.1859, *Kirk* (K). MS : Chimoio, Garuso, fl. 21.iii.1948, *Barbosa* 1223 (BM ; LISC). SS : Salela, 10 km. E. of Inhambane, fl. & fr. x.1935, *Gomes e Sousa* 1658 (COI). LM : Maputo, Goba, fl. & fr. 23.xii.1942, *Mendonça* 1624 (BM ; LISC).

Eastern tropical Africa from Ethiopia to the Transvaal and in the Belgian Congo. Sporadic throughout our region in various habitats up to 1300 m.

A form with long racemes (up to 18 cm. long) occurs in the south of Mozambique (*Barbosa* 696, *Torre* 7641 A, etc.) and a very small-flowered form, *Faulkner* 320 (PRE), has been collected at Namagoa.

P. filifera Chod., from the SW. of Angola, is very close but has smaller caruncular appendages. At present it seems unwise to unite it with *P. sphenoptera* in view of this difference and some discontinuity in distribution.

I have not been able to find the type (*Rüppell*) but it seems highly probable that *P. sphenoptera* Fresen. is the oldest name for the species usually known as *P. quartiniana* A. Rich. or *P. ukambica* Chod. and I have followed Cufodontis (loc. cit.) in using this name.

The species as delimited here is very variable in leaf-size and length of raceme, and somewhat variable in flower-size and indumentum.

11. **Polygala erioptera** DC., Prodr. **1** : 326 (1824).—Chod. in Mém. Soc. Phys. Hist. Nat. Genève, **31**, 2 : 342, t. 28 fig. 1–4 (1893).—Eyles in Trans. Roy. Soc. S. Afr. **5** : 391 (1916).—Burtt Davy, F.P.F.T. **1** : 133 (1926).—Norlindh in Bot. Notis. **1935** : 361 (1935).—Exell & Mendonça, C.F.A. **1**, 1 : 95 (1937).—Martineau, Rhod. Wild Fl. : 43 (1953).—Cufod. in Bull. Jard. Bot. Brux. **26** : 407 (1956).— E. Petit, F.C.B. **7** : 255 (1958). Syntypes from Senegal and Egypt.
　　Polygala linearis R. Br. in Salt., Voy. Abyss. App. : 95 (1814) *nom. nud.*
　　Polygala triflora sensu Oliv., F.T.A. **1** : 128 (1868).

Annual herb, 10–40 cm. high, usually much branched from near the base with arcuate-ascending branches densely crisped- or occasionally subpatent-pubescent. Leaves 8–30 × 1·5–6 mm., linear-oblong-elliptic to narrowly oblong-elliptic, crisped-pubescent. Flowers greenish-white, pedicels up to 2 mm. long, in short few-flowered lateral racemes up to about 1 cm. long or solitary ; bracts and bracteoles persistent, about 1 mm. long. Posterior sepal 3 × 1 mm., keel-shaped ; wing sepals up to 5 × 2·5 mm., greenish-white with a pronounced green stripe down the centre, obliquely elliptic, densely pubescent ; anterior sepals 2·5 × 0·5 mm., free. Upper petals 3 × 1 mm., obliquely narrowly obovate-elliptic ; carina 3·5 × 1·2 mm., crest 1 mm. long. Stamens 8. Capsule up to 4 × 2·8 mm., ovate-oblong in outline, pubescent at the margins, marginal wing obsolete. Seeds 3·5 × 1·2 mm., flattened-cylindric, brown-sericeous ; caruncle sharply conical, without appendages.

N. Rhodesia. N : Lake Tanganyika, Mbulu I., 760 m., fl. 17.ii.1955, *Richards* 4529 (BM ; K). **S. Rhodesia.** N : Urungwe, Zambezi Valley, 370 m., fl. & fr. 24.ii.1953, *Wild* 4058 (BM ; SRGH). W : Bulawayo, fl. v.1898, *Rand* 290 (BM). E : Inyanga near Cheshire, 1300 m., fl. & fr. 3.ii.1931, *Norlindh & Weimarck* 4790 (BM). **Nyasaland.** N : Nyika Plateau, fl. ii.1903, *McClounie* 43 (K). S : west of Fort Johnston, fl. 14.iii.1955, *E.M. & W.* 861 (BM ; LISC ; SRGH). **Mozambique.** LM : Impamputo, fl. & fr. 23.ii.1955, *E.M. & W.* 549 (BM ; LISC ; SRGH).

Widespread in tropical Africa, Transvaal, Egypt and tropical Asia. Grassland, sandy or muddy roadsides, and as a sporadic and not very common weed of cultivated ground, up to 1300 m., but usually at lower altitudes.

For an account of the numerous varieties of this species which have been described see Chodat (loc. cit.) and Cufodontis (loc. cit.).

12. **Polygala sadebeckiana** Gürke in Engl., Pflanzenw. Ost-Afr. **C** : 233 (1895). TAB. **56** fig. 9. Type from Tanganyika.
　　Polygala polygoniiflora Chod. in Journ. of Bot. **34** : 200 (1896). Type : Nyasaland, Shire (?) or Mlanje (?), *Scott Elliot* 8670 (BM, holotype ; K, isotype).

Perennial herb or shrublet 10–30 cm. tall sending up annual shoots after burning ; stems slender, crisped-pubescent, rarely branched. Leaves 1–5 × 0·5–2 cm., narrowly elliptic-oblong, pubescent on the margins and veins and sparsely so elsewhere, tips blunt to subacute, mucronate. Flowers greenish-white, pedicels 2–3 mm. long, in short somewhat congested 5–15-flowered lateral racemes 1–2·5 cm. long ; bracts and bracteoles persistent, 0·8–1·2 mm. long. Posterior sepal keel-shaped, 3 mm. long, pubescent ; wing sepals 5–6 × 3–3·5 mm., greenish-white, obliquely elliptic, glabrous ; anterior sepals 2·5 mm. long, keel-shaped, free. Upper petals 2 × 1·5 mm., obliquely ovate ; carina 4 mm. long, crest 1·5 mm. long. Stamens 8. Capsule 3–4 mm. in diam., suborbicular in outline, ciliolate at the margin, marginal wing about 0·5 mm. broad. Seeds 2·5 × 1·5 mm., sericeous ; caruncular appendages 0·5 mm. long.

Nyasaland. S : Zomba, fl. & fr. ix.1891, *Whyte* (BM). **Mozambique.** Z : Namagoa, 60–120 m., *Faulkner* 287 (K ; PRE).
Also in Tanganyika. Open woodland on hillsides and on cultivated ground, 60–1220 m.
An apparently rare species or one which has escaped notice.

13. **Polygala nyikensis** Exell in Bol. Soc. Brot., Sér. 2, **31** : 10 (1957). Type : Nyasaland, Nyika Plateau, *Benson* 1392 (BM, holotype).

Perennial with short, crisped-pubescent stems, 6–14 cm. tall, produced annually from a woody rootstock. Leaves 5–22 × 2–6 mm., very narrowly to narrowly elliptic-oblong, somewhat pubescent when young but soon becoming glabrous, apex obtuse to acute, mucronulate. Flowers magenta or pale mauve, pedicels up to 5 mm. long, in short, somewhat congested, terminal racemes 2–5 cm. long, bracts and bracteoles 1·5–2 mm. long, persistent. Posterior sepal 3 mm. long ; wing sepals 6–7 × 4·5–5 mm., obliquely elliptic, glabrous ; anterior sepals 2·5 mm. long, free. Upper petals 5·5 × 3 mm., broadly falcate ; carina 6 × 3 mm., crest 2·5 mm. long. Stamens 8. Ovary 1·5 × 1·2 mm., obovate in outline, style 5·5–6 mm. long. Capsule 5·5 × 4·5 mm., suborbicular, slightly pubescent. Seed 3 × 1·5 mm., ellipsoid, sericeous ; caruncular appendages 0·7 mm. long.

N. Rhodesia. E : between Mts. Kongula and Kangampanda and the Nyasaland border, near Mt. Mwanda, fl. 18.xi.1952, *Temperley* (BM). **Nyasaland.** N : Nyika Plateau, fl. xi.1903, *Henderson* (BM).
N. Rhodesia and Nyasaland. In submontane grassland up to 2300 m.

The flowers are comparatively large for such a small species. The type shows clearly the burning off effect of fire and none of the specimens so far collected shows evidence of more than one year's growth. The form of the plant might be considerably modified under fire-free conditions.

14. **Polygala ohlendorfiana** Eckl. & Zeyh., Enum. Pl. Afr. Austr. Extra-trop. : 22 (1834).—Harv. in Harv. & Sond., F.C. **1** : 91 (1860).—Chod. in Mém. Soc. Phys. Hist. Nat. Genève, **31,** 2 : 395, t. 30 fig. 15–16 (1893).—Burtt Davy, F.P.F.T. **1** : 134 (1926).—Norlindh in Bot. Notis. **1935** : 364 (1935). Type from S. Africa (Cape Province).

Perennial with slender crisped-pubescent and sparsely pilose, prostrate or sprawling stems, 6–20 cm. long from a woody rootstock. Leaves 5–15 × 3–9 mm., ovate to suborbicular or elliptic to broadly elliptic, obtuse to rounded at the apex, rounded or obtuse to very slightly cordate at the base, pilose or crisped-pubescent especially on the margins and midrib. Flowers purple or pink sometimes streaked with yellow, with slender pedicels up to 5 mm. long, in few-flowered, terminal racemes 1·5–5 cm. long ; bracts and bracteoles about 1 mm. long, persistent. Posterior sepal 2·5 mm. long, keel-shaped, wing sepals 5–7 × 3·5–4·5 mm., obliquely broadly elliptic, glabrous ; anterior sepals 2 mm. long, free, keel-shaped. Upper petals 3·5 × 2 mm., obovate-spathulate ; carina 4 × 2·8 mm., crest 1·5 mm. long. Stamens 8. Capsule 4·5 × 4 mm., suborbicular in outline, with a narrowly winged margin 0·5 mm. broad. Seeds 2 × 1·5 mm., broadly ovoid, sericeous-pubescent ; caruncular appendages 0·8 mm. long.

S. Rhodesia. E : Mt. Inyangani, 2300 m., fl. 7.xii.1930, *F.N. & W.* 3562 (BM). **Nyasaland.** N : Nyika Plateau, fl. xi.1903, *Henderson* (BM). C : Mt. Dedza 2080 m., fl. & fr. 15.x.1937, *Longfield* 37 (BM). S : Mt. Mlanje, Chambe Plateau, 1670 m., fl. & fr. 4.ix.1956, *Newman & Whitmore* 661 (BM ; SRGH). **Mozambique.** MS : Banti Forest, 1980 m., fl. 4.ii.1955, *E.M. & W.* 218 (BM ; LISC ; SRGH).
From S. Tanganyika to S. Africa (Cape Province). In submontane grassland from 1800–2300 m.

A montane species with a well-marked N–S distribution.

15. **Polygala wilmsii** Chod. in Engl., Bot. Jahrb. **48**: 329 (1912).—Burtt Davy, F.P.F.T.
1 : 134 (1926).—Norlindh in Bot. Notis. **1935** : 365 (1955).—Martineau, Rhod.
Wild Fl. : 44 (1953). TAB. **56** fig. 7. Type from the Transvaal.

Perennial with slender, pilose, erect or subdecumbent stems 5–25 cm. long produced annually from a woody rootstock. Leaves 6–20 × 1–5 mm., very narrowly
elliptic or very narrowly oblong-elliptic to elliptic or oblong-elliptic, laxly pilose.
Flowers purple or purple and yellow with pedicels 1 mm. long in congested
terminal racemes up to 5 cm. long (sometimes subcapitate) ; bracts and bracteoles
1–1·2 mm. long, persistent. Posterior sepal 1·8 mm. long ; wing sepals 3–4 × 1·5–
1·8 mm., obliquely elliptic, rather markedly unguiculate, glabrous ; anterior sepals
1·5 mm. long, free. Upper petals 3 × 1·2 mm., obliquely and irregularly oblanceolate ; carina 3 × 1·5 mm., crest 1·2 mm. long. Stamens 8. Capsule c. 2·1 × 2·1
mm., suborbicular in outline, with a very narrow marginal wing, glabrous. Seeds
1·5 × 0·9 mm., ellipsoid, rather sparsely white-pubescent ; caruncular appendages
c. 1 mm. long, narrow.

S. Rhodesia. E : Banti Forest, 1830 m., fl. 4.ii.1955, *E.M. & W.* 182 (BM ; LISC ;
SRGH) ; Mt. Inyangani, 2000 m., fl. & fr. 6.xii.1930, *F.N. & W.* 3495 (BM). **Mozam-
bique.** MS : Tsetsera, 2140 m., fl. 7.ii.1955, *E.M. & W.* 256 (BM ; LISC ; SRGH).
LM : Libombos, near Namaacha, Mt. Mponduim, 800 m., fl. & fr. 22.ii.1955, *E.M. & W.*
516 (BM ; LISC ; SRGH).
Also in the Transvaal. Submontane grassland up to 2140 m.

A mainly montane species but found at 800 m. in the Libombos.
The name *P. wilmsii* var. *subcapitata* Chod. (loc. cit.), type from the Transvaal, was
applied to the form with subcapitate racemes. Specimens from our area show every
gradation.

16. **Polygala virgata** Thunb., Prodr. Pl. Cap. **2** : 120 (1800).—Chod. in Mém. Soc. Phys.
Hist. Nat. Genève, **31,** 2 : 403, t. 30 fig. 33–35 (1893).—Bak. f. in Journ. Linn. Soc.,
Bot. **40** : 24 (1911).—Eyles in Trans. Roy. Soc. S. Afr. **5** : 391 (1916).—Milne-
Redh. in Mem. N.Y. Bot. Gard. **8,** 3 : 220 (1953).—Martineau, Rhod. Wild Fl. :
143, t. 14 fig. 4 (1953).—E. Petit, F.C.B. **7** : 266 (1958). Type from S. Africa.
Var. **decora** (Sond.) Harv. in F.C. **1** : 85 (1860).—Chod., tom. cit. : 404 (1893).—Eyles,
loc. cit. TAB. **56** fig. 14, TAB. **58** fig. A. Type from Natal.
Polygala decora Sond. in Linnaea, **23** : 14 (1850). Type as above.
Polygala ourolopha Chod. in Engl., Bot. Jahrb. **48** : 331 (1912). Type : Mozam-
bique, Gorongosa, *Carvalho* (B † ; COI, lectotype).
Polygala speciosa sensu Burtt Davy, F.P.F.T. **1** : 136 (1926) excl. syn. *P. virgata*
var. *speciosa* Harv.

Shrub 1·5–3 m. tall, virgately branched ; stems sparsely to fairly densely crisped-
pubescent, eventually glabrous. Leaves 2–9 × 0·3–2 cm., linear to narrowly
elliptic, pubescent when young, becoming glabrous, acute at the apex and cuneate
at the base. Flowers large for the genus, varying from deep purple to pale lilac,
with slender pedicels up to 7 mm. long, in terminal many-flowered racemes 3–15
cm. long with pubescent rhachis, deciduous bracts 3 mm. long and bracteoles
2 mm. long. Posterior sepal 4–5 mm. long, keel-shaped ; wing sepals 10–17 × 9–13
mm., very broadly obliquely elliptic to suborbicular, glabrous ; anterior sepals
3·5 mm. long, free. Upper petals 7 × 3–3·5 mm., broadly and irregularly spathulate ;
carina 3 × 4–5 mm., margins minutely ciliate, crest 4–5 mm. long, purple. Style
strongly curved, 10–12 mm. long, ciliate on the inside of the curve. Stamens 8.
Capsule 10 × 8 mm., obliquely obovate in outline, glabrous, winged margin 1 mm.
broad. Seeds 4·5 × 1·5 mm., cylindric, with rather short brownish pubescence ;
caruncle without appendages.

N. Rhodesia. C : Broken Hill, fl. & fr. xi.1928, *van Hoepen* 1342 (PRE). E : Fort
Young, fl. ix.1896, *Nicholson* (K). **S. Rhodesia.** N : Mazoe, 1280 m., fl. xii.1905,
Eyles 198 (BM). C : Salisbury, fl. vii.1898, *Rand* 437 (BM). E : Vumba Mts., source
of Impodzi R., fl. & fr. 29.vii.1956, *Chamberlain* 14 (BM) ; near Chirinda, 1200 m., fl.
xii.1903, *Swynnerton* 294 (BM). **Nyasaland.** N : Nyika Plateau, 2300 m., fl. 13.viii.1946,
Brass 17206 (BM ; K). C : Dedza Mt., 1830 m., fl. 20.iii.1955, *E.M. & W.* 1074 (BM ;
LISC ; SRGH). S : Mt. Mlanje, 1950 m., fl. & fr. 16.vii.1956, *Newman & Whitmore* 25
(BM ; SRGH). **Mozambique.** N : Vila Cabral, fl. & fr. 30.x.1934, *Torre* 446 (BM ;
COI). Z : Guruè, 1200 m., fl. & fr. 30.ix.1941, *Torre* 3534 (BM ; LISC). MS : Manica,
Mavita, Rotanda, fl. & fr. 8.viii.1948, *Barbosa* 1710 (BM ; LISC).
Also in the Belgian Congo, Tanganyika and S. Africa. Submontane grasslands and
forest edges up to 2300 m.

Most of the Nyasaland specimens have narrower and more grasslike leaves than those from the rest of the area, and slightly smaller flowers with a tendency for the bracteoles to persist. This leads to the suspicion that there may have been some hybridization with *P. macrostigma*, which is common in the same area.

Var. *decora* can be fairly satisfactorily distinguished from the type variety by its more acute leaves and (usually) larger flowers and from var. *speciosa* (Sims) Harv. by the pubescent inflorescences. Many other varieties have been described from S. Africa but it is difficult to separate more than the three mentioned (see Norlindh, loc. cit.). Only var. *decora* has been recorded from tropical Africa.

This species is one of the most handsome in the genus and has been grown in greenhouses in Britain. Known as " Pride of Manicaland " in Southern Rhodesia.

17. **Polygala gomesiana** Welw. ex. Oliv., F.T.A. **1** : 126 (1868) pro parte excl. specim. Kirk.—Welw. in Trans. Linn. Soc. **27** : 14, t. 4 (1869).—Chod. in Mém. Soc. Phys. Hist. Nat. Genève, **31**, 2 : 336, t. 27 fig. 32–34 (1893).—Exell & Mendonça, C.F.A., **1**, 1 : 95 (1937). Type from Angola (Huila).

Perennial herb up to about 2 m. with woody rootstock with usually a single, longitudinally furrowed, glabrous or nearly glabrous stem. Leaves very variable in size (reaching a maximum of 10 × 2·5 cm. in Angolan material), elliptic to very narrowly elliptic acute or rounded at the apex, cuneate at the base, glabrous or nearly so. Flowers bright purple, pedicels slender, pilose, up to 12 mm. long, in many flowered terminal racemes up to 40 cm. long, branched or simple, rhachis pubescent, with persistent, purplish scarious, ovate or ovate-acuminate bracts and bracteoles up to about 7 (9) mm. long. Posterior sepal 6 mm. long, keel-shaped, pilose ; wing sepals up to 12 × 11 mm., obliquely suborbicular, glabrous ; anterior sepals c. 5·5 mm. long, free, keel-shaped. Upper petals 6 × 3 mm. laterally deeply notched ; carina 7 × 4 mm., crest 3 mm. long. Stamens 8. Style 9–10 mm. long, strongly curved, glabrous. Capsule 5–6 × 4·5–5·5 mm., broadly obovate in outline, sparsely pilose on the margin with marginal wing 0·5–1 mm. broad. Seeds 2·5 × 1·8 mm., ovoid, white-silky-pubescent ; caruncular appendages absent.

N. Rhodesia. W : Mwinilunga Distr., Dobeka Bridge, fl. 3.i.1938, *Milne-Redhead* 3932 (BM ; K).
Also in Angola. In boggy grassland.

This handsome, purple-flowered species is abundant and widespread in Angola but only enters our area in NW. Rhodesia. Here the distribution overlaps that of *P. exelliana* and some specimens appear to be hybrids.

18. **Polygala macrostigma** Chod. in Engl., Bot. Jahrb. **48** : 319 (1912).—Exell & Mendonça, C.F.A. **1**, 1 : 96 (1937).—E. Petit, F.C.B. **7** : 260 (1958). TAB. **56** fig. 1. Type from the Belgian Congo.
Polygala gomesiana sensu Milne-Redh. in Mem. N.Y. Bot. Gard. **8**, 3 : 219 (1953).

Annual (always ?) herb up to 3 m. tall with erect, sparsely pilosulose or glabrous stems. Leaves up to 25 cm. long and usually not more than 4–5 mm. broad, but occasionally up to 16 mm. broad (in Belgian Congo material), usually linear, grasslike, occasionally very narrowly elliptic, slightly rough with short, somewhat bristly hairs mainly on the upper surface. Flowers purple or pale purple fading to white (usually almost white in dried specimens), pedicels slender, pilose, up to 14 mm. long, in terminal many-flowered racemes up to 30 cm. long, with pilose rhachis and persistent bracts 5–9 mm. long with ovate base and filiform apex, and similar bracteoles. Posterior sepal 5 mm. long, keel-shaped, pilose ; wing sepals 9–14 × 9–11 mm., obliquely suborbicular, sometimes very minutely ciliate on the margin, otherwise glabrous ; anterior sepals 4 mm. long, free, keel-shaped, pilose. Upper petals 7 mm. long, bent and contorted, ciliate at the base ; carina 8 × 4 mm., crest 4 mm. long. Stamens 8. Capsule 8 × 5·5–6 mm., obovate-oblong, in outline, marginal wing 0·5–0·8 mm. broad. Seeds 3 × 1·8 mm., cylindric, with long white silky hairs ; caruncular appendages absent.

Nyasaland. N : Mzimba Distr., Mbawa, fl. & fr. 10.iii.1954, *Jackson* 1293 (BM). C : Mt. Nchisi, 1400 m., fl. & fr. 25.vii.1946, *Brass* 16939 (K). S : Zobwe, 900 m., fl. 8.x.1944, *Benson* 509 (K). **Mozambique.** N : Metónia, Litunde, fl. & fr. 9.x.1942, *Mendonça* 705 (BM ; LISC).
Also in Angola, Belgian Congo and Tanganyika. Woodland, wooded grassland and submontane grasslands up to 1900 m.

Africans in Nyasaland chew the roots as a remedy for coughs.

19. **Polygala exelliana** Troupin in Bull. Jard. Bot. Brux. **19** : 206 (1948) ("*exelleana* ").
—E. Petit, F.C.B. **7** : 262 (1958). TAB. **56** fig. 2. Type from the Belgian Congo (Katanga).
Polygala gomesiana sensu R.E.Fr. in Wiss. Ergebn. Schwed. Rhod.-Kongo-Exped. **1** : 112 (1914).

Shrub or perennial herb up to 2·5 m. in height, stems densely crisped- or patent-pubescent. Leaves 2·5–7 × 0·4–1·4 cm., linear-elliptic to narrowly elliptic, rounded or acute at the apex, cuneate at the base, densely pubescent. Flowers red or purple, pedicels slender pilose up to 8 mm. long, in congested terminal racemes usually 4–5 cm. long (up to 10 cm.) with densely pubescent rhachis and persistent, opaque, dark red or greenish bracts and bracteoles up to about 3 mm. long, pointed but without long filiform tips. Posterior sepal 5 mm. long ; wing sepals 9–11 × 7–8·5 mm., obliquely suborbicular, pubescent or nearly glabrous with a few hairs towards the base ; anterior sepals 4·5 mm. long, free. Upper petals 8 mm. long, bent nearly at right angles and laterally notched ; carina 11 × 5 mm., crest 3 mm. long. Stamens 8. Capsule 6–9 × 4–5 mm., obovate-elliptic in outline, shiny, with very narrowly winged pilose margin. Seed 4·5 × 1·5 mm., cylindric, with rather short, white, silky-pubescent indumentum ; caruncular appendages absent.

N. Rhodesia. N : Lungunwa Valley, 1520 m., fl. 28.iii.1955, *E.M. & W.* 1259 (BM ; LISC ; SRGH). W : Mwinilunga, 1370–1520 m., fl. & fr. vi.1929, *Marks* E24 (K).
Nyasaland. N : Mugesse Forest Reserve, fl. 30.viii.1952, *Chapman* 11 (BM).
Also in the Belgian Congo and Tanganyika. In submontane grassland, marshes, sides of streams and along the edges of montane forest up to 1640 m.

I accept Petit's decision (F.C.B. **7** : 262–263 (1958)) to separate this species from *P. bakerana* Chod. but I am not convinced that the latter species exists in N. Rhodesia.
The Africans use an infusion of the roots as cough medicine.

20. **Polygala arenaria** Willd. in L., Sp. Pl. ed. 4, **3** : 880 (1802).—Chod. in Mém. Soc. Phys. Hist. Nat. Genève, **31**, 2 : 337, t. 27 fig. 35–6 (1893).—Exell & Mendonça, C.F.A., **1**, 1 : 101 (1937).—E. Petit, F.C.B. **7** : 248 (1958).—TAB. **57** fig. 17, TAB. **58** fig. C. Type from " Guinea ".

Annual 5–25 cm. tall with spreading habit, dichotomously or trichotomously branched ; stem apparently terminating in an inflorescence and growth continuing by lateral branches. Stems crisped-pubescent. Leaves 1–6 × 0·2–2 cm., linear-elliptic to elliptic, fairly densely pubescent to nearly glabrous. Flowers white, dull orange or pale purple, deflexed, with pedicels c. 3 mm. long, in very congested terminal racemes usually 1·5–3 cm. long, often globose, bracts and bracteoles persistent, 1·5 mm. long. Posterior sepal 2–2·5 mm. long ; wing sepals 5–7 × 3–4 mm., obliquely elliptic, usually somewhat pilose towards the base ; anterior sepals 1·5–2 mm. long, free. Upper petals 2 × 0·9 mm., obovate-elliptic, constricted below the middle ; carina 2·5 × 1 mm., crest 1·5 mm. long. Stamens 8. Capsule 3·5 × 2·5 mm., broadly obovate-elliptic in outline, wing less than 0·5 mm. broad, ciliate on the margin ; caruncle 0·5 mm. long, appendages very small.

Nyasaland. N : Karonga, fl. & fr. vii. 1896, *Whyte* (K). S : Malosa, fl. & fr. 10.iii.1955, *E.M. & W.* 787 (BM ; LISC ; SRGH). **Mozambique.** Z : Mocuba, Namagoa, 60–120 m., fl. & fr. ii–iii. 1943, *Faulkner* 161 (K ; PRE).
Widespread in tropical Africa. Roadsides and cultivated ground at low altitudes.

This species can usually be distinguished from *P. albida* by its peculiar branching and by the usually more hairy flowers and leaves but it may well hybridize with the latter as some specimens seem to be intermediate. It is comparatively rare in our area and has been collected only on cultivated ground and along roadsides so that there is a possibility that it is not indigenous.

21. **Polygala albida** Schinz in Verh. Bot. Verein. Brand. **29** : 53 (1888).—Burtt Davy, F.P.F.T. **1** : 133 (1926).—Norlindh in Bot. Notis. **1935** : 361 (1935).—Exell & Mendonça, C.F.A. **1**, 1 : 101 (1937).—Milne-Redh. in Mem. N.Y. Bot. Gard. **8**, 3 : 220 (1953). Type from SW. Africa.

Inflorescences dense-flowered ; wings 6–8 mm. long   -   -   -   - var. *albida*
Inflorescences laxer ; wings up to 6 mm. long   -   -   -   - var. *angustifolia*

Var. albida. TAB. **57** fig. 21.
Polygala livingstoniana Chod in Mém. Soc. Phys. Hist. Nat. Genève, **31**, 2 : 339 t. 27 fig. 38 (1893).—Eyles in Trans. Roy. Soc. S. Afr. **5** : 391 (1916).
Polygala arenaria sensu Bak. f. in Journ. Linn. Soc., Bot. **40** : 24 (1911).—Eyles, loc. cit.

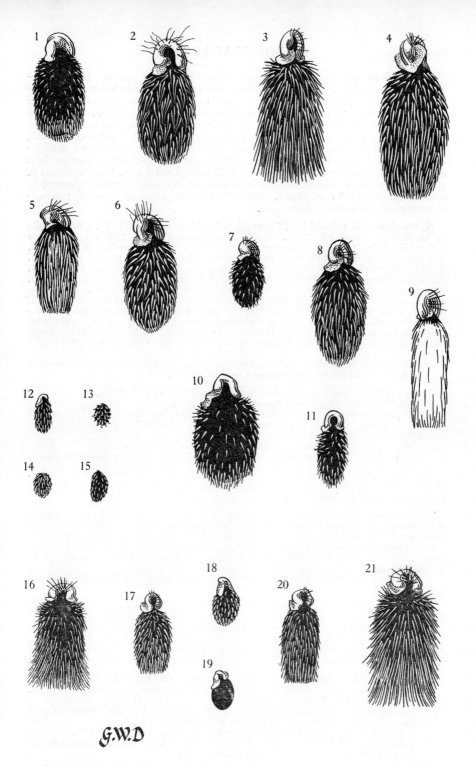

G.W.D

Tab. 57. Seeds of POLYGALA (all ×8) 1. P. STENOPETALA ; 2. P. PRODUCTA ; 3. P. SCHIN-
ZIANA ; 4. P. MARENSIS ; 5. P. PETITIANA VAR. PARVIFLORA ; 6. P. BAUMII ; 7. P. DEWEVREI ;
8. P. REHMANNII ; 9. P. UNCINATA ; 10. P. WESTII ; 11. P. RIVULARIS ; 12. P. MYRIANTHA ;
13. P. AFRICANA ; 14. P. SPICATA ; 15. P. CAPILLARIS ; 16. P. PERSICARIIFOLIA ; 17.
P. ARENARIA ; 18. P. PYGMAEA ; 19. P. MELILOTOIDES ; 20. P. ALBIDA VAR. ANGUSTIFOLIA ;
21. P. ALBIDA VAR. ALBIDA.

Annual 6–60 cm. tall with slender, crisped-pubescent stems. Leaves 1·5–10 × 0·2–1·5 cm., linear to narrowly elliptic, sometimes minutely hairy on the midrib and margins and pubescent towards the base, otherwise glabrous. Flowers white, greenish-white, pale pink or pale blue, pedicels deflexed, 3–4 mm. long, in dense terminal and lateral racemes; bracts 2 mm. long; bracteoles persistent, 1·5 mm. long. Posterior sepal 4 mm. long, keel-shaped ; wing sepals 6–8 × 6 mm., obliquely obovate-elliptic, sometimes with a few hairs towards the base ; anterior sepals 3 mm. long, free, keel-shaped. Upper petals 1·5 × 2 mm., bent at right angles and notched on one side ; carina 4 × 2 mm., crest 2 mm. long. Stamens 8. Capsule very broadly elliptic-oblong in outline, 4·5 × 4 mm., marginal wing ciliate towards the apex, 0·5 mm. broad. Seeds 3 × 1·5 mm., ellipsoid-cylindric ; caruncular appendages scarcely developed.

N. Rhodesia. N : Lumi R., 1520 m., fl. 9.ii.1955, *Richards* 4372 (K). W : Ndola, fl. 14.vii.1932, *Young* 105 (BM). C : Lusaka, Leopard's Hill, 1220 m., *Veterinary Officer* CRS 276 (PRE). S : Livingstone, fl. & fr., v.1911, *Gardner* 564 (K). **S. Rhodesia.** N : Miami, fl. & fr. iv.1926, *Rand* 16 (BM). W : Matobo, Besna Kobila, 1460 m., fl. ii.1955, *Miller* 2673 (BM ; SRGH). C : Salisbury, fl. & fr. vii.1898, *Rand* 463 b (BM). E : Umtali Distr., Engwa, 1600 m., fl. 1.ii.1955, *E.M. & W.* 25 (BM ; LISC ; SRGH). **Nyasaland.** N : Mzimba Distr., Mlawa, fl. 10.iii.1954, *Jackson* 1227 (BM). C : Lilongwe, fl. 31.iii.1955, *Jackson* 1533 (BM). S : Ncheu Distr., Lower Kirk Range, Chipusiri, fl. & fr. 17.iii.1955, *E.M. & W.* 931 (BM ; LISC ; SRGH). **Mozambique.** N : between Nampula and Ribauè, fl. & fr. 24.iv.1937, *Torre* 1464 (COI ; LISC). Z : Alto Molocuè, fl. & fr. 30.iv.1952, *Pereira* 27 (COI). T : Angónia, Vila Mousinho, fl. 12.v.1948, *Mendonça* 4172 (LISC). MS : Manica, Mavita, fl. & fr. 6.iv.1948, *Barbosa* 1374 (BM ; LISC).

Also in Angola, SW. Africa and the Transvaal. In *Brachystegia* woodland, wooded grassland, cultivated ground and roadsides ; up to 1600 m.

Var. **angustifolia** (Chod.) Exell in Bol. Soc. Brot., Sér. 2, **31** : 11 (1957).
    TAB. **57** fig. 20. Type from Angola (Cuanza Norte).
    Polygala stanleyana Chod. in Mém. Soc. Phys. Hist. Nat. Genève, **31,** 2 : 340, t. 27 fig. 39 (1893).—Exell & Mendonça, C.F.A. **1,** 1 : 102 (1937).—E. Petit, F.C.B. **7** 250 (1958). Type from Angola.
    Polygala stanleyana var. *angustifolia* Chod., loc. cit. Type as above.
    Polygala stanleyana var. *latifolia* Chod., loc. cit. Type from the Belgian Congo.

Flowers slightly smaller than in var. *albida* (above) and inflorescences laxer. Wing sepals 4–6 mm. long. Otherwise as in var. *albida* but flowers slightly smaller in all dimensions.

N. Rhodesia. N : Abercorn, Chitimbwa, fl. & fr. 30.v.1955, *Nash* N 116 (BM). **S. Rhodesia.** C : Marandellas, fl. & fr. 26.ii.1942, *Dehn* 76, 576 (BM ; SRGH). **Nyasaland.** N : Mzimba Distr., Mlawa, fl. & fr. 10.iii.1954, *Jackson* 1223 (BM). S : Ncheu Distr., Lower Kirk Range, Chipusiri, 1460 m., fl. 17.iii.1955, *E.M. & W.* 930 (BM ; LISC ; SRGH). **Mozambique.** N : Vila Cabral, fl. & fr. v–vi.1934, *Torre* 447 (BM ; COI ; LISC).

Also in the Cameroons, Belgian Congo, Angola and Uganda. In *Brachystegia* woodland, grassland, old cultivation and roadsides.

The choice of the epithet *angustifolia* is forced on nomenclatural grounds although the intention is to reduce the whole of *P. stanleyana* to varietal rank under *P. albida*.

Var. *angustifolia* has a more northern distribution than var. *albida* but where the ranges overlap there are so many intermediates that it seems preferable to regard these two taxa as varieties rather than species. *E.M. & W.* 930 and 931 cited above were collected in the same locality to illustrate the difference between the two varieties.

22. **Polygala persicariifolia** DC., Prodr. **1** : 326 (1824).—Chod. in Mém. Soc. Phys. Hist. Nat. Genève, **31,** 2 : 331, t. 27 fig. 22–23 (1893).—E. Petit, F.C.B. **7** : 254 (1958). TAB. **57** fig. 16. Type from Nepal.
    Polygala persicariifolia var. *latifolia* Oliv., F.T.A. **1** : 129 (1868). Type : Mozambique, Manganja Hills, fl. & fr. 8.iii.1862, *Kirk* (K).
    Polygala persicariifolia var. *densiflora* Chod., tom. cit. : 333 (1893). Type : Nyasaland, Shire Highlands, " leg. Blantyre " (sic). One of the specimens collected by Buchanan may have been intended as the type.

Annual up to 60 cm. or more tall ; stems slender, often arcuate-ascending, crisped-pubescent. Leaves 2–8 × 0·5–2 cm., elliptic to very narrowly elliptic, somewhat acuminate, with rather sparse, short, bristly hairs. Flowers greenish, pink or purplish, pedicels 3–4 mm. long, in many-flowered, terminal and lateral racemes up to 10 cm. long ; rhachis crisped-pubescent ; bracts 1·5–2 mm. long ; bracteoles up to 0·8 mm. long, persistent. Posterior sepal 3–3·5 mm. long, keel-shaped ; wing sepals 5–7 × 3·5–6 mm., very broadly elliptic to suborbicular with rather prominent venation, glabrous ; anterior sepals 2·5–3 mm. long, free, keel-shaped. Upper petals 3 mm. long, obliquely and irregularly elliptic, notched on one side ; carina 4 × 1·5 mm., crest 1·5 mm. long, purple. Stamens 8. Capsule 5 × 4·5 mm., very broadly elliptic to suborbicular in outline, ciliate on the margin, with a narrow wing less than 0·5 mm. broad. Seeds 3 × 1·5 mm., oblong-ellipsoid, with white silky hairs ; caruncular appendages scarcely developed.

N. Rhodesia. N : Abercorn, Kalambo Falls, fl. & fr. 30.iii.1952, *Richards* 1298 (K). W : Ndola, fl. 27.ii.1954, *Fanshawe* 877 (K). **Nyasaland.** S : Namwera Escarpment, Jalasi, 1120 m., fl. 15.iii.1955, *E.M. & W.* 897 (BM ; LISC ; SRGH). **Mozambique.** N : Massangulo, 1100 m., fl. iii.1933, *Gomes e Sousa* 1325 (COI). T : Manganja Hills, fl. and fr. 8.iii.1862, *Kirk* (K).
Tropical Asia and from the Sudan and Ethiopia to Mozambique. Grasslands and roadsides up to c. 900–1250 m.

Owing to the erroneous synonymy given by Oliver (loc. cit.) this species has been much confused with the quite distinct *P. sphenoptera* Fresen. (*P. quartiniana* A. Rich.)

23. **Polygala pygmaea** Gürke in Engl., Pflanzenw. Ost-Afr. **C** : 234 (1895).—Exell & Mendonça, C.F.A. **1**, 1 : 103 (1937).—E. Petit, F.C.B. **7** : 248, fig. 7B (1958). TAB. 57 fig. 18. Type from Tanganyika.

Annual 6–15 cm. tall, usually of dwarfed, tufted appearance ; stems crisped-pubescent. Leaves 15–60 × 1–7 mm., linear or linear-elliptic to narrowly elliptic, glabrous or very minutely pubescent. Flowers white, lilac or pinkish, pedicels 2–3 mm. long, in terminal racemes with dichotomous cymose branching much as in *P. arenaria* but the whole inflorescence much congested in conformity with the dwarf habit. Posterior sepal 1·5 mm. long ; wing sepals 2–2·5 × 2–3 mm., suborbicular ; anterior sepals, 1·5 mm. long, free. Upper petals 1·5 × 1·5 mm., rounded at the apex with a notch on one side and a projection on the other ; carina 2·5 mm. long, crest 1 mm. long. Stamens 8. Capsule 2·5 × 2·2 mm., suborbicular in outline, glabrous. Seeds 2 × 1 mm., ellipsoid, white-silky-pubescent ; caruncle about 0·5 mm. long with appendages about the same length.

N. Rhodesia. B : below Gonye Falls, 1030 m., fl. & fr. 28.iv.1925, *Pocock* 154 (PRE). W : Mwinilunga Distr. near source of Matonchi R., fl. & fr., 16.ii.1938, *Milne-Redhead* 4602 (BM ; K). S : Mapanza, 1060 m., fl. & fr. 26.ii.1955, *Robinson* 1121 (K). **S. Rhodesia.** W : Matopos, World's View, fl. & fr. 5.viii.1929, *Rendle* 373 (BM). C : Marandellas, Grasslands, fl. & fr. 20.iii.1948, *Corby* in GHS. 20900 (BM ; SRGH). E : Umtali, Dora R., fl. & fr. 20.v.1935, *Eyles* 8429 (K).
Also in Uganda, Tanganyika, Belgian Congo and Angola. Usually on sand, but little ecological information is available.

This species is very close to *P. welwitschii* Chod. (1893) and may eventually have to be united with it. *P. pygmaea* has a more dwarfed habit and slightly smaller flowers. *Milne-Redhead* 4602 is of less compact habit than the other specimens seen but may have been growing in shade.
The species is either rare in our area or may have been overlooked.

24. **Polygala melilotoides** Chod. in Engl., Bot. Jahrb. **48** : 320 (1912).—Exell & Mendonça, C.F.A. **1**, 1 : 103 (1937).—E. Petit, F.C.B. **7** : 246, fig. 7A (1958). TAB. 57 fig. 19. Type from the Belgian Congo (Katanga).

Annual 3–10 cm. tall, similar in habit and in the morphology of the flower to the preceding species but with usually relatively broader leaves, 10–20 × 5–8 mm., elliptic to broadly elliptic, and slightly smaller, glabrous seeds, 1·8 × 0·8 mm.

N. Rhodesia. N : Abercorn, Lucheche R., 1670 m., fl. & fr. 2.v.1952, *Siame* 180 A (BM). W : Ndola, fl. & fr. 14.vii.1932, *Young* 68 (BM). C : Kapiri Mposhi, fl. & fr. 13.vii.1932, *Young* 39 (BM). **Nyasaland.** S : Lower Kasupe, 700 m., fl. & fr. 13.iii.1955, *E.M. & W.* 818 (BM ; LISC ; SRGH).
Also in Angola, Belgian Congo and Tanganyika. On sand in *Brachystegia* woodland at 700–1670 m.

This is the only species of *Polygala* with completely glabrous seeds found in our area.

25. **Polygala mendoncae** E. Petit in Bull. Jard. Bot. Brux. **26** : 259 (1956) ; F.C.B. **7** : 264 (1958). TAB. 56 fig. 16. Type from Angola.
Polygala sp.—Exell & Mendonça, C.F.A. **1**, 2 : 367 (1951).

Annual 1–1·5 m. tall. Stems with short, crisped pubescence interspersed with longer patent hairs. Leaves 2–6 × 0·2–1·2 cm., linear or linear-lanceolate to very narrowly elliptic becoming broader towards the base of the stem, crisped-pubescent. Flowers white, pink, crimson or mauve, pedicels 3–5 mm. long, in elongated, terminal racemes up to 40 cm. long, with densely pubescent rhachis and persistent bracts 2–3 mm. long and bracteoles 1–1·5 mm. long. Posterior sepal 4 mm. long, keel-shaped ; wing sepals 7–8 × 6·5–7 mm. (up to 9 × 8 mm. in Angolan material), broadly obovate to suborbicular ; anterior sepals 3–4 mm. long, free, keel-shaped. Upper petals 5–6 × 1–2 mm., ligulate, curved almost at right angles ; carina 7–8 mm. long, crest 3–4 mm. long. Stamens 8. Capsule 5–6 × 3–4 mm., obovate in outline, ciliate on the margin. Seeds 3 × 1 mm., cylindric, with long, white, silky hairs ; caruncular appendages scarcely developed.

N. Rhodesia. W : Mwinilinga Distr., source of Matonchi R., fl. & fr. 19.xii.1937, *Milne-Redhead* 3744 (BM ; K).
Also in Angola and the Belgian Congo.

Mixed evergreen-deciduous woodland.

26. **Polygala usafuensis** Gürke in Engl., Bot. Jahrb. **30** : 337 (1900).—R.E.Fr. in Wiss. Ergebn. Schwed. Rhod.-Kongo-Exped. **1** : 113 (1914).—Exell & Mendonça, C.F.A. **1,** 1 : 99 (1937).—E. Petit, F.C.B. **7** : 257, fig. 7E (1958). TAB. **56** fig. 15. Type from Tanganyika.
  ? *Polygala engleranum* Busc. & Muschl. in Engl., Bot. Jahrb. **49** : 476 (1913) pro parte. Type (destroyed) from the Belgian Congo (Bangweolo region).

Annual 1·4–2 m. tall ; stems crisped-pubescent. Leaves 1–7 × 0·2–0·8 cm., linear to very narrowly elliptic, pubescent. Flowers salmon-pink to orange-red, pedicels 2–3·5 mm. long, in elongated, terminal racemes up to 30 cm. long with densely pubescent rhachis and persistent bracts 2 mm. long and bracteoles 1 mm. long. Posterior sepal 3 mm. long ; wing sepals 5–8 × 3–6 mm., broadly elliptic to suborbicular, pubescent ; anterior sepals 2 mm. long, free. Upper petals 4·5 × 2·5 mm., irregularly obovate with a finger-like projection at the apex and one margin bent over ; carina 8 × 2·8 mm., minutely ciliate on the upper margins, crest 4–5 mm. long. Stamens 8. Capsule 5–6 × 2·2–3 mm., narrowly obovate-oblong in outline, ciliate on the margin. Seeds 2·5 × 1 mm., cylindric, densely sericeous ; caruncular appendages scarcely developed.

N. Rhodesia. N : Abercorn, 1830 m., fl. & fr. 21.iv.1931, *Gamwell* 45 (BM). W : Ndola, fl. & fr. 14.vii.1932, *Young* 80 (BM). C : Serenje, Kaombi, 1370 m., fl. & fr. iv.1930, *Lloyd* (BM). E : Lunkwakwa Valley, fl. & fr. 23.iii.1955, *E.M. & W.* 1141 (BM ; LISC ; SRGH). **Nyasaland.** N : Mzimba, Mlawa, fl. 7.iv.1955, *Jackson* 1611 (BM ; LISC). C : Lilongwe, Dzalanyama Forest Reserve, fl. & fr. 27.iv.1958, *Jackson* 2210 (BM ; SRGH).
Also in Angola, Belgian Congo and Tanganyika.

The plate of *P. engleranum* Busc. & Muschl. in the Brussels Herbarium looks more like *P. friesii* Chod. than *P. usafuensis* but there are serious discrepancies between the plate and the description. The latter seems to apply rather to *P. usafuensis*. In view of the well-known unreliability of the Buscalioni and Muschler descriptions it seems best to leave *P. engleranum* as a doubtful name.

27. **Polygala friesii** Chod. apud R.E.Fr. in Wiss. Ergebn. Schwed. Rhod.-Kongo-Exped. **1** : 113 (1914). Type : N. Rhodesia, near Fort Rosebery, Monglobi, fl. ix.1911, *Fries* 606 (UPS, holotype).
  ? *Polygala engleranum* Busc. & Muschl. in Engl., Bot. Jahrb. **49** : 476 (1913) pro parte. Type (destroyed) from the Belgian Congo (Bangweolo). See note under *P. usafuensis*, above.

Annual 0·5–0·8 m. tall ; stems pubescent at first becoming nearly glabrous. Leaves poorly developed, 10–25 × 0·5–1 mm., linear to almost acicular, occasionally pubescent, but usually glabrous, with needle-like points. Flowers pale to deep blue, densely pubescent, pedicels 4–5 mm. long, in terminal racemes usually not more than about 10 cm. long, inflorescence often considerably branched, rhachis patent-pubescent ; bracts 2·5 mm. long ; bracteoles 1·5 mm. long, persistent. Posterior sepal 3 mm. long ; wing sepals 9–11 × 5–6 mm., obliquely elliptic ; anterior sepals 2·5 mm. long, free. Upper petals 6·5 × 3·5 mm., irregularly oblong-elliptic, lobed at one side ; carina 9 × 3·5 mm., minutely ciliate on the upper margins, crest attached by a very broad base, 5 mm. long. Stamens 8. Neither capsules nor seeds seen.

N. Rhodesia. N : Abercorn, Lake Chila, 1750 m., fl. 9.i.1952, *Nash* 71 (BM). W : Katubo Stream, fl. 30.xii.1907, *Kassner* 2256 (BM).
Known at present only from N. Rhodesia. In swamps, peat bogs and marshy grassland at 1500–1750 m.

This species has long been confused with *P. usafuensis* to which it is very near. It is distinguishable by its larger flowers of a beautiful blue colour, longer pedicels, shorter inflorescences and poorly developed leaves and by the fact that it is always found in swampy ground or even growing in water.

28. **Polygala britteniana** Chod. in Journ. of Bot. **34** : 198 (1896). Type : N. Rhodesia, Stevenson Road, *Scott Elliot* 8256 (BM, holotype ; K, isotype).

Perennial herb up to 2·5 m. tall, with wiry, minutely appressed-pubescent stems. Leaves 20–80 × 0·5–1 mm., linear, with needle-like points, glabrous. Flowers pale blue or pale mauve and white, pedicels 2·5–3 mm. long, in elongated terminal racemes up to 20 cm. long with minutely appressed-pubescent rhachis and persistent bracts 2 mm. long and bracteoles 1·5 mm. long. Posterior sepal 5 mm. long ; wing sepals 8 × 6 mm., obliquely broadly obovate in outline, glabrous ; anterior sepals 4 mm. long, free. Upper petals 5·5–6 × 3 mm., obliquely oblong, lobed at one side ; carina 5·5 mm. long, crest 4 mm. long. Stamens 8. Capsule 7 × 2·8 mm., narrowly oblanceolate-oblong in outline, ciliate on the margin. Seeds 2 × 1 mm., densely sericeous ; caruncle c. 1 mm. long, sagittate, appendages scarcely developed.

N. Rhodesia. N : Abercorn, 8 km. south on the Kasama Road, fl. & fr. 19.x.1947, *Greenway & Brenan* 8237 (K).
Known only from N. Rhodesia (and one doubtful record from Angola). Riverine forest and sandy roadsides at 1200–1700 m.

29. **Polygala nambalensis** Gürke in Warb., Kunene-Samb.-Exped. Baum : 276 (1903). —Exell & Mendonça, C.F.A. **1**, 1 : 99 (1937).—E. Petit, F.C.B. **7** : 258, fig. 7F (1958). TAB. **56** fig. 11. Type from Angola (Huila).
   Polygala psammophila Gürke, tom. cit. : 279 (1903) non Chod. & Huber (1901). Type from Angola (Bié).
   Polygala guerkei Chod. in Engl., Bot. Jahrb. **48** : 318 (1912). Type as above, a new name for *P. psammophila* Gürke.

Annual herb 0·5–2 m. tall with slender, wiry, green, minutely appressed-pubescent or glabrous stems. Leaves 10–20 × 0·5–1 mm., linear with needle-like points and incurved margins, sparsely pubescent or glabrous. Flowers pale blue or yellowish, pedicels 1·5–2 mm. long, in elongated, one-sided terminal racemes up to 25 cm. long with glabrous or nearly glabrous rhachis and persistent bracts and bracteoles 1 mm. long. Posterior sepal 3·5 mm. long ; wing sepals 6–7 × 4–4·5 mm., obliquely broadly elliptic, glabrous except for the occasional presence of a few hairs towards the base ; anterior sepals 3 mm. long, free. Upper petals 5 × 3·5 mm., obliquely ovate-lanceolate, notched at one side ; carina 8 × 3 mm., slightly exceeding the wings at maturity, crest 4·5 mm. long. Stamens 8. Capsule 6 × 2·5–3 mm. narrowly elliptic in outline, ciliate on the margin, marginal wing extremely narrow. Seeds 4 × 1·2 mm., shortly cylindric, nearly 2 mm. of the length consisting of a caruncle large in relation to the size of the seed and without appendages.

N. Rhodesia. B : between Lukona and the Angolan frontier, 1120 m., fl. & fr. 9.v.1925, *Pocock* 224 (PRE). W : Kamwedzi R., fl. & fr. 19.vi.1953, *Fanshawe* 100 (K).
Also in the Belgian Congo and Angola. Edges of swamps (also recorded from grasslands and woodland in Angola).

30. **Polygala hottentotta** C. Presl in Abh. Böhm. Ges. Wiss., Ser. 5, **3** : 445 (1845).—Chod. in Mém. Soc. Phys. Hist. Nat. Genève, **31**, 2 : 399, t. 30 fig. 25 (1893).—Bak. f. in Journ. Linn. Soc., Bot. **40** : 24 (1911).—Burtt Davy, F.P.F.T. **1** : 135, fig. 13 I–J (1926). TAB. **56** fig. 17. Type from S. Africa.
   Polygala abyssinica sensu Gibbs in Journ. Linn. Soc., Bot. **37** : 419 (1906).— Eyles in Trans. Roy. Soc. S. Afr. **5** : 390 (1916).—Norlindh in Bot. Notis. **1935** : 364 (1935).—Merxm. in Proc. & Trans. Rhod. Sci. Ass. **43** : 109 (1951).—Wild, Guide Fl. Vict. Falls : 143 (1953).

Perennial (sometimes annual ?) herb or shrublet 30–60 cm. tall with slender, wiry, minutely pubescent or glabrous stems. Leaves 10–40 × 0·5–2 mm., lorate to linear, glabrous or very sparsely pubescent. Flowers dull yellowish-white, pale purple or pinkish, pedicels 2–3 mm. long, in elongated, terminal racemes up to 20 cm. long with sparsely pubescent or glabrous rhachis and caducous bracts 1·4 mm. long and bracteoles about 1–1·2 mm. long. Posterior sepal 2·5–3 mm. long ; wing sepals 5–8 × 3–4·5 mm., obovate, obovate-elliptic or suborbicular, glabrous ; anterior sepals 2–2·5 mm. long, free. Upper petals c. 2·2 × 2·2 mm., almost square, with a short stalk ; carina 6 × 2·5–3 mm., crest 2·5–3 mm. long. Stamens 8. Capsule 4–5 × 3–3·5 mm., obliquely obovate-elliptic, with very narrowly winged margin, glabrous. Seeds 4 × 2 mm., ellipsoid, with white silky hairs ; caruncle oblique, appendages scarcely developed.

S. Rhodesia. W : banks of Inseze R., fl. & fr. xii.1929, *Cheesman* 113 (BM). C : Salisbury, fl. ix.1898, *Rand* 599 (BM). E : Umtali Distr., Inyamatshira Mts., 1830 m., fl. & fr. 29.x.1950, *Chase* 3081 (BM ; COI ; LISC ; SRGH). S : Fort Victoria, fl. & fr., 1909, *Monro* 835 (BM). **Mozambique.** LM : Goba, Maputo, fl. & fr. 13.xii.1947, *Barbosa* 713 (BM ; LISC).
Also in S. Africa. Grassland and woodland from low altitudes to 1830 m.

The flowers are similar to those of *P. abyssinica* Fresen. but the caruncle is much larger and of a different shape.

31. **Polygala seminuda** Harv. in Harv. & Sond., F.C. **1** : 86 (1860).—Exell & Mendonça, C.F.A. **1,** 1 : 98 (1937). Type from S. Africa.

Perennial herb up to about 30 cm. tall, with slender, greyish, crisped-pubescent stems. Leaves linear to linear-oblong, 10–13 × 1–1·5 mm., sparsely pubescent, tips recurved and mucronate. Flowers white, pedicels 1–2 mm. long, in elongated, terminal racemes up to 10 cm. long with sparsely pubescent rhachis and caducous bracts and bracteoles. Posterior sepal 2 mm. long ; wing sepals narrowly obovate-elliptic, 6 × 2·5 mm., glabrous, rounded at the apex ; anterior sepals free, 2 mm. long. Upper petals 3 mm. long ; carina 3 mm. long, crest 1 mm. long. Stamens 8. Capsule 5 × 3·5 mm., narrowly obovate in outline, glabrous, margin narrowly winged. Seeds cylindric, 3 × 1·5 mm., with dense, white, silky hairs ; caruncular appendages about 0·5 mm. long.

Bechuanaland Prot. N : Ngamiland, near Kwebe, fl. & fr. xii.1896, *Lugard* 84 (K). Also in Angola, SW. Africa and Cape Province (Namaqualand).

Clearly very near to *P. hottentotta* and perhaps no more than a semi-desert ecotype.

32. **Polygala spicata** Chod. in Mém. Soc. Phys. Hist. Nat. Genève, **31,** 2 : 221, t. 23 fig. 36–7 (1893).—R.E.Fr. in Wiss. Ergebn. Schwed. Rhod.-Kongo-Exped. **1** : 112 (1914).—Exell & Mendonça, C.F.A. **1,** 1 : 97 (1937).—E. Petit, F.C.B. **7** : 270 (1958). TAB. **57** fig. 14. Type from Angola (Huila).
    Polygala capillaris var. *angolensis* Oliv., Fl. Trop. Afr. **1** : 131 (1868). Type from Angola (Huila).
    Polygala schlechteri Schinz in Viert. Naturf. Ges. Zür. **55** : 237 (1910).—Burtt Davy, F.P.F.T. **1** : 133 (1926). Type from the Transvaal.

Annual herb up to 40 cm. tall with slender, wiry, glabrous stems. Leaves 5–8 × 0·5–1 mm., linear to lorate, glabrous, often scarcely developed, sometimes a little broader at the extreme base of the stem. Flowers small, white or pale greenish-yellow, pedicels 0·5–1 mm. long, in many-flowered terminal racemes up to 10 cm. long with glabrous rhachis and threadlike, caducous bracts 1–1·5 mm. long and bracteoles 0·8 mm. long. Posterior sepal 1 mm. long ; wing sepals 2–2·5 × 1·3–1·5 mm., broadly elliptic to rotund, glabrous ; anterior sepals 1 mm. long, free. Upper petals 1·3 × 0·8 mm., broadly elliptic ; carina 1·5 × 0·8 mm. crest 0·6 mm. long. Stamens 8. Capsules 1 × 1 mm., suborbicular in outline, glabrous. Seeds 0·5–0·8 × 0·4–0·5 mm., ellipsoid, with white or brownish hairs ; caruncle absent.

N. Rhodesia. N : Abercorn Distr., Old Tsanga Road, 1520 m., fl. & fr. 11.i.1952, *Richards* 379 (K). W : Mwinilunga Distr., near source of Musangila R., fl. & fr. 14.x.1937, *Milne-Redhead* 2771 (BM ; K). **S. Rhodesia.** C : Marandellas, Diggleford, fl. & fr. 15.iii.1948, *Corby* 32 (BM ; SRGH).
    Also in Angola and the Transvaal. Boggy upland grasslands.

33. **Polygala filicaulis** Baill. in Bull. Soc. Linn. Par. **1** : 608 (1886).—Chod. in Mém. Soc. Phys. Hist. Nat. Genève, **31,** 2 : 221, t. 23 fig. 34–35 (1893). Type from Madagascar.

Annual herb up to 40 cm. tall with slender, wiry or filiform, glabrous stems. Leaves 5–12 × 0·5–1 mm., linear, with needle-like tips. Flowers small, purplish, pedicels very short, in terminal, congested racemes 2–5 cm. long with glabrous rhachis and very small, caducous bracts and bracteoles. Posterior sepal 1·2 mm. long ; wing sepals 1·8–2·2 × 0·6–0·9 mm., elliptic to broadly elliptic ; anterior sepals 1 mm. long, free. Upper petals 0·6 mm. long ; carina 1 × 0·8 mm., crest 0·5 mm. long. Capsule 1 × 0·8 mm. Seeds 0·8 × 0·5 mm., ellipsoid, with a few minute white hairs and no caruncle.

Mozambique. Z : Maganja da Costa, fl. & fr. 14.ix.1944, *Mendonça* 2064 (BM ; LISC).
Also in Madagascar. Coastal swamps.

34. **Polygala claessensii** Chod. in Bull. Soc. Bot. Genève, Sér. 2, **5** : 190 (1913).—Exell & Mendonça, C.F.A. **1,** 1 : 98 (1937).—E. Petit, F.C.B. **7** : 256, fig. 7D (1958). TAB. **56** fig. 8. Type from the Belgian Congo.

Annual herb up to 50–60 cm. tall with slender, densely patent-pubescent stems. Leaves 12–20 × 0·5–1·5 mm., linear to lorate, occasionally shorter and broader and somewhat crowded near the base of the stem. Flowers yellow, pedicels 1·5–2·5 mm. long, in elongated, terminal racemes up to 12 cm. long with patent-pubescent rhachis, persistent bracts 1 mm. long and bracteoles 0·6 mm. long. Posterior sepal 2 mm. long ; wing sepals 4–5 × 3·5–4·5 mm., obliquely rotund to sub-orbicular, densely pubescent ; anterior sepals 2 mm. long, free. Upper petals 2·5 × 1·5 mm., irregularly elliptic ; carina 5 × 2·5 mm., crest 1·5 mm. long. Stamens 8. Capsule 2 × 2 mm., suborbicular in outline, pubescent with a very narrow marginal wing. Seeds 1·5 × 0·8 mm., ellipsoid, with short, white, silky hairs ; caruncular appendages very short.

N. Rhodesia. N : Shiwa Ngandu, 1670 m., fl. 19.vii.1938, *Greenway* 5402 (BM ; K). W : Mwinilunga Distr., Kasompa R., fl. 31.x.1937, *Milne-Redhead* 3025 (BM ; K). Also in Angola and the Belgian Congo. In bogs in upland grassland.

35. **Polygala ukirensis** Gürke in Engl., Bot. Jahrb. **14** : 310 (1891).—E. Petit, F.C.B. **7** : 256 (1958). TAB. **56** fig. 12. Type from Tanganyika.

Annual herb up to 50–60 cm. tall with very slender, wiry, appressed-pubescent stems. Leaves 10–25 × 0·5–1 mm., linear, glabrous. Flowers golden-yellow or brownish tinged with smoky blue, pedicels 1·5–2 mm. long, in elongated, terminal racemes up to 15 cm. long with pubescent rhachis and persistent bracts 1·5 mm. long and bracteoles 1 mm. long. Posterior sepal 2·5 × 1·3 mm. ; wing sepals 4·5–5·5 × 3·2–3·7 mm., broadly elliptic, nearly glabrous, rounded or slightly emarginate at the apex ; anterior sepals 2 × 1 mm., upper petals 3 × 1·7 mm., irregularly obovate, with a short, stump-like projection at the apex ; carina 4–5 × 2·8 mm., yellow, crest 2 mm. long. Stamens 8. Capsule 4 × 2·5 mm., narrowly obovate, pubescent. Seeds 2 × 1 mm., with long, white, silky hairs ; caruncular appendages scarcely developed.

N. Rhodesia. N : Lunzua Valley, Kafakula Village, 910 m., fl. 5.iii.1953, *Richards* 4777 (BM ; K). W : Solwezi Distr., Mutanda Bridge, fl. & fr. 20.vi.1930, *Milne-Redhead* 544 (K). **Nyasaland.** N : Mzimba Distr., Mlawa, fl. 7.iv.1955, *Jackson* 1614 (BM ; LISC).
Also from the Belgian Congo and Tanganyika. Wet places, sandy soil.
The crushed roots smell of wintergreen.

36. **Polygala myriantha** Chod. in Engl., Bot. Jahrb. **48** : 321 (1912).—Exell in Journ. of Bot. **64** : 302 (1926).—Exell & Mendonça, C.F.A. **1,** 1 : 96 (1937).—E. Petit, F.C.B. **7** : 267, fig. 7H (1958). TAB. **57** fig. 12. Type from the Cameroons.

Annual herb up to 40 cm. tall usually branched from near the base with slender crisped-pubescent stems. Leaves 7–30 × 1–3 mm., linear to lorate or linear-elliptic, sparsely pubescent or glabrous, mucronate at the apex. Flowers mauve or lilac (usually drying yellowish-white), pedicels 0·5–1 mm. long, in elongated, terminal racemes up to 12 cm. long with crisped-pubescent or almost glabrous rhachis and caducous bracts 1·8 mm. long and bracteoles 1 mm. long. Posterior sepal 2 × 0·8 mm. ; wing sepals 3 × 1·2 mm., obliquely, narrowly obovate, marginally ciliolate ; anterior sepals 1–1·2 mm. long, free. Upper petals 1 × 0·8 mm., irregularly and obliquely oblong-ovate ; carina 1·8 × 0·8 mm., crest 0·8 mm. long. Stamens 8. Capsule 2 × 1·2 mm., obovate-oblong in outline, margin very narrowly winged. Seed 1·5 × 0·6 mm., cylindric, with very short appressed hairs ; caruncle c. 0·3 mm. long with very short appendages.

N. Rhodesia. W : Mwinilunga, near Matonchi R., fl. 12.i.1938, *Milne-Redhead* 4545 (K). C : 25 miles SW. of Serenje Corner, 1610 m., fl. & fr. 15.vii.1930, *Hutchinson & Gillett* 3702 (K). S : Mumbwa, fl. & fr. 1911, *Macaulay* 678 (K). **S. Rhodesia.** N : Karoi Exp. Farm, fl. & fr. 15.vi.1946, *Wild* 1160 (BM ; SRGH). **Nyasaland.** N : Mzimba Distr., Mlawa, fl. & fr. iii.1954, *Jackson* 1233 (BM). S : Lower Kasupe, 700 m., fl. & fr. 13.iii.1955, *E.M. & W.* 820 (BM ; LISC ; SRGH).

Also in the Cameroons, Belgian Congo, Angola, Uganda and Tanganyika. In *Brachystegia* woodland.

37. **Polygala youngii** Exell in Journ. of Bot. **73** : 227 (1935).—Exell & Mendonça, C.F.A. **1**, 1 : 101 (1937). Type from Angola (Lunda).

Annual herb 40–60 cm. tall with very slender, crisped-pubescent, pilose or nearly glabrous stems. Leaves 10–12 × 0·5 cm., linear, with needle-like tips. Flowers yellowish, pink or blue, pedicels 1·5 mm. long, in terminal racemes 4–15 cm. long with glabrous rhachis, caducous bracts 1 mm. long and bracteoles 0·5 mm. long. Posterior sepal 3 mm. long ; wing sepals 5–6 × 3–3·5 mm., oblong-elliptic, glabrous ; anterior sepals 2 mm. long, free. Upper petals 5–6 × 2·5 mm., obliquely oblong ; carina 3·5 mm. long, crest 5 mm. long. Stamens 6 fertile. Capsule 3 × 2 mm., obovate in outline, marginal wing very narrow, sparsely ciliate. Seed 1·5–1·6 × 0·8–0·9 mm., cylindric, with white, silky hairs ; caruncular appendages scarcely developed.

N. Rhodesia. W : Mwinilunga Distr., SW. of Dobeka Bridge, fl. 9.xi.1937, *Milne-Redhead* 3324 (BM ; K).

Also in Angola. Boggy upland or grassland.

38. **Polygala africana** Chod. in Mém. Soc. Phys. Hist. Nat. Genève, **31**, 2 : 168, t. 21 fig. 20–21 (1893).—Eyles in Trans. Roy. Soc. S. Afr. **5** : 390 (1916) pro parte excl. syn.—Burtt Davy, F.P.F.T. **1** : 133 (1926).—Norlindh in Bot. Notis. **1935** : 360 (1935).—Exell & Mendonça C.F.A. **1**, 1 : 96 (1937).—Martineau, Rhod. Wild. Fl. : 44, t. 14 fig. 2 (1954).—E. Petit, F.C.B. **7** : 268 (1958). TAB. **57** fig. 13, TAB. **58** fig. B. Type from Angola.

Polygala capillaris E. Mey. ex Harv. in Harv. & Sond., F.C. **1** : 93 (1860) pro parte quoad specim. " Burke and Zeyher ".—Milne-Redh. in Mem. N.Y. Bot. Gard. **8**, 3 : 219 (1953).

Small annual herb usually about 8–12 cm. tall, often much branched with threadlike, glabrous stems. Leaves up to about 5 × 0·5 mm., linear or filiform, with sharp, often uncinate, tips. Flowers pink, pedicels very short, in dense, terminal racemes about 1–2 cm. long with glabrous rhachis and very small caducous bracts and bracteoles. Posterior sepal 0·7 mm. long ; wing sepals 1–1·5 × 1–1·2 mm., suborbicular, margins minutely ciliate ; anterior sepals 0·5 mm. long, free. Upper petals 0·8 mm. long, oblong to lanceolate ; carina 1 mm. long, crest 0·6 mm. long. Capsule 1·2 × 1·2 mm. Seeds 0·6 × 0·5–0·6 mm., subglobose, with minute glochidiate hairs ; caruncle absent.

N. Rhodesia. B: between Kassassa and the Angolan frontier, 1120 m., fl. & fr. 10.v.1925, *Pocock* 247 (BM ; PRE). N : Abercorn Distr., Old Boma gardens, 1670 m., fl. & fr. 10.v.1955, *Richards* 5632 (BM ; K). W : Mwinilunga Distr., Kalenda Dambo, fl. & fr. 1.i.1938, *Milne-Redhead* 3910 (BM ; K). S : Victoria Falls, fl. & fr. 25.iv.1932, *St. Clair-Thompson* 1317 (K). **S. Rhodesia.** W : Matopos, near World's View, fl. & fr. 14.iv.1955, *E.M. & W.* 1514 (BM ; LISC ; SRGH). C : Salisbury, 1520 m., fl. & fr. ii.1919, *Eyles* 1480 (BM). E : Penhalonga and Vumba (according to Martineau, loc. cit.). S : Fort Victoria, *Monro* 1770 (BM). **Nyasaland.** C : Kota Kota, fl. & fr. 1.ix.1946, *Brass* 17472 (K). **Mozambique.** N : R. Chitope, Massangulo, fl. & fr. iii.1933, *Gomes e Sousa* 1272 (COI).

Also in Angola, Belgian Congo, Tanganyika and the Transvaal. Annual herb of boggy grassland up to 1670 m.

While the seeds of nearly all *Polygala* spp. in our area have well-developed caruncles and long silky hairs, the latter greatly facilitating their passage " head first " but obstructing any backward movement, in this species, usually found in swamps (dambos, vleis), there is no caruncle and the hairs are glochidiate. In most species the seeds are probably distributed by ants, which are said to drag them away by the caruncle ; but in the swampy conditions in which *P. africana* is found this would be difficult or impossible. The hairs thus appear to have become modified for adhesion to the coats of animals and the now useless caruncle has disappeared. Field observations of seed distribution in this and other species would be interesting.

39. **Polygala capillaris** E. Mey. [ex Drège, Zwei Pflanz.-Docum.: 212 (1843) *nom. nud.*] ex Harv. in Harv. & Sond., F.C. **1** : 93 (1860) emend. excl. specim. " Burke and Zeyher ".—Chod. in Mém. Soc. Phys. Hist. Nat. Genève, **31**, 2 : 220, t. 23 fig. 32–33 (1893).—Exell & Mendonça, C.F.A. **1**, 1 : 97 (1937).—E. Petit, F.C.B. **7** : 269 (1958). TAB. **57** fig. 15. Type from S. Africa.

Annual herb up to 50 cm. tall with filiform, glabrous stems. Leaves 3–7 × 0·5–1 mm., linear to lorate. Flowers greenish-white or purplish with very short pedicels, in elongated, terminal racemes up to 6–7 cm. long with glabrous rhachis and very small, caducous bracts and bracteoles. Posterior sepal 1·5 mm. long ; wing sepals 1·5–2 × 1 mm., elliptic, glabrous ; anterior sepals 0·8 mm. long, free. Upper petals 1·5 mm. long, lanceolate ; carina 1·5 mm. long, crest 0·5 mm. long. Capsule 1·5 × 1·2 mm., glabrous. Seeds 1 × 0·6 mm., ellipsoid, with minute, white, appressed hairs ; caruncle absent.

N. Rhodesia. S : 24 km. N. of Mapanza, 1000 m., fl. & fr. 21.i.1954, *Robinson* 482 (K). **Nyasaland.** S : Nyambi, Sinjani Dambo, fl. & fr. 26.iv.1955, *Jackson* 1646 (BM). **Mozambique.** N : Mecubúri, fl. & fr. 21.xi.1936, *Torre* 1079 (COI ; LISC). MS : Sofala, 40 km. from the road along the R. Save towards the R. Buzi, fl. & fr. 2.ix.1942, *Mendonça* 104 (BM ; LISC). SS : Inhambane, fl. & fr. xii.1898, *Schlechter* 12088 (BM ; K).

Also in Angola, Belgian Congo, Tanganyika and S. Africa (Natal and Cape Prov.). Swamps from sea-level to 1600 m. (in SE. Africa).

40. **Polygala homblei** Exell in Journ. of Bot. **70** : 186 (1932).—E. Petit, F.C.B. **7** : 273 (1958). Type from the Belgian Congo.

Perennial sending up annual shoots 10–16 cm. tall from a woody rootstock and forming large tufts ; stems pubescent. Leaves 5–25 × 1·5–3·5 mm., linear-elliptic to narrowly elliptic, with mucronate tips. Flowers dull mauve, shaded greenish outside and cream-coloured inside with glabrous pedicels 4 mm. long, in short, terminal racemes with puberulous rhachis, caducous, scarious bracts 1·5–2 mm. long and bracteoles 1 mm. long. Posterior sepal 4 mm. long ; wing sepals 10–11 × 7–9 mm., broadly elliptic to ovate, glabrous or almost so ; anterior sepals 3·5 mm. long, connate almost to the apex. Upper petals 9–10 × 7 mm., obovate-spathulate, glabrous ; carina 11 × 5 mm., crest 2–3 mm. long. Stamens 8. Capsule 8–10 × 7–8 mm., broadly oblong-elliptic, glabrous, with a winged margin 1·5 mm. broad. Seeds (immature) with long, white, silky hairs ; caruncle well-developed.

N. Rhodesia. W : Solwezi Distr., Lumwana R., fl. & fr. 14.viii.1930, *Milne-Redhead* 1136 (BM ; K).

Also in the Belgian Congo. In open, stony ground (on termite mounds in the Belgian Congo).

41. **Polygala marensis** Burtt Davy, F.P.F.T. **1** : 48 (1926). TAB. **57** fig. 4. Type from the Transvaal.

Polygala rogersii Burtt Davy, tom. cit. : 133, 134 (1926). Same type as above.

Prostrate shrublet of greyish appearance with densely pubescent branchlets. Leaves oblanceolate or spathulate, normally 5–25 × 1–6 mm. (extremely diminutive leaves are occasionally found). Flowers blue, pedicels 2–2·5 mm. long, in terminal racemes up to 15 cm. long with pubescent rhachis, caducous bracts 2 mm. long and bracteoles 1 mm. long. Posterior sepal 2·5–3 mm. long ; wing sepals 6–8 × 4 mm., glabrous ; anterior sepals 2–2·5 mm. long, connate. Upper petals 4·5 × 3 mm., spathulate or obovate ; carina 8 × 4 mm., crest 1·5–2 mm. long. Stamens 8. Capsule oblong, oblique, 7 × 3·5–4 mm., glabrous, apical lobes pointed, marginal wing very narrow except at the apex. Seeds 5 × 1·5 mm., cylindric, with a dense indumentum of long, white, silky hairs ; caruncle well-developed but appendages minute.

Mozambique. LM : Maputo, Bela Vista, fl. 16.xi.1944, *Mendonça* 2885 (BM ; LISC).

Also in the Transvaal. In clearings in secondary bush on white sand and in sandveld in the Transvaal.

Tab. 58. A.—POLYGALA VIRGATA VAR. DECORA. A1, flowering stem (× ⅔) ; A2, upper
petal (× 2) ; A3, carina and crest (× 2) ; A4, androecium (× 2) ; A5, ovary with
style and stigma (× 2) ; A6, capsule (× 2). All from *Eyles* 1688. B.—POLYGALA
AFRICANA, flowering stem (× ⅔) *E.M. & W.* 1514. C.—POLYGALA ARENARIA,
flowering stem (× ⅔) *E.M. & W.* 787.

Burtt Davy probably intended to call this *P. rogersii* but on discovering the existence of *P. rogersiana* Bak. f. he altered the name accompanying the latin diagnosis on p. 48 (loc. cit.) but forgot to make the alterations in the general text. It is preferable in every respect to choose the name *P. marensis*.

42. **Polygala schinziana** Chod. in Mém. Soc. Phys. Hist. Nat. Genève, **31,** 2 : 364, t. 28 fig. 38 (1893).—Exell & Mendonça, C.F.A. **1,** 1 : 103 (1937).—TAB. **57** fig. 3. Type from SW. Africa (Amboland).

Polygala sp.—Eyles in Trans. Roy. Soc. S. Afr. **5** : 392 (1916).

Perennial with woody rootstock sending up annual shoots; stems densely patent-pubescent. Leaves 5–30 × 2–7 mm., narrowly oblong-elliptic to narrowly elliptic or occasionally almost suborbicular near the base of the stem, pubescent, blunt at the apex. Flowers pinkish or mauve, pedicels 2 mm. long, in terminal racemes up to 12 cm. long with patent-pubescent rhachis and small, pubescent, scarious or subscarious caducous bracts 1·5–2 mm. long and bracteoles 0·8–1 mm. long. Posterior sepal 2 mm. long ; wing sepals 4–4·5 × 2·5 mm., obliquely elliptic, glabrous ; anterior sepals 1·5 mm. long, connate. Upper petals 2·5 × 2 mm., irregularly obovate ; carina 3·5 mm. long, crest 1·5 mm. long. Capsule 6–7 × 3–4·5 mm., oblong-elliptic, ciliolate on the margins, apical lobes rounded, marginal wing very narrow except at the apex. Seeds 4 × 1·2 mm., cylindric, with a very dense white silky indumentum ; caruncular appendages minute.

S. Rhodesia. W : Victoria Falls, fl. & fr. 12.ii.1912, *Rogers* 5698 (BM ; K). Also in Angola and SW. Africa.

Distinguished from the preceding species by the smaller flowers and more rounded apical lobes of the capsule.

43. **Polygala rehmannii** Chod. in Mém. Soc. Phys. Hist. Nat. Genève, **31,** 2 : 362, t. 28 fig. 33–34 (1893) emend. excl. var. *parviflora.*—Burtt Davy, F.P.F.T. **1** : 134 (1926) pro parte excl. specim. *Quintas* 15.—Norlindh in Bot. Notis. **1935** : 361, fig. 1 (left) (1935) pro parte excl. var. *latipetala.* TAB. **57** fig. 8. Type from the Transvaal.

Polygala gracilenta Burtt Davy, tom. cit. : 48 (1926). A new name for *P. tenuifolia* sensu Harv. non Willd.

Perennial 10–30 cm. tall with a woody rootstock ; stems glabrous or nearly so, slightly winged. Leaves 10–25 × 1·5–4 mm., linear-lanceolate or linear-elliptic, sometimes suborbicular towards the base of the stem, glabrous, usually rounded at the apex, but sometimes acute, mucronate. Flowers blue or greenish-blue, pedicels 1·5–2·5 mm. long, in terminal or lateral racemes usually 3–7 cm. long, occasionally longer, with glabrous rhachis, small caducous bracts 1 mm. long and bracteoles 0·5 mm. long. Posterior sepal 3 mm. long ; wing sepals 6–8 × 3–4 mm., elliptic, glabrous ; anterior sepals connate, 2 mm. long. Upper petals comparatively large, 6 × 5 mm., obovate, unguiculate ; carina 7 × 2·5 mm., crest 1·5 long. Capsule 7·5 × 4·5 mm., oblong-elliptic, marginal wing very narrow. Seeds 5 × 2 mm., elliptic, with very dense, white, silky indumentum ; caruncle well developed but appendages minute.

S. Rhodesia. C : Makoni Distr., Dunedin, 1800 m., fl. 9.ii.1931, *Norlindh & Weimarck* 4945 (BM). E : Umtali, Inyamatshira, 1220 m., fl. 2.xi.1951, *Chase* 4264 (BM ; LISC ; SRGH). Also in the Transvaal. Submontane grasslands up to 2300 m.

44. **Polygala stenopetala** Klotzsch in Peters Reise Mossamb. Bot. **1** : 114, t. 23 (" stenophylla ") (1861). TAB. **57** fig. 1. Type : Mozambique, Inhambane, *Peters* (B, holotype †, not seen).

Polygala rarifolia sensu Chod. in Mém. Soc. Phys. Hist. Nat. Genève, **31,** 2 : 367 (1893) pro parte quoad syn. *P. stenophylla* Klotzsch.

Polygala viminalis Gürke in Engl., Pflanzenw. Ost-Afr. C : 234 (1895).— R.E.Fr. in Wiss. Ergebn. Schwed. Rhod.-Kongo-Exped. **1** : 113 (1914).— Exell & Mendonça, C.F.A. **1,** 1 : 104 (1937).—Milne-Redh. in Mém. N.Y. Bot. Gard. **8,** 3 : 220 (1953). Type from Tanganyika.

Polygala viminalis forma *brachyptera* Chod. apud R.E.Fr., tom. cit. : 114 (1914). Type : N. Rhodesia, Abercorn, *Fries* 1257 (UPS, holotype).

Perennial herb (sometimes annual ?) or shrublet up to 1·5 m. tall with slender, erect, slightly winged, glabrous stems. Leaves 15–50 × 1–5 mm., linear, grasslike, pointed at the apex, glabrous. Flowers blue or greenish with purple-brown veining, pedicels 3–4 mm. long, in elongated, terminal racemes up to 20 cm. long

Polygala asperifolia Chod., tom. cit.: 323 (1912). Type from the Transvaal (*Galpin* 844, the same gathering as the lectotype of *P. producta*).
with glabrous rhachis and small caducous bracts 1 mm. long and bracteoles 0·8 mm. long. Posterior sepal 2 mm. long ; wing sepals 5–6·5 × 4–5·5 mm., obliquely suborbicular or rotund ; anterior sepals 1·8 mm. long, connate. Upper petals 3 × 2·5 mm., obliquely and irregularly obovate ; carina 6·5 × 3·5 mm., crest 2 mm. long. Stamens 8. Capsule 5–6 × 3·5–4·5 mm., obliquely oblong, glabrous. Seeds 2·5 × 1·5 mm., ellipsoid, with rather short, brownish, appressed hairs ; caruncular appendages absent.

N. Rhodesia. N : Upper Luangwa R., fl. & fr. x.1897, *Nicholson* (K). W : Solwezi Dambo, fl. 20.ix.1930, *Milne-Redhead* 1161 (K). **Nyasaland.** N : Vipya, fl. & fr. iv.1956, *Chapman* 279 (BM). S : Zomba Mt., fl. & fr. 22.ii.1956, *Banda* 179 (BM ; LISC). **Mozambique.** N : Metónia, fl. & fr. 10.x.1942, *Mendonça* 706 (BM ; LISC). Z : Quelimane Distr., Mgulumi Mission, *Faulkner* 62 (K).
Also in Angola and Tanganyika. In *Brachystegia* woodland, wooded grassland and submontane grassland.

The record from N. Rhodesia (W), *Milne-Redhead* 1161, is doubtful as the specimen is an immature one from a dense tuft which has been much burnt. It has, however, the characteristic, slightly winged stems. Klotzsch (loc. cit.) employed the epithet " stenopetala " in his text but " stenophylla " on his figure. *P. stenophylla* would have been a more appropriate name and may have been originally intended ; but Klotzsch no doubt discovered that it was preoccupied by *P. stenophylla* A. Gray (1854) and made the minimum alteration in his text but was perhaps unable to, or forgot to, alter the name on the figure.

45. **Polygala uncinata** E. Mey. ex Meisn. in Hook., Lond. Journ. Bot. **1** : 468 (1842). TAB. **57** fig. 9. Type from S. Africa (Port Natal).
Polygala tenuifolia var. *uncinata* (E. Mey. ex Meisn.) Harv. in Harv. & Sond., F.C. **1** : 88 (1860). Type as above.
Polygala rigens sensu Chod. in Mém. Soc. Phys. Hist. Nat. Genève, **31**, 2 : 361 (1893) pro parte quoad syn. *P. uncinata*.—Gibbs in Journ. Linn. Soc., Bot. **37** : 430 (1906).
Polygala latipetala N.E.Br. in Kew Bull. **1906** : 98 (1906).—Eyles in Trans. Roy. Soc. S. Afr. **5** : 391 (1916). Type : S. Rhodesia, between Umtali and Salisbury, *Cecil* 45 (K, holotype).
Polygala rarifolia sensu Eyles, loc. cit.
Polygala hamata Burtt Davy, F.P.F.T. **1** : 48 (1926). Same type as *P. uncinata*.
Polygala rehmannii var. *latipetala* (N.E.Br.) Norlindh in Bot. Notis. **1935** : 362, fig. 1 (right) (1935). Same type as *P. latipetala*.

Perennial sending up annual shoots from a woody rootstock or shrublet with slender, minutely pubescent, angled or slightly winged stems. Leaves 3–15 × 0·5–2·5 mm., linear to narrowly elliptic or linear-spathulate, minutely pubescent or glabrous, usually rounded at the apex with a recurved mucro. Flowers purple or blue, pedicels 2–2·5 mm. long, in terminal racemes 3–10 cm. long with minutely pubescent rhachis and minute, caducous bracts and bracteoles. Posterior sepal 2·5 mm. long ; wing sepals 5 × 3 mm., oblong-elliptic, blunt or slightly retuse at the apex, glabrous ; anterior sepals 2 mm. long, connate. Upper petals 4·5 × 2·5 mm., obovate-spathulate ; carina 5 × 2·5 mm., crest 1·5 nm. long. Stamens 8. Capsule 5·5 × 3·5–4 mm., broadly oblong-elliptic, glabrous, with a narrow marginal wing. Seeds 3·5 × 1·2 mm., elliptic, with dense, white, silky indumentum ; caruncle 1 mm. long, appendages absent.

S. Rhodesia. N : Mazoe, 1310 m., fl. & fr. xii.1906, *Eyles* 484 (BM). W : Matobo Distr., Besna Kobila, fl. & fr. i.1954, *Miller* 2071 (BM ; SRGH). C : 10 km. N. of Rusape, Maidstone, 1450 m., fl. & fr. 6.i.1931, *Norlindh & Weimarck* 4136 (BM). E : Chimanimani, 1520 m., fl. 26.ii.1907, *Johnson* 212 (K). S : Fort Victoria, *Monro* 1354 (BM). **Mozambique.** SS : Inhambane, Govuro, Mambone, fl. & fr. 2.ix.1942, *Mendonça* 102 (BM ; LISC). LM : Inhaca I., fl. 13.vii.1934, *Weintroub* (PRE).
Also in S. Africa (Transvaal, Orange Free State, Cape Prov.). Grasslands and edges of swamps from sea-level to 1520 m.

46. **Polygala producta** N.E.Br. in Kew Bull. **1895** : 142 (1895) emend. Burtt Davy in Kew Bull. **1924** : 225 (1924) ; F.P.F.T. **1** : 134 (1926). TAB. **57** fig. 2. Type from the Transvaal (*Galpin* 844 selected by Burtt Davy, loc. cit., 1924).
Polygala rehmannii var. *parviflora* Chod. in Mém. Soc. Phys. Hist. Nat. Genève, **31**, 2 : 362 (1893). Type from the Transvaal (*Rehmann* 4563).
Polygala rehmannii var. *gymnoptera* Chod. in Engl., Bot. Jahrb. **48** : 321 (1912). Type : Mozambique, Lourenço Marques, *Quintas* 15 (B † ; COI, lectot

Shrublet (or sometimes an annual herb ?) up to 1·5 m. tall with slender, minutely crisped-pubescent, slightly winged stems. Leaves 10–30 × 1–5·5 mm., linear to lorate, usually sparsely minutely hairy, somewhat scabrid on the edges and upper surface, rounded or subacute at the apex, and usually mucronate. Flowers pink or purple, pedicels 2·5–3·5 mm. long, in elongated terminal racemes up to 20 cm. long with minutely pubescent or almost glabrous rhachis, small caducous bracts 0·8 mm. long and bracteoles 0·5 mm. long. Posterior sepal 2 mm. long ; wing sepals 4·5–5 × 2·5–3 mm., obliquely ovate, glabrous except for minute cilia round the margin ; anterior sepals 1·5 mm. long, connate. Upper petals 4 × 2 mm., irregularly obovate-spathulate ; carina 4 × 1·8 mm., crest 1·5 mm. long. Stamens 8. Capsule 5 × 4 mm., broadly elliptic, glabrous, marginal wing very narrow. Seeds 3 × 1·8 mm., oblong-elliptic ; caruncle 0·8 mm. long, appendages very small or absent.

S. Rhodesia. E : Inyanga, Pungwe River Valley, fl. & fr. 17.vii.1948, *Chase* 844 (BM ; SRGH). S : Old Bikita, 1310 m., fl. & fr. 16.xii.1953, *Wild* 4400 (BM ; SRGH). **Mozambique.** MS : Chimoio, Monte de Belas, fl. & fr. 2.iii.1948, *Garcia* 466 (BM ; LISC). LM : Namaacha Falls, fl. & fr. 22.ii.1955, *E.M. & W.* 539 (BM ; LISC ; SRGH).
Also in the Transvaal. In patches of evergreen forest and in secondary thickets, grassland and abandoned cultivation from sea-level to about 1300 m.

P. producta can be distinguished from *P. rehmannii* by its smaller flowers and more elongated racemes. Specimens which appear to be hybrids between these two species have been collected in the Transvaal but in our area, although there is some overlap, *P. rehmannii* usually occurs at much higher altitudes.

47. **Polygala rivularis** Gürke in Warb., Kunene-Samb.-Exped. Baum : 278 (1903).— Exell & Mendonça, C.F.A. **1**, 1 : 105 (1937). TAB. **57** fig. 11. Type from Angola (Bié).

Erect, annual herb, up to 100 cm. tall with glabrous, often almost leafless stems. Leaves 5–25 × 0·3–0·8 mm., linear to filiform, glabrous, with needle-like tips. Flowers blue, purplish or white, often with purplish-brown-veined wings and white or bluish petals, pedicels 1·5–2 mm. long, in elongated, terminal racemes up to 15 cm. long, with glabrous rhachis, caducous bracts 1·2 mm. long and bracteoles 0·6 mm. long. Posterior sepal 1·6 mm. long ; wing sepals 2·5–3 × 1·8–2·5 mm., obliquely obovate to suborbicular, glabrous ; anterior sepals 1·5 mm. long, connate. Upper petals 2·5 × 1·5 mm., obovate-spathulate ; carina 3 × 1·5 mm., crest 1·5 mm. long. Stamens 8. Capsule 3 × 2 mm., oblong, glabrous, apical lobes pointed, marginal wing scarcely developed. Seeds 4 × 2 mm., ellipsoid ; caruncle 1·2 mm. long, appendages scarcely developed.

N. Rhodesia. N : Kasama-Abercorn road, 72 km. from Abercorn, 1520 m., fl. 30.iii.1955, *Richards* 5216 (BM ; K). W : Mwinilunga Distr., Dobeka Bridge, fl. 11.xii.1937, *Milne-Redhead* 3584a (K).
Also in Angola.

48. **Polygala xanthina** Chod. in Engl., Bot. Jahrb. **48** : 325 (1912).—E. Petit, F.C.B. **7** : 276 (1958). Type from the Belgian Congo (Katanga).

Annual with glabrous, ridged, arcuate-ascending stems, up to 50 cm. tall. Leaves 2–8 × 0·4–1·4 cm., linear-elliptic to very narrowly elliptic, slightly scabrid on the margins, otherwise glabrous. Flowers greenish-purple, pedicels 1·5–2 mm. long, in elongated, terminal racemes up to 20 cm. long with glabrous rhachis, caducous bracts 1·3–2 mm. long and bracteoles 0·5–0·8 mm. long. Posterior sepal 3 mm. long ; wing sepals 3·5–5 × 2·5–3 mm., ovate-elliptic, unguiculate, glabrous ; anterior sepals 2 mm. long, connate. Upper petals 4·5 × 3 mm., obliquely cuneate, slightly emarginate at the apex, salmon-pink or purple-pink ; carina 5 mm. long, greenish or pinkish-purple tipped with blue ; crest absent. Stamens 8. Capsule 4·5–5 × 3·5–4 mm., oblong, glabrous, margin narrowly winged. Seeds 3 × 1·2 mm., ellipsoid, with dense, white, silky hairs ; caruncle 0·8 mm. long, appendages scarcely developed.

N. Rhodesia. N : Abercorn Distr., above Howser's Farm, 1830 m., fl. 23.iii.1955, *Richards* 5075 (BM ; K). **Nyasaland.** C : Lilongwe, Chankhandwe Dambo, fl. & fr. 11.iv.1956, *Jackson* 1836 (BM ; COI ; LISC). S : Ncheu Distr., Lower Kirk Range, 1460 m., fl. & fr. 17.iii.1955, *E.M. & W.* 952 (BM ; LISC ; SRGH).
Also in the Belgian Congo. In submontane grassland and seasonal swamps.

49. **Polygala petitiana** A. Rich., Tent. Fl. Abyss. **1** : 37 (1847).—Chod. in Mém. Soc. Phys. Hist. Nat. Genève, **31**, 2 : 370, t. 29 fig. 12–13 (1893).—Eyles in Trans. Roy. Soc. S. Afr. **5** : 391 (1916).—Milne-Redh. in Mem. N.Y. Bot. Gard. **8,** 3 : 220 (1953).—E. Petit, F.C.B. **7** : 278, t. 29 (1958).   Type from Ethiopia.

Annual herb up to 90 cm. tall with slender, glabrous or nearly glabrous stems. Leaves 20–40 × 0·5–3 mm., linear to very narrowly elliptic, with needle-like tips, glabrous.   Flowers blue, white or yellow, pedicels 1–4 mm. long, in elongated terminal racemes up to 15 cm. long with glabrous rhachis, caducous bracts up to 2 mm. long and bracteoles up to 1 mm. long. Posterior sepal 1·5–3 mm. long ; wing sepals 2–5 × 0·8–3 mm., elliptic ; anterior sepals 1–1·5 mm. long, connate. Upper petals 1–2 × 0·8–1·5 mm., obovate-spathulate ; carina 2–4 mm. long, crest absent.   Stamens 8.   Capsule 3–4 × 2–2·5 mm., oblong-elliptic, very narrowly winged.   Seeds 1·8–2 × 0·8–1 mm., ellipsoid, with long, white, silky hairs ; caruncle 0·8 mm. long, appendages absent.

Pedicels up to 3 mm. long ; wings 4–5 × 2–3 mm. ;
   carina 3·5–4·5 mm. long   -   -   -   -   -   -   -   var. *calceolata*
Pedicels 1–2 mm. long ; wings 2–3 × 0·8–1·5 mm. ;
   carina 2–2·5 mm. long -   -   -   -   -   -   -   var. *parviflora*

Var. **calceolata** Norlindh in Bot. Notis. **1935** : 364 (1935).   Type : S. Rhodesia, Inyanga, 1300 m., *Norlindh & Weimarck* 4354 (BM, isotype ; K, isotype).

Annual herb up to 90 cm. tall.   Leaves linear-elliptic to very narrowly elliptic. Pedicels up to 3 mm. long.   Wing sepals 4–5 × 2–3 mm., elliptic ; carina 3·5–4·5 mm. long.   Capsule up to 4 × 2·5 mm.

N. Rhodesia.   W : Mwinilunga Distr., S. of Matonchi Farm, fl. & fr. 3.ii.1938, *Milne-Redhead* 4447 (BM ; G ; K).   C : Lusaka, 1220 m., fl. & fr. 5.iii.1955, *Best* 63 (BM ; K).   **S. Rhodesia.**   N : Miami, fl. & fr. iv.1926, *Rand* 28 (BM).   W : Matobo Distr., fl. & fr. ii.1957, *Garley* 141 (BM ; SRGH).   C : Marandellas, Digglefold, fl. & fr. 26.ii.1948, *Corby* 23 (BM ; SRGH).   E : Umtali, Fern Valley, 1060 m., fl. & fr. 21.ii.1954, *Chase* 5197 (BM ; SRGH).   S : Fort Victoria, fl. & fr. 1909–12, *Monro* 1354 A (BM). **Nyasaland.**   C : Lilongwe, Civuwo Dambo, fl. & fr. 13.iv.1956, *Jackson* 1840 (BM). S : Ncheu Distr., Lower Kirk Range, Chipusiri, 1460 m., fl. & fr. 17.iii.1955, *E.M. & W.* 953 (BM ; LISC ; SRGH).   **Mozambique.**   N : Manganja Hills. 12.xii.1865, *Kirk* (K).   MS : R. Mossurizi, fl. & fr. 15.ii.1907, *Johnson* 114 (K).

Also in Angola, Belgian Congo, Uganda and Tanganyika.   In *Brachystegia* woodland, upland grassland or swamps, usually on sandy soil.

Var. **parviflora** Exell in Bol. Soc. Brot. Sér. 2, **31** : 12 (1957).—Norlindh, tom. cit. : 363 (1935) (as " *petitiana* ").   TAB. **57** fig. 5.   Type : Nyasaland, Lower Kasupe, 700 m., fl. & fr. 13.iii.1955, *E.M. & W.* 819 (BM, holotype : LISC ; SRGH).
   Polygala liniflora sensu Exell in Journ. of Bot. **64** : 302 (1926).   The specimen cited, originally *Rand* 28 (in part), was later renumbered 28 A.

Annual herb up to 75 cm. tall.   Leaves linear.   Pedicels 1–2 mm. long.   Wing sepals 2–3 × 0·8–1·5 mm., elliptic, yellow (sometimes orange ?) ; carina 2–2·5 mm. long.   Capsule 3 × 2 mm.

N. Rhodesia.   N : Chilila, 1520 m., fl. & fr. 11.ii.1953, *Richards* 4453 (BM ; K). W : Ndola, fl. & fr. 10.v.1954, *Fanshawe* 1181 (K ; SRGH).   **S. Rhodesia.**   N : Miami, fl. & fr. 1.iv.1926, *Rand* 28 A (BM).   E : Inyanga, c. 1700 m., fl. 20.i.1921, *Norlindh & Weimarck* 4482 (BM ; G).   **Nyasaland.**   N : Mzimba Distr., Mlawa, fl. & fr. 10.iii.1954, *Jackson* 1226 (BM).   S : Kirk Range, Zonze Hill, 1550 m., fl. & fr. 17.iii.1955, *E.M. & W.* 981 (BM ; LISC ; SRGH).   **Mozambique.**   N : Metónia, Vila Cabral, fl. & fr. 27.vii.1934, *Torre* 749 (COI ; LISC).   T : Angónia, between Metengo Bolane and Vila Mousinho, fl. & fr. 11.v.1948, *Mendonça* 4156 (BM ; LISC).
Also in the Belgian Congo and probably in Tanganyika.   In *Brachystegia* woodland.

The two varieties are quite distinct in our area and could well be separated specifically but further north they tend to merge or rather, if Ethiopia be the place of origin, as it seems to be, the taxa tend to differentiate more distinctly the further one goes from that region.
Fanshawe 1181 has orange flowers and may eventually be separable as another variety.

50. **Polygala westii** Exell in Bol. Soc. Brot., Sér. 2, **31** : 13 (1957).   TAB. **57** fig. 10. Type : S. Rhodesia, Matobo Distr., *West* 2702 (BM, holotype).

Herb 10–12 cm. tall, stems minutely crisped-pubescent, branchlets arcuate-ascendant.   Leaves 10–45 × 1·5–4 mm., linear-elliptic, glabrous.   Flowers greenish, pedicels 1·5 mm. long, in lateral racemes 3·5 cm. long with almost glabrous rhachis and caducous bracts and bracteoles. Posterior sepal 2 mm. long ; wing sepals 4 × 2 mm., elliptic, with minutely ciliolate margin otherwise glabrous, apparently greenish-white in colour ; anterior sepals apparently connate almost to the apex but coming apart rather readily, 1·5 mm. long. Capsule 4·5–5 × 3·5–4 mm. elliptic-oblong, clearly broader than the wing sepals, margin with a narrow wing 0·2–0·5 mm. broad. Seeds 2 × 1·2 mm., cylindric, with white, silky hairs ; caruncle 0·7 mm. long with appendages 0·3 mm. long.

S. Rhodesia. W : Matobo Distr., Westacre Creek, fr. 26.ii.1948, *West* 2702 (BM ; SRGH).

Very little material was available for examination. The anterior sepals seem to be truly joined ; otherwise the affinity might be rather with the first four species enumerated here as the capsule is broader than the wings.

51. **Polygala kalaxariensis** Schinz in Verh. Bot. Verein. Brand. **29** : 52 (1888).— Exell & Mendonça, C.F.A. **1**, 1 : 106 (1937).—E. Petit, F.C.B. **7** : 274 (1958). Type from SW. Africa.

Procumbent, perennial herb or shrublet with minutely pubescent or nearly glabrous stems and branches. Leaves 8–15 × 1–3 mm. (often much broader in Angolan specimens), linear-oblong to narrowly oblong, minutely pubescent or nearly glabrous, blunt and mucronate at the apex. Flowers blue, pedicels rather stout up to 6 mm. long, in short, lateral racemes up to 2 cm. long or solitary, with pubescent rhachis and caducous bracts and bracteoles. Posterior sepal 2·5–3 mm. long ; wing sepals 6–7 × 3·5–5 mm., greenish, obliquely elliptic or ovate-elliptic ; anterior sepals 2·5 mm. long, connate. Upper petals 5 × 4 mm., obcuneate ; carina 7 × 3 mm., minutely pubescent on the margin towards the base, crest 2 mm. long. Stamens 6 fertile and 2 staminodes. Capsule 6 × 3·5 mm., elliptic-oblong, minutely pubescent near the margins, margin very narrowly winged. Seeds 3 × 1·6 mm., cylindric, sparsely white-sericeous ; caruncle 1·3 mm. long with very small appendages.

N. Rhodesia. B : near Kassassa, 1120 m., fl. & fr. 10.v.1925, *Pocock* 244 (PRE). **S. Rhodesia.** W : Bulawayo, fl. xii.1897, *Rand* 15 (BM). C : Marandellas, fl. & fr. 7.iv.1948, *Corby* 60 (BM ; SRGH).
Also in SW. Africa, Angola and the Belgian Congo. In upland grasslands on sandy soils, up to 1520 m.

52. **Polygala baumii** Gürke in Warb., Kunene-Samb.-Exped. Baum : 276 (1903).— Exell & Mendonça, C.F.A. **1**, 1 : 106 (1937). TAB. **57** fig. 6. Type from Angola (Bié).

Shrublet up to 30–40 cm. tall of ericoid appearance with glabrous, slightly winged stems. Leaves 8–15 × 0·5–2 mm., linear to very narrowly oblanceolate, glabrous, apex often uncinate, margin usually incurved. Flowers greenish or yellowish with blue crest, pedicels up to 3 mm. long, in dense, subcapitate, lateral racemes about 2 cm. in diam., with glabrous rhachis, bracts 1·5–2 mm. long and bracteoles 1 mm. long, persistent for a time then caducous. Posterior sepal 3–3·5 mm. long ; wing sepals 7–10 × 5–5·5 mm., obliquely ovate-elliptic, ciliolate on the margin ; anterior sepals 2·5–3 mm. long, connate. Upper petals 4 × 2·5 mm., obovate-cuneate, contorted ; carina 6 × 2 mm., crest 2–2·5 mm. long. Stamens 6 fertile with 2 staminodes. Capsule 4·5 × 2·5 mm., oblong-elliptic, glabrous, with very narrowly winged margin. Seeds 2·8–3 × 1·2–1·7 mm., ellipsoid, with white, silky hairs ; caruncle 1 mm. long with small appendages.

N. Rhodesia. W : Kitwe, fl. & fr. 10.v.1955, *Fanshawe* 2266 (BM ; K). C : Kapiri Mposhi, fl. & fr. 13.vii.1932, *Young* 32 (BM).
Also in Angola. No information as to habitat but found in grassland with shrubs in Angola.

53. **Polygala gossweileri** Exell & Mendonça in Journ. of Bot. **74** : 133 (1936) ; C.F.A. **1**, 1 : 107 (1937). Type from Angola (Bié).

Perennial herb with woody base, about 15 cm. tall. Stems slender, striate, glabrous. Leaves 8–14 × 0·5–1 mm., linear, with needle-like points, glabrous. Flowers blue, pedicels 5–7 mm. long, solitary, or in few-flowered, lateral racemes about 1·5 cm. long. Posterior sepal 2·2 mm. long ; wing sepals 5 × 2·5 mm., obliquely elliptic, acute, with ciliolate margins ; anterior sepals 2 mm. long, connate. Upper petals 4 × 3·3 mm., obovate. Stamens 6 fertile with 2 staminodes. Capsule 2·5–3 × 2–2·5 mm., margin narrowly winged. Seeds 2 × 0·8 mm., cylindric, white-sericeous ; caruncle 0·7 mm. long with a very short appendage.

N. Rhodesia. W : Chingola, fl. 18.i.1956, *Fanshawe* 2744 (K). Also in Angola and S. Tanganyika.

54. **Polygala dewevrei** Exell in Journ. of Bot. **70** : 186 (1932).—Exell & Mendonça, C.F.A. **1**, 2 : 366 (1951).—E. Petit, F.C.B. **7** : 271 (1958). TAB. **57** fig. 7. Type from the Belgian Congo.

Much branched annual herb up to 60 cm. tall with slender, glabrous or almost glabrous stems. Leaves 10–20 × 0·3–1 mm., linear to almost acicular, with incurved margins. Flowers yellowish-green and blue, pedicels 2–3 mm. long, in subcapitate, lateral racemes about 1 cm. in diam. with glabrous rhachis and small caducous bracts and bracteoles c. 1 mm. long. Posterior sepal 1·5 mm. long ; wing sepals 4–5 × 2–2·5 mm., obliquely elliptic, acute at the apex, glabrous ; anterior sepals 1·2 mm. long, connate. Upper petals 1·8 × 1 mm., broadly oblong ; carina 4·5 × 2 mm., crest 1·5 mm. long. Stamens 6 fertile and 2 staminodes. Capsule 2–2·5 × 1·8–2 mm., suborbicular in outline, glabrous, marginal wing 0·5 mm. broad. Seeds (somewhat immature) 1·5 × 0·8 mm., ellipsoid, with white, silky hairs ; caruncle 0·5 mm. long, appendages scarcely developed. (Capsule and seeds described from Angolan material).

N. Rhodesia. W : Mwinilunga Distr., Sinkabolo Dambo, fl. 21.xii.1937, *Milne-Redhead* 3772 (BM ; K). Also in Angola and the Belgian Congo. On termite mounds (N. Rhodesia) and in swamps (Angola).

55. **Polygala arenicola** Gürke in Warb., Kunene-Samb.-Exped. Baum : 273 (1903).— Exell & Mendonça, C.F.A. **1**, 1 : 107 (1937). Type from Angola (Bié).

Spreading annual or perennial herb or shrublet with thick, woody rootstock ; stems slender, slightly winged, glabrous. Leaves 8–12 × 0·3–0·6 mm., linear to acicular, glabrous, margins incurved. Flowers pale blue, pedicels 2·5–3 mm. long, in terminal racemes 2–4 cm. long with glabrous rhachis and caducous bracts and bracteoles. Posterior sepal 1·5–1·8 mm. long ; wing sepals 4–5 × 2 mm., greenish, obliquely elliptic, acute at the apex, glabrous. Upper petals 3 × 1·5–2 mm., obcuneate ; carina 5 × 2 mm., crest 1·5 mm. long. Stamens 6 fertile and 2 staminodes. Capsule 3·5–4 × 2–2·5 mm., elliptic-oblong, sometimes minutely puberulous when young, glabrescent. Seeds 2·5 × 1–1·2 mm., ellipsoid, with rather white, appressed, silky hairs ; caruncle 1 mm. long, appendages minute. (Seed described from Angolan material).

N. Rhodesia. W : between Mwinilunga and Matonchi, fl. 26.i.1937, *Milne-Redhead* 4364 (BM ; K). C : Chilanga, 910 m., fl. & fr. 10.ix.1909. *Rogers* 8525 (BM ; K). S : Choma, 1310 m., fl. 28.iii.1955, *Robinson* 1212 (BM ; K ; SRGH). Also in the Belgian Congo and Angola. Edges of woodland in bush and in swamps ; mainly on Kalahari Sand.

56. **Polygala** sp. aff. **amatymbica** Eckl. & Zeyh.
Polygala amatymbica sensu Eyles in Trans. Roy. Soc. S. Afr. **5** : 300 (1916).

Small perennial herb with woody rootstock ; stems 4–12 cm. tall, crisped-pubescent. Leaves up to 15 × 2·5 mm., very broadly obovate towards the base of the stem, becoming narrowly oblanceolate towards the apex, apex blunt or rounded and mucronate, crisped-pubescent. Wing sepals 3·5 × 1·5 mm., obliquely elliptic, acute at the apex, greenish except at the margins ; anterior sepals free. Carina 2·5 mm. long, crest 1·5 mm. long. Capsule 3·5 × 2·5 mm., oblong-elliptic, puberulous, notched at its apex. Seeds 2·8 × 1·3 mm., ellipsoid, densely sericeous ; caruncular appendages c. 0·4 mm. long.

S. Rhodesia. W : Bulawayo, fl. & fr. iii.1914, *Rogers* 13731 (BM).

I have seen only one very small specimen (cited above) and I have not succeeded in finding *Rogers* 13547 also cited by Eyles (loc. cit.) as *P. amatymbica* Eckl. & Zeyh. The wings appear smaller and rather different in shape from those of *P. amatymbica* but the material is so meagre that I hesitate to express any opinion.